CONTROL STRATEGY FOR TIME-DELAY SYSTEMS

Part I: Concepts and Theories

Edited by

MOHAMMAD-HASSAN KHOOBAN

Department of Engineering - Electrical and Computer Engineering
Aarhus University
Aarhus N, Denmark

TOMISLAV DRAGICEVIC

Department of Electrical Engineering
Center for Electric Power and Energy
Technical University of Denmark
Copenhagen, Denmark

ELSEVIER

ACADEMIC PRESS
An imprint of Elsevier

Academic Press is an imprint of Elsevier
125 London Wall, London EC2Y 5AS, United Kingdom
525 B Street, Suite 1650, San Diego, CA 92101, United States
50 Hampshire Street, 5th Floor, Cambridge, MA 02139, United States
The Boulevard, Langford Lane, Kidlington, Oxford OX5 1GB, United Kingdom

Notices

Knowledge and best practice in this field are constantly changing. As new research and experience broaden our understanding, changes in research methods, professional practices, or medical treatment may become necessary.

Practitioners and researchers must always rely on their own experience and knowledge in evaluating and using any information, methods, compounds, or experiments described herein. In using such information or methods they should be mindful of their own safety and the safety of others, including parties for whom they have a professional responsibility.

To the fullest extent of the law, neither the Publisher nor the authors, contributors, or editors, assume any liability for any injury and/or damage to persons or property as a matter of products liability, negligence or otherwise, or from any use or operation of any methods, products, instructions, or ideas contained in the material herein.

Library of Congress Cataloging-in-Publication Data
A catalog record for this book is available from the Library of Congress

British Library Cataloguing-in-Publication Data
A catalogue record for this book is available from the British Library

ISBN: 978-0-12-820599-0

For information on all Academic Press publications
visit our website at https://www.elsevier.com/books-and-journals

Publisher: Mara Conner
Acquisitions Editor: Sonnini R. Yura
Editorial Project Manager: Charlotte Rowley
Production Project Manager: Prasanna Kalyanaraman
Designer: Alan Studholme

Typeset by VTeX

Contents

CONTROL STRATEGY FOR TIME-DELAY SYSTEMS

List of contributors

Khalid Abidi
Newcastle University (Singapore Campus), SIT Building @ Nanyang Polytechnic, Ang Mo Kio, Singapore

Francisco Javier Bejarano
SEPI, ESIME Ticomán, Instituto Politécnico Nacional, Mexico City, Mexico

Wenjian Cai
School of Electrical and Electronic Engineering, Nanyang Technological University, Singapore

Eugenio B. Castelan
Department of Automation and Systems Engineering, PPGEAS — DAS/CTC/UFSC, Florianópolis, SC, Brazil

Niraj Choudhary
Department of Electrical Engineering, Indian Institute of Technology Delhi, New Delhi, India

Carla de Souza
Department of Automation and Systems Engineering, PPGEAS — DAS/CTC/UFSC, Florianópolis, SC, Brazil

S.H. Hashemipour
Department of Electrical Engineering, Roudsar and Amlash Branch, Islamic Azad University, Roudsar, Iran

S. Janardhanan
Department of Electrical Engineering, Indian Institute of Technology Delhi, New Delhi, India

Indra Narayan Kar
Department of Electrical Engineering, Indian Institute of Technology Delhi, New Delhi, India

A. Khaki Sedigh
Department of Electrical Engineering, Khaje Nasir Toosi University of Technology, Tehran, Iran

Valter J.S. Leite
Department of Mechatronics Engineering, *Campus* Divinópolis — CEFET–MG, Divinópolis, MG, Brazil

Magdi S. Mahmoud
Distributed Control Research Lab, Systems Engineering Department, KFUPM, Dhahran, Saudi Arabia

Majdeddin Najafi
Research institute for Avionics, Isfahan University of Technology, Isfahan, Iran

Libor Pekař
Faculty of Applied Informatics, TBU in Zlín, Zlín, Czechia

Spandan Roy
Robotics Research Center, International Institute of Information Technology Hyderabad (IIIT-H), Hyderabad, India

Harsh Shukla
Robotics Research Center, International Institute of Information Technology Hyderabad (IIIT-H), Hyderabad, India

Luis F.P. Silva
Department of Mechatronics Engineering, *Campus* Divinópolis — CEFET–MG, Divinópolis, MG, Brazil

Hang Jian Soo
Technology Centre for Offshore and Marine, Singapore (TCOMS), Prince George's Park, Singapore

Long Teng
Department of Materials and Production, Aalborg University, Aalborg, Denmark

N. Vasegh
Department of Electrical Engineering, Shahid Rajaee Teacher Training University, Tehran, Iran

Youyi Wang
School of Electrical and Electronic Engineering, Nanyang Technological University, Singapore

Fen Wu
Department of Mechanical and Aerospace Engineering, North Carolina State University, Raleigh, NC, United States

Xiangming Xue
Department of Mechanical and Aerospace Engineering, North Carolina State University, Raleigh, NC, United States

Chengzhi Yuan
Department of Mechanical, Industrial and Systems Engineering, University of Rhode Island, Kingston, RI, United States

Gang Zheng
Defrost, INRIA - Lille Nord Europe, Lille, France

Preface

In a general sense, a system is considered as a time-delay system when the rate of change in system states is related to its previous states. Delays repeatedly appear in different real-world engineering problems. It has been proved that the presence of delays in controlled plants (e.g., communication delays in networked control systems) may have a "*vague*" feature: they highly threaten the stability of the system and deteriorate the system performance, which imposes great difficulty in the control system design and stability analysis. This necessitates the implementation of particular methodologies and design methods in the control theory to effectively stabilize the time-delayed systems.

The research on control design of time-delay system dates back to the 1960s and has received considerable attention over the two past decades. Over these years, the study of such systems has been reported at many international congresses and organized conferences and published in high-qualified journals of control engineering, control theory mathematics, and numerical problems. It is obvious that any developments in the control theory and numerical schemes, particularly the laws of robust stability and performance evaluation, will greatly help a lot in making progress.

This book is dedicated to present an up-to-date framework in the field of the advanced control strategy for time-delay systems along with an overview of the most recent studies in this context. An attempt is carried out to provide a specific perspective, which is concentrated on the robust design, stability analysis, adaptive control, and synchronization of time-delay systems for graduate students studying robust control schemes and/or their application to systems with delays. The usefulness of the suggested schemes and their effectiveness over other prevalent ones are verified theoretically and described by means of various case-studies. The different chapters of this book introduce the latest advances in the active research field to the readers and acquaint them with a state-of-the-art methodology for investigating time-delay systems.

In Chapter 1 the stability and stabilization of time-delay systems are reviewed with emphasis on complex models of partial differential equations, the Lyapunov-based tools, and systems with input or output delays and state delays.

In Chapter 2 a class of sequential subpredictors (SSP) is provided to solve a stability problem for the systems with a large input delay, where the system states are predicted over a large range of delay times.

Chapter 3 proposes a discrete-time version of adaptive posicast controller (APC) for dealing with plant uncertainties that are not well characterized. To avoid the numerical complexity related to the implementation of time-delay compensation on sampled-data plants, the suggested controller scheme is developed and analyzed in discrete time.

Chapter 4 proposes an application of the quasipolynomial meromorphic functions (R_{QM}) ring and elegant algebraic parlance to stabilize the controller structures with the capability of their further parameterization for time-delayed systems.

Chapter 5 proposes a model predictive control of discrete-time nonlinear systems with unknown time-varying delay via the T-S fuzzy modeling and control approach. In this application, the Lyapunov–Razumikhin function (LRF) is employed such that new stabilization conditions are provided.

In Chapter 6 the observability and observer design for a general class of commensurate time-delayed systems affected by unknown inputs is presented. In this endeavor, sufficient conditions ensuring the observability of the system (a precise definition is given further) is provided, which also allows for recovering the state variables of the system by two different ways.

Chapter 7 addresses a new adaptive-robust time delay control (ARTDC) for tracking control of a class of systems under arbitrarily varying input delay and state-dependent uncertainty.

Chapter 8 discusses the problem of adaptive control of the decentralized model reference for a class of uncertain large-scale systems with constant state and input delays and varying delays in the interconnected terms.

Chapter 9 proposes the Lambert W function (LWF) based approach for design and analysis of linear time-invariant (LTI) multidimensional TDSs with a single time delay.

Chapter 10 focuses on the stabilization problem for the class of discrete-time state-delayed systems subjected to saturating actuators and state constraints.

Chapter 11 is concerned with the robust output-feedback delay control synthesis problem for a class of linear systems with time-varying state delay under the integral quadratic constraints (IQCs) framework.

CHAPTER ONE

An overview of time-delay control systems

Magdi S. Mahmoud
Distributed Control Research Lab, Systems Engineering Department, KFUPM, Dhahran, Saudi Arabia

Contents

Control Strategy for Time-Delay Systems
https://doi.org/10.1016/B978-0-12-820599-0.00006-9

1.1. Introduction

The last few decades have witnessed a rapid development of networked control systems due to their significant advantages, and they have been applied to variant industrial areas, such as unmanned surface vehicles, unmanned space vehicles, smart grids, waste-water treatment processes, Internet-based teleoperation, intelligent transportation systems, and so on. However, time-delays are ubiquitous in networked control systems due to limited network bandwidth and possibly degrade the closed-loop system performance.

The occurrence of time-delay phenomenon appears to prevail many real-world systems and engineering applications. This takes place in either the state, the control input side, or the measurements side. It turns out that delays are strongly involved in challenging areas of communication and information technologies including stabilization of networked controlled systems and high-speed communication networks. In terms of mathematical representation, time-delay systems are often described by *functional differential equations* (FDEs), which are of infinite dimension [1], [2], [3]. It turns out that a wide variety of dynamical systems can be modeled as time-delay systems [4]. Stability analysis and robust control of time-delay systems (TDSs) are of theoretical and practical importance [5–9]. A great deal of the basic results is reported in [10–18]. Broadly speaking, stability conditions for time-delay systems can be broadly classified into two categories, delay-independent and delay-dependent stability conditions. Much attention was paid to the study of delay-dependent stability conditions as they yield less conservative design results. Recent works presented in [19–35] cover alternative issues pertaining to stability and stabilization of dynamical systems with time-delays.

The primary objective of this chapter is to wider familiarize the readers with TDSs and provide a systematic treatment of modern ideas and techniques for researchers. It presents a guided tour from basic classical results to recent developments on Lyapunov-based analysis and design with applications to the attractive topics of network-based control and interconnected time-delay control systems.

The primary objective of this chapter is to

- wider familiarize the readers with TDSs,
- provide a systematic treatment of modern ideas and techniques for researchers,
- present a guided tour from basic classical results to recent developments on Lyapunov-based analysis and design, and
- present applications to the attractive topics of resilient delay-dependent stability and stabilization, network-based control, and interconnected TDSs.

Essentially, it provides an overview on the mathematical modeling and technical progress of stability and stabilization of TDSs. Particular emphases will be placed on issues concerned with the conservatism and the computational complexity of the results. For simplicity in exposition, the discussions are limited to linear or linearizable systems. Some methods and techniques used to derive stability conditions for time-delay systems are reviewed. Several future research directions on this topic are also discussed.

The remaining part of the chapter is organized as follows: In Section 1.2, many applications are presented to illuminate the sources of delay phenomena. Section 1.3 is devoted to the models and solutions of time-delay systems. After recalling the notion of stability of TDS in Section 1.4, a comprehensive theoretical study of the methods of Lyapunov are provided in Section 1.5 to establish pertinent theorems leading to stability conditions. Specializing to linear class of TDS, Section 1.6 includes stability analysis under different delay patterns where the ensuing results/conditions are formulated as linear matrix inequalities for both continuous-time and discrete-time systems. This provides a convenient computational tool for control research and investigations. In Section 1.7, we present delay-dependent stability results using model transformations. In Section 1.8, we move to study the stabilization problem of TDSs and establish appropriate design techniques. A typical application of TDSs is the cart-pole systems presented in Section 1.9. Complete results of resilient delay-dependent stability and stabilization are developed in Section 1.10. On a different research line, we provide in Section 1.11 a concise treatment of Kalman filtering for continuous and discrete TDSs. Finally, Section 1.12 is devoted to brief account of networked control systems followed by a class of interconnected TDSs in Section 1.13. Pertinent concluding remarks are outlined in Section 1.14. Several numerical simulations are provided throughout the chapter.

Notations: Let $\mathbb{R}^n$ denote the n-dimensional Euclidean space equipped with norm $||\cdot||$. We use W^t, W^{-1}, $\lambda_m(W)$, $\lambda_M(W)$ to denote, respectively, the transpose, the inverse, the minimum eigenvalue, the maximum eigenvalue of any square matrix W, and $W > 0$ ($W < 0$) stands for a symmetrical and positive-definite (negative-definite) matrix W; I stands for the unit matrix of appropriate dimension; $||\alpha||_2^2 = \sum_{k=0}^{\infty} \alpha^t(k)\alpha(k)$; δH denotes the first difference of H. We denote by $\mathbb{R}^+$ the set of nonnegative real numbers and by $\mathbb{C}_n = \mathbb{C}([-h, 0], \mathbb{R}^n)$ the Banach space of continuous functions $\varphi : [-h, 0] \to \mathbb{R}^n$ with associated norm $||\varphi||_c = \sup_{-h\le s\le 0} ||\varphi||$. We let $\mathsf{N} = \{1, \ldots, N\}$. Matrices, if their dimensions are not explicitly stated, are assumed to be compatible for algebraic operations. In symmetric block matrices, we use the symbol $\bullet$ to represent the terms induced by symmetry. Sometimes the arguments of a function will be omitted when no confusion can arise.

The following facts are provided in [6].

Fact 1.1. Let Σ_1, Σ_2, Σ_3, and $0 < R = R^t$ be real constant matrices of compatible dimensions, and let $H(t)$ be a real matrix function satisfying $H^t(t)H(t) \le I$. Then for any $\rho > 0$ satisfying $\rho\Sigma_2^t\Sigma_2 < R$, we have the following matrix inequality:

$$\begin{aligned}&(\Sigma_3 + \Sigma_1 H(t)\Sigma_2)R^{-1}(\Sigma_3^t + \Sigma_2^t H^t(t)\Sigma_1^t)\\ &\le \rho^{-1}\Sigma_1\Sigma_1^t + \Sigma_3\left(R - \rho\Sigma_2^t\Sigma_2\right)^{-1}\Sigma_3^t.\end{aligned}$$

Fact 1.2. For any real matrices Σ_1, Σ_2, and Σ_3 of appropriate dimensions such that $\Sigma_3^t\,\Sigma_3 \le I$, it follows that

$$\Sigma_1\Sigma_3\Sigma_2 + \Sigma_2^t\Sigma_3^t\Sigma_1^t \le \alpha^{-1}\,\Sigma_1\Sigma_1^t + \alpha\,\Sigma_2^t\Sigma_2,\ \forall\,\alpha > 0.$$

Lemma 1.1 (Finsler's lemma, [18]). *Let* $X \in \mathbb{R}^n$, $P = P^t \in \mathbb{R}^{n\times n}$, *and* $H \in \mathbb{R}^{m\times n}$ *of rank* $(H) = r < n$. *The following statements are equivalent:*

- $x^t P x < 0 \; \forall Hx = |0, \; x \neq 0;$
- $(H^{\perp})^t P (H^{\perp}) < 0;$
- $\exists N \in \mathbb{R}^{n\times m} : P + NH + H^t N^t < 0;$
- $\exists \lambda \in \mathbb{R} : P - \lambda H^t H < 0.$

Lemma 1.2. *For given two vectors* $\alpha \in \mathbb{R}^n$ *and* $\beta \in \mathbb{R}^m$ *and matrix* $\mathbb{N} \in \mathbb{R}^{n\times m}$ *defined over a prescribed interval* Ω, *it follows for any matrices* $X \in \mathbb{R}^{n\times n}$, $Y \in \mathbb{R}^{n\times m}$, *and* $Z \in \mathbb{R}^{m\times m}$, *we have the following inequality:*

$$
\begin{aligned}
-2\,\alpha^t\, \mathbb{N}\, \beta &\leq \begin{bmatrix} \alpha \\ \beta \end{bmatrix}^t \begin{bmatrix} X & Y - \mathbb{N} \\ Y^t - \mathbb{N}^t & Z \end{bmatrix} \begin{bmatrix} \alpha \\ \beta \end{bmatrix} \\
&= \alpha^t X \alpha + \beta^t (Y^t - \mathbb{N}^t)\alpha + \alpha^t (Y - \mathbb{N})\beta + \beta^t Z \beta
\end{aligned}
$$

subject to

$$
\begin{bmatrix} X & Y \\ Y^t & Z \end{bmatrix} \geq 0.
$$

1.2. Sources of time-delays

There are many applications where time-delay phenomena appear quite naturally. This includes, but not limited to:

A. *Automotive:* Combustion model (ignition delay), electromechanical brakes (actuator delay).

B. *Electrical networks.*

C. *Epidemics:* Understanding the dynamics of biological processes and epidemics is a challenge for health workers engaged in managing treatment strategies. The underlying mechanisms can be revealed by considering epidemics and diseases as dynamical processes, for which the hematology dynamics can be modeled by

$$
\dot{y}(t) = -\lambda y(t) + F[y(t - \tau)], \tag{1.1}
$$

which formulates the circulating cell populations in one compartment, where y represents the circulating cell population, λ is the cell-loss rate, and the monotone function F (describing a feedback mechanism) denotes the flux of cells from the previous compartment. The delay τ represents the average length of time required to go through the compartment.

D. *Glucose-insulin model:* Let $G(t)$ and $I(t)$ represent the levels of plasma glycemia and insulinemia. Then

$$\dot{G}(t) = K_{xgi}\, G(t)I(t) + \frac{T_{gh}}{V_G},$$
$$\dot{I}(t) = -K_{xi}\, I(t) + \frac{T_{iG_{\max}}}{V_I} f[G(t-\tau_g)], \tag{1.2}$$

where K_{xgi} is the rate of glucose uptake by tissues (insulin-dependent) per pM of plasma insulin concentration, T_{gh} is the net balance between hepatic glucose output and insulin-independent zero-order glucose tissue uptake (mainly by the brain), V_G is the apparent distribution volume for glucose, K_{xi} is the apparent first-order disappearance rate constant for insulin, $T_{iG_{\max}}$ is the maximal rate of second-phase insulin release, V_I is the apparent distribution volume for insulin, τ_g is the apparent delay with which the pancreas varies secondary insulin release in response to varying plasma glucose concentrations, and f is the nonlinear function that models the Insulin Delivery Rate.

E. *Heat exchanger:* distributed delay due to conduction in a tube.

F. *Hydraulic networks:* the transport phenomenon of water is modeled as a varying time-delay

G. *Intelligent building:* time-delay due to wireless transmission of sensor data

H. *Manufacturing process:* The metal cutting process on a lathe can be described as

$$m\, \ddot{y}(t) + c\, \dot{y}(t) + ky(t) = -F_t[f + y(t) - y(t-\tau)]. \tag{1.3}$$

The study of this model is critical in understanding the regenerative chattering phenomenon.

I. *Marine robotics:* transport delay due to sonar measurement of depth.

J. *Neutral delay systems:* Arise, for instance, in the analysis of the coupling between transmission lines and population dynamics: evolution of forests. The model is based on a refinement of the delay-free logistic (or the Pearl–Verhulst equation), where effects as soil depletion and erosion have been introduced:

$$\dot{x}(t) = r\,x(t)\left[1 - \frac{x(t-\tau) + c\dot{x}(t-\tau)}{K}\right], \tag{1.4}$$

where x is the population, r is the intrinsic growth rate, and K is the environmental carrying capacity.

K. *Population dynamics:* Predator–prey model based on the Volterra model with predator (y) and prey (x) populations (t is the life time of prey):

$$\begin{aligned}\dot{x}(t) &= r\,x\left(1 - \frac{x(t-\tau)}{K}\right) - \alpha\,x\,y,\\ \dot{y}(t) &= -c\,y + \beta\,x\,y.\end{aligned} \tag{1.5}$$

1.3. Models and solutions

A general model of TDSs can be expressed as

$$\begin{aligned}\dot{x}(t) &= f(x_t, t, u_t), \quad t \geq t_0,\\ y(t) &= g(x_t; t; u_t),\end{aligned} \tag{1.6}$$

where

$$\begin{aligned}x_t(\theta) &= x(t+\theta),\ -h \leq \theta \leq 0, \qquad u_t(\theta) = u(t+\theta),\ -h \leq \theta \leq 0,\\ x(\theta) &= \varphi(\theta),\ t_0 - h \leq \theta \leq t_0.\end{aligned} \tag{1.7}$$

1.3.1 Retarded systems

It is quite natural to consider, as a state space, the set $\mathbb{S} = \mathbb{S}([-h; 0]; \mathbb{R}^n)$ of continuous functions mapping the interval $[-h; 0] \rightarrow \mathbb{R}^n$ with the topology of uniform convergence. The initial condition $\varphi(\theta)$ must be prescribed as $\phi : [-h; 0] \rightarrow \mathbb{R}^n$. Observe that ($\phi \in \mathbb{C}$, or ϕ may involve bounded jumps at some discontinuity instants). The nature of the solution (and of its initial value) distinguishes FDE from ODE.

Definition 1.1 ([3]). A function x is said to be a solution on $[\sigma - h; \sigma + a]$ of the retarded functional differential equation (RDE)

$$\begin{aligned}\dot{x}(t) &= f(t, x_t),\\ f &: \Omega \subset \mathbb{R} \times \mathbb{S} \rightarrow \mathbb{R}^n,\end{aligned} \tag{1.8}$$

if there are $\sigma \in \mathbb{R}$ and $a > 0$ such that $x \in \mathbb{S}([\sigma - h; \sigma + h]; \mathbb{R}^n)$, $(t, x_t) \in \Omega$, and x(t) satisfies (1.8) for $t \in [\sigma - h; \sigma + a]$. For given $\sigma \in \mathbb{R}$; $\varphi \in \mathbb{S}$, we say that $x(\sigma; \varphi; f)$ is a solution of (1.8) with initial value φ at σ or simply a solution through $(\sigma; \varphi)$ if there is $a > 0$ such that $x(\sigma; \varphi; f)$ is a solution of (1.8) on $t \in [\sigma - h; \sigma + a]$ and $x_\sigma(\sigma; \varphi; f) = \varphi$.

Supposing that Ω is open and $f \in \mathbb{S}(\Omega, \mathbb{R}^n)$, a function $x \in \mathbb{S}([\sigma - h - \alpha; \sigma]; \mathbb{R}^n)$, $\alpha > 0$, is referred to as a backward continuation of the solution through $(\sigma; \varphi)$ if $x_\sigma = \varphi$ and for any $\sigma_1 \in [\sigma - \alpha, \sigma]$, we have $(\sigma_1; x_{\sigma_1}) \in \Omega$, and x is a solution of (1.8) on $(\sigma_1 - h; \sigma)$ through $(\sigma_1; x_{\sigma_1})$.

The interested reader is referred to [3] for further useful discussions.

1.3.2 Models for linear time-invariant systems

In the linear time-invariant case (LTI), the corresponding general time-delay model is

$$\begin{aligned} \dot{x}(t) &= \sum_{\ell=1}^{p} D_\ell \dot{x}(t-\omega_\ell) + \sum_{j=0}^{q} [A_j x(t-h_j) + B_j u(t-h_j)] \\ &\quad + \sum_{m=1}^{r} \int_{t-\tau_m}^{t} [G_m(\theta)x(\theta) + H_m(\theta)u(\theta)]d\theta \end{aligned} \tag{1.9}$$

$$y(t) = \sum_{j=0}^{q} C_j x(t-h_j) + \sum_{m=1}^{r} \int_{t-\tau_m}^{t} N_m(\theta)x(\theta)d\theta, \tag{1.10}$$

where

- $h_0 = 0$ and A_0 is a constant instantaneous matrix;
- Constant matrices A_j, $j > 0$, represent discrete-delay phenomena;
- The sum of integrals corresponds to distributed delay effects, weighted by G_m over the time intervals $[t - \tau_m; t]$;
- Matrices D_ℓ account for the neutral part;
- Matrices B_j and $H_m(s)$ are input matrices;
- $h = \max_{j,m,\ell}\{h_j; \tau_m; \omega_\ell\}$.

Note that in (1.10), $y(t) \in \mathbb{R}^p$ represents the output description, with discrete C_j and distributed $N_m(\theta)$ delayed parts as well. The particular case of (1.9)–(1.10)

$$\begin{aligned} \dot{x}(t) &= \sum_{j=0}^{q} [A_j x(t-h_j) + B_j u(t-h_j)], \\ h_0 &= 0 < h_1 < \cdots < h_{q-1} < h_q, \end{aligned} \tag{1.11}$$

has been investigated extensively in the literature.

1.4. Stability notion

We further recall the following stability notion for time-delay system (1.5).

Definition 1.2. If for any $t_0 \in \mathbb{R}$ and any $\varepsilon > 0$, there exists $\delta = \delta(t_0, \varepsilon) > 0$ such that $||x_{t_0}||_c < \delta$ implies $||x(t)|| < \varepsilon$ for all $t \geq t_0$, then the trivial solution of time-delay system (1.5) is called stable.

The following properties are readily recognized:

- If the trivial solution of time-delay system (1.5) is *stable* and if δ can be chosen independently of t_0, then the trivial solution of time-delay system (1.5) is called *uniformly stable.*
- If the trivial solution of time-delay system (1.5) is stable and if for any $t_0 \in \mathbb{R}$ and any $\varepsilon > 0$, there exists $\delta_a = \delta_a(t_0, \varepsilon) > 0$ such that $||x_{t_0}||_c < \delta_a$ implies $\lim_{t\to\infty} x(t) = 0$, then the trivial solution of time-delay system (1.5) is called *asymptotically stable.*
- If the trivial solution of time-delay system (1.5) is uniformly stable and there exists $\delta_a > 0$ such that $||x_{t_0}||_c < \delta_a$ implies $||x(t)|| < \eta$ for $t \geq t_0 + T$ and $t_0 \in \mathbb{R}$, then the trivial solution of time-delay system (1.5) is called *uniformly asymptotically stable.*
- If the trivial solution of time-delay system (1.5) is (uniformly) asymptotically stable and if δ_a can be an arbitrarily large finite number, then the trivial solution of time-delay system (1.5) is called *globally (uniformly) asymptotically stable*

1.5. Stability theorems

In the study of stability analysis of time-delay systems, the methods of Lyapunov functions and Lyapunov–Krasovskii functionals play important roles. Two Lyapunov methods are often used:

A. *Lyapunov–Krasovskii functional (LKF)* method,
B. *Lyapunov–Razumikhin function (LRF)* method.

It is significant to observe that the LKF method deals with functionals that essentially have scalar values, whereas the LRF method involves only functions rather than functionals.

In this section, we review these two methods; see [6] for details.

Consider the following time-delay system

$$\dot{x}(t) = f(t, x_t), \quad t \geq t_0, \tag{1.12}$$

where

- $x_t = x(t+\theta)$, $-\theta_m \le \theta \le 0$,
- $f; \mathbb{R} \times \mathbb{C}_n \to \mathbb{R}^n$ is continuous and Lipschitz in x_t,
- $f(t, 0) = 0$.

In what follows, we let $x_t(s, \varphi)$ be the solution of (1.12) at time t with initial condition $x_s = \varphi$. Let $\mathbb{G}$ be a bounded subset of $\mathbb{C}_n$, and let $\mathbb{H}$ be a bounded subset of $\mathbb{R}^n$.

The *Lyapunov–Krasovskii stability method* is provided by the following theorem.

Theorem 1.1. *Let $f : \mathbb{R} \times \mathbb{G} \to \mathbb{H}$, and let $u, v, w : \mathbb{R}^+ \to \mathbb{R}^+$ be continuous nondecreasing functions with $u(0) = v(0) = 0$ and $u(\beta) > 0$ and $v(\beta) > 0$ for $\beta > 0$. Suppose that there exists a continuous functional $\mathbb{V} : \mathbb{R} \times \mathbb{C}_n \to \mathbb{R}$ such that*

1. $u(||\varphi(0)||) \le \mathbb{V}(t, \varphi) \le v(||\varphi(0)||)$;
2. $\dot{\mathbb{V}}(t, \varphi) \le -w(||\varphi(0)||)$,

where

$$\dot{\mathbb{V}}(t, \varphi) = \lim_{\Delta t \to 0^+} \frac{1}{\Delta t} (\mathbb{V}(t + \Delta t, x_{t+\Delta t}(t, \varphi)) - \mathbb{V}(t, \varphi)).$$

Then the trivial solution of time-delay system (1.5) *is uniformly stable. If $w(\beta) > 0$ for $\beta > 0$, then the trivial solution of time-delay system* (1.5) *is uniformly asymptotically stable. Additionally, if $\lim_{\beta\to\infty} u(\beta) \to \infty$, then the trivial solution of time-delay system* (1.5) *is globally uniformly asymptotically stable.*

In some cases, the LKF involving terms depending on the state derivatives $\dot{x}_t$ are quite effective in the derivation of the stability conditions. This will in turn requires the modification of the conditions in Theorem 1.1. See [8] for details.

The *Lyapunov–Razumikhin stability method* is provided by the following theorem.

Theorem 1.2. *Let $f : \mathbb{R} \times \mathbb{G} \to \mathbb{H}$, let $u, v, w : \mathbb{R}^+ \to \mathbb{R}^+$ be continuous nondecreasing functions with $u(0) = v(0) = 0$ and $u(\beta) > 0$ and $v(\beta) > 0$ for $\beta > 0$, and let v be strictly increasing. Suppose that there exists a continuous functional $\mathbb{V} : \mathbb{R} \times \mathbb{R}^n \to \mathbb{R}$ such that*

1. $u(||x||) \le \mathbb{V}(t, x) \le v(||x||)$;
2. $\dot{\mathbb{V}}(t, \varphi) \le -w(||\varphi(0)||)$ *if*

$$[\mathbb{V}(t+\theta, x(t+\theta)) \le \mathbb{V}(t, x(t))] \, \textit{for} \, \theta \in [-h, 0],$$

where

$$\dot{\mathbb{V}}(t,\ x(t)) = \frac{d}{dt}\mathbb{V}(t,\ x(t)) = \frac{\partial \mathbb{V}(t,\ x(t))}{\partial t} + \frac{\partial \mathbb{V}(t,\ x(t))}{\partial x} f(t,\ x_t).$$

Then the trivial solution of time-delay system (1.5) *is uniformly stable. If* $w(\beta) > 0$ *for* $\beta > 0$, *then there exits a continuous nondecreasing function* $q(\beta) > 0$ *for* $\beta > 0$. *If the foregoing condition* 2) *is strengthened to* $\dot{\mathbb{V}}(t,\ x(t)) \leq -w(||x(t)||)$ *if*

$$(\mathbb{V}(t+\theta,\ x(t+\theta)) \leq q(\mathbb{V}(t,\ x(t)))) \, for \, \theta \in [-h,\ 0],$$

then the trivial solution of time-delay system (1.5) *is uniformly asymptotically stable. Additionally, if* $\lim_{\beta \to \infty} u(\beta) \to \infty$, *then the trivial solution of time-delay system* (1.5) *is globally uniformly asymptotically stable.*

The following Halanay result [7] also plays an important role in the stability analysis of time-delay systems.

Theorem 1.3. *Suppose that constant scalars* k_1 *and* k_2 *satisfy* $k_1 > k_2 > 0$ *and that* $y(t)$ *is a nonnegative continuous function on* $[t_0 - \tau,\ t_0]$ *satisfying*

$$\begin{aligned} \dot{y}(t) &\leq -k_1 y(t) + k_2 \bar{y}(t), \quad t \geq t_0, \\ \bar{y}(t) &= \sup_{t-\tau \leq s \leq t} \{y(s)\}, \quad \tau \geq 0. \end{aligned} \tag{1.13}$$

Then, for $t \geq t_0$, *we have*

$$y(t) \leq \bar{y}(t_0) \exp(-\kappa(t - t_0)),$$

where $\kappa > 0$ *is the unique solution to the equation*

$$\kappa = k_1 - k_2 \exp(-\kappa\tau).$$

Remark 1.1. Theorems 1.1–1.3 can be used to derive stability conditions for the case where the delay is time-varying, continuous, but not necessarily differentiable.

Remark 1.2. In what follows, stability conditions for time-delay systems can be broadly classified into two types:

1. Delay-independent stability (DIS) conditions, which do not include information about the delay. Generally speaking, DIS conditions are simpler to apply.

2. Delay-dependent stability (DDS) conditions, which involve information on the size and pattern of the delay. DDS conditions are less conservative, especially in the case where the time-delay is small.

In what follows, this chapter focuses on the delay-dependent stability problem, and the objective is twofold:

A. to develop delay-dependent conditions to provide a maximal allowable delay as large as possible,

B. to develop delay-dependent conditions by using as few as possible decision variables while keeping the same maximal allowable delay.

Alternative approaches were proposed in the literature to obtain DDS conditions, among which the linear matrix inequality (LMI) approach is the most popular [18].

1.6. Stability results for linear delay systems

For simplicity, we consider the following linear system with a single discrete delay:

$$\dot{x}(t) = A\,x(t) + A_d(t - \tau(t)), \quad t \geq t_0, \tag{1.14}$$

where $x(t) \in \mathbb{R}^n$ is the state vector, A and A_d are system matrices of appropriate dimensions, and $\tau(t)$ is the time-delay factor. There are several classes of time-delay patterns considered in the literature.

Class A Constant delay,

$$\tau(t) = d, \ \forall\, t.$$

Class B Unknown-but-bounded delay,

$$0 < \tau(t) \leq \tau_M, \ \forall\, t.$$

Class C Bounded time-varying delay,

$$\tau_m \leq \tau(t) \leq \tau_M, \ \forall\, t.$$

Class D Bounded time-varying delay with bounded derivative:

$$\tau_m \leq \tau(t) \leq \tau_M, \ \forall\, t,$$
$$d_m \leq \dot{\tau}(t) \leq d_M, \ \forall\, t.$$

1.6.1 Constant delay

When the time-delay is constant, the system described by (1.14) can be rewritten as

$$\dot{x}(t) = A_0\, x(t) + A_d x(t-d), \quad t \geq t_0. \tag{1.15}$$

Natural extensions of the quadratic Lyapunov functions can be particularly used to study in the framework of LTI delay systems (1.15) and the functional

$$\mathbb{V}_0(x_t) = x^t(t)\mathcal{P}x(t) + \int_{-d}^{0} x^t(t+\theta)\mathcal{Q}x(t+\theta)d\theta. \tag{1.16}$$

We obtain sufficient conditions by the following theorem.

Theorem 1.4. *The time-delay system* (1.15) *is asymptotically stable for any* $d \geq 0$ *if there exist matrices* $\mathcal{P} > 0$, $\mathcal{Q} > 0$, *and* $\mathcal{R}$ *satisfying*

$$\mathcal{P}A_0 + A_0^t\mathcal{P} + \mathcal{P}A_d\mathcal{Q}^{-1}A_d^t\mathcal{P} + \mathcal{Q} + \mathcal{R} = 0 \tag{1.17}$$

or, equivalently, the LMI

$$\begin{bmatrix} \mathcal{P}A_0 + A_0^t\mathcal{P} + \mathcal{Q} & \mathcal{P}A_d \\ \bullet & -\mathcal{Q} \end{bmatrix} < 0. \tag{1.18}$$

It is significant to observe in the delay-free case ($A_d = 0$) that (1.18) provides the link with the Lyapunov equation for ODE. Nevertheless, in the delayed case $A_d \neq 0$, this sufficient condition is far from being necessary. From here many generalizations were proposed, involving different alternative terms:

$$\begin{aligned}
\mathbb{V}_1(x_t) &= x^t(t)\mathcal{P}x(t), \\
\mathbb{V}_2(x_t) &= x^t(t)\int_{-d_j}^{0} \mathcal{Q}_j x(t+\theta)d\theta, \\
\mathbb{V}_3(x_t) &= \int_{-d_j}^{0} x^t(t+\theta)\mathcal{S}_j x(t+\theta)d\theta, \\
\mathbb{V}_4(x_t) &= \int_{-\tau_j}^{0}\int_{t+\theta}^{0} x^t(\theta)\mathcal{R}_j x(\theta)d\theta\, ds, \\
\mathbb{V}_5(x_t) &= x^t(t)\int_{-d_j}^{0} \mathcal{P}_j(\eta)x(t+\eta)d\eta,
\end{aligned} \tag{1.19}$$

$$\mathbb{V}_6(x_t) = \int_{-d_j}^{0}\int_{-d_j}^{0} x^t(t+\eta)\mathcal{P}(\eta,\theta)x(t+\theta)d\eta d\theta.$$

The following points are worth noting:

1. Loosely speaking, the terms $\mathbb{V}_2$ and $\mathbb{V}_3$ are used for the delay-independent stability of discrete delays.
2. The term V_4 is meant for distributed delays or discrete-delay dependent stability. Considering system (1.15) along with

$$\mathbb{V}(x_t) = \mathbb{V}_1(x(t)) + \mathbb{V}_4(x_t) + \mathbb{V}_4(x_{t-h}),$$

standard manipulation leads, with R_1 for $V_4(x_t)$ and R_2 for $V_4(x_t - h)$, to the following delay-dependent LMI condition:

$$\begin{bmatrix} \mathcal{P}(A_0 + A_d) + (A_0 + A_d)^t\mathcal{P} + d\mathcal{R}_1 + d\mathcal{R}_2 & d\mathcal{P}A_dA_0 & h\mathcal{P}A_d^2 \\ \bullet & -d\mathcal{R}_1 & 0 \\ \bullet & \bullet & -d\mathcal{R}_2 \end{bmatrix} < 0. \tag{1.20}$$

3. Although the terms V_5 and V_6 appear, in a general form, in necessary and sufficient schemes (see [10], [11], [12]), the general computation of the time-varying matrices is excessively burden. To avoid such computational limitations, a discretization scheme incorporating piecewise-constant functions P_j was introduced in [16,17].

1.6.2 Time-varying delay

In what follows, we will review the LMI techniques in deriving DDS results for the single-delay case. Extension to the multiple-delay case is a straightforward task. We consider the class of time-delay systems (class B) in which the delay factor is continuous but bounded:

$$\dot{x}(t) = A_0\, x(t) + A_d x(t-\tau(t)), \;\; x(t) = \varphi(t), \;\; t \in [-\tau_M,\, 0]. \tag{1.21}$$

Similar to (1.16), we consider the LKF of the form

$$\widehat{V}_0(x_t) = x^t(t)\mathcal{P}x(t) + \int_{t-\tau(t)}^{t} x^t(t+\theta)\mathcal{Q}x(t+\theta)d\theta. \tag{1.22}$$

Since the time-varying delay $\tau(t)$ may be nondifferentiable, we introduce the following equalities for any matrices $\mathcal{Y}$, $\mathcal{W}$, and $\mathcal{S}$ of appropriate dimensions:

$$\begin{aligned}
&\dot{x}^t(t)\mathcal{Y}[A_0\, x(t) + A_d x(t-\tau(t)) - \dot{x}(t)] = 0, \\
&x^t(t)\mathcal{W}[A_0\, x(t) + A_d x(t-\tau(t)) - \dot{x}(t)] = 0, \\
&\dot{x}^t(t-\tau(t))\mathcal{S}[A_0\, x(t) + A_d x(t-\tau(t)) - \dot{x}(t)] = 0.
\end{aligned} \tag{1.23}$$

The following theorem summarizes the main result.

Theorem 1.5. *The time-delay system* (1.21) *is asymptotically stable if there exist matrices* $\mathcal{P} > 0$, $\mathcal{Y}$, $\mathcal{W}$, *and* $\mathcal{S}$ *such that*

$$\begin{bmatrix} \mathcal{W}A_0 + A_0^t\mathcal{W}^t & \mathcal{W}A_d + A_0^t\mathcal{S}^t & A_0^t\mathcal{Y}^t + \mathcal{P} - \mathcal{W} \\ \bullet & \mathcal{S}A_d + A_d^t\mathcal{S}^t & A_d^t\mathcal{Y}^t - \mathcal{S} \\ \bullet & \bullet & -\mathcal{Y} - \mathcal{Y}^t \end{bmatrix} < 0. \tag{1.24}$$

Consider the time-delay system

$$\begin{aligned}
&\dot{x}(t) = A_0\, x(t) + A_d x(t-\tau(t)), \;\; x(t) = \varphi(t), \;\; t \in [-\tau_M, 0], \\
&\quad 0 \le \tau_m \le \tau(t) \le \tau_M, \;\; \sigma \le \dot{\tau}(t) \le \mu.
\end{aligned} \tag{1.25}$$

According to the Lyapunov–Razumikhin stability method (Theorem 1.2), we can obtain the following stability condition.

Theorem 1.6. *The time-delay system* (1.25) *is asymptotically stable if there exist a matrix* $\mathcal{P} > 0$ *and a scalar* $\sigma > 0$ *such that*

$$\begin{bmatrix} \mathcal{P}A_0 + A_0^t\mathcal{P} + \sigma\mathcal{P} & \mathcal{P}A_d \\ \bullet & -\sigma\mathcal{P} \end{bmatrix} < 0. \tag{1.26}$$

On choosing the LKF (1.16), a delay-independent stability condition can be derived in the following form.

Theorem 1.7. *The time-delay system* (1.25) *is asymptotically stable if there exist matrices* $\mathcal{P} > 0$ *and* $\mathcal{Q} > 0$ *such that*

$$\begin{bmatrix} \mathcal{P}A_0 + A_0^t\mathcal{P} + \mathcal{Q} & \mathcal{P}A_d \\ \bullet & -(1-\mu)\mathcal{Q} \end{bmatrix} < 0. \tag{1.27}$$

Remark 1.3. Note that Theorem 1.7 is independent of the time-delay and therefore is very conservative, especially when the time-delay is small.

When the delay is constant, $\tau(t) \equiv d$, from the Schur complements it follows that (1.24) is equivalent to

$$\mathcal{P}A_0 + A_0^t\mathcal{P} + \mathcal{Q} + \mathcal{P}A_d\mathcal{Q}^{-1}A_d\mathcal{P} < 0.$$

In turn, this implies that

$$\mathcal{P}(A_0 + A_d) + (A_0 + A_d)^t\mathcal{P} < 0,$$

which is a necessary and sufficient condition for the stability of system (1.25) with zero delay.

In the literature the following Lyapunov functional is often used to derive delay-dependent results:

$$\mathbb{V}(t, x_t) = x^t(t)\mathcal{P}x(t) + \int_{t-\tau(t)}^{t} x^t(s)\mathcal{Q}x(s)ds + \int_{-\tau_M}^{0}\int_{t+\theta}^{t} \dot{x}^t(s)\mathcal{Z}\dot{x}(s)dsd\theta. \tag{1.28}$$

It was first introduced in [44,45]. Using the free-weighting [54], the following DDS condition can be derived based on the LKF (1.28).

Theorem 1.8. *The time-delay system* (1.25) *is asymptotically stable if there exist matrices* $\mathcal{P} > 0$, $\mathcal{Q} > 0$, $\mathcal{Z} > 0$, $\begin{bmatrix} X_{11} & X_{12} \\ \bullet & X_{22} \end{bmatrix} \geq 0$, *and any matrices* M, N *of appropriate dimensions such that*

$$\begin{bmatrix} \mathbb{L}_{11} & \mathbb{L}_{12} & \tau_M A_0^t\mathcal{Z} \\ \bullet & \mathbb{L}_{22} & \tau_M A_d^t\mathcal{Z} \\ \bullet & \bullet & -\tau_M\mathcal{Z} \end{bmatrix} < 0 \quad \text{and} \quad \begin{bmatrix} X_{11} & X_{12} & M \\ \bullet & X_{22} & N \\ \bullet & \bullet & \mathcal{Z} \end{bmatrix} \geq 0, \tag{1.29}$$

where

$$\mathbb{L}_{11} = \mathcal{P}A_0 + A_0^t\mathcal{P} + M + M^t + \mathcal{Q} + \tau_M X_{11},$$
$$\mathbb{L}_{12} = \mathcal{P}A_d + M + N^t + \tau_M X_{12},$$
$$\mathbb{L}_{22} = -N - N^t - (1-\mu)\mathcal{Q} + \tau_M X_{22}.$$

1.6.3 Augmented Lyapunov functional

Recalling that the first term in most LKFs is $x^t(t)Px(t)$, which involves the current state $x(t)$ only and does not reflect the delayed state. Hence an augmented Lyapunov functional was proposed in [55] for system described by (1.15):

$$\mathbb{V}(t,x_t)=\xi^t(t)\mathcal{P}\xi(t)+\int_{t-d}^{t}\varrho^t(s)\mathcal{Q}\varrho(s)ds+\int_{-d}^{0}\int_{t+\theta}^{t}\varrho^t(s)\mathcal{Z}\varrho(s)dsd\theta, \quad (1.30)$$

$$\xi^t(t)=\begin{bmatrix}x^t(t) & x^t(t-d) & \int_{t-d}^{t}x^t(s)ds\end{bmatrix},\ \varrho^t(s)=\begin{bmatrix}x^t(s) & \dot{x}^t(s)\end{bmatrix}. \quad (1.31)$$

Remark 1.4. Compared with the Lyapunov functional (1.28), the augmented Lyapunov functional can lead to less conservative results. Additionally, it is also applicable for systems with time-varying delay, which can been seen in [56] and references therein.

1.6.4 Triple integral Lyapunov functional

On examining the LKFs (1.16) and (1.30), we can see that the Lyapunov functional often contains integral terms, single $\int_{t-\tau(t)}^{t}x^t(t+\theta)\mathcal{Q}x(t+\theta)d\theta$ and double $\int_{-d}^{0}\int_{t+\theta}^{t}\varrho^t(s)\mathcal{Z}\varrho(s)dsd\theta$, in order to bring the effect of time-delays.

A natural question arose whether introducing triple integral terms in the Lyapunov functional would yield improvement in the stability behavior. This question is addressed in [57], [58] by extending the LKF (1.30)–(1.31) and incorporating a triple integral term to yield the form

$$\begin{aligned}\mathbb{V}(t,x_t)&=\xi^t(t)\mathcal{P}\xi(t)+\int_{t-d}^{t}\varrho^t(s)\mathcal{Q}\varrho(s)ds+\int_{-d}^{0}\int_{t+\theta}^{t}\varrho^t(s)\mathcal{Z}\varrho(s)dsd\theta\\&+\int_{-d}^{0}\int_{\theta}^{0}\int_{t+\beta}^{t}\dot{x}^t(s)\mathcal{R}\dot{x}(s)dsd\beta d\theta. \quad (1.32)\end{aligned}$$

Remark 1.5. It is reported in [57], [58] by simulation results that the Lyapunov functional containing triple integral terms is quite effective in reduction of the conservatism of the stability conditions.

1.6.5 Newton–Leibniz formula

An alternative route can be pursued by using the Newton–Leibniz formula

$$\begin{aligned}x(t-d)&=x(t)-\int_{t-d}^{t}\dot{x}(\alpha)d\alpha\\&=x(t)-\int_{t-d}^{t}[A_0x(\alpha)+A_dx(\alpha-d)]d\alpha \quad (1.33)\end{aligned}$$

and recalling (1.15) to yield

$$\dot{x}(t) = [A_0 + A_d]\, x(t) - A_d \int_{t-d}^{t} [A_0\, x(\alpha) + A_d x(\alpha - d)] d\alpha. \tag{1.34}$$

Remark 1.6. It should be clear that the asymptotic stability of the time-delay system in (1.34) implies that of system (1.15).

Following [6], we proceed to study the DDS of system (1.34) using the following LKF candidate:

$$\begin{aligned}\mathbb{V}(t, x_t) &= x^t(t)\mathcal{P}^{-1}x(t) + \int_{-h}^{0}\int_{t+\alpha}^{t} x^t(\theta)A_d^t\mathcal{Q}_1^{-1}A_d x(\theta)\, d\theta\, d\alpha \\ &+ \int_{-h}^{0}\int_{t-h+\alpha}^{t} x^t(\theta)A_d^t\mathcal{Q}_2^{-1}A_d x(\theta) d\theta\, d\alpha, \ \ \mathcal{P} > 0, \ \mathcal{Q}_1 > 0, \ \mathcal{Q}_2 > 0.\end{aligned} \tag{1.35}$$

Define

$$\Gamma = \mathcal{P}(A_0 + A_d) + (A_0 + A_d)^t\mathcal{P} + A_d(\mathcal{Q}_1 + \mathcal{Q}_2)A_d^t.$$

The main stability result is established by the following theorem.

Theorem 1.9. *The time-delay system* (1.34) *is asymptotically stable for any delay satisfying* $0 < d \le d_M$ *if there exist matrices* $\mathcal{P} > 0$, $\mathcal{Q}_1$, *and* $\mathcal{Q}_2$ *such that*

$$\begin{bmatrix} \Gamma & d_M\mathcal{P}A_0^t & d_M\mathcal{P}A_d^t \\ \bullet & -\mathcal{Q}_1 & 0 \\ \bullet & \bullet & -\mathcal{Q}_2 \end{bmatrix} < 0. \tag{1.36}$$

Remark 1.7. The technique by using the Newton–Leibniz formula to transform the time-delay system to an appropriate one for DDS analysis is quite useful. However, still a different route of writing (1.34) would be

$$\dot{x}(t) = [A_0 + A_d]\, x(t) - A_d \int_{t-d}^{t} \dot{x}(d\alpha) d\alpha \tag{1.37}$$

and

$$\frac{d}{dt}\left[x(t) + A_d \int_{t-d}^{t} x(\alpha) d\alpha \right] = (A_0 + A_d)x(t). \tag{1.38}$$

However, all the transformed time-delay systems by using the Newton–Leibniz formula introduce additional dynamics, which may cause conservatism as the delay-dependent conditions derived based on the transformed systems.

1.6.6 Bounding techniques

In studying delay-dependent stability for time-delay systems, it is desirable to find methods that yield stability conditions with reduced conservatism. A wide class of early methods rely on generating improved bounds on some weighted cross products arising in the analysis of the delay-dependent stability problem. This class of methods was by using the well-known algebraic inequality

$$-2\,\alpha^t\,\beta \le \alpha^t\,\Xi\,\alpha + \beta^t\,\Xi^{-1}\,\beta, \tag{1.39}$$

where the vectors $\alpha, \beta \in \mathbb{R}^n$ and the matrix $\Xi \in \mathbb{R}^{n\times n}$. An integral bounding inequality is as follows.

Lemma 1.3 ([39]). *Let $a(\alpha) \in \mathbb{R}^{n_a}$ and $b(\alpha) \in \mathbb{R}^{n_b}$ be given for $\alpha \in \Omega$. Then, for any $0 < X \in \mathbb{R}^{n_a\times n_a}$ and any matrix $M \in \mathbb{R}^{n_a\times n_a}$, we have*

$$\begin{aligned} &-\int_{\Omega} a^t(\alpha)\, b(\alpha)\, d\alpha \\ &\le \int_{\Omega} \begin{bmatrix} a(\alpha) \\ b(\alpha) \end{bmatrix}^t \begin{bmatrix} X & XM \\ \bullet & (M^tX+I)X^{-1}(M^tX+I)^t \end{bmatrix} \begin{bmatrix} a(\alpha) \\ b(\alpha) \end{bmatrix} d\alpha. \end{aligned} \tag{1.40}$$

Which, when applied to time-delay systems of type (1.15), yields the following:

Theorem 1.10. *The time-delay system (1.15) is asymptotically stable for any delay satisfying $0 < h \le h_M$ if there exist matrices $\mathcal{P} > 0$, $\mathcal{Q}$, $\mathcal{V}$, and $\mathcal{W}$ such that*

$$\begin{bmatrix} \Phi & -\mathcal{W}^t A_d^t & A_0^t A_d^t \mathcal{V} & h_M(\mathcal{P}+\mathcal{W}^t) \\ \bullet & -\mathcal{Q} & A_d^t A_d^t \mathcal{V} & 0 \\ \bullet & \bullet & -\mathcal{V} & 0 \\ \bullet & \bullet & \bullet & -\mathcal{V} \end{bmatrix} < 0.$$

$$\Phi = \mathcal{P}(A_0 + A_d) + (A_0 + A_d)^t\mathcal{P} + \mathcal{W}^t A_d + A_d^t \mathcal{W}_1 + \mathcal{Q} \tag{1.41}$$

An improved version of Lemma 1.3 is expressed by the following:

Lemma 1.4 ([40]). *Let $a(\alpha) \in \mathbb{R}^{n_a}$, $b(\alpha) \in \mathbb{R}^{n_b}$, and $\mathcal{N}(\alpha) \in \mathbb{R}^{n_a \times n_b}$ be given for $\alpha \in \Omega$. Then, for any $0 < X \in \mathbb{R}^{n_a \times n_a}$ and any matrix $M \in \mathbb{R}^{n_a \times n_a}$, we have*

$$-\int_{\Omega} a^t(\alpha)\,\mathcal{N}(\alpha)\,b(\alpha)\,d\alpha \leq \int_{\Omega} \begin{bmatrix} a(\alpha) \\ b(\alpha) \end{bmatrix}^t \begin{bmatrix} X & Y - \mathcal{N}(\alpha) \\ \bullet & Z \end{bmatrix} \begin{bmatrix} a(\alpha) \\ b(\alpha) \end{bmatrix} d\alpha, \tag{1.42}$$

where

$$\begin{bmatrix} X & Y \\ \bullet & Z \end{bmatrix} \geq 0. \tag{1.43}$$

By considering the LKF

$$V(t, x_t) = x^t(t)\mathcal{P}x(t) + \int_{t-d}^{t} x^t(\alpha)\mathcal{Q}x(\alpha)\,d\alpha + \int_{-d}^{0}\int_{t+\beta}^{t} \dot{x}^t(\alpha)Z\dot{x}(\alpha)\,d\alpha\,d\beta \tag{1.44}$$

and applying Lemma 1.4, we obtain the following delay-dependent stability theorem.

Theorem 1.11. *The time-delay system* (1.15) *is asymptotically stable for any delay satisfying $0 < d \leq d_M$ if there exist matrices $\mathcal{P} > 0$, $\mathcal{Q}$, X, Y, and Z such that*

$$\begin{bmatrix} \Lambda & \mathcal{P}A_d - Y & d_M A_0^t Z \\ \bullet & -\mathcal{Q} & d_M A_d^t Z \\ \bullet & \bullet & -d_M Z \end{bmatrix} < 0 \quad \textit{and} \quad \begin{bmatrix} X & Y \\ \bullet & Z \end{bmatrix} \geq 0, \ \textit{where}$$

$$\Lambda = \mathcal{P}(A_0 + A_d) + (A_0 + A_d)^t\mathcal{P} + \mathcal{W}^t A_d + A_d^t \mathcal{W}_1 + \mathcal{Q}. \tag{1.45}$$

On the other hand, deploying Lemma 1.1 together with Lemma 1.4, a different delay-dependent stability criterion is provided by the following theorem.

Theorem 1.12 ([41]). *The time-delay system* (1.15) *is asymptotically stable for any delay satisfying $0 < d \leq d_M$ if there exist matrices $\mathcal{P}_1 > 0$, $\mathcal{S}$, P_2, P_3, P_4, Y_1, Y_2, Z_1, Z_2, Z_3, and $R > 0$ such that the following LMIs hold:*

$$\begin{bmatrix} \Theta_1 & \Theta_2 & \Theta_3 \\ \bullet & \Theta_4 & \Theta_5 \\ \bullet & \bullet & \Theta_6 \end{bmatrix} < 0, \qquad \begin{bmatrix} R & Y_1 & Y_2 \\ \bullet & Z_1 & Z_2 \\ \bullet & \bullet & Z_3 \end{bmatrix} \geq 0, \tag{1.46}$$

$$\begin{aligned}
\Theta_1 &= A_0^t P_2 + P_2^t A_0 + Y_1 + Y_1^t + \mathcal{S} + d_M Z_1,\\
\Theta_2 &= \mathcal{P}_1 - P_2^t + A_0^t P_3 + Y_2 + d_M Z_2,\\
\Theta_3 &= P_2^t A_d - Y_1^t + A_0^t P_4, \;\; \Theta_4 = d_M(R + Z_3) - P_3^t - P_3,\\
\Theta_5 &= P_3^t A_d - Y_2^t - P_4, \;\; \Theta_6 = A_d^t P_4 + P_4^t A_d - \mathcal{S}.
\end{aligned} \tag{1.47}$$

Remark 1.8. The inequality in Lemma 1.4 is more general than both inequalities (1.39) and (1.40), and for this reason, it was extensively used in dealing with various issues related to time-delay systems to obtain delay-dependent results.

Now we present another important inequality, which is also effective in the derivation of DDS conditions.

Lemma 1.5 ([42]). *For any constant matrix $0 < \mathcal{M} \in \mathbb{R}^{m\times m}$, scalars $b > a$, and a vector function $\omega : [a, b] \to \mathbb{R}^m$, we have*

$$\begin{aligned}
(b-a)&\int_a^b \omega^t(\alpha)\,\mathcal{M}\,\omega(\alpha)\,d\alpha\\
&\geq \int_a^b \left[\int_a^b \omega^t(\alpha)\,d\alpha\right]^t \mathcal{M} \left[\int_a^b \omega^t(\alpha)\,d\alpha\right],
\end{aligned} \tag{1.48}$$

provided that the integrals are well-defined.

Using Lemma 1.5 and selecting the LKF

$$\begin{aligned}
V(t, x_t) &= x^t(t)\mathcal{P}x(t) + \int_{t-d}^t x^t(\theta)\mathcal{Q}x(\theta)d\theta\\
&+ d\int_{-d}^0\int_{t+\alpha}^t \dot{x}^t(\theta)\mathcal{Z}\dot{x}(\theta)d\theta\,d\alpha, \;\; \mathcal{P} > 0, \; \mathcal{Q} > 0, \; \mathcal{Z} > 0,
\end{aligned} \tag{1.49}$$

we obtain the following stability result.

Theorem 1.13 ([6]). *The time-delay system (1.15) is asymptotically stable for any delay satisfying $0 < d \leq d_M$ if there exist matrices $\mathcal{P} > 0$, $\mathcal{Q} > 0$, and $\mathcal{Z} > 0$ such that*

$$\begin{bmatrix}
\mathcal{P}A_0 + A_0^t\mathcal{P} + \mathcal{Q} - \mathcal{Z} & \mathcal{P}A_d + \mathcal{Z} & d_M A_0^t \mathcal{Z}\\
\bullet & -\mathcal{Q} - \mathcal{P} & d_M A_d^t \mathcal{Z}\\
\bullet & \bullet & -\mathcal{Z}
\end{bmatrix} < 0. \tag{1.50}$$

Alternatively, selecting the LKF

$$\mathbb{V}(t, x_t) = x^t(t)\mathcal{P}x(t) + \tau_M \int_{-\tau_M}^{0} \int_{t+\alpha}^{t} \dot{x}^t(\theta)\mathcal{Z}\dot{x}(\theta)d\theta\, d\alpha, \;\; \mathcal{P} > 0, \;\; \mathcal{Z} > 0, \tag{1.51}$$

we obtain the following stability result.

Theorem 1.14 ([6]). *The time-delay system* (1.21) *is asymptotically stable for all continuous delays* $\tau(t)$ *satisfying* $0 < \tau(t) \le \tau_M$ *if there exist matrices* $\mathcal{P} > 0$ *and* $\mathcal{Z} > 0$ *such that*

$$\begin{bmatrix} \mathcal{P}A_0 + A_0^t\mathcal{P} - \mathcal{Z} & \mathcal{P}A_d + \mathcal{Z} & \tau_M A_0^t \mathcal{Z} \\ \bullet & -\mathcal{Z} - \mathcal{P} & \tau_M A_d^t \mathcal{Z} \\ \bullet & \bullet & -\mathcal{Z} \end{bmatrix} < 0. \tag{1.52}$$

A useful result is summarized by the following lemma.

Lemma 1.6 (The integral inequality, [59]). *For any constant matrix* $0 < \Sigma \in \mathbb{R}^{n\times n}$, *a scalar* $\tau_* < \tau(t) < \tau^+$, *and a vector function* $\dot{x} : [t - \tau^+, t - \tau_*] \to \mathbb{R}^n$, *we have*

$$\begin{aligned} &-(\tau^+ - \tau_*) \int_{t-\tau^+}^{t-\tau_*} \dot{x}^t(s)\, \Sigma\, \dot{x}(s)\, ds \\ &\le -[x(t - \tau_*) - x(t - \tau^+)]^t\, \Sigma\, [x(t - \tau_*) - x(t - \tau^+)], \end{aligned} \tag{1.53}$$

provided that the integral is well-defined.

Lemma 1.6 is frequently called the "integral inequality" and is derived from Jensen's inequality [38].

Remark 1.9. It is significant to observe that Theorem 1.13 establishes that the time-delay system (1.15) is asymptotically stable for any delay d satisfying $0 < d \le d_M$ when the LMI (1.50) attains a feasible solution, which implies that for d satisfying $0 < d \le d_M/2$, the time-delay system (1.15) is asymptotically stable as well. Then, introducing the half delay into the time-delay system (1.15) will take more information on the system and thus may tend to reduce the conservatism in Theorem 1.13. For further elaboration on this argument, see [60].

1.6.7 Discrete-time systems

Less attention has been paid to discrete-time systems with a time-delay because a linear discrete-time system with constant time-delay can be transformed into a delay-free system by means of a state-augmentation approach.

However, this approach is not suitable for systems with either unknown or time-varying delays. For small time-varying delays, the descriptor model transformation approach was employed in [46].

Consider the class of discrete-time systems with state-delay is represented by

$$x(k+1) = A_o x(k) + D_o x(k - d(k)), \tag{1.54}$$

where for $k \in \mathbf{Z}_+ \triangleq \{0, 1, \dots\}$, $x(k) \in \mathbb{R}^n$ is the state, and $A_o \in \mathbb{R}^{n\times n}$ and $D_o \in \mathbb{R}^{n\times n}$ are constant matrices. The delay factor $d(k)$ is unknown but bounded of the form

$$0 < d_m \leq d(k) \leq d_M, \quad d_s = d_M - d_m + 1, \tag{1.55}$$

where the scalars d_m and $_M$ represent the lower and upper bounds, respectively, and d_s denotes the number of samples within the delay interval.

Remark 1.10. By setting $d(k) == \equiv 0$ in (1.54) it is readily seen that $|\lambda(A_o + D_o)| < 1$ is a necessary condition for stability of system (1.54). From all studies on discrete-time delay systems it is assumed that this is always the case.

Remark 1.11. The class of systems (1.54) represents a nominally linear model emerges in many areas dealing with applications to functional difference equations or delay-difference equations. These applications include cold rolling mills, decision-making processes, and manufacturing systems.

Related results for a class of discrete-time systems with time-varying delays can be found in [47], where delay-dependent stability and stabilization conditions were derived. It should be stressed that although we consider only the case of single time-delay, extension to multiple time-delay systems can be easily attained using an augmentation procedure.

Intuitively, if we associate with system (1.54) a positive-definite Lyapunov–Krasovskii functional $V(k, x(k)) > 0$ and find that its first difference $\Delta V(k, x(k)) = V(k+1, x(k+1)) - V(k, x(k))$ is negative-definite along the solutions of (1.54), then the origin of system (1.54) is globally asymptotically stable. Formally, we present the following theorem for discrete-time systems of the type (1.54).

Theorem 1.15. *The equilibrium* $\mathbf{0}$ *of the discrete-time system*

$$x(k+1) = h(x(k)) \tag{1.56}$$

is globally asymptotically stable if there is a function $V : \{0, 1, 2, \ldots\} \times \mathbb{R}^n \rightarrow \Re$ *such that*

- $V(k, x(k))$ *is a positive-definite function, decreasing, and radially unbounded,*
- $\Delta V(k, x(k)) = V(k+1, x(k+1)) - V(k, x(k))$ *is negative-definite along the solutions of system* (1.54).

For arbitrary value of $d(k)$, denote

$$z(k) = \left[x^t(k) \;\; \vdots \;\; x^t(k - d(k)) \right]^t .$$

We have

$$z(k+1) = \begin{bmatrix} A_o & 0 & \ldots & 0 & D_o \\ I & 0 & \ldots & 0 & 0 \\ \vdots & \ddots & \ldots & \vdots & \vdots \\ 0 & 0 & \ldots & I & 0 \end{bmatrix} z(k). \tag{1.57}$$

It is obvious that system (1.54) is globally asymptotically stable if and only if system (1.57) is globally asymptotically stable. For system (1.57), we define

$$\hat{V}(k, z(k)) = z^t(k)\text{diag}\left[\mathcal{P} \quad \mathcal{Q} \quad \ldots \quad \mathcal{Q} \right] z(k), \tag{1.58}$$

where $\mathcal{P} > 0$ and $\mathcal{Q} > 0$. It is easy to see that $\hat{V}(k, z(k)) > 0$ is decreasing and radially unbounded, and hence system (1.57) and thus system (1.54) are globally asymptotically stable.

By selecting the Lyapunov–Krasovskii functional

$$V(k) = x^t(k)\mathcal{P}x(k) + \sum_{m=k-d(k)}^{k-1} x^t(m)\mathcal{Q}x(m), \;\; 0 < \mathcal{P}, \; 0 < \mathcal{Q}, \tag{1.59}$$

and invoking the Lyapunov–Krasovskii theorem, the following stability condition can be derived.

Theorem 1.16. *The discrete-delay system* (1.54) *is asymptotically stable if there exist matrices* $\mathcal{P} > 0$ *and* $\mathcal{Q} > 0$ *such that*

$$\begin{bmatrix} -(\mathcal{P} - \mathcal{Q}) & 0 & A_o^t\mathcal{P} \\ \bullet & -\mathcal{Q} & D_o^t\mathcal{P} \\ \bullet & \bullet & -\mathcal{P} \end{bmatrix} < 0. \tag{1.60}$$

We stress that LMI (1.60) is virtually delay-independent since it is satisfied no matter the size of delay $d(k)$ is.

Next, we give sufficient delay-dependent LMI-based stability conditions. The approach used here does not introduce any dynamics and leads to a product separation between the matrices of the system and those from the Lyapunov–Krasovskii functional. The following theorem provides some LMI conditions depending on the values d_m and d_M.

Theorem 1.17. *Given the delay sample number d_s, system* (1.54) *subject to* (1.55) *is delay-dependent asymptotically stable if one of the following equivalent conditions is satisfied:*

A. *there exist matrices $0 < \mathcal{P} \in \mathbb{R}^{n\times n}$ and $0 < \mathcal{Q} \in \mathbb{R}^{n\times n}$ such that*

$$\Xi_a = \begin{bmatrix} A_o^t \mathcal{P} A_o + d_s \mathcal{Q} - \mathcal{P} & A_o^t \mathcal{P} D_o^t \\ \bullet & D_o^t \mathcal{P} D_o - \mathcal{Q} \end{bmatrix} < 0; \qquad (1.61)$$

B. *there exist matrices $0 < \mathcal{P} \in \mathbb{R}^{n\times n}$, $0 < \mathcal{Q} \in \mathbb{R}^{n\times n}$, $\mathcal{X} \in \mathbb{R}^{n\times n}$, $\mathcal{Y} \in \mathbb{R}^{n\times n}$, and $\mathcal{Z} \in \mathbb{R}^{n\times n}$ such that*

$$\Xi_c = \begin{bmatrix} \mathcal{P} + \mathcal{X} + \mathcal{X}^t & \mathcal{Y} - \mathcal{X} A_o & \mathcal{Z} - \mathcal{X} D_o \\ \bullet & \Gamma_v & -A_o^t \mathcal{Z}^t - \mathcal{Y} D_o \\ \bullet & \bullet & \Gamma_w \end{bmatrix} < 0, \qquad (1.62)$$

where

$$\Gamma_v = -A_o^t \mathcal{Z}^t - \mathcal{Y} A_o + d_s \mathcal{Q} - \mathcal{P},$$
$$\Gamma_v = -\mathcal{Q} - \mathcal{Z} D_o - D_o^t \mathcal{Z}^t.$$

In this case, the Lyapunov–Krasovskii functional (LKF)

$$\tilde{V}(k) = x^t(k)\mathcal{P}x(k) + \sum_{m=k-d(k)}^{k-1} x^t(m)\mathcal{Q}x(m) + \sum_{s=2-d_M}^{1-d_m} \sum_{m=k+s-1}^{k-1} x^t(m)\mathcal{Q}x(m) > 0 \qquad (1.63)$$

is such that

$$\Delta \tilde{V}(k) < 0, \ \forall \begin{bmatrix} x^t(k) & x^t(k-d(k)) \end{bmatrix}^t \neq \mathbf{0}. \qquad (1.64)$$

The result of Theorem 1.17 was developed in [47–49].

Next, we consider the following discrete-time piecewise linear systems with infinite distributed delays [50]:

$$x(k+1) = A_\ell x(k) + D_\ell \sum_{d=1}^{\infty} \mu_d x(k-d) + B_\ell u(k), \tag{1.65}$$

where $x(k) \in \mathbb{R}^n$ is the state, $\{S_\ell\}_{\ell \in L}$ denotes a partition of the state-space into a number of closed polyhedral subspaces, L is the index set of subspaces, $u(k) \in \mathbb{R}^m$ is the control input, and A_ℓ, D_ℓ, and B_ℓ are constant matrices of appropriate dimensions corresponding to the ℓth local model of the systems. When the state of the system transits from one region to another at time k, the dynamics is governed by the local model of the former one; the convergence constants $\mu_d \geq 0$ satisfy the following condition:

$$\bar{\mu} \triangleq \sum_{d=1}^{\infty} \mu_d \leq \sum_{d=1}^{\infty} d\, \mu_d < +\infty. \tag{1.66}$$

Distributed time delays have been widely recognized and intensively studied for continuous-time systems [51]. However, the corresponding results for discrete-time systems have been very few due mainly to the difficulty in formulating the distributed delays in a discrete-time domain. The distributed delay term $\sum_{d=1}^{\infty} \mu_d x(k-d)$ can be regarded as a discretization of the infinite integral form $\int_{-\infty}^{t} k(t-s)x(s)ds$ for the continuous-time system. The following result is recalled [51].

Lemma 1.7. *Let $0 \leq M \in \mathbb{R}^{n\times n}$, $x_j \in \mathbb{R}^n$, and $a_j > 0$, $j = 1, 2, \ldots$. If the series concerned is convergent, then we have*

$$\Big(\sum_{j=1}^{\infty} a_j x_j\Big)^t M \Big(\sum_{j=1}^{\infty} a_j x_j\Big) \leq \Big(\sum_{j=1}^{\infty} a_j\Big) \sum_{j=1}^{\infty} a_j x_j^t M x_j. \tag{1.67}$$

Introduce the following Lyapunov–Krasovskii functional candidate:

$$V(k) = x^t(k)\mathcal{P}_\ell x(k) + \sum_{d=1}^{\infty} \mu_d \sum_{m=k-d}^{k-1} x^t(m)\mathcal{Q}x(m). \tag{1.68}$$

By setting $\bar{\mathcal{Q}} = \bar{\mu}_d \mathcal{Q}$ and invoking Theorem 1.16 we obtain the following result.

Theorem 1.18. *Consider the piecewise linear system* (1.65) *with $u \equiv 0$. This system is asymptotically stable if there exist matrices $X_\ell > 0$ and $\bar{Q} > 0$ such that* *the following linear matrix inequalities hold for* $(\ell, j) \in \Omega \triangleq \{\ell, j | x(k) \in S_\ell,\ x(k+1) \in S_j,\ j \neq \ell\}$

$$\begin{bmatrix} -\mathcal{P} + \bar{\mathcal{Q}} & 0 & A_o^t \mathcal{P} \\ \bullet & -\frac{\bar{\mathcal{Q}}}{\bar{\mu}_d} & D_o^t \mathcal{P} \\ \bullet & \bullet & -\mathcal{P} \end{bmatrix} < 0 \tag{1.69}$$

We emphasize that Theorem 1.18 was established in [50].

1.7. Model transformations

We must recall that in the prototype system (1.15), the independent of delay (IOD) stability demands matrix A_0 to be Hurwitz, which, coherently, can be found in condition (1.18). On the other hand, the criteria ensuring delay-dependent stability for $h \in [0, h_M)$ require the matrix $A_0 + A_d$ to be Hurwitz as evident in condition (1.20). On this basis, several results concerning delay-dependent stability were derived from the formula

$$\int_{t-d}^{t} \dot{x}(s)\, ds = x(t) - x(t-d).$$

Consider the change of variables

$$A_j x(t-d) = [A_j - L_j] x(t-d_j) + L_j \big[x(t) - \int_{t-d_j}^{t} \dot{x}(s)\, ds\big]. \tag{1.70}$$

This will transform the multiple-delay system, with possibly $d_1 = 0$,

$$\dot{x}(t) = \sum_{j=1}^{m} A_j x(t-d_j) \tag{1.71}$$

into the system with augmented delay $h = \max(h_j + h_k)$

$$\dot{x}(t) = [\sum_{j=1}^{m} L_j] x(t) + \sum_{j=1}^{m} A_j x(t-d_j) + \sum_{j=1,k=1}^{m} \int_{t-d_j}^{t} L_j A_k x(s-d_j)\, ds. \tag{1.72}$$

Model (1.72) guarantees that the unstable nondelayed part A_d in system (1.15) is absorbed in the stable part L_1. Indeed, such a decomposition can be conveniently handled using LMI tools. It is shown in [5,52] that the foregoing system can be written in the three following forms:

$$\dot{x}(t) = Ax(t) - \sum_{j,k=1}^{m} A_{jk} \int_{t-d_{jk}}^{t-d_k} x(s)\, ds, \tag{1.73}$$

$$\dot{x}(t) = Ax(t) - \sum_{j=1}^{m} A_j \int_{t-d_j}^{t} \dot{x}(s)\, ds, \tag{1.74}$$

$$\frac{d}{dt}\left[x(t) + \sum_{j=1}^{m} A_j \int_{t-d_j}^{t} x(s)\, ds\right] = Ax(t), \tag{1.75}$$

$$A = \sum_{j=1}^{m} A_j, \;\; A_{jk} = A_j A_k, \;\; d_{jk} = d_j + d_k. \tag{1.76}$$

It turns out that each of these formulations can be studied by using specific Lyapunov–Krasovskii functionals (1.19) leading to three different Riccati equations [5]:

$$\Pi + md\mathcal{R} + \mathcal{P} \sum_{j,k=1}^{m} d_j A_{jk} \mathcal{R}^{-1} B_{jk}^t \mathcal{P} = -\mathcal{Q}, \tag{1.77}$$

$$\Pi + \sum_{j=1}^{m} (d_j \mathcal{P} A_j \mathcal{R}^{-1} B_j^t \mathcal{P} + md A_j^t \mathcal{R} A_j) = -\mathcal{Q}, \tag{1.78}$$

$$\Pi + \sum_{j=1}^{m} d_j \mathcal{R}_j + \sum_{j,k=1}^{m} d_j A^t \mathcal{P} A_j \mathcal{R}_j^{-1} A_j^t \mathcal{P} A = -\mathcal{Q}, \tag{1.79}$$

$$\Pi = \mathcal{P}A + A^t\mathcal{P} \tag{1.80}$$

Remark 1.12. In the literature, there were other different methods to develop delay-dependent stability criteria. These methods include the discretized LKF approach [4], the descriptor system approach [43], and the delay-partitioning projection approach [61]. Declaring the stability result as conservative or not requires well-defined quantitative measures. More importantly, it must be pointed out that the issue of computational complexity and the associated number of manipulated matrices deserve a serious investigation.

1.8. Delay-dependent stabilization

Extending the time-delay system (1.15) for stabilization studies, we start with the form

$$\dot{x}(t) = A_0\, x(t) + A_d x(t-d) + B_0 u(t), \;\; t \geq t_0, \tag{1.81}$$

where $u(t)$ is the control input, and B_0 is the input matrix with the pair A_0, B_0 being controllable. We seek to design a state-feedback controller

$$u(t) = \mathbb{K}_0\, x(t) \tag{1.82}$$

such that the closed-loop system

$$\dot{x}(t) = A_c\, x(t) + A_d x(t-d), \;\; A_c = A_0 + B_0 \mathbb{K}_0, \tag{1.83}$$

is asymptotically stable [13]. This is attained by convex analysis [18] leading to the following theorem.

Theorem 1.19. *The closed-loop time-delay system* (1.81) *is asymptotically stable for any* $d \geq 0$ *if there exist matrices* $\mathcal{X} > 0,\ \mathcal{Y},\ \mathcal{W} > 0$ *satisfying*

$$\begin{bmatrix} A_0\mathcal{X} + \mathcal{X}A_0^t + B_0\mathcal{Y} + \mathcal{Y}^t B_0^t + \mathcal{W} & A_d\mathcal{X} \\ \bullet & -\mathcal{W} \end{bmatrix} < 0. \tag{1.84}$$

1.8.1 A class of nonlinear systems

One of the standard classes of nonlinear time-delay systems is given by

$$\dot{x}(\sigma) = A_o x(\sigma) + A_d x(\sigma - \sigma_d) + B_o u(\sigma) + h(t, x, x(t-d)) \tag{1.85}$$

in the dimensionless coordinates σ, where $x(\sigma)$ is the state vector, $u(t)$ is the control input, and A_o, B_o, A_d are known real constant matrices. The nonlinear vector function $h(\cdot, \cdot)$ is a piecewise-continuous function in its arguments. In the discussions that follow, we assume that this uncertain function satisfies the quadratic inequality

$$h^t(t, x, x(t-d))\, h(t, x, x(t-d)) \leq \alpha^2\, x^t H_o^t H_o x + \theta^2\, x^t(t-d) H_d^t H_d x(t-d), \tag{1.86}$$

where $\alpha > 0$ and $\theta > 0$ are the bounding parameters. The matrices $H_o \in \mathbb{R}^{r\times n}$ and $H_d \in \mathbb{R}^{p\times n}$ are constant and characterize the upper bound on system nonlinearities.

For stability purposes, we let $\alpha^{-2} = \gamma$ and $\theta^{-2} = \psi$. The following convex optimization result holds.

Theorem 1.20. *Nonlinear system* (1.85) *with* $u \equiv 0$ *is robustly stable if the following LMI feasibility problem is solvable:*

$$\min \;\; \gamma\, ,\; \psi$$

$$\textit{subject to} \quad \mathcal{X} > 0,$$

$$\begin{bmatrix} A_o\mathcal{X} + \mathcal{X}A_o^t & \mathcal{X}\mathcal{W} & \mathcal{X}H_o^t & I & A_d & 0 \\ \bullet & -\mathcal{W} & 0 & 0 & 0 & 0 \\ \bullet & \bullet & -\gamma I & 0 & 0 & 0 \\ \bullet & \bullet & \bullet & -I & 0 & 0 \\ \bullet & \bullet & \bullet & \bullet & -W & H_d^t \\ \bullet & \bullet & \bullet & \bullet & \bullet & -\psi \end{bmatrix} < 0. \tag{1.87}$$

Given that the pair (A_o, B_o) is stabilizable, we achieve state feedback stabilization in two-stages as follows:

S1. Let the linear state feedback be $u(t) = K_o\, x(t)$. Then the closed-loop system becomes

$$\dot{x}(t) = A_k x(t) + A_d x(t-d) + h(t, x, x(t-d)), \quad A_k = (A_o + B_o K_o). \tag{1.88}$$

This establishes the following theorem.

Theorem 1.21. *Nonlinear system* (1.85) *is robustly stabilized by control law* $u(t) = K_o\, x(t)$ *if the following LMI problem has a feasible solution:*

$$\min \quad \gamma\,,\ \psi$$

subject to $\quad \mathcal{X} > 0\ ,\ \mathcal{Y},$

$$\begin{bmatrix} A_o\mathcal{X} + \mathcal{X}A_o^t + B_o\mathcal{Y} + \mathcal{Y}^t B_o^t & \mathcal{X}\mathcal{W} & \mathcal{X}H_o^t & I & A_d & 0 \\ \bullet & -\mathcal{W} & 0 & 0 & 0 & 0 \\ \bullet & \bullet & -\gamma I & 0 & 0 & 0 \\ \bullet & \bullet & \bullet & -I & 0 & 0 \\ \bullet & \bullet & \bullet & \bullet & -W & H_d^t \\ \bullet & \bullet & \bullet & \bullet & \bullet & -\psi \end{bmatrix} < 0. \tag{1.89}$$

S2. Next, to include bounds for the gain matrix K_o, we set the bounding relations

$$M^t M < \mu I,\ \mu > 0,\ X^{-1} < \varphi I,\ \varphi > 0.$$

Moreover, to guarantee desired values $\{\bar{\alpha}, \bar{\theta}\}$ of the bounding factors $\{\alpha, \theta\}$, we enforce $\alpha^{-2} = \gamma$ and $\theta^{-2} = \psi$. The following theorem summarizes the main result.

Theorem 1.22. *Nonlinear system* (1.85) *is robustly stabilized by the control law* $u(t) = K_o\, x(t)$ *with constrained feedback gains if following convex optimization problem over LMIs has a feasible solution:*

$$\min \quad \gamma + \psi + \mu + \varphi$$
$$\textit{subject to} \quad \mathcal{X} > 0, \ \mathcal{Y},$$
$$\begin{bmatrix} A_o\mathcal{X} + \mathcal{X}A_o^t + B_o\mathcal{Y} + \mathcal{Y}^t B_o^t & \mathcal{X}\mathcal{W} & \mathcal{X}H_o^t & I & A_d & 0 \\ \bullet & -\mathcal{W} & 0 & 0 & 0 & 0 \\ \bullet & \bullet & -\gamma I & 0 & 0 & 0 \\ \bullet & \bullet & \bullet & -I & 0 & 0 \\ \bullet & \bullet & \bullet & \bullet & -W & H_d^t \\ \bullet & \bullet & \bullet & \bullet & \bullet & -\psi \end{bmatrix} < 0, \quad (1.90)$$
$$\gamma - \frac{1}{\bar{\alpha}^2} < 0, \ \psi - \frac{1}{\bar{\sigma}^2} < 0,$$
$$\begin{bmatrix} -\mu I & \mathcal{Y}^t \\ \bullet & -I \end{bmatrix} < 0, \ \begin{bmatrix} -\varphi I & I \\ \bullet & -X \end{bmatrix} < 0. \quad (1.91)$$

Remark 1.13. We can address the performance deterioration issue by considering that the actual linear state-feedback controller has the form $u(t) = [K_o + \Delta K_o]\, x(t)$, where $K_o \in \mathbb{R}^{m\times n}$ is a constant gain matrix, and ΔK_o is a gain perturbation matrix.

1.9. Resilient delay-dependent control

Now we direct attention to delay-dependent methodologies for continuous- and discrete-time systems by exhibiting the delay-dependence dynamics in the design procedure.

1.9.1 Continuous-time systems

We consider the class of time-delay systems by

$$(\Sigma_J): \quad \dot{x}(t) = A_o x(t) + A_d x(t - \tau(t)) + B_o u(t) + \Gamma w(t),$$
$$x(s) = \phi(s), \ t \in [-\tau, 0], \quad (1.92)$$
$$y(t) = C_o x(t) + D_o u(t), \quad (1.93)$$
$$z(t) = G_o x(t) + F_o u(t) + \Phi w(t), \quad (1.94)$$

where $x(t) \in \mathbb{R}^n$ is the state vector, $u(t) \in \mathbb{R}^m$ is the control input, $w(t) \in \mathbb{R}^q$ is the disturbance input, which belongs to $\mathcal{L}_2[0, \mathcal{T}]$, $y(t) \in \mathbb{R}^p$, $z(t) \in \mathbb{R}^p$ is the controlled output, $\tau \in [\tau_0, \tau^*]$ is a constant delay factor, and A_o, B_o, A_d, C_o, D_o, G_o, F_o, Γ, and Φ are known real constant matrices of appropriate dimensions, which describe the nominal system of (Σ_J). The initial condition is specified as $\langle x(0), x(s)\rangle = \langle x_o, \phi(s)\rangle$, where $\phi \in \mathcal{L}_2[-\tau, 0]$.

Remark 1.14. Our purpose further is developing methods for determining delay-dependent stabilizing controllers that minimize the upper bound of an $\mathcal{H}_2$ performance measure while guaranteeing that a prescribed upper bound on an $\mathcal{H}_\infty$ performance is attained for all possible $w(t) \in \mathcal{L}_2[0,\infty]$. It will be shown that this problem is related to the existence of a feasible solution of linear matrix inequalities (LMIs), thereby providing a clear key to designing the state feedback controller.

1.9.2 Descriptor transformation

We employ the descriptor system approach to exhibit the delay-dependence dynamics in the design procedure. Thus we rewrite system (Σ_J) in the form

$$\begin{aligned}(\Sigma_D): \ \dot{x}(t) &= \sigma(t),\\ 0 &= -\sigma(t) + [A_o + A_d]x(t) + B_o u(t)\\ &\quad - A_d \int_{t-\tau(t)}^{t} \sigma(s)ds + \Gamma w(t), \qquad (1.95)\\ y(t) &= C_o x(t) + D_o u(t), \qquad (1.96)\\ z(t) &= G_o x(t) + F_o u(t) + \Phi w(t). \qquad (1.97)\end{aligned}$$

Under the state-feedback control law

$$u(t) = K_o\, x(t) \qquad (1.98)$$

in system (Σ_{Do}), it becomes

$$\begin{aligned}(\Sigma_{DK}): \quad \dot{x}(t) &= \sigma(t),\\ 0 &= -\sigma(t) + A_{Kd}x(t) - A_d \int_{t-\tau(t)}^{t} \sigma(s)ds + \Gamma w(t), \qquad (1.99)\\ y(t) &= C_K x(t), \qquad (1.100)\\ z(t) &= G_K x(t) + \Phi w(t), \qquad (1.101)\end{aligned}$$

where

$$\begin{aligned}&C_K = C_o + D_o K_o,\ G_K = G_o + F_o K_o,\ A_{od} = A_o + A_d,\\ &A_{Kd} = A_o + B_o K_o + A_d = A_{od} + B_o K_o. \qquad (1.102)\end{aligned}$$

In the discussions to follow, we develop tractable synthesis procedures to compute the feedback gain K_o.

1.9.3 Nominal $\mathcal{H}_2$ design

In this regard, we introduce the $\mathcal{H}_2$ performance measure

$$\mathsf{J}_2 \triangleq \left[\int_0^{\infty} y^t(s) y(s)\, ds \right]. \tag{1.103}$$

The problem of $\mathcal{H}_2$ state-feedback control can be stated as follows:

Given system Σ_J, determine the control law (1.98) **that achieves the minimal value of $\mathcal{H}_2$ performance measure** (1.103)**.**

For convenience, we introduce the matrices

$$\begin{aligned} \bar{P} &= \mathsf{U}\,\mathbb{P}\ ;\ \mathsf{U} = \begin{bmatrix} I & 0 \\ 0 & 0 \end{bmatrix}\ ;\ \mathbb{P} = \begin{bmatrix} P_x & 0 \\ P_d & P_\sigma \end{bmatrix}, \\ E_1 &= \begin{bmatrix} I \\ 0 \end{bmatrix},\ E_2 = \begin{bmatrix} 0 \\ I \end{bmatrix},\ Y = \mathbb{P}^{-1} = \begin{bmatrix} Y_x & 0 \\ Y_d & Y_\sigma \end{bmatrix}, \\ Y_x &= P_x^{-1},\ Y_\sigma = P_\sigma^{-1},\ Y_d = -Y_x P_d Y_\sigma, \end{aligned} \tag{1.104}$$

$$\mathcal{A}_{Kd} = \begin{bmatrix} 0 & I \\ A_{Kd} & -I \end{bmatrix},\ \bar{A}_d = \begin{bmatrix} 0 \\ A_d \end{bmatrix},\ \bar{C}_K = [C_K \ \ 0] \tag{1.105}$$

and provide the following stability result.

Theorem 1.23. *In the absence of input disturbance $w(t) \equiv 0$, controller* (1.98) *is an $\mathcal{H}_2$-optimal controller for system (Σ_{DK}) minimizing the $\mathcal{H}_2$-performance measure* (1.103) *if, given a matrix sequence $Q_\sigma = Q_\sigma^t > 0$, $i \in \mathcal{S}$, there exist matrices $P_x = P_x^t > 0$, $P_\sigma = P_\sigma^t > 0$, $P_d = P_d^t > 0$, $i \in \mathcal{S}$, satisfying the system of LMIs*

$$\Pi_t \triangleq \begin{bmatrix} \Omega_1 + C_K^t C_K & \Omega_2 & \tau P_d^t A_d \\ \bullet & -\Omega_3 & \tau P_\sigma^t A_d \\ \bullet & \bullet & -Q_\sigma \end{bmatrix} < 0, \tag{1.106}$$

where

$$\begin{aligned} \Omega_1 &= P_d^t \mathcal{A}_{Kd} + \mathcal{A}_{Kd}^t P_d\ ,\ \Omega_2 = P_x^t - P_d^t + \mathcal{A}_{Kd}^t P_\sigma, \\ \Omega_3 &= P_\sigma^t + P_\sigma - \tau\, Q_\sigma. \end{aligned} \tag{1.107}$$

An upper bound on the $\mathcal{H}_2$ performance measure is given by

$$\mathsf{J}_2 \le \mathsf{J}^+ \triangleq \left[x^t(0) P_x x(0) + \tau_o \int_{-\tau_o}^{0} \sigma^t(s)\, Q_\sigma\, \sigma(s)\, ds \right]^{1/2}. \tag{1.108}$$

Remark 1.15. It should be noted that Theorem 1.23 establishes an LMI-based sufficient condition for the existence of state-feedback controller (1.98), and hence it depends on K_o.

The following theorem provides a method for computing the feedback gains.

Theorem 1.24. *The feedback gain associated with the $\mathcal{H}_2$-optimal controller for system (Σ_{DK}) is given by $K_o = Z\, Y_x^{-1}$, where the matrices $Y_x = Y_x^t > 0$, $Y_\sigma = Y_\sigma^t > 0$, $Y_d = Y_d^t > 0$ satisfy the system of LMIs*

$$\begin{bmatrix} \Lambda_1 & \Lambda_2 & \tau Y_d^t & Y_x^t C_o^t + Z^t D_o^t \\ \bullet & -\Lambda_3 & \tau Y_\sigma^t & 0 \\ \bullet & \bullet & -\tau Q_\sigma & 0 \\ \bullet & \bullet & \bullet & -I \end{bmatrix} < 0, \qquad (1.109)$$

where

$$\Lambda_1 = Y_d^t + Y_d \ , \ \Lambda_2 = Y_\sigma - Y_d^t + Y_x^t A_{od}^t + Z^t B_o^t,$$
$$\Lambda_3 = Y_\sigma^t + Y_\sigma - \tau A_d Q_\sigma^{-1} A_d. \qquad (1.110)$$

1.9.4 Resilient $\mathcal{H}_2$ design

In the course of controller implementation, it turns out that the controllers are very sensitive to errors in the controller coefficients. It is considered beneficial that the designed (nominal) controllers should be capable of tolerating some level of controller gain variations. This illuminates the controller fragility problem, for which some relevant results are given in [6]. Looked at in this light, we consider the actual (implemented) control law

$$u(t) = [K_o + \Delta K]\, x(t) \ , \ \Delta K = H \Delta_c(t) E, \qquad (1.111)$$

where $\Delta_c(t)$ is the uncertainty matrix represented by

$$\Delta_c(t) \in \mathbf{\Delta_C(t)} \stackrel{\Delta}{=} \{\Delta_c(t) : \Delta_c^t(t)\Delta_c(t) \le I\}. \qquad (1.112)$$

The closed-loop system becomes

$$(\Sigma_{DK}): \ \dot{x}(t) = \sigma(t),$$

$$0 = -\sigma(t) + A_{\Delta K d} x(t) - A_d \int_{t-\tau(t)}^{t} \sigma(s) ds + \Gamma w(t), \qquad (1.113)$$

$$y(t) = \bar{C}_K x(t), \qquad (1.114)$$

where

$$A_{\Delta Kd} = A_o + B_oK_o + A_d + B_oH\Delta_cE = A_{od} + B_oK_o + B_oH\Delta_cE,$$
$$\hat{C}_K = C_o + D_oK_o + D_oH\Delta_cE. \quad (1.115)$$

The following summarizes the main result.

Theorem 1.25. *The perturbed feedback gain* (1.111) *associated with the* $\mathcal{H}_2$-*optimal controller for system* (Σ_{DK}) *is given by* $K_o = Z\,Y_x^{-1}$ *for some scalars* $\varepsilon > 0$, $\varrho > 0$. *where the matrices* $Y_x = Y_x^t > 0$, $Y_\sigma = Y_\sigma^t > 0$, $Y_d = Y_d^t > 0$ *satisfy the system of LMIs*

$$\begin{bmatrix} \Lambda_1 & \Lambda_2 & \tau Y_d^t & \begin{matrix} Y_x^t C_o^t \\ +Z^t D_o^t \end{matrix} & \varepsilon Y_d^t B_o H & \mathcal{Y} & \begin{matrix} Y_x^t C_o^t \\ +Z^t D_o^t \end{matrix} \\ \bullet & -\Lambda_3 & \tau Y_\sigma^t & 0 & \varepsilon Y_\sigma^t B_o H & 0 & 0 \\ \bullet & \bullet & -\tau Q_\sigma & 0 & 0 & 0 & 0 \\ \bullet & \bullet & \bullet & -I & 0 & 0 & 0 \\ \bullet & \bullet & \bullet & \bullet & -\varepsilon I & 0 & 0 \\ \bullet & \bullet & \bullet & \bullet & \bullet & -\Lambda_e & 0 \\ \bullet & \bullet & \bullet & \bullet & \bullet & \bullet & \begin{matrix} -I+ \\ \varrho D_o H H^t D_o^t \end{matrix} \end{bmatrix} < 0 \quad (1.116)$$

for all admissible gain perturbations, where

$$\Lambda_e = [\varepsilon I, \ \ \varrho I], \ \mathcal{Y} = [Y_x^t E^t, \ \ Y_x^t E^t].$$

1.9.5 Nominal $\mathcal{H}_\infty$ design

For a prescribed $\gamma > 0$, we introduce the performance measure

$$J_\infty \triangleq \left[\int_0^\infty \{z^t(s)z(s) - \gamma^2 w^t(s)w(s)\}\, ds \right]. \quad (1.117)$$

The problem of $\mathcal{H}_\infty$ state-feedback control could be phrased as follows:

Given system Σ_J, **determine the control law** (1.98) **that guarantees that** $\mathcal{H}_\infty$ **performance measure** (1.117) **is bounded by** γ **for all** $w(t) \in \mathcal{L}_2[0, \infty]$

The following two theorems summarize the main results pertaining to the $\mathcal{H}_\infty$ performance:

Theorem 1.26. *Given a prescribed scalar $\gamma > 0$, state feedback controller* (1.98) *renders system* (Σ_{DK}) *exponentially stable with a disturbance attenuation level γ for all $w(t) \in \mathcal{L}_2[0, \infty)$ if, given a matrix sequence $Q_x = Q_x^t > 0$, there exist matrices $P_x = P_x^t > 0$, $P_\sigma = P_\sigma^t > 0$, $P_d = P_d^t > 0$ satisfying the system of LMIs*

$$\Pi_h \triangleq \begin{bmatrix} \Omega_1 + G_K^t G_K & \Omega_2 & \tau P_d^t A_d & \mathbb{P}^t \bar{\Gamma} + \bar{G}_K^t \Phi \\ \bullet & -\Omega_3 & \tau P_\sigma^t A_d & 0 \\ \bullet & \bullet & -Q_\sigma & 0 \\ \bullet & \bullet & \bullet & -\gamma^2 I + \Phi^t \Phi \end{bmatrix} < 0. \tag{1.118}$$

Moreover,

$$||z(t)|| \leq \Bigg[\gamma^2 ||w(t)||^2 + x^t(0) P_x(\eta_o) x(0) + \tau_o \int_{-\tau_o}^{0} \sigma^t(s) Q_\sigma(\eta_o) \sigma(s) ds\Bigg]^{1/2}. \tag{1.119}$$

Theorem 1.27. *Given a prescribed constant $\gamma > 0$, the feedback gain associated with the $\mathcal{H}_\infty$-controller for system* (Σ_{DK}) *is given by $K_o = Z\, Y_x^{-1}$, where the matrices $Y_x = Y_x^t > 0$, $Y_\sigma = Y_\sigma^t > 0$, $Y_d = Y_d^t > 0$ satisfy the system of LMIs*

$$\begin{bmatrix} \Lambda_1 & \Lambda_2 & \tau Y_d^t & Y_x^t G_o^t + Z^t F_o^t & \Gamma + Y^t G_o^t \Phi \\ \bullet & -\Lambda_3 & \tau Y_\sigma^t & 0 & 0 \\ \bullet & \bullet & -\tau Q_\sigma & 0 & 0 \\ \bullet & \bullet & \bullet & -I & 0 \\ \bullet & \bullet & \bullet & \bullet & -\gamma^2 I + \Phi^t \Phi \end{bmatrix} < 0. \tag{1.120}$$

1.9.6 Resilient $\mathcal{H}_\infty$ design

Following the procedure of resilient $\mathcal{H}_2$ design, we develop hereafter a tractable method for resilient $\mathcal{H}_\infty$ design. The following summarizes the main result.

Theorem 1.28. *Given a prescribed scalar $\gamma > 0$, the perturbed feedback gain* (1.111) *associated with the $\mathcal{H}_\infty$-optimal controller for system* (Σ_{DK}) *is given by $K_o = Z\, Y_x^{-1}$ for some scalars $\varepsilon > 0$, $\varrho > 0$, where the matrices $Y_x = Y_x^t > 0$, $Y_\sigma = Y_\sigma^t > 0$, $Y_d = Y_d^t > 0$, satisfy the system of LMIs*

$$\begin{bmatrix} \Lambda_1 & \Lambda_2 & \tau Y_d^t & \begin{matrix} Y_x^t G_o^t + \\ Z^t F_o^t \end{matrix} & \begin{matrix} \Gamma + \\ Y^t G_o^t \Phi \end{matrix} & \varepsilon Y_d^t B_o H & \mathcal{Y} & \begin{matrix} Y_x^t C_o^t + \\ Z^t D_o^t \end{matrix} \\ \bullet & -\Lambda_3 & \tau Y_\sigma^t & 0 & 0 & \varepsilon Y_\sigma^t B_o H & 0 & 0 \\ \bullet & \bullet & -\tau Q_\sigma & 0 & 0 & 0 & 0 & 0 \\ \bullet & \bullet & \bullet & -I & 0 & 0 & 0 & 0 \\ \bullet & \bullet & \bullet & \bullet & \begin{matrix} -\gamma^2 I + \\ \Phi^t \Phi \end{matrix} & 0 & 0 & 0 \\ \bullet & \bullet & \bullet & \bullet & \bullet & -\varepsilon I & 0 & 0 \\ \bullet & \bullet & \bullet & \bullet & \bullet & \bullet & -\Lambda_e & 0 \\ \bullet & \bullet & \bullet & \bullet & \bullet & \bullet & \bullet & \begin{matrix} -I + \\ \varrho D_o H H^t D_o^t \end{matrix} \end{bmatrix} < 0 \tag{1.121}$$

for all admissible gain perturbations.

1.9.7 Simultaneous nominal $\mathcal{H}_2/\mathcal{H}_\infty$ design

Having developed two separate nominal design methods, we obviously tempted to combine both methods in one way or another. This can successfully be achieved by a simultaneous $\mathcal{H}_2/\mathcal{H}_\infty$ design approach. Adopting one approach, the problem of simultaneous $\mathcal{H}_2/\mathcal{H}_\infty$ state-feedback control could be phrased as follows:

Given system Σ_J, determine a control law of type (1.98) **that achieves the minimal value of $\mathcal{H}_2$ performance measure while guaranteeing that $\mathcal{H}_\infty$ performance measure is bounded by γ for all** $w(t) \in \mathcal{L}_2[0, \infty]$.

Remark 1.16. Observe that the objective of the simultaneous $\mathcal{H}_2/\mathcal{H}_\infty$ feedback control under consideration is minimizing the energy of the output $y(t)$ while simultaneously satisfying the prescribed $\mathcal{H}_\infty$-norm bound of the controlled system $w(t) \to z(t)$.

By exploiting the benefits of the foregoing results, the solution to the simultaneous nominal $\mathcal{H}_2/\mathcal{H}_\infty$ control problem is established by the following theorem.

Theorem 1.29. *Given a prescribed scalar $\gamma > 0$, the feedback controller* (1.98) *with gain $K_o = Z\, Y_x^{-1}$ is a simultaneous $\mathcal{H}_2/\mathcal{H}_\infty$ controller satisfying the performance measure* (1.117) *for system* (Σ_{DK}) *if there exist matrices $Y = Y^t > 0$, $Z, L = L^t > 0$, $W = W^t > 0$ such the system of generalized eigenvalue problems*

$$
\min \Big[\lambda + Tr(W) \Big]
$$

subject to (1.109), (1.120), *and*

$$
\begin{bmatrix} -L & I \\ I & -Q_x \end{bmatrix} < 0\,, \begin{bmatrix} -\lambda & \phi^t(0) \\ \phi(0) & -Y_x \end{bmatrix} < 0\,, \begin{bmatrix} -W & X^t \\ X & -L \end{bmatrix} < 0 \tag{1.122}
$$

has a feasible solution.

1.9.8 Simultaneous resilient $\mathcal{H}_2/\mathcal{H}_\infty$ design

Taking gain perturbations into account, we have the following result.

Theorem 1.30. *Given a prescribed scalar $\gamma > 0$, the feedback controller* (1.111) *with gain $K_o = Z\, Y_x^{-1}$ is a simultaneous resilient $\mathcal{H}_2/\mathcal{H}_\infty$ controller satisfying the performance measure* (1.117) *for system* (Σ_{DK}) *if there exist matrices $Y = Y^t > 0$, $Z, L = L^t > 0$, $W = W^t > 0$ such that the system of generalized eigenvalue problems*

$$
\min \Big[\lambda + Tr(W) \Big]
$$

subject to (1.116), (1.121), *and*

$$
\begin{bmatrix} -L & I \\ I & -Q_x \end{bmatrix} < 0\,, \begin{bmatrix} -\lambda & \phi^t(0) \\ \phi(0) & -Y_x \end{bmatrix} < 0\,, \begin{bmatrix} -W & X^t \\ X & -L \end{bmatrix} < 0 \tag{1.123}
$$

has a feasible solution.

1.9.9 Numerical simulation 1

To illustrate Theorem 1.29, we consider a pilot-scale single-reach water quality system, which can fall into the type (1.92)–(1.94) with the following associated data:

$$
A_o = \begin{bmatrix} -0.2 & 0 \\ 0 & -0.1 \end{bmatrix}, \; A_d = \begin{bmatrix} -0.1 & 0 \\ -0.1 & -0.1 \end{bmatrix}, \; \Gamma = \begin{bmatrix} 1 \\ 2 \end{bmatrix},
$$

$$
B_o = \begin{bmatrix} 1 & 0 \\ 0 & 1 \end{bmatrix}, \; C_o = \begin{bmatrix} 2 & 0 \\ 0 & 2 \end{bmatrix}, \; D_o = \begin{bmatrix} 1 & 0 \\ 0 & 1 \end{bmatrix},
$$

$$
G_o = [0.2 \;\; 0.1]\,, \; \Phi = 0.5\,, \; F_o = [2 \;\; 1].
$$

Invoking the Matlab®-LMI software environment, we solve the system of LMIs (1.122) using

$$Q_\sigma = \begin{bmatrix} 2 & 0 \\ 0 & 2 \end{bmatrix}, \quad \phi = \begin{bmatrix} e^{0.4\,t} \\ e^{0.4\,t} \end{bmatrix}, \quad \tau = 0.6.$$

The feasible solution is given by

$$Y_x = \begin{bmatrix} 2.2354 & 1.1532 \\ 1.1532 & 4.0422 \end{bmatrix}, \quad Y_\sigma = \begin{bmatrix} 0.9765 & 0.2240 \\ 0.2240 & 3.1367 \end{bmatrix},$$

$$Y_d = \begin{bmatrix} 2.0065 & 1.1213 \\ 1.1213 & 3.2524 \end{bmatrix},$$

$$Z = \begin{bmatrix} 1.0124 & -0.2341 \\ -0.4125 & 1.0332 \end{bmatrix}, \quad W = \begin{bmatrix} 5.2054 & 2.1443 \\ 2.1443 & 4.7805 \end{bmatrix},$$

$$\lambda = 1.2154, \quad \gamma = 2.1275.$$

The feedback gain and the associated performance bound are

$$K = \begin{bmatrix} 0.5661 & -0.2194 \\ -0.3710 & 0.3614 \end{bmatrix}, \quad \mathcal{J} = 11.2013.$$

This shows that the resilient controllers render the water quality model asymptotically with a disturbance attenuation level of $\gamma = 2.1275$.

1.9.10 Numerical simulation 2

Here we try another operating point of the model examined in Numerical simulation 1. It is described by the following information:

$$A_o = \begin{bmatrix} -2 & -1 \\ 0 & -2 \end{bmatrix}, \quad A_d = \begin{bmatrix} 0 & 1 \\ 1 & 0 \end{bmatrix}, \quad \Gamma = \begin{bmatrix} 2 \\ 1 \end{bmatrix},$$

$$B_o = \begin{bmatrix} 1 & 0 \\ 0 & 1 \end{bmatrix}, \quad C_o = \begin{bmatrix} 1 & 0 \\ 0 & 2 \end{bmatrix}, \quad D_o = \begin{bmatrix} 1 & 0 \\ 0 & 1 \end{bmatrix},$$

$$G_o = [0.1 \;\; 0.1], \quad \Phi = 0.3, \quad F_o = [1 \;\; 2].$$

Similarly, we solve the system of LMIs (1.122) using

$$Q_\sigma = \begin{bmatrix} 1 & 0 \\ 0 & 2 \end{bmatrix}, \quad \phi = \begin{bmatrix} e^{0.4\,t} \\ e^{0.4\,t} \end{bmatrix}, \quad \tau = 0.65.$$

This yields the feasible solution

$$Y_x = \begin{bmatrix} 5.2064 & 1.4423 \\ 1.4423 & 3.8605 \end{bmatrix}, \; Y_\sigma = \begin{bmatrix} 1.3264 & 0.7443 \\ 0.7443 & 2.8987 \end{bmatrix},$$

$$Y_d = \begin{bmatrix} 1.6554 & 1.4142 \\ 1.4142 & 6.8055 \end{bmatrix},$$

$$Z = \begin{bmatrix} 1.4564 & -0.6214 \\ -0.7250 & 1.2112 \end{bmatrix}, \; W = \begin{bmatrix} 5.2054 & 2.1443 \\ 2.1443 & 4.7805 \end{bmatrix},$$

$$\lambda = 1.3461 \; , \; \gamma = 3.6345.$$

The feedback gains and the associated performance bound are

$$K = \begin{bmatrix} 0.3618 & -0.2961 \\ -0.2533 & 0.4080 \end{bmatrix} \; , \; \mathcal{J} = 13.0542.$$

1.9.11 Discrete-time systems

Consider the following class of discrete-time state-delay systems with parametric uncertainties:

$$x_{k+1} = A_{\Delta o} x_k + B_o u_k + A_{\Delta d} x_{k-d_k} \; , \; x_0 = \psi_0, \tag{1.124}$$

where $x_k \in \mathbb{R}^n$ is the state vector, $u_k \in \mathbb{R}^p$ is the control input, d_k is a positive number representing the delay such that $d_k \le d^+$, and the uncertain matrices $A_\Delta \in \mathbb{R}^{n\times n}$ and $A_{\Delta d} \in \mathbb{R}^{n\times n}$ are represented by

$$[A_{\Delta o} \;\; A_{\Delta d}] = [A_o \;\; A_d] + M \Delta_k [N_a \;\; N_d], \tag{1.125}$$

where $A_o \in \mathbb{R}^{n\times n}$, $B_o \in \mathbb{R}^{n\times p}$, $C_o \in \mathbb{R}^{q\times n}$, $A_d \in \mathbb{R}^{n\times n}$, $M \in \mathbb{R}^{n\times \alpha}$, $N_a \in \mathbb{R}^{\beta\times n}$, and $N_d \in \mathbb{R}^{\beta\times n}$ are known real constant matrices with bounded matrix Δ_k of uncertainties such that $\Delta_k \in \mathbf{\Delta_k} \stackrel{\Delta}{=} \{\Delta_k^t : \Delta_k \Delta_k < I, \; \forall \, k\}$. For the time being, we set $u_k \equiv 0$ to yield the free uncertain system

$$x_{k+1} = A_{\Delta o} x_k + A_{\Delta d} x_{k-d_k}, \; x_0 = \psi_0. \tag{1.126}$$

By setting $d_k = 0$ we readily see that $|\lambda(A_o + A_d)| < 1$ is a necessary condition for the stability of system (1.126). From now on, we assume that this is the case. We further analyze two problems.

Problem A, in which we consider $d_k = d$ as unknown constant satisfying $0 \le d_k \le d^+$.

Problem B, in which we consider $d_k = d$ as time-varying unknown satisfying $0 \le d_k \le d^+$ and $\Delta_k \in \mathbf{\Delta_k} \triangleq \{\Delta_k^t : \Delta_k \Delta_k < I \,, \; \forall\, k\}$.

We study stability in two stages. In the first stage, we consider the nominal case by setting $\Delta_k \equiv 0$, and in the second stage, we incorporate the norm-bounded parametric uncertainties into the system dynamics.

1.9.12 Descriptor model transformation

Let y_k denote the state increment, that is,

$$y_k = x_{k+1} - x_k. \tag{1.127}$$

Then in line with Section 1.9.11, system (1.126) with $\Delta_k \equiv 0, \; u_k \equiv 0$ can be represented by the following descriptor form:

$$\begin{bmatrix} x_{k+1} \\ 0 \end{bmatrix} = \begin{bmatrix} y_k + x_k \\ -y_k + A_o\, x_k - x_k + A_d x_{k-d_k} \end{bmatrix}. \tag{1.128}$$

Recall by successive iterations on (1.127) that

$$x_{k-d_k} = x_k - \sum_{j=k-d_k}^{k-1} y_j,$$

and letting

$$\xi_k \triangleq \begin{bmatrix} x_k \\ y_k \end{bmatrix}, \quad E^t = \begin{bmatrix} I \\ 0 \end{bmatrix}, \quad A_{od} = A_o + A_d,$$

it follows that

$$\begin{aligned} E\,\xi_{k+1} &= \begin{bmatrix} I & I \\ A_{od} - I & -I \end{bmatrix} \xi_k - \begin{bmatrix} 0 \\ A_d \end{bmatrix} \sum_{j=k-d_k}^{k-1} y_j \\ &= \bar{A}_\xi\, \xi_k - \bar{A}_d \sum_{j=k-d_k}^{k-1} y_j, \end{aligned} \tag{1.129}$$

where the initial conditions are characterized by

$$\xi_0 = \begin{bmatrix} x_0 \\ y_0 \end{bmatrix} = \begin{bmatrix} \psi_0 \\ (A_o - I)\psi_0 - A_d \psi_{-d_0} \end{bmatrix}. \tag{1.130}$$

In short, if x_k is a solution of system (1.126) with $\Delta_k \equiv 0$ and $u_k \equiv 0$, then $\xi_k = \{x_k, y_k\}$ is a solution of the free descriptor system (1.129) subject to (1.130), and the reverse is true. This is the essence of descriptor transformation.

Theorem 1.31. *Consider system* (1.126) *without uncertainties* $\Delta_k \equiv 0$. *If there exists a continuous functional*

$$V_k \triangleq V(x_{k-d^+}, \ldots., x_k, y_{k-d^+}, \ldots., y_{k-1}) \tag{1.131}$$

such that

$$0 \leq V_k \leq \omega \max\left\{ \max_{k-d^+ \leq j \leq k} |x_j|^2 \, , \max_{k-d^+ \leq j \leq k-1} |y_j|^2 \right\}, \tag{1.132}$$

$$V_{k+1} - V_k \leq -\kappa \, |x_k|^2 \tag{1.133}$$

for x_k *and* y_k *satisfying* (1.129), *then system* (1.126) *is asymptotically stable.*

Proof. Summing up (1.133), it follows that

$$\sum_{j=0}^{k} V_{j+1} - V_j = V_{k+1} - V_0 \leq -\kappa \sum_{j=0}^{k} |x_j|^2.$$

Since x_k and y_k satisfy (1.129), (1.132) implies that

$$\begin{aligned} |x_k|^2 &\leq \sum_{j=0}^{k} |x_j|^2 \leq \kappa^{-1} V_0 \\ &\leq \kappa^{-1}\omega \max\left\{ \max_{-d^+ \leq j \leq 0} |x_j|^2 \, , \max_{-d^+ \leq j \leq -1} |y_j|^2 \right\} \; \forall k \geq 0. \end{aligned} \tag{1.134}$$

If x_k is a solution of (1.126) and y_k is defined by (1.127), then $\{x_k, y_k\}$ satisfies (1.128)–(1.130), and hence (1.134) holds.

Note that (1.134) implies that if $\max_{j\in[-d^+,0]} |\psi_{-j}|^2$ is sufficiently small, then $|x_k|^2$ is sufficiently small, and subsequently $\sum_{j=0}^{\infty} |x_j|^2 < \infty$. Therefore $|x_j|^2 \to 0$ as $j \to \infty$. □

To derive tractable conditions for stability, we introduce the Lyapunov–Krasovskii functional

$$\begin{aligned} V_k &= V_{k,a} + V_{k,b} + V_{k,c}, \\ V_{k,a} &= x_k^t \mathbb{P}_x x_k = \xi_k^t E^t \mathbb{P} E \xi_k \\ &= \xi_k^t E^t \begin{bmatrix} \mathbb{P}_x & 0 \\ \mathbb{P}_f & \mathbb{P}_d \end{bmatrix} E \xi_k \; , \; \mathbb{P}_x > 0, \end{aligned} \tag{1.135}$$

$$V_{k,b} = \sum_{p=-d^+}^{-1} \sum_{j=k+p}^{k-1} y_j^t \mathcal{Q} y_j \ , \ \mathcal{Q} > 0,$$

$$V_{k,c} = \sum_{j=k-d}^{k-1} x_k^t \mathcal{W} x_k \ , \ \mathcal{W} > 0. \tag{1.136}$$

We observe that V_k has a general form and is constructed from three terms: $V_{k,a}$ signifies necessary and sufficient conditions for the stability of discrete descriptor system without delay, $V_{k,b}$ corresponds to delay-dependent criteria [6], and $V_{k,c}$ is common for delay-independent stability conditions. For simplicity, we introduce the following matrix expressions:

$$\mathcal{W}_q = \begin{bmatrix} \mathcal{W} & 0 \\ \bullet & d^+\mathcal{Q} \end{bmatrix} , \ \begin{bmatrix} \mathcal{Z} & \mathcal{Y} \\ \bullet & \mathcal{Q} \end{bmatrix} \geq 0 \, , \ \mathcal{Z} \in \mathbb{R}^{2n\times 2n} \ , \ \mathcal{Y} \in \mathbb{R}^{2n\times n}. \tag{1.137}$$

We now address Problem A. The following theorem establishes LMI-based sufficient conditions for the asymptotic stability of system (1.126).

Theorem 1.32. *Consider system* (1.126) *without uncertainties* $\Delta_k \equiv 0$ *with the delay factor* $d_k = d$ *being an unknown constant satisfying* $0 \leq d_k \leq d^+$. *Given matrices* $0 < \mathcal{Q} = \mathcal{Q}^t \in \mathbb{R}^{n\times n}$ *and* $0 < \mathcal{W} = \mathcal{W}^t \in \mathbb{R}^{n\times n}$, *this system is asymptotically stable if there exist matrices* $0 < \mathbb{P} = \mathbb{P}^t \in \mathbb{R}^{2n\times 2n}$, $\mathcal{Y} \in \mathbb{R}^{2n\times n}$, *and* $\mathcal{Z} \in \mathbb{R}^{2n\times 2n}$ *satisfying* (1.137) *and the LMI*

$$\Upsilon(d^+) \triangleq \begin{bmatrix} \begin{array}{c} \bar{A}_\xi^t \mathbb{P} \bar{A}_\xi - E^t \mathbb{P} E + \\ \mathcal{W}_q + d^+ \mathcal{Z} + \mathcal{Y} E + E^t \mathcal{Y}^t \end{array} & \mathcal{Y} - \bar{A}_\xi^t \mathbb{P} \bar{A}_d \\ \bullet & -\mathcal{W} + \bar{A}_d^t \mathbb{P} \bar{A}_d \end{bmatrix} < 0. \tag{1.138}$$

Proof. Let

$$\zeta_k^t = [x_k^t \ \ 0] \ , \ \pi_k = x_k - x_{k-d_k}. \tag{1.139}$$

We consider V_k and evaluate the first difference of the functionals $V_{k,a}$, $V_{k,b}$, and $V_{k,c}$. Thus using (1.129), we have

$$\begin{aligned} V_{k+1,a} - V_{k,a} &= \zeta_{k+1}^t E^t \mathbb{P} E \zeta_{k+1} - \zeta_k^t E^t \mathbb{P} E \zeta_k \\ &= \zeta_{k+1}^t \mathbb{P} \zeta_{k+1} - \xi_k^t E^t \mathbb{P} E \xi_k \\ &= \left[\bar{A}_\xi \, \xi_k - \bar{A}_d \sum_{j=k-d_k}^{k-1} y_j \right]^t \mathbb{P} \left[\bar{A}_\xi \, \xi_k - \bar{A}_d \sum_{j=k-d_k}^{k-1} y_j \right] \\ &\quad - \xi_k^t E^t \mathbb{P} E \xi_k \end{aligned}$$

$$= \xi_k^t \Big[\bar{A}_\xi^t \mathbb{P} \bar{A}_\xi - E^t \mathbb{P} E \Big] \xi_k + \pi_k^t \bar{A}_d^t \mathbb{P} \bar{A}_d \pi_k$$
$$- 2 \sum_{j=k-d}^{k-1} \xi_k^t \bar{A}_\xi^t \mathbb{P} \bar{A}_d \gamma_j. \tag{1.140}$$

Similarly,

$$\begin{aligned} V_{k+1,b} - V_{k,b} &= d^+ \, \gamma_k^t \mathcal{Q} \gamma_k - \sum_{j=k-d^+}^{k-1} \gamma_j^t \mathcal{Q} \gamma_j \\ &= \xi_k^t \begin{bmatrix} 0 & 0 \\ 0 & d^+ \mathcal{Q} \end{bmatrix} \xi_k - \sum_{j=k-d^+}^{k-1} \gamma_j^t \mathcal{Q} \gamma_j, \end{aligned} \tag{1.141}$$

$$\begin{aligned} V_{k+1,c} - V_{k,c} &= x_k^t \mathcal{W} x_k - x_{k-d}^t \mathcal{W} x_{k-d} \\ &= \xi_k \begin{bmatrix} \mathcal{W} & 0 \\ 0 & 0 \end{bmatrix} \xi_k - x_{k-d}^t \mathcal{W} x_{k-d}. \end{aligned} \tag{1.142}$$

It follows from (1.139)–(1.142) that

$$\begin{aligned} V_{k+1} - V_k = \xi_k^t \Big[\bar{A}_\xi^t \mathbb{P} \bar{A}_\xi - E^t \mathbb{P} E + \mathcal{W}_q \Big] \xi_k + \pi_k^t \bar{A}_d^t \mathbb{P} \bar{A}_d \pi_k \\ - 2 \sum_{j=k-d}^{k-1} \xi_k^t \bar{A}_\xi^t \mathbb{P} \bar{A}_d \gamma_j \\ - x_{k-d}^t \mathcal{W} x_{k-d} - \sum_{j=k-d^+}^{k-1} \gamma_j^t \mathcal{Q} \gamma_j. \end{aligned} \tag{1.143}$$

Using Lemma 1.2 with

$$\mathbb{N} = \bar{A}_\xi^t \mathbb{P} \bar{A}_d \,, \; \beta = \gamma_j \,, \; \alpha = \xi_k,$$

we obtain

$$\begin{aligned} -2 \sum_{j=k-d}^{k-1} \xi_k^t \bar{A}_\xi^t \mathbb{P} \bar{A}_d \gamma_j &\le \sum_{j=k-d}^{k-1} \begin{bmatrix} \xi_k \\ \gamma_j \end{bmatrix}^t \begin{bmatrix} \mathcal{Z} & \mathcal{Y} - \bar{A}_\xi^t \mathbb{P} \bar{A}_d \\ \bullet & \mathcal{Q} \end{bmatrix} \begin{bmatrix} \xi_k \\ \gamma_j \end{bmatrix} \\ &= \sum_{j=k-d}^{k-1} \gamma_j^t \mathcal{Q} \gamma_j + d^+ \, 2 \xi_k^t \mathcal{Z} \xi_k + 2 \Big[\mathcal{Y} - \bar{A}_\xi^t \mathbb{P} \bar{A}_d \Big] x_k \\ &\quad - 2 \, \xi_k^t \Big[\mathcal{Y} - \bar{A}_\xi^t \mathbb{P} \bar{A}_d \Big] x_{k-d}. \end{aligned} \tag{1.144}$$

Substituting (1.144) into (1.143), applying Schur complements, and arranging terms, we get

$$V_{k+1} - V_k = \begin{bmatrix} \xi_k \\ x_{k-d} \end{bmatrix}^t \Upsilon(d^+) \begin{bmatrix} \xi_k \\ x_{k-d} \end{bmatrix}.$$

Based on LMI (1.138), we infer that $V_{k+1} - V_k < 0$, and, more importantly, from (1.135)–(1.137) it follows that $V_k \geq 0$ and $V_{k+1} - V_k < -\kappa|x_k|^2$ for all $k \geq 0$. By Theorem 1.31 the asymptotic stability is therefore guaranteed. □

Remark 1.17. Consider the nominal delayless system

$$x_{k+1} = A_{od}\, x_k \ , \ \ x_0 = \psi_0, \tag{1.145}$$

obtained from (1.126) by setting $\Delta_k \equiv 0$, $d_k \equiv 0$, and $u_k \equiv 0$. If $|\lambda(A_o + A_d)| < 1$, then system (1.145) is asymptotically stable, and, equivalently, its descriptor form

$$\xi_{k+1} = \bar{A}_\xi\, \xi_k$$

is asymptotically stable, that is, the LMI

$$\bar{A}_\xi^t \mathbb{P} \bar{A}_\xi - E^t \mathbb{P} E < 0$$

has a feasible solution $\mathbb{P} > 0$. This is also true for all possible and sufficiently small $d^+ > 0$. Since the results of Theorem 1.32 depend on the bound d^+, the stability condition is called weakly delay-dependent. In the particular case of

$$\mathcal{Z} = \upsilon I \ , \ \ \mathcal{Q} = \upsilon I \ , \ \ \mathcal{Y} = \upsilon[0, \ \ I] \ , \ \ \upsilon \longrightarrow 0,$$

LMI (1.138) reduces to

$$\Upsilon \triangleq \begin{bmatrix} \bar{A}_\xi^t \mathbb{P} \bar{A}_\xi - E^t(\mathbb{P} - \mathcal{W})E & A_\xi^t \mathbb{P} \bar{A}_d \\ \bullet & -\mathcal{W} + \bar{A}_d^t \mathbb{P} \bar{A}_d \end{bmatrix} < 0, \ \mathbb{P} > 0, \tag{1.146}$$

which provides an LMI characterization of delay-independent stability.

1.9.13 Robust stability

When we considered system (1.126) and applied the descriptor model transformation, we have obtained the descriptor system

$$E\,\xi_{k+1} = \bar{A}_{\Delta\xi}\,\xi_k - \bar{A}_{\Delta d}\sum_{j=k-d_k}^{k-1} \gamma_j, \tag{1.147}$$

where

$$\begin{aligned}
\bar{A}_{\Delta\xi} &= \begin{bmatrix} I & I \\ A_{\Delta od} - I & -I \end{bmatrix} = \begin{bmatrix} I & I \\ A_{od} - I & -I \end{bmatrix} + \begin{bmatrix} 0 & 0 \\ M\Delta_k[N_a + N_d] & 0 \end{bmatrix} \\
&= \bar{A}_\xi + \bar{M}\Delta_k\bar{N}_{ad}\ , \quad \bar{N}_{ad} = [N_a + N_d \ \ 0] = [N_{ad} \ \ 0], \\
\bar{A}_{\Delta d} &= \bar{A}_d + \bar{M}\Delta_k N_d\ , \quad \bar{M}^t = [0 \ \ M^t].
\end{aligned} \tag{1.148}$$

By Theorem 1.32 it follows that the uncertain system (1.126) is asymptotically stable if, given matrices $0 < \mathcal{Q} = \mathcal{Q}^t \in \mathbb{R}^{n\times n}$ and $0 < \mathcal{W} = \mathcal{W}^t \in \mathbb{R}^{n\times n}$, there exist matrices $0 < \mathbb{P} = \mathbb{P}^t \in \mathbb{R}^{2n\times 2n}$, $\mathcal{Y} \in \mathbb{R}^{2n\times n}$, and $\mathcal{Z} \in \mathbb{R}^{2n\times 2n}$ satisfying (1.137) and the LMI

$$\Upsilon_\Delta \triangleq \begin{bmatrix} \begin{array}{c} \bar{A}^t_{\Delta\xi}\mathbb{P}\bar{A}_{\Delta\xi} - E^t\mathbb{P}E + \\ \mathcal{W}_q + d^+\mathcal{Z} + \mathcal{Y}E + E^t\mathcal{Y}^t \end{array} & \mathcal{Y} - \bar{A}^t_{\Delta\xi}\mathbb{P}\bar{A}_{\Delta d} \\ \bullet & -\mathcal{W} + \bar{A}^t_{\Delta d}\mathbb{P}\bar{A}_{\Delta d} \end{bmatrix} < 0. \tag{1.149}$$

In terms of (1.148), we rewrite the LMI (1.149) in the form

$$\begin{aligned}
\Upsilon_\Delta \triangleq &\begin{bmatrix} \begin{array}{c} -E^t\mathbb{P}E + \mathcal{W}_q + \\ d^+\mathcal{Z} + \mathcal{Y}E + E^t\mathcal{Y}^t \end{array} & \mathcal{Y} \\ \bullet & -\mathcal{W} \end{bmatrix} \\
&+ \left(\begin{bmatrix} \bar{A}^t_{\Delta\xi} \\ -\bar{A}^t_{\Delta d} \end{bmatrix} \mathbb{P} + \begin{bmatrix} \bar{N}^t_{ad} \\ N^t_d \end{bmatrix} \Delta^t_k \bar{M}^t\mathbb{P} \right) \mathbb{P}^{-1} \\
&\left(\mathbb{P} \begin{bmatrix} \bar{A}^t_{\Delta\xi} \\ -\bar{A}^t_{\Delta d} \end{bmatrix}^t + \mathbb{P}\bar{M}\Delta_k \begin{bmatrix} \bar{N}^t_{ad} \\ N^t_d \end{bmatrix}^t \right) < 0.
\end{aligned} \tag{1.150}$$

A robust stability result is now established by the following theorem.

Theorem 1.33. *Consider system* (1.126) *with the delay factor $d_k = d$ being an unknown constant satisfying $0 \le d_k \le d^+$. Given matrices $0 < \mathcal{Q} = \mathcal{Q}^t \in \mathbb{R}^{n\times n}$ and $0 < \mathcal{W} = \mathcal{W}^t \in \mathbb{R}^{n\times n}$, this system is robustly asymptotically stable if there exist matrices $0 < \mathbb{P} = \mathbb{P}^t \in \mathbb{R}^{2n\times 2n}$, $\mathcal{Y} \in \mathbb{R}^{2n\times n}$, and $\mathcal{Z} \in \mathbb{R}^{2n\times 2n}$ and scalar $\varrho > 0$*

satisfying (1.137) *and the LMI*

$$\begin{bmatrix} \begin{array}{c} -E^t\mathbb{P}E+\mathcal{W}_q+ \\ d^+\mathcal{Z}+\mathcal{Y}E+E^t\mathcal{Y}^t \\ +\varrho\bar{N}^t_{ad}\bar{N}_{ad} \end{array} & \mathcal{Y}+\varrho\bar{N}^t_{ad}N_d & \bar{A}^t_\xi\mathbb{P} & 0 \\ \bullet & -\mathcal{W}+\varrho N^t_d N_d & -\bar{A}_d\mathbb{P} & 0 \\ \bullet & \bullet & -\mathbb{P} & \mathbb{P}\bar{M} \\ \bullet & \bullet & \bullet & -\varrho I \end{bmatrix} < 0. \quad (1.151)$$

Extending on Remark 1.17, we have the following result.

Theorem 1.34. *Consider system* (1.126) *with the delay factor* $d_k = d$ *being an unknown constant satisfying* $0 \le d_k \le d^+$. *Given a matrix* $0 < \mathcal{W} = \mathcal{W}^t \in \mathbb{R}^{n\times n}$, *this system is robustly asymptotically stable independently of delay if there exist matrices* $0 < \mathcal{P} = \mathcal{P}^t \in \mathbb{R}^{n\times n}$ *and scalars* $0 < \vartheta < 1$, $\sigma > 1$, *and* $\varrho > 0$ *satisfying the LMI*

$$\begin{bmatrix} \begin{array}{c} -E^t(\mathbb{P}-\mathcal{W})E \\ +\varrho\bar{N}^t_{ad}\bar{N}_{ad} \end{array} & \varrho\bar{N}^t_{ad}N_d & \bar{A}^t_\xi\mathbb{P} & 0 \\ \bullet & -\mathcal{W}+\varrho N^t_d N_d & -\bar{A}^t_d\, P & 0 \\ \bullet & \bullet & -\mathbb{P} & \mathbb{P}\bar{M} \\ \bullet & \bullet & \bullet & -\varrho I \end{bmatrix} < 0. \quad (1.152)$$

1.9.14 Nominal feedback stabilization

Now we consider the state stabilization problem for system (1.126). Initially, we use the nominal feedback controller

$$u_k = K_o\, x_k. \quad (1.153)$$

Applying controller (1.153) to system (1.124) with $\Delta_k \equiv 0$ yields the nominal closed-loop system

$$\begin{aligned} x_{k+1} &= [A_o + B_o K_o]x_k + A_d x_{k-d_k}\ , \quad x_0 = \psi_0 \\ &= A_{Ko}x_k + A_d x_{k-d_k}. \end{aligned} \quad (1.154)$$

With reference to (1.129), it is readily evident that the corresponding descriptor system matrix takes the form

$$\bar{A}_{K\xi} = \begin{bmatrix} I & I \\ A_{Kod}-I & -I \end{bmatrix}, \quad A_{Kod} = A_{Ko} + A_d. \quad (1.155)$$

Now, given matrices $0 < \mathcal{Q} = \mathcal{Q}^t \in \mathbb{R}^{n\times n}$ and $0 < \mathcal{W} = \mathcal{W}^t \in \mathbb{R}^{n\times n}$, it follows from Theorem 1.32 that system (1.154) is asymptotically stable if there exist matrices $0 < \mathbb{P} = \mathbb{P}^t \in \mathbb{R}^{2n\times 2n}$, $\mathcal{Y} \in \mathbb{R}^{2n\times n}$, and $\mathcal{Z} \in \mathbb{R}^{2n\times 2n}$ satisfying (1.137) and the LMI

$$\Upsilon_K(d^+) \triangleq \begin{bmatrix} \begin{array}{c} \bar{A}^t_{K\xi}\mathbb{P}\bar{A}_{K\xi} - E^t\mathbb{P}E + \\ \mathcal{W}_q + d^+\mathcal{Z} + \mathcal{Y}E + E^t\mathcal{Y}^t \end{array} & \mathcal{Y} - \bar{A}^t_{K\xi}\mathbb{P}\bar{A}_d \\ \bullet & -\mathcal{W} + \bar{A}^t_d\mathbb{P}\bar{A}_d \end{bmatrix} < 0. \tag{1.156}$$

Performing a Schur complement operation, LMI (1.156) becomes equivalent to

$$\begin{bmatrix} \begin{array}{c} -E^t\mathbb{P}E + \mathcal{W}_q + \\ d^+\mathcal{Z} + \mathcal{Y}E + E^t\mathcal{Y}^t \end{array} & \mathcal{Y} & \bar{A}^t_{K\xi}\mathbb{P} \\ \bullet & -\mathcal{W} & -\bar{A}^t_d\mathbb{P} \\ \bullet & \bullet & -\mathbb{P} \end{bmatrix} < 0. \tag{1.157}$$

The feedback stabilization result is readily established by the following theorem.

Theorem 1.35. *Consider system (1.124) with $\Delta_k \equiv 0$ and the delay factor $d_k = d$ being an unknown constant satisfying $0 \le d_k \le d^+$. Given matrices $0 < \mathcal{Q} = \mathcal{Q}^t \in \mathbb{R}^{n\times n}$ and $0 < \mathcal{W} = \mathcal{W}^t \in \mathbb{R}^{n\times n}$, this system is asymptotically stable if there exist matrices $0 < \mathbb{L}_x = \mathbb{L}^t_x \in \mathbb{R}^{n\times n}$, $0 < \mathbb{L}_d = \mathbb{L}^t_d \in \mathbb{R}^{n\times n}$, $\mathbb{L}_f \in \mathbb{R}^{n\times n}$, $\mathbb{L}_a \in \mathbb{R}^{n\times n}$, $\mathbb{L}_b \in \mathbb{R}^{n\times n}$, $\mathbb{L}_g \in \mathbb{R}^{n\times n}$, $\mathbb{L}_h \in \mathbb{R}^{n\times n}$, $\mathbb{B}_x \in \mathbb{R}^{m\times n}$, $\mathbb{B}_d \in \mathbb{R}^{m\times n}$, $\mathbb{M}_x \in \mathbb{R}^{n\times n}$, and $\mathbb{M}_d \in \mathbb{R}^{n\times n}$ satisfying the LMI*

$$\begin{bmatrix} -\mathbb{L}_x & \mathbb{L}_f + \mathbb{M}_x & \mathbb{L}_a & \mathbb{L}_x & \begin{array}{c} \mathbb{L}_x(A^t_{od} - I) \\ +\mathbb{B}^t_x B^t_o \end{array} & \mathbb{L}_g \\ \bullet & -\mathbb{M}_d & \mathbb{L}_b & \mathbb{L}_f + \mathbb{L}_d & \begin{array}{c} \mathbb{L}_f(A^t_{od} - I) \\ -\mathbb{L}_d + \mathbb{B}^t_d B^t_o \end{array} & \mathbb{L}_h \\ \bullet & \bullet & -\mathcal{W} & 0 & -A^t_d & 0 \\ \bullet & \bullet & \bullet & -\mathbb{L}_x & 0 & 0 \\ \bullet & \bullet & \bullet & -\mathbb{L}_f & -\mathbb{L}_d & 0 \\ \bullet & \bullet & \bullet & \bullet & \bullet & -\mathbb{H}_x \end{bmatrix} < 0. \tag{1.158}$$

Moreover, the nominal gain is given by $K_o = \mathbb{B}_x\, \mathbb{L}_x^{-1}$.

Considering the uncertain case $\Delta_k \neq 0$, we would obtain the closed-loop system

$$x_{k+1} = A_{\Delta Ko}x_k + A_d x_{k-d_k}\ , \ x_0 = \psi_0,$$

$$A_{\Delta Ko} = [A_{Ko} + M\Delta N_k]. \tag{1.159}$$

In a similar way, the descriptor system matrix takes the form

$$\begin{aligned} \bar{A}_{\Delta K\xi} &= \begin{bmatrix} I & I \\ A_{\Delta Kod} - I & -I \end{bmatrix} = \begin{bmatrix} I & I \\ A_{Kod} - I & -I \end{bmatrix} + \begin{bmatrix} 0 & 0 \\ M\Delta[N_a + N_d] & 0 \end{bmatrix} \\ &= \bar{A}_{K\xi} + \bar{M}\Delta_k\bar{N}_{ad} \ , \quad A_{Kod} = A_{od} + B_o K_o. \end{aligned} \tag{1.160}$$

By Theorem 1.33 it follows that system (1.159) is asymptotically stable if, given matrices $0 < \mathcal{Q} = \mathcal{Q}^t \in \mathbb{R}^{n\times n}$ and $0 < \mathcal{W} = \mathcal{W}^t \in \mathbb{R}^{n\times n}$, there exist matrices $0 < \mathbb{P} = \mathbb{P}^t \in \mathbb{R}^{2n\times 2n}$, $\mathcal{Y} \in \mathbb{R}^{2n\times n}$, and $\mathcal{Z} \in \mathbb{R}^{2n\times 2n}$ and scalar $\varrho > 0$ satisfying (1.137) and the LMI

$$\begin{bmatrix} \begin{array}{c} -E^t\mathbb{P}E + \mathcal{W}_q + \\ d^+\mathcal{Z} + \mathcal{Y}E + E^t\mathcal{Y}^t \\ +\varrho\bar{N}_{ad}^t\bar{N}_{ad} \end{array} & \mathcal{Y} + \varrho\bar{N}_{ad}^t N_d & \bar{A}_{K\xi}^t\mathbb{P} & 0 \\ \bullet & -\mathcal{W} + \varrho N_d^t N_d & -\bar{A}_d^t\mathbb{P} & 0 \\ \bullet & \bullet & -\mathbb{P} & \mathbb{P}\bar{M} \\ \bullet & \bullet & \bullet & -\varrho I \end{bmatrix} < 0. \tag{1.161}$$

Applying Schur complements convert LMI (1.161) into the form

$$\begin{bmatrix} \begin{array}{c} -E^t\mathbb{P}E + \mathcal{W}_q + \\ d^+\mathcal{Z} + \mathcal{Y}E + E^t\mathcal{Y}^t \end{array} & \mathcal{Y} & \bar{A}_{K\xi}^t\mathbb{P} & \varrho\bar{N}_{ad}^t \\ \bullet & -\mathcal{W} & -\bar{A}_d^t\mathbb{P} & \varrho N_d^t \\ \bullet & \bullet & -\mathbb{P} + \varrho^{-1}\mathbb{P}\bar{M}\bar{M}^t\mathbb{P} & 0 \\ \bullet & \bullet & \bullet & -\varrho I \end{bmatrix} < 0. \tag{1.162}$$

The robust feedback stabilization result is readily established by the following theorem.

Theorem 1.36. *Consider system (1.124) with the delay factor $d_k = d$ being an unknown constant satisfying $0 \le d_k \le d^+$. Given matrices $0 < \mathcal{Q} = \mathcal{Q}^t \in \mathbb{R}^{n\times n}$ and $0 < \mathcal{W} = \mathcal{W}^t \in \mathbb{R}^{n\times n}$, this system is robustly asymptotically stable if there exist matrices $0 < \mathbb{L}_x = \mathbb{L}_x^t \in \mathbb{R}^{n\times n}$, $0 < \mathbb{L}_d = \mathbb{L}_d^t \in \mathbb{R}^{n\times n}$, $\mathbb{L}_f \in \mathbb{R}^{n\times n}$, $\mathbb{L}_a \in \mathbb{R}^{n\times n}$, $\mathbb{L}_b \in \mathbb{R}^{n\times n}$, $\mathbb{L}_g \in \mathbb{R}^{n\times n}$, $\mathbb{L}_h \in \mathbb{R}^{n\times n}$, $\mathbb{B}_x \in \mathbb{R}^{m\times n}$, $\mathbb{B}_d \in \mathbb{R}^{m\times n}$, $\mathbb{M}_x \in \mathbb{R}^{n\times n}$, and*

$\mathbb{M}_d \in \mathbb{R}^{n\times n}$ *and scalars* $\mu > 0$ *satisfying the LMI*

$$\begin{bmatrix} -\mathbb{L}_x & \mathbb{L}_f+\mathbb{M}_x & \mathbb{L}_a & \mathbb{L}_x & \begin{matrix}\mathbb{L}_x(A^t_{od}-I)\\ +\mathbb{B}^t_x B^t_o\end{matrix} & \mathbb{L}_g & \mathbb{L}_x N^t_{ad} \\ \bullet & -\mathbb{M}_d & \mathbb{L}_b & \mathbb{L}_f+\mathbb{L}_d & \begin{matrix}\mathbb{L}_f(A^t_{od}-I)\\ -\mathbb{L}_d+\mathbb{B}^t_d B^t_o\end{matrix} & \mathbb{L}_h & \mathbb{L}_f N^t_{ad} \\ \bullet & \bullet & -\mathcal{W} & 0 & -A^t_d & 0 & N^t_d \\ \bullet & \bullet & \bullet & -\mathbb{L}_x & 0 & 0 & 0 \\ \bullet & \bullet & \bullet & -\mathbb{L}_f & -\mathbb{L}_d+\mu MM^t & 0 & 0 \\ \bullet & \bullet & \bullet & \bullet & \bullet & -\mathbb{H}_x & 0 \\ \bullet & \bullet & \bullet & \bullet & \bullet & \bullet & -\mu I \end{bmatrix} < 0. \tag{1.163}$$

Moreover, the nominal gain is given by $K_o = \mathbb{B}_x\,\mathbb{L}_x^{-1}$.

1.9.15 Resilient feedback stabilization

We now consider the practical case where the controller gain K_o is subject to perturbations due to various technical reasons. Thus the actual implemented control law would be

$$u_k = [K_o + \Delta K_o]x_k = [K_o + H\Delta_c N_c]x_k, \tag{1.164}$$

which implies that the gain perturbations are of additive type. Extension to multiplicative type can be derived in the manner of the foregoing chapters. In (1.164), $H \in \mathbb{R}^{m\times\alpha_k}$ and $N_c \in \mathbb{R}^{\beta_k\times n}$ are known real constant matrices with bounded matrix Δ_c of uncertainties such that $\Delta_c \in \mathbf{\Delta_c} \stackrel{\Delta}{=} \{\Delta^t_c : \Delta_c\Delta_c < I\ ,\ \forall\, k\}$.

Applying controller (1.164) to system (1.124) yields the perturbed closed-loop system

$$\begin{aligned} x_{k+1} &= [A_o + M\Delta_k N_a]x_k + B_o[K_o + H\Delta_c N_c]x_k + [A_d + M\Delta_k N_d]x_{k-d_k} \\ &= [A_{Ko} + M\Delta_k N_a + B_o H\Delta_c N_c] \\ &= \bar{A}_{\Delta Ko}x_k + A_{\Delta d}x_{k-d_k}\ ,\ x_0 = \psi_0. \end{aligned} \tag{1.165}$$

Now the perturbed descriptor system matrix takes the form

$$\begin{aligned} \hat{A}_{\Delta K\xi} &= \begin{bmatrix} I & I \\ \bar{A}_{\Delta Kod} - I & -I \end{bmatrix} \\ &= \begin{bmatrix} I & I \\ A_{Kod} - I & -I \end{bmatrix} + \begin{bmatrix} 0 & 0 \\ M\Delta[N_a + N_d] & 0 \end{bmatrix} + \begin{bmatrix} 0 & 0 \\ B_o H\Delta N_c & 0 \end{bmatrix} \end{aligned}$$

$$
\begin{aligned}
&= \bar{A}_{K\xi} + \bar{M}\Delta_k\bar{N}_{ad} + \bar{H}\Delta_c\bar{N}_c, \\
&\bar{H}^t = [0 \quad H^t B_o^t],\ \bar{N}_c = [N_c \quad 0].
\end{aligned} \tag{1.166}
$$

Again, by Theorem 1.33 it follows that system (1.165) is asymptotically stable if, given matrices $0 < \mathcal{Q} = \mathcal{Q}^t \in \mathbb{R}^{n\times n}$ and $0 < \mathcal{W} = \mathcal{W}^t \in \mathbb{R}^{n\times n}$, there exist matrices $0 < \mathbb{P} = \mathbb{P}^t \in \mathbb{R}^{2n\times 2n}$, $\mathcal{Y} \in \mathbb{R}^{2n\times n}$, and $\mathcal{Z} \in \mathbb{R}^{2n\times 2n}$ and scalars $\varrho > 0$ and $\psi > 0$ satisfying (1.137) and the LMI

$$
\begin{bmatrix}
\begin{array}{c} -E^t\mathbb{P}E + \mathcal{W}_q + \\ d^+\mathcal{Z} + \mathcal{Y}E + E^t\mathcal{Y}^t \\ +\varrho\bar{N}_{ad}^t\bar{N}_{ad} + \psi\bar{N}_c^t\bar{N}_c \end{array} & \mathcal{Y} + \varrho\bar{N}_{ad}^t N_d & \bar{A}_{K\xi}^t\mathbb{P} & 0 & \mathbb{P}\bar{M} \\
\bullet & -\mathcal{W} + \varrho N_d^t N_d & -\bar{A}_d^t\mathbb{P} & 0 & 0 \\
\bullet & \bullet & -\mathbb{P} & \mathbb{P}\bar{M} & 0 \\
\bullet & \bullet & \bullet & -\varrho I & 0 \\
\bullet & \bullet & \bullet & \bullet & -\psi I
\end{bmatrix} < 0. \tag{1.167}
$$

Applying Schur complements converts LMI (1.167) into the form

$$
\begin{bmatrix}
\begin{array}{c} -E^t\mathbb{P}E + \mathcal{W}_q + \\ d^+\mathcal{Z} + \mathcal{Y}E + E^t\mathcal{Y}^t \\ +\psi\bar{N}_c^t\bar{N}_c \end{array} & \mathcal{Y} & \bar{A}_{K\xi}^t\mathbb{P} & \varrho\bar{N}_{ad}^t & \mathbb{P}\bar{H} \\
\bullet & -\mathcal{W} & -\bar{A}_d^t\mathbb{P} & \varrho N_d^t & 0 \\
\bullet & \bullet & -\mathbb{P} + \varrho^{-1}\mathbb{P}\bar{M}\bar{M}^t\mathbb{P} & 0 & 0 \\
\bullet & \bullet & \bullet & -\varrho I & 0 \\
\bullet & \bullet & \bullet & \bullet & -\psi I
\end{bmatrix} < 0. \tag{1.168}
$$

The resilient feedback stabilization result is readily established by the following theorem.

Theorem 1.37. *Consider system* (1.124) *with the delay factor* $d_k = d$ *being an unknown constant satisfying* $0 \leq d_k \leq d^+$ *and subject to the perturbed feedback control* (1.164). *Given matrices* $0 < \mathcal{Q} = \mathcal{Q}^t \in \mathbb{R}^{n\times n}$ *and* $0 < \mathcal{W} = \mathcal{W}^t \in \mathbb{R}^{n\times n}$, *this system is robustly asymptotically stable if there exist matrices* $0 < \mathbb{L}_x = \mathbb{L}_x^t \in \mathbb{R}^{n\times n}$, $0 < \mathbb{L}_d = \mathbb{L}_d^t \in \mathbb{R}^{n\times n}$, $\mathbb{L}_f \in \mathbb{R}^{n\times n}$, $\mathbb{L}_a \in \mathbb{R}^{n\times n}$, $\mathbb{L}_b \in \mathbb{R}^{n\times n}$, $\mathbb{L}_g \in \mathbb{R}^{n\times n}$, $\mathbb{L}_h \in \mathbb{R}^{n\times n}$, $\mathbb{B}_x \in \mathbb{R}^{m\times n}$, $\mathbb{B}_d \in \mathbb{R}^{m\times n}$, $\mathbb{M}_x \in \mathbb{R}^{n\times n}$, *and* $\mathbb{M}_d \in \mathbb{R}^{n\times n}$ *and scalars*

$\mu > 0$ and $\psi > 0$ satisfying the LMI

$$\begin{bmatrix} -\mathbb{L}_x & \mathbb{L}_f + \mathbb{M}_x & \mathbb{L}_a & \mathbb{L}_x & \begin{matrix}\mathbb{L}_x(A^t_{od} - I) \\ +\mathbb{B}^t_x B^t_o\end{matrix} & \mathbb{L}_g & \mathbb{L}_x N^t_{ad} & 0 \\ \bullet & -\mathbb{M}_d & \mathbb{L}_b & \mathbb{L}_f + \mathbb{L}_d & \begin{matrix}\mathbb{L}_f(A^t_{od} - I) \\ -\mathbb{L}_d + \mathbb{B}^t_d B^t_o\end{matrix} & \mathbb{L}_h & \mathbb{L}_f N^t_{ad} & B_o H \\ \bullet & \bullet & -W & 0 & -A^t_d & 0 & N^t_d & 0 \\ \bullet & \bullet & \bullet & -\mathbb{L}_x & 0 & 0 & 0 & 0 \\ \bullet & \bullet & \bullet & -\mathbb{L}_f & -\mathbb{L}_d + \mu M M^t & 0 & 0 & 0 \\ \bullet & \bullet & \bullet & \bullet & \bullet & -\mathbb{H}_x & 0 & 0 \\ \bullet & \bullet & \bullet & \bullet & \bullet & \bullet & -\mu I & 0 \\ \bullet & \bullet & \bullet & \bullet & \bullet & \bullet & 0 & -\psi I \end{bmatrix} < 0$$

Moreover, the nominal gain is given by $K_o = \mathbb{B}_x \mathbb{L}_x^{-1}$.

1.9.16 Numerical simulation 3

We consider a second-order system with the following data

$$A_o = \begin{bmatrix} 0.81 & 0.1 \\ 0 & 0.96 \end{bmatrix}, \; A_d = \begin{bmatrix} -0.1 & 0 \\ -0.1 & -0.1 \end{bmatrix}.$$

Let $d_k = d$. It is required to determine the maximum value of d^+ guaranteeing the asymptotic stability of the system. The delay-dependent method of [6] gives $d^+ = 11$, whereas application of Theorem 1.32 yields $d^+ = 17$.

1.9.17 Numerical simulation 4

Consider the system of type (1.126) with

$$A_o = \begin{bmatrix} 0.81 & 0.09 \\ 0 & 0.97 \end{bmatrix}, \; A_d = \begin{bmatrix} -0.1 & 0.01 \\ -0.1 & -0.1 \end{bmatrix},$$

$$M = \begin{bmatrix} 0.2 & 0 \\ 0 & 0.01 \end{bmatrix}, \; N_a = \begin{bmatrix} 1.1 & 0 \\ 0 & 1.1 \end{bmatrix}, \; N_d = \begin{bmatrix} 0.6 & 0 \\ 0 & 0.4 \end{bmatrix}.$$

Applying Theorem 1.33, we obtain that the maximum value of d^+ guaranteeing the asymptotic stability is 6.

1.9.18 Numerical simulation 5

A discrete-time system of type (1.126) has

$$A_o = \begin{bmatrix} 1 & 0.02 \\ 0.02 & 0.49 \end{bmatrix}, \quad A_d = \begin{bmatrix} 0.1 & 0 \\ 0 & 0.1 \end{bmatrix}, \quad B_o = \begin{bmatrix} 1 \\ 0 \end{bmatrix},$$
$$M = \begin{bmatrix} 0.1 & 0 \\ 0 & 0.1 \end{bmatrix}, \quad N_a = \begin{bmatrix} 0.1 & 0.05 \\ -0.03 & 0.1 \end{bmatrix}, \quad N_d = \begin{bmatrix} 0.1 & 0 \\ -0.02 & 0.2 \end{bmatrix}.$$

Barring the uncertainties, we apply Theorem 1.35 and obtain the nominal feedback controller

$$u_k = [-0.9442 \quad -0.7614]\, x_k.$$

Including the uncertainties, we apply Theorem 1.36 and obtain the nominal feedback controller

$$u_k = [-0.9805 \quad -0.8135]\, x_k.$$

Taking the gain perturbations into account with $H = 0.9$ and $N_c = [0.1 \quad 0.7]$, the feasible solution gives the actual gain

$$K_o = [-0.6985 \quad -1.1025].$$

1.10. The cart-pole system

Consider a pendulum-cart system that consists of a pole mounted on a cart by means of a pivot allowing the pole to swing freely in the xy plane. Neglecting the friction in the pivot, the dynamics of the system can be described by

$$(m_1 + m_2)\ddot{x} + m_2\ell_2 cos(\theta)\ddot{\theta} - m_2\ell_2 \sin(\theta)\dot{\theta}^2 = F(t), \tag{1.169}$$
$$m_2\ell_2 \cos(\theta)\ddot{x} + (m_2\ell_2^2 + I_2)\ddot{\theta} - m_2 g\ell_2 \sin(\theta) = 0, \tag{1.170}$$

where x and θ are the positions of the cart and the unactuated joint of the pendulum, respectively, and $F(t)$ is the horizontal force applied to the cart to swing the pendulum until it reaches the desired unstable position $(\theta = 0,\ \dot{\theta} = 0,\ \dot{x} = 0)$.

Since the cart-pole system is underactuated, partial linearization technique is needed among others to reduce the design complexity of the controller. Following the approach in [8] and taking into consideration possible processing delays $\tau(t)$, a new control input u is chosen in terms of the cart acceleration designed to stabilize the cart-pole states at the origin. In this regard, the control input $F(t)$ is selected as

$$F(t-\tau(t)) = [m_2\ell_2\cos(\theta) - \frac{(m_1+m_2)(m_2\ell_2^2+I_2)}{m_2\ell_2}]u(t-\tau(t))$$
$$+ (m_1+m_2)g\tan(\theta) - m_2\ell_2\sin(\theta)\dot{\theta}^2$$

The cart-pole dynamics reduce to

$$\begin{aligned}\ddot{\theta}(t) &= u(t-\tau(t)),\\ \ddot{x}(t) &= g\,\tan(\theta) - \frac{(m_2\ell_2^2+I_2)}{m_2\ell_2}\cos(\theta)u(t-\tau(t)),\end{aligned} \tag{1.171}$$

which is of the strict feedforward form.

Seeking general finite-time stability and boundedness results, we cast model (1.171) into am LTI continuous-time system

$$\dot{x}(t) = Ax(t) + Cu(t-\tau) + B\omega(t), \tag{1.172}$$

$$x(\theta) = \varphi(\theta),\ \theta \in [-\tau, 0], \tag{1.173}$$

where $x(t) \in \mathbb{R}^n$ denotes the state vector, $u(t) \in \mathbb{R}^m$ denotes the control input vector, A, C, B are constant matrices of appropriate dimensions, $\varphi(\theta)$ is a differentiable vector-valued initial function on $[-\tau, 0]$; $\omega(t)$, meaning the time-varying exogenous disturbance, and satisfies

$$\int_0^{T_f} \omega^T(t)\,\omega(t)\,dt \le \Omega,$$

where x_0 is the initial state, t_0 is the initial time, and $\tau(t)$ is a differentiable time-varying delay satisfying

$$0 \le \tau(t) \le \tau_m,\quad \dot{\tau} \le \tau_M \le 1,$$

where τ_m and τ_M are upper bounds of the time-varying delay and its derivative, respectively. Assume that system (1.172) without delay and $\omega(t) \equiv 0\omega(t)$ is stabilizable. Then, for small enough τ_m, there exists a linear state-feedback $u(t) = Kx(t)$ that exponentially stabilizes system (1.172) for all piecewise-continuous $\tau(t) \in [0, \tau_m]$. With $D = CK$, the closed-loop takes the form

$$\dot{x}(t) = Ax(t) + Dx(t-\tau) + B\omega(t), \tag{1.174}$$

$$x(\theta) = \varphi(\theta),\ \theta \in [-\tau, 0]. \tag{1.175}$$

The following definitions and lemmas will be useful in the subsequent analysis.

Definition 1.3 (Finite-time boundedness). Given four positive constants c_1, c_2, h, T_f with $c_1 < c_2$ and a positive-definite matrix R, if

$$x^T(\bar{t_0})\,R\,x(\bar{t_0}) \le c_1 \Rightarrow x^T(t)\,R\,x(t) < c_2,\ \ \forall\, t \in [0,\ T_f]$$
$$\forall \int_0^{T_f} \omega^T(t)\,\omega(t)\,dt \le h, \tag{1.176}$$

where $x^T(\bar{t_0})\,R\,x(\bar{t_0}) = \sup_{-\tau\le s\le 0}\{x^T(s)\,R\,x(s),\ \dot{x}^T(s)\,R\,\dot{x}(s)\}$, then system (1.174) is said to be finite-time bounded with respect to $(c_1,\ c_2,\ h,\ T_f)$.

Remark 1.18. Note that the concept of finite-time boundedness is more general than that of finite-time stability. The definition of finite-time boundedness recovers the definition of finite-time stability as a particular case.

Lemma 1.8 (The integral inequality). *For any constant matrix $0 < \Sigma \in \mathbb{R}^{n\times n}$, a scalar $\tau_* < \tau(t) < \tau^+$, and a vector function $\dot{x} : [t-\tau^+, t-\tau_*] \to \mathbb{R}^n$ such that the following integration is well-defined, we have*

$$-(\tau^+ - \tau_*)\int_{t-\tau^+}^{t-\tau_*} \dot{x}^t(s)\ \Sigma\ \dot{x}(s)\ ds$$
$$\le -[x(t-\tau_*) - x(t-\tau^+)]^t\ \Sigma\ [x(t-\tau_*) - x(t-\tau^+)].$$

Now we choose a Lyapunov–Krasovskii functional candidate of the form

$$V(t) = V_1(t) + V_2(t) + V_3(t) + V_4(t), \tag{1.177}$$

where

$$V_1(t) = x^T(t)Px(t),$$
$$V_2(t) = \int_{t-\tau(t)}^{t} e^{-\alpha_0(s-t)}x^T(s)Qx(s)\ dt,$$

$$V_3(t) = \int_{t-\tau_m}^{t} e^{-\alpha_0(s-t)} x^T(s) W x(s)\, dt,$$

$$V_4(t) = \tau_m \int_{-\tau_m}^{0} \int_{t+\theta}^{t} e^{-\alpha_0(s-t)} \dot{x}^T(s) S \dot{x}(s)\, dt. \tag{1.178}$$

Let us introduce the extended vector $\eta^T(t) = [x^T(t),\ x^T(t - -\tau(t)),\ x^T(t - -\tau_m),\ \omega^T(t)]$.

Now we are in the position to discuss the finite-time boundedness of delay system (1.172). Evaluating the derivative of $V_1(t), \ldots, V_4(t)$ along the trajectories of system (1.172) leads to

$$\dot{V}_1(t) = 2x^T(t) P \dot{x}(t), \tag{1.179}$$

$$\begin{aligned}\dot{V}_2(t) &= \alpha_0 V_2 + x^T(t) Q x(t) - (1 - \dot{\tau}) e^{-\alpha_0 \tau(t)} x^T(t-\tau) Q x(t-\tau)\\ &\le \alpha_0 V_2 + x^T(t) Q x(t) - (1 - \tau_M) e^{-\alpha_0 \tau_m} x^T(t-\tau) Q x(t-\tau),\end{aligned} \tag{1.180}$$

$$\dot{V}_3(t) = \alpha_0 V_3 + x^T(t) W x(t) - e^{-\alpha_0 \tau_m} x^T(t-\tau_m) W x(t-\tau_m), \tag{1.181}$$

$$\begin{aligned}\dot{V}_4(t) &= \alpha_0 V_4 + \tau_m^2 \dot{x}^T(t) S \dot{x}(t) - \tau_m \int_{-\tau_m}^{0} e^{-\alpha_0 \tau_m} \dot{x}^T(t+\theta) S \dot{x}(t+\theta)\, d\theta\\ &= \alpha_0 V_4 + \tau_m^2 \dot{x}^T(t) S \dot{x}(t) - \tau_m \int_{t-\tau_m}^{t} e^{-\alpha_0 \tau_m} \dot{x}^T(s) S \dot{x}(s)\, ds.\end{aligned} \tag{1.182}$$

By Lemma 1.6, (1.182) is bounded by

$$\begin{aligned}\dot{V}_4(t) &\le \alpha_0 V_4 + \tau_m^2 \dot{x}^T(t) S \dot{x}(t) - x^T(t) S x(t)\\ &\quad - x^T(t-\tau_m) S x(t-\tau_m) + x^T(t-\tau_m) S x(t) + x^T(t) S x(t-\tau_m).\end{aligned} \tag{1.183}$$

It follows from (1.178)–(1.183) that

$$\begin{aligned}\dot{V}(t) &\le \alpha_0 V + x^T(t)[Q + W - S - \alpha_0 P] x(t) + 2x^T(t) P \dot{x}(t)\\ &\quad - x^T(t-\tau_m)[S + e^{-\alpha_0 \tau_m} W] x(t-\tau_m)\\ &\quad - (1-\tau_M) e^{-\alpha_0 \tau_m} x^T(t-\tau) Q x(t-\tau) + x^T(t-\tau_m) S x(t)\\ &\quad + x^T(t) S x(t-\tau_m) + \tau_m^2 \dot{x}^T(t) S \dot{x}(t).\end{aligned} \tag{1.184}$$

1.11. Kalman filtering

The seminal Kalman filtering algorithm [83] is the optimal estimator over all possible linear ones and gives unbiased estimates of the unknown state vectors under the condition that the system and measurement noise processes are mutually independent Gaussian distributions. We consider the state-estimation problem for a class of linear continuous time-lag systems with norm-bounded parameter uncertainties.

1.11.1 A class of continuous time-lag systems

We consider a class of uncertain time-delay systems represented by

$$\begin{aligned} \dot{x}(t) &= [A(t) + \Delta A(t)]x(t) + A_d(t)x(t-\tau) + w(t) \\ &= A_\Delta(t)x(t) + A_d(t)x(t-\tau) + w(t), \end{aligned} \tag{1.185}$$

$$\begin{aligned} y(t) &= [C(t) + \Delta C(t)]x(t) + v(t) \\ &= C_\Delta(t)x(t) + v(t), \end{aligned} \tag{1.186}$$

where $x(t) \in \mathbb{R}^n$ is the state, $y(t) \in \mathbb{R}^m$ is the measured output, and $w(t) \in \mathbb{R}^n$ and $v(t) \in \mathbb{R}^m$ are, respectively, the process and measurement noises. In (1)–(2), $A(t) \in \mathbb{R}^{n\times n}$, $A_d(t) \in \mathbb{R}^{n\times n}$, and $C(t) \in \mathbb{R}^{m\times n}$ are piecewise-continuous matrix functions. Here τ is a constant scalar representing the amount of time-lag in the state. The matrices $\Delta A(t)$ and $\Delta C(t)$ represent time-varying parametric uncertainties of the form

$$\begin{bmatrix} \Delta A(t) \\ \Delta C(t) \end{bmatrix} = \begin{bmatrix} H(t) \\ H_c(t) \end{bmatrix} \Delta(t)\, E(t), \tag{1.187}$$

where $H(t) \in \mathbb{R}^{n\times\alpha}$, $H_c(t) \in \mathbb{R}^{m\times\alpha}$, and $E(t) \in \mathbb{R}^{\beta\times n}$ are known piecewise-continuous matrix functions, and $\Delta(t) \in \mathbb{R}^{\alpha\times\beta}$ is an unknown matrix with Lebesgue measurable elements satisfying

$$\Delta^t(t)\Delta(t) \leq I \quad \forall\, t. \tag{1.188}$$

The initial condition is specified as $\langle x(0), x(s)\rangle = \langle x_o, \phi(s)\rangle$, where $\phi \in \mathcal{L}_2[-\tau, 0]$, which is assumed to be a zero-mean Gaussian random vector. We recall the following standard assumptions on noise statistics.

Assumption 1.1. $\forall t, s \geq 0$,

$$\textbf{(a)} \quad \mathbb{E}[w(t)] = 0\ ;\ \mathbb{E}[w(t)w^t(s)] = W(t)\delta(t-s)\ ;\ W(t) > 0, \tag{1.189}$$

$$\textbf{(b)} \quad \mathbb{E}[v(t)] = 0\ ;\ \mathbb{E}[v(t)v^t(s)] = V(t)\delta(t-s)\ ;\ V(t) > 0, \tag{1.190}$$

$$\textbf{(c)} \quad \mathbb{E}[x(0)w^t(t)] = 0\,;\, \mathbb{E}[x(0)v^t(t)] = 0, \tag{1.191}$$

$$\textbf{(d)} \quad \mathbb{E}[w(t)v^t(s)] = 0\,;\, \mathbb{E}[x(0)x^t(0)] = R_o, \tag{1.192}$$

where $\mathbb{E}$ stands for the mathematical expectation, and δ is the Dirac function.

1.11.2 Robust Kalman filtering

Our objective is to design a stable state estimator of the form

$$\dot{\hat{x}}(t) = G(t)\,\hat{x}(t) + K(t)\,y(t), \quad \hat{x}(0) = 0, \tag{1.193}$$

where $G(t) \in \mathbb{R}^{n\times n}$ and $K(t) \in \mathbb{R}^{n\times m}$ are piecewise-continuous matrices to be determined such that there exists a matrix $\Psi \geq 0$ satisfying

$$\mathbb{E}[(x-\hat{x})(x-\hat{x})^t] \leq \Psi\ , \ \forall\Delta : \Delta^t(t)\Delta(t) \leq I. \tag{1.194}$$

Note that (1.189) implies

$$\mathbb{E}[(x-\hat{x})^t(x-\hat{x})] \leq tr(\Psi)\ , \ \forall\Delta : \Delta^t(t)\Delta(t) \leq I. \tag{1.195}$$

In this case, estimator (1.193) is said to provide a guaranteed cost (GC) matrix Ψ.

Examination of the proposed estimator proceeds by analyzing the estimation error

$$e(t) = x(t) - \hat{x}(t). \tag{1.196}$$

Substituting (1.186) and (1.193) into (1.196), we express the dynamics of the error in the form

$$\begin{aligned}\dot{e}(t) &= G(t)\,e(t) + [A(t) - G(t) - K(t)C(t)]x(t) \\ &+ [\Delta A(t) - K(t)\Delta C(t)]x(t) \\ &+ A_d(t)x(t-\tau) + [w(t) - K(t)v(t)].\end{aligned} \tag{1.197}$$

By introducing the extended state-vector

$$\xi(t) = \begin{bmatrix} x(t) \\ e(t) \end{bmatrix} \in \mathbb{R}^{2n} \tag{1.198}$$

it follows from (1.185)–(1.186) and (1.197) that

$$\begin{aligned}\dot{\xi}(t) &= [\widehat{A}(t) + \widehat{H}(t)F(t)\widehat{E}(t)]\xi(t) + \widehat{D}(t)\xi(t-\tau) + \widehat{B}(t)\eta(t) \\ &= \widehat{A}_\Delta(t)\xi(t) + \widehat{D}(t)\xi(t-\tau) + \widehat{B}(t)\eta(t),\end{aligned} \tag{1.199}$$

where $\eta(t)$ is a stationary zero-mean noise signal with identity covariance matrix, and

$$\widehat{A}(t) = \begin{bmatrix} A(t) & 0 \\ A(t) - G(t) - K(t)C(t) & G(t) \end{bmatrix}, \tag{1.200}$$

$$\widehat{H}(t) = \begin{bmatrix} H(t) \\ H(t) - K(t)H_c(t) \end{bmatrix}, \quad \widehat{E}(t) = [E(t) \ \ 0], \tag{1.201}$$

$$\widehat{B}\widehat{B}^t(t) = \begin{bmatrix} W(t) & W(t) \\ W(t) & W(t) + K(t)V(t)K^t(t) \end{bmatrix}, \tag{1.202}$$

$$\widehat{D}(t) = \begin{bmatrix} A_d(t) & 0 \\ A_d(t) & 0 \end{bmatrix}, \quad \eta = \begin{bmatrix} w(t) \\ v(t) \end{bmatrix}. \tag{1.203}$$

Definition 1.4. Estimator (1.188) is said to be a quadratic estimator (QE) associated with a matrix $\Omega(t) > 0$ for system (1.185) if there exist a scalar $\lambda(t) > 0$ and a matrix

$$0 < \Omega(t) = \begin{bmatrix} \Omega_1(t) & \Omega_3(t) \\ \Omega_3^t(t) & \Omega_2(t) \end{bmatrix} \tag{1.204}$$

satisfying the algebraic inequality:

$$\begin{aligned} &- \dot{\Omega}(t) + \widehat{A}_\Delta(t)\,\Omega(t) + \Omega(t)\,\widehat{A}_\Delta^t(t) + \lambda(t)\,\Omega(t-\tau) \\ &+ \lambda^{-1}(t)\,\widehat{D}(t)\,\Omega(t-\tau)\,\widehat{D}^t(t) + \widehat{B}(t)\widehat{B}^t(t) \le 0. \end{aligned} \tag{1.205}$$

The next result shows that if (1.189) is QE for system (1.185)–(1.186) with cost matrix $\Omega(t)$, then $\Omega(t)$ defines an upper bound for the filtering error covariance, that is,

$$\mathbb{E}[e(t)e^t(t)] \le \Omega_2(t) \quad \forall\ t$$

for all admissible uncertainties satisfying (1.187)–(1.188).

Theorem 1.38. *Consider the time-delay* (1.185)–(1.186) *satisfying* (1.187)–(1.188) *with known initial state. Suppose there exists a solution* $\Omega(t) \ge 0$ *to inequality (21) for some* $\lambda(t) > 0$ *and for all admissible uncertainties. Then estimator* (1.193) *provides an upper bound for the filtering error covariance, that is,*

$$\mathbb{E}[e(t)e^t(t)] \le \Omega_2(t).$$

We employ hereafter a Riccati equation approach to solve the robust Kalman filtering for time-delay systems. To this end, we define piecewise matrices $P(t) = P^t(t) \in \mathbb{R}^{n\times n}$ and $L(t) = L^t(t) \in \mathbb{R}^{n\times n}$ as the solutions of the Riccati differential equations (RDEs)

$$\begin{aligned} \dot{P}(t) &= A(t)P(t) + P(t)A^t(t) + \lambda(t)P(t-\tau) + \hat{W}(t) \\ &+ \lambda^{-1}(t)A_d(t)P(t-\tau)A_d^t(t) + \mu(t)P(t)E^t(t)E(t)P(t), \end{aligned}$$

$$P(t-\tau)=0 \quad \forall t \in [0,\tau], \tag{1.206}$$

$$\begin{aligned}\dot{L}(t) &= A(t)L(t)+L(t)A^t(t)+\lambda(t)L(t-\tau)+\hat{W}(t)\\ &+\lambda^{-1}(t)A_d(t)P(t-\tau)A_d^t(t)+\mu(t)L(t)E^t(t)E(t)L(t)\\ &-[L(t)C^t(t)+\mu^{-1}(t)H(t)H_c^t(t)]\hat{V}^{-1}(t)\\ &[C(t)L(t)+\mu^{-1}(t)H_c(t)H^t(t)],\end{aligned}$$

$$L(t-\tau)=0 \quad \forall t \in [0,\tau], \tag{1.207}$$

where $\lambda(t)>0$ and $\mu(t)>0$ for all t are scaling parameters, and the matrices $\hat{A}(t)$, $\hat{V}(t)$, and $\hat{W}(t)$ are given by

$$\hat{W}(t)=W(t)+\mu^{-1}(t)H(t)H^t(t), \tag{1.208}$$

$$\hat{V}(t)=V(t)+\mu^{-1}(t)H_c(t)H_c^t(t), \tag{1.209}$$

$$\begin{aligned}\hat{A}(t) &= A(t)+\delta A(t)\\ &= A(t)+\mu^{-1}(t)L^t(t)E^t(t)E(t).\end{aligned} \tag{1.210}$$

Let the (λ,μ)-parameterized estimator be expressed as

$$\dot{\hat{x}}(t)=\left\{A(t)+\mu^{-1}(t)L^t(t)E^t(t)E(t)\right\}\hat{x}(t)+K(t)\left\{y(t)-C(t)\hat{x}(t)\right\}, \tag{1.211}$$

where the gain matrix $\mathbb{K}_e(t)\in\mathbb{R}^{n\times m}$ is to be determined. The following theorem summarizes the main result.

Theorem 1.39. *Consider system* (1.185)–(1.186) *satisfying the uncertainty structure* (1.187)–(1.188) *with zero initial condition. Suppose the process and measurement noises satisfy Assumption 1.1. For some $\mu(t)>0$ and $\lambda(t)>0$, let $P(t)=P^t(t)$ and $L(t)=L^t(t)$ be the solutions of RDEs* (1.210)–(1.211), *respectively. Then the (λ,μ)-parameterized estimator* (1.211) *is a QE estimator with GC such that*

$$\mathbb{E}[\{x(t)-\hat{x}(t)\}^t\{x(t)-\hat{x}(t)\}]\le tr[L(t)]. \tag{1.212}$$

Moreover, the gain matrix $\mathbb{K}(t)$ is given by

$$\mathbb{K}_e(t)=\left\{L(t)C^t(t)+\mu^{-1}(t)H(t)H_c^t(t)\right\}\hat{V}^{-1}(t). \tag{1.213}$$

Further details can be found in [84].

1.11.3 Numerical simulation 6

To illustrate the developed theory, we focus on the steady-state Kalman filtering and proceed to determine the estimator gains. Essentially, we seek to solve (1.206)–(1.211) in the steady state (as $\dot{P}(t) \rightarrow 0$ and $\dot{L}(t) \rightarrow 0$) when $\lambda \in [\lambda_1 \rightarrow \lambda_2]$, $\mu \in [\mu_1 \rightarrow \mu_2]$, where λ_1, λ_2, μ_1, μ_2 are given constants. Initially, we observe that (1.206) depends on P only, and it is not of the standard form of AREs. On the other hand, (1.207) depends on both L and P, and it can be put into the standard ARE form. For numerical simulation, we employ a Kronecker product-like technique to reduce (1.206) into a system of nonlinear algebraic equations of the form

$$f(\alpha) = G\alpha + h(\alpha) + q, \tag{1.214}$$

where $\alpha \in \mathbb{R}^{n(n+1)/2}$ is a vector of the unknown elements of the P-matrix. Then the algebraic equation (1.71) can be solved using an iterative Newton–Raphson technique according to the rule

$$\alpha_{(i+1)} = \alpha_{(i)} - \gamma_{(i)}[G + \nabla_\alpha h(\alpha_{(i)}]^{-1} f(\alpha_{(i)}), \tag{1.215}$$

where i is the iteration index, $\alpha_{(o)} = 0$, $\nabla_\alpha h(\alpha)$ is the Jacobian of $h(\alpha)$, and the step-size $\gamma_{(i)}$ is given by $\gamma_{(i)} = 1/[||f(\alpha(i))|| + 1]$.

Given the solution of (1.206), we proceed to solve (1.207) using a standard Hamiltonian/eigenvector method. All the computations are carried out using the Matlab software system. As a typical case, consider a time-delay system of type (1.185)–(1.186) with

$$A = \begin{bmatrix} -2 & 0.5 \\ 1 & -3 \end{bmatrix}, \quad A_d = \begin{bmatrix} -0.2 & -0.1 \\ 0.1 & 0.4 \end{bmatrix},$$
$$W = \begin{bmatrix} 1 & 0 \\ 0 & 1 \end{bmatrix}, \quad H = \begin{bmatrix} 0.5 \\ 0.5 \end{bmatrix}$$
$$C = [1 \ \ -3], \quad E = [0.5 \ \ 1], \quad H_c = 2, \quad V = 1.$$

A summary of the computational results is presented in Table 1.1, from which we observe the following:

1. For a given $\lambda \in [0.1 - 0.9]$, increasing μ by 50% results in 0.3% increase in $||K||$ (for small λ) and about 1.12% increase in $||K||$ when λ is relatively large.
2. For a given μ, increasing λ from 0.1 to 0.9 causes $||K||$ to increase by about 5.35%.
3. For $\mu < 0.6$ and $\lambda \in [0.1, 0.9]$, the estimator is unstable.
4. Increasing (λ, μ) beyond $(1, 1)$ yields unstable estimator.

Table 1.1 Summary of some of the computational results.

λ	μ	P		L		δA		K^t		$\lambda(\hat{A})$	
0.1	0.6	0.574	0.175	0.566	0.176	0.382	0.784	0.314	0.301	−3.227	−0.472
		0.175	0.457	0.176	0.464	0.459	0.919				
0.1	0.8	0.535	0.156	0.500	0.136	0.242	0.483	0.314	0.301	−3.249	−0.897
		0.156	0.431	0.136	0.421	0.306	0.612				
0.1	0.9	0.525	0.151	0.480	0.124	0.202	0.404	0.315	0.302	−3.258	−1.017
		0.151	0.425	0.124	0.409	0.261	0.523				
0.2	0.6	0.564	0.206	0.554	0.210	0.406	0.811	0.317	0.295	−3.279	−0.498
		0.206	0.374	0.210	0.386	0.409	0.818				
0.2	0.8	0.525	0.183	0.488	0.167	0.257	0.514	0.318	0.301	−3.293	−0.913
		0.183	0.350	0.167	0.346	0.269	0.537				
0.2	0.9	0.515	0.177	0.468	0.153	0.215	0.430	0.318	0.294	−3.298	−1.031
		0.177	0.344	0.153	0.334	0.228	0.456				
0.4	0.6	0.603	0.242	0.576	0.238	0.438	0.877	0.324	0.296	−3.302	−0.450
		0.242	0.357	0.238	0.367	0.405	0.810				
0.4	0.8	0.564	0.217	0.508	0.192	0.278	0.557	0.325	0.301	−3.311	−0.880
		0.217	0.335	0.192	0.328	0.265	0.530				
0.4	0.9	0.555	0.210	0.486	0.177	0.233	0.466	0.326	0.295	−3.415	−1.002
		0.210	0.330	0.177	0.316	0.225	0.449				
0.6	0.6	0.665	0.276	0.612	0.259	0.471	0.942	0.331	0.300	−3.311	−0.382
		0.276	0.369	0.259	0.372	0.418	0.836				
0.6	0.8	0.629	0.252	0.540	0.210	0.300	0.600	0.333	0.299	−3.318	−0.834
		0.252	0.348	0.210	0.333	0.274	0.548				
0.6	0.9	0.622	0.247	0.517	0.195	0.252	0.503	0.335	0.299	−3.321	−0.962
		0.247	0.344	0.195	0.321	0.232	0.464				
0.8	0.6	0.751	0.319	0.656	0.281	0.507	1.015	0.340	0.304	−3.318	−0.300
		0.319	0.392	0.281	0.384	0.437	0.874				
0.8	0.8	0.723	0.298	0.580	0.229	0.324	0.649	0.343	0.304	−3.324	−0.779
		0.298	0.374	0.229	0.344	0.287	0.574				
0.8	0.9	0.726	0.298	0.556	0.213	0.273	0.545	0.345	0.304	−3.326	−0.914
		0.298	0.372	0.213	0.332	0.244	0.487				
0.9	0.6	0.806	0.346	0.681	0.292	0.527	1.055	0.344	0.307	−3.321	−0.255
		0.346	0.407	0.292	0.392	0.448	0.897				
0.9	0.8	0.790	0.330	0.603	0.239	0.338	0.676	0.349	0.307	−3.326	−0.747
		0.330	0.392	0.239	0.352	0.294	0.589				
0.9	0.9	0.805	0.336	0.579	0.222	0.284	0.569	0.351	0.307	−3.328	−0.887
		0.336	0.393	0.222	0.339	0.250	0.501				

Therefore we conclude that:

A. The stable-estimator gains are practically insensitive to the (λ, μ)-parameters.

B. There is a finite range for (λ, μ) that guarantees a stable performance of the developed Kalman filter.

Remark 1.19. Had we considered a class of uncertain time-delay systems represented by

$$\begin{aligned} x_{k+1} &= [A_k + \Delta A_k]\, x_k + D_k\, x_{k-\tau} + w_k \\ &= A_{k,\Delta}\, x_k + D_k\, x_{k-\tau} + w_k, \end{aligned} \tag{1.216}$$

$$\begin{aligned} y_k &= [C_k + \Delta C_k]\, x_k + v_k \\ &= C_{k,\Delta}\, x_k + v_k, \end{aligned} \tag{1.217}$$

$$z_k = C_{1,k}\, x_k, \tag{1.218}$$

where $x_k \in \mathbb{R}^n$ is the state, $y_k \in \mathbb{R}^m$ is the measured output, $z_k \in \mathbb{R}^p$ is a linear combination of the state variables to be estimated, and $w_k \in \mathbb{R}^r$ and $v_k \in \mathbb{R}^m$ are, respectively, the process and measurement noise sequences, following parallel development to the continuous-case, we would be able to generate a robust discrete-time Kalman filter.

1.11.4 Numerical simulation 7

Consider the following discrete-time delay system:

$$\begin{aligned} x_{k+1} &= \left(\begin{bmatrix} 0.2 & -0.1 & 0 \\ 0.004 & 0.4 & 0.1 \\ 0 & 0.1 & 0.6 \end{bmatrix} + \begin{bmatrix} 0.1 \\ 0.1 \\ 0.1 \end{bmatrix} \Delta_k [0.5\ 0.4\ 0.2] \right) x_k \\ &\quad + \begin{bmatrix} -0.1 & 0 & 0 \\ 0.05 & -0.2 & 0.1 \\ 0 & 0 & -0.1 \end{bmatrix} x_{k-3} + w_k, \end{aligned} \tag{1.219}$$

$$y_k = \left(\begin{bmatrix} 1.0 & 0 & 0 \\ 0 & 1.0 & 0 \end{bmatrix} + \begin{bmatrix} 0.2 \\ 0.3 \end{bmatrix} \Delta_k [0.5\ 0.4\ 0.2] \right) x_k + v_k, \tag{1.220}$$

which is of type (1.216)–(1.218) with three units of delay. We further assume that $W = I$ and $V = 0.2I$. To determine the Kalman gains, we solve the discrete ARE equivalent to (1.206)–(1.211) for selected values of λ, μ. The numerical computation is basically of the form of iterative schemes, and the results for a typical case of $\mu = 0.7$ and $\lambda = 0.3$ are given by

$$P = \begin{bmatrix} 0.141 & 0.005 & 0.003 \\ 0.005 & 0.255 & 0.175 \\ 0.003 & 0.175 & 0.501 \end{bmatrix}, \quad S = 10^{-4} \begin{bmatrix} 0.284 & -1.17 & -2.966 \\ -1.17 & 4.813 & 12.208 \\ -2.966 & 12.208 & 30.962 \end{bmatrix},$$

$$K_o = 10^{-6}\begin{bmatrix} -0.841 & -1.309 \\ 3.463 & 5.388 \\ 8.782 & 13.665 \end{bmatrix}, \quad \hat{A} = \begin{bmatrix} 0.331 & -0.019 & -0.034 \\ -0.277 & 0.254 & 0.175 \\ -1.641 & -0.897 & 0.961 \end{bmatrix}.$$

The developed estimator is indeed asymptotically stable since

$$\lambda(\hat{A}) = \{0.302, 0.48, 0.765\} \in (0, 1).$$

The guaranteed cost is 36.059×10^{-4}. Several simulation studies have been carried out to examine the performance of the steady-state Kalman filter. In Table 1.1 the guaranteed cost is presented for selected values of the scaling parameters (μ, λ).

It is readily evident that the scaling parameters (μ, λ) have a crucial impact on the optimality of the guaranteed cost. This is equally true for specified μ while changing λ or given λ and varying μ.

For the purpose of comparison, a standard Kalman filter was designed for the nominal delay-less system by setting $\Delta_k \equiv 0$ and $x_{k-3} \equiv 0$. Then we applied the developed robust Kalman filter and the standard Kalman filter with $\Delta_k = 0$, $\Delta_k = 0.2$, and $\Delta_k = -0.2$ and retained the delayed state. The resulting filtering costs for both filtering schemes are provided in Table 1.2. Again, it is clearly shown that the developed robust Kalman filter outperforms the standard nominal Kalman filter in the presence of uncertainty and delay factor.

Together, Tables 1.2 and 1.3 demonstrate the superior performance of the developed robust Kalman filter.

1.12. Networked control systems

Typically in process industries, a *network* used at the lowest level of a process/factory communication hierarchy is called a *fieldbus*. Fieldbuses are intended to replace the traditional wiring between sensors, actuators, and controllers. In distributed control system applications a feedback control loop is often closed through the network, which is called a network-based control system (NBCS); see details in [38,63–74]. A generic setup of NCS is depicted in Fig. 1.1. In the NBCS, various delays with variable lengths are occurred due to sharing a common network medium, which are called *network-induced delays*. These delays depend on configurations of the network and given system. Those make the NBCS unstable.

In feedback control systems, it is significant that sampled data must be transmitted within a sampling period and stability of control systems should

Table 1.2 The guaranteed cost (GC) vs the scaling parameters (μ, λ).

$\lambda = 0.2$

μ	0.2	0.45	0.7	1.1	1.5	2.3
$GC \times 10^{-4}$	219.241	141.706	37.029	27.583	65.742	120.333

$\lambda = 0.4$

μ	0.2	0.45	0.7	1.1	1.5	2.3
$GC \times 10^{-4}$	173.601	121.016	31.125	25.113	60.142	101.471

$\lambda = 0.8$

μ	0.2	0.45	0.7	1.1	1.5	2.3
$GC \times 10^{-4}$	96.711	101.315	29.451	24.881	51.371	89.116

$\lambda = 1.4$

μ	0.2	0.45	0.7	1.1	1.5	2.3
$GC \times 10^{-4}$	109.345	117.996	38.222	24.003	61.332	110.541

$\lambda = 2.7$

μ	0.2	0.45	0.7	1.1	1.5	2.3
$GC \times 10^{-4}$	165.124	122.236	46.113	35.723	72.119	121.171

$\lambda = 3.4$

μ	0.2	0.45	0.7	1.1	1.5	2.3
$GC \times 10^{-4}$	259.943	165.176	51.152	41.907	77.139	151.454

Table 1.3 Comparison between the nominal Kalman filter.

	Actual cost		
Filter	$\Delta_k = -0.8$	$\Delta_k = 0$	$\Delta_k = 0.8$
Nominal Kalman filter	72.034×10^{-2}	45.147×10^{-4}	235.654×10^{-2}
Robust Kalman filter	66.802×10^{-4}	61.147×10^{-4}	67.113×10^{-4}

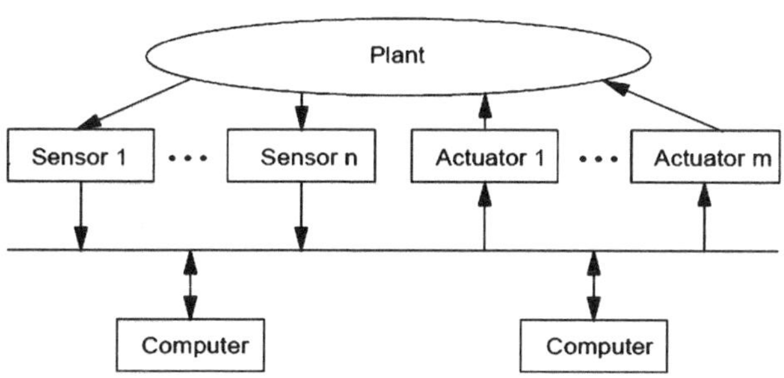

Figure 1.1 Generic setup of NCS.

be guaranteed. Although a shorter sampling period is preferable in most control systems, for some cases, it can be lengthened up to a certain bound within which stability of the system is guaranteed in spite of the performance degradation. This certain bound is called a *maximum allowable delay bound* (MADB). The MADB depends only on parameters and configurations of the given plant and the controller.

In addition, a faster sampling is said to be desirable in sampled-data systems because the performance of the discrete-time system controller can approximate that of the continuous-time system. But in NBCS the high sampling rate can increase network load, which in turn results in longer delay of the signals. Thus finding a sampling rate that can both tolerate the network-induced delay and achieve desired system performance is of fundamental importance in the NBCS design.

1.12.1 State-feedback stabilization

Consider the plant model described as

$$\begin{aligned}\dot{x}(t) &= Ax(t) + Bu(t), \ t \in [kh + \tau_k, \ (k+1)h + \tau_{k+1}), \\ y(t) &= Cx(t), \\ u(t^+) &= -K_i x(t - \tau_k), \ t \in \{kh + \tau_k, \ k = 0, 1, 2, \ldots\}. \end{aligned} \quad (1.221)$$

Sampling this system with period h and defining $z(kh) = [x^T(h), \ u^T((k-1)h)]^T$ yielded the following closed-loop system:

$$z((k+1)h) = \tilde{\Phi}(K_i)z(kh) \quad \forall i = 1, 2, \ldots, p. \quad (1.222)$$

A recent survey of the stabilization methods is reported in [38].

1.12.2 Observer-based feedback stabilization

An observer-based stabilizing controller can be designed for networked systems involving both random measurement and actuation delays; see Fig. 1.2. The LTI plant under consideration was assumed to be of the form

$$\begin{aligned} x_p(k+1) &= Ax_p + Bu_p, \\ y_p &= Cx_p, \end{aligned} \quad (1.223)$$

where $x_p(k) \in \mathbb{R}^n$ is the state vector, and $u_p(k) \in \mathbb{R}^m$ and $y_p(k) \in \mathbb{R}^p$ are the control input and output vectors of the plant, respectively. The measure-

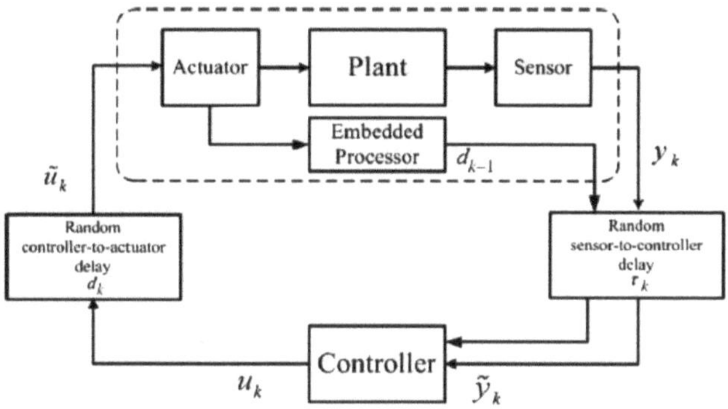

Figure 1.2 Setup of NCS.

ment subjected to random communication delay is given by

$$y_c(k) = (1 - \delta(k))y_p(k) + \delta(k)y_p(k - \tau_k^m), \tag{1.224}$$

where τ_k^m is the measurement delay whose occurrence is governed by the Bernoulli distribution, and $\delta(k)$ is a Bernoulli distributed sequence with

$$\begin{aligned} Prob\{\delta(k) = 1\} &= \mathbb{E}\{\delta(k)\} = \bar{\delta}, \\ Prob\{\delta(k) = 0\} &= 1 - \mathbb{E}\{\delta(k)\} = 1 - \bar{\delta}. \end{aligned} \tag{1.225}$$

The following observer-based controller is designed when the full state vector is not available:

$$\text{Observer:} \begin{cases} \hat{x}(k+1) = A\hat{x} + Bu_c(k) + L(y_c(k) - \hat{y}_c(k)), \\ \hat{y}_c(k) = (1 - \bar{\delta})C\hat{x}(k) + \bar{\delta}C\hat{x}(k - \tau_k^m), \end{cases} \tag{1.226}$$

$$\text{Controller:} \begin{cases} u_c(k) = K\hat{x}(k), \\ u_p = (1 - \alpha)u_c(k) + \alpha u_c(k - \tau_k^a), \end{cases} \tag{1.227}$$

where $\hat{x}(k) \in \mathbb{R}^n$ is the estimate of system (1.223), $\hat{y}_c(k) \in \mathbb{R}^p$ is the observer output, and $L \in \mathbb{R}^{n\times p}$ and $K \in \mathbb{R}^{m\times n}$ are the observer and controller gains, respectively. The stochastic variable α, mutually independent of δ, is also a Bernoulli distributed white sequence with

$$\begin{aligned} Prob\{\alpha(k) = 1\} &= \mathbb{E}\{\alpha(k)\} = \bar{\alpha}, \\ Prob\{\alpha(k) = 0\} &= 1 - \mathbb{E}\{\alpha(k)\} = 1 - \bar{\alpha}, \end{aligned} \tag{1.228}$$

where τ_k^a is the actuation delay. It is assumed that τ_k^a and τ_k^m are time varying and have the following bounded condition:

$$\underline{d}_m \leq \tau_k^m \leq \overline{d}_m, \quad \underline{d}_a \leq \tau_k^a \leq \overline{d}_a. \tag{1.229}$$

The estimation error is defined by

$$e(k) = x_p(k) - \hat{x}(k). \tag{1.230}$$

This yields

$$\begin{aligned} x_p(k+1) = {} & [A + (1-\bar{\alpha})BK]x_p(k) - (1-\bar{\alpha})BKe(k) \\ & + \bar{\alpha}BKx_p(k - \tau_k^a) - \bar{\alpha}BKe(k - \tau_k^a) \\ & - (\alpha - \bar{\alpha})BKx_p(k) + (\alpha - \bar{\alpha})BKe(k) \\ & + (\alpha - \bar{\alpha})BKx_p(k - \tau_k^a) - (\alpha - \bar{\alpha})BKe(k - \tau_k^a), \end{aligned} \tag{1.231}$$

$$\begin{aligned} e(k+1) = {} & [A - (1-\bar{\delta})LC]e(k) - \bar{\delta}LCe(k - \tau_k^m) \\ & + (\delta - \bar{\delta})LCx_p(k) - (\delta - \bar{\delta})LCx_p(k - \tau_k^m). \end{aligned} \tag{1.232}$$

Systems (1.231) and (1.232) are equivalent to the following compact form:

$$\varepsilon(k+1) = (\bar{A} + \tilde{A})\varepsilon(k) + (\bar{B} + \tilde{B})\varepsilon(k - \tau_k^m) + (\bar{C} - \tilde{C})\varepsilon(k - \tau_k^a), \tag{1.233}$$

where

$$\begin{aligned} \varepsilon(k) &= [x_p^T(k) \quad e^T(k)]^T, \\ \bar{A} &= \begin{bmatrix} A + (1-\bar{\alpha})BK & -(1-\bar{\alpha})BK \\ 0 & A - (1-\bar{\delta})LC \end{bmatrix}, \\ \tilde{A} &= \begin{bmatrix} -(\alpha - \bar{\alpha})BK & (\alpha - \bar{\alpha})BK \\ (\delta - \bar{\delta})LC & 0 \end{bmatrix}, \\ \bar{B} &= \begin{bmatrix} 0 & 0 \\ 0 & -\bar{\delta}LC \end{bmatrix}, \quad \tilde{B} = \begin{bmatrix} 0 & 0 \\ -(\delta - \bar{\delta})LC & 0 \end{bmatrix}, \\ \bar{C} &= \begin{bmatrix} \bar{\alpha}BK & -\bar{\alpha}BK \\ 0 & 0 \end{bmatrix}, \quad \tilde{C} = \begin{bmatrix} (\alpha - \bar{\alpha})BK & -(\alpha - \bar{\alpha})BK \\ 0 & 0 \end{bmatrix}. \end{aligned}$$

Remark 1.20. Note that a majority of the existing works on the stability of NCS (in the framework of time-delay approach) are reduced to some Lyapunov-based analysis of systems with uncertain and bounded time-varying delays; see [38]. In the following sections, we present alternative approaches that lead to improved results.

1.12.3 Lyapunov-based sampled-data stabilization

Three main approaches have been used to the sampled-data control and later to the Networked Control Systems (NCS), where the plant is controlled via communication network:

A. The first one is based on discrete-time models [67], [31]. This approach is not applicable to the performance analysis (like the exponential decay rate) of the resulting continuous-time closed-loop system.

B. The second one is a time-delay approach, where the system is modeled as a continuous-time system with time-varying sawtooth delay in the control input [8], [75], [76], [77]. The time-delay approach via time-independent Lyapunov–Krasovskii functionals or Lyapunov–Razumikhin functions leads to linear matrix inequalities (LMIs) for analysis and design of linear uncertain NCS.

C. The third approach is based on the representation of the sampled-data system in the form of impulsive model [75], [76]. Recently, the impulsive model approach was extended to the case of uncertain sampling intervals [34] by employing a discontinuous Lyapunov function method, which improved the existing Lyapunov-based results. Recently, the latter result was recovered via an input–output approach by application of the vector extension of Wirtinger inequality [37].

Consider the continuous-time system

$$\begin{aligned} \dot{x}(t) &= Ax(t) + Bu(t) + Ew(t), \\ z(t) &= C_1 x(t) + D_1 u(t), \end{aligned} \tag{1.234}$$

where $x(t) \in \mathbb{R}^n$ is the state, $z(t) \in \mathbb{R}^r$ is the signal to be controlled or estimated, $w(t) \in \mathbb{R}^q$ is the disturbance, $u(t) \in \mathbb{R}^m$ is the control input, and $A(t) \in \mathbb{R}^{n\times n}$, $B(t) \in \mathbb{R}^{n\times m}$, $C_1 \in \mathbb{R}^{p\times n}$, $D_1 \in \mathbb{R}^{p\times m}$, and $E \in \mathbb{R}^{n\times q}$ are system matrices.

In (1.234) the sampler is time-driven, whereas the controller and the zero-order hold (ZOH) are event-driven (in the sense that the controller and the ZOH update their outputs as soon as they receive a new sample). For simplicity, we assume that the measurement output $y(s_k) \in \mathbb{R}^p$ is available at discrete sampling instants

$$0 = s_0 < s_1 < \cdots < s_k < \cdots, \quad \lim_{k\to\infty} s_k = \infty,$$

and it may be corrupted by a measurement noise signal $v(s_k)$:

$$y(t) = Cx(t) + Du(t). \tag{1.235}$$

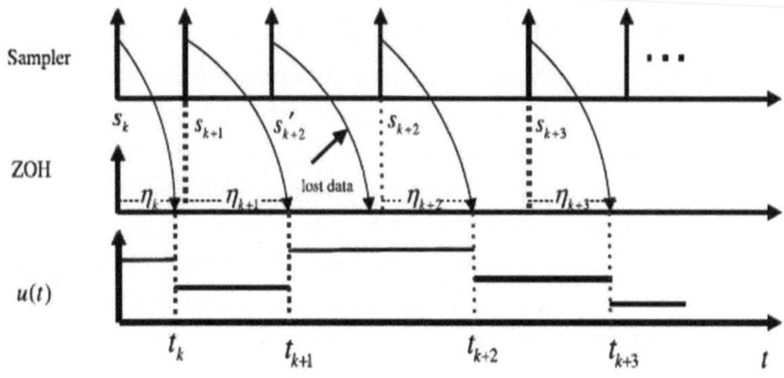

Figure 1.3 NCS timing diagram.

By considering a nonuniform sampling data packet dropouts can be accommodated. In this respect, $y(s_k)$, $k = 0, 1, 2, \ldots$ correspond to the measurements that are not lost. The timing diagram of the considered NCS with both delay and packet dropout is shown in Fig. 1.3, where $s_k = t_k - \lambda_k$ accounts for the sampling time of the data that has not been lost. In this setup, t_k denotes the updating instant time of the ZOH. Suppose that the updating signal at the instant t_k has experienced a signal transmission delay λ_k. Adopting the approach of [34], we allow the delays λ_k to grow larger than $s_{k+1} - s_k$, provided that the sequence of input update times t_k remains strictly increasing. This implies that if an old sample gets to the destination after the most recent one, then it should be dropped.

The static output-feedback controller has the form

$$u(t_k) = K_o y(t_k - \lambda_k), \quad t_k \leq t \leq t_{k+1},$$

where K_o is the controller gain, and t_{k+1} is the next updating instant time of the ZOH after t_k. It is known that

$$t_{k+1} - t_k + \lambda_k \leq \tau_M, \qquad 0 \leq \lambda_k \leq \lambda_M, \;\; k = 0, 1, 2, \ldots,$$

where λ_M is a known upper bound on the network-induced delays λ_k, and τ_M denotes the maximum time span between the time $s_k = t_k - \lambda_k$ at which the state is sampled, and the time t_{k+1} at which next update arrives at the ZOH. Observe that the sampling intervals and numbers of successive packet dropouts are uniformly bounded.

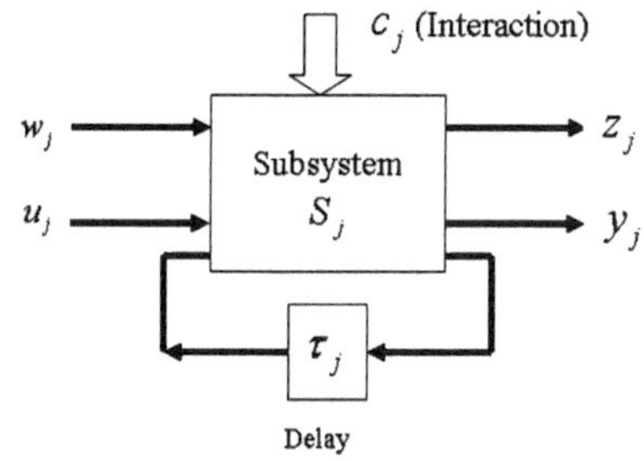

Figure 1.4 Subsystem model.

Within the foregoing representation, exponential stability, state-feedback, and static output feedback results are developed in [78]. More elaborate results can be found in [77].

1.13. Interconnected systems

We consider a class of linear systems S structurally composed of n_s coupled subsystems S_j depicted in Fig. 1.4 and modeled by the state-space model:

$$\begin{aligned}
\dot{x}_j(t) &= A_{j\Delta}x_j(t) + A_{dj\Delta}x_j(t-\tau_j) + B_{j\Delta}u_j(t) + c_j(k) + \Gamma_{j\Delta}w_j(t),\\
z_j(t) &= G_{j\Delta}x_j(t) + G_{dj\Delta}x(t-\tau_j) + D_{j\Delta}u_j(t) + \Phi_{j\Delta}w_j(t),\\
y_j(t) &= C_{j\Delta}x_j(t) + C_{dj\Delta}x(t-\tau_j),\\
c_j(t) &= \sum_{k=1}^{n_s} F_{jk\Delta}x_k(t) + \sum_{k=1}^{n_s} E_{jk\Delta}x_k(t-\eta_{jk}(t)),
\end{aligned} \tag{1.236}$$

whose matrices contain uncertainties that belong to a real convex bounded polytopic model of the type

$$\begin{aligned}
&\begin{bmatrix} A_{j\Delta} & A_{dj\Delta} & B_{j\Delta} & \Gamma_{j\Delta} \\ G_{j\Delta} & G_{dj\Delta} & D_{j\Delta} & \Phi_{j\Delta} \\ C_{j\Delta} & C_{dj\Delta} & E_{jk\Delta} & F_{jk\Delta} \end{bmatrix} \in \Pi_\lambda \triangleq \left\{ \begin{bmatrix} A_{j\lambda} & A_{dj\lambda} & B_{j\lambda} & \Gamma_{j\lambda} \\ G_{j\lambda} & G_{dj\lambda} & D_{j\lambda} & \Phi_{j\lambda} \\ C_{j\lambda} & C_{dj\lambda} & E_{jk\lambda} & F_{jk\lambda} \end{bmatrix} \right. \\
&\left. = \sum_{s=1}^{N} \lambda_s \begin{bmatrix} A_{js} & A_{djs} & B_{js} & \Gamma_{js} \\ G_{js} & G_{djs} & D_{js} & \Phi_{js} \\ C_{js} & C_{djs} & E_{jks} & F_{jks} \end{bmatrix}, \lambda_s \in \Lambda \right\},
\end{aligned} \tag{1.237}$$

where Λ is the unit simplex

$$\Lambda \triangleq \left\{(\lambda_1, \cdots, \lambda_N) : \sum_{j=1}^{N} \lambda_j = 1 \,,\, \lambda_j \geq 0\right\}. \tag{1.238}$$

Define the vertex set $\mathcal{N} = \{1, \ldots, N\}$. We use $\{A, \ldots, \Phi\}$ to imply generic system matrices and $\{A_j, \ldots, \Phi_j, j \in \mathcal{N}\}$ to represent the respective values at the vertices.

In the absence of uncertainties, system (1.236) reduces to the nominal state-space model

$$\begin{aligned}
\dot{x}_j(t) &= A_j x_j(t) + A_{dj} x_j(t - \tau_j(t)) + B_j u_j(t) + c_j(t) + \Gamma_j w_j(t), \\
z_j(t) &= G_j x_j(t) + G_{dj} x_j(t - \tau_j(t)) + D_j u_j(t) + \Phi_j w_j(t), \\
y_j(t) &= C_j x_j(t) + C_{dj} x_j(t - \tau_j(t)), \\
c_j(t) &= \sum_{k=1}^{n_s} F_{jk} x_k(t) + \sum_{k=1}^{n_s} E_{jk} x_k(t - \eta_{jk}(t)),
\end{aligned} \tag{1.239}$$

where for $j \in \{1, \ldots, n_s\}$, $x_j(t) \in \mathbb{R}^{n_j}$ is the state vector, $u_j(t) \in \mathbb{R}^{m_j}$ is the control input, $y_j(t) \in \mathbb{R}^{p_j}$ is the measured output, $w_j(t) \in \mathbb{R}^{q_j}$ is the disturbance input belonging to $\mathcal{L}_2[0, \infty)$, $z_j(t) \in \mathbb{R}^{q_j}$ is the performance output, $c_j(t) \in \mathbb{R}^{n_j}$ is the coupling vector, and τ_j, η_{jk}, $j, k \in \{1, \ldots, n_s\}$, are unknown time-delay factors satisfying

$$\begin{aligned}
0 \leq \tau_j(t) \leq \varrho_j, \quad \dot{\tau}_j(t) \leq \mu_j, \\
0 \leq \eta_{jk}(t) \leq \varrho_{jk}, \quad \dot{\eta}_{jk}(t) \leq \mu_{jk},
\end{aligned} \tag{1.240}$$

where the bounds ϱ_j, ϱ_{jk}, μ_j, μ_{jk} are known constants to guarantee smooth growth of the state trajectories. The matrices $A_j \in \mathbb{R}^{n_j \times n_j}$, $B_j \in \mathbb{R}^{n_j \times m_j}$, $D_j \in \mathbb{R}^{q_j \times m_j}$, $A_{dj} \in \mathbb{R}^{n_j \times n_j}$, $\Phi_j \in \mathbb{R}^{q_j \times q_j}$, $\Gamma_j \in \mathbb{R}^{n_j \times q_j}$, $C_j \in \mathbb{R}^{p_j \times n_j}$, $C_{dj} \in \mathbb{R}^{p_j \times n_j}$, $G_j \in \mathbb{R}^{q_j \times n_j}$, $G_{dj} \in \mathbb{R}^{q_j \times n_j}$, $F_{jk} \in \mathbb{R}^{n_j \times n_k}$, and $E_{jk} \in \mathbb{R}^{n_j \times n_k}$ are real and constant. The initial condition $\langle x_j(0), x_j(r) \rangle = \langle x_{oj}, \phi_j \rangle$, $j \in \{1, \ldots, n_s\}$, where $\phi_j \in \mathcal{L}_2[-\tau_j^*, 0]$, $j \in \{1, \ldots, n_s\}$. The inclusion of the terms $A_{dj} x_j(t - \tau_j(t))$ and $E_{jk} x_k(t - \eta_{jk}(t))$ emphasizes the delays within each subsystem (local delay) and among the subsystems (coupling delay), respectively.

We develop new criteria for LMI-based characterization of delay-dependent asymptotic stability and $\mathcal{L}_2$ gain analysis, which requires only subsystem information thereby assuring decentralization. The criteria include some parameter matrices and aim at expanding the range of applicability of the developed conditions. The following theorem establishes the main result for subsystem $\mathbf{S}_j$.

Theorem 1.40. *Given $\varrho_j > 0$, $\mu_j > 0$, $\varrho_{jk} > 0$, and $\mu_{jk} > 0$, $j, k = 1, \ldots, n_s$. The family of nominal subsystems $\{\mathsf{S}_j\}$ with $u_j(\cdot) \equiv 0$, where S_j is described by (1.239), is delay-dependent asymptotically stable with $\mathcal{L}_2$-performance bound γ_j, $j = 1, \ldots, n_s$, if there exist positive-definite matrices $\mathcal{P}_j$, $\mathcal{Q}_j$, $\mathcal{W}_j$, $\mathcal{Z}_{kj}$, $k, j = 1, \ldots, n_s$, parameter matrices Θ_j and Υ_j satisfying the following LMIs for $j = 1, \ldots, n_s$:*

$$\begin{bmatrix} \Xi_{oj} & \Xi_{aj} & -\varrho_j\Theta_j & \Xi_{cj} & \mathcal{P}_j\Gamma_j & G_j^t & \varrho_j A_j^t \mathcal{W}_j \\ \bullet & -\Xi_{mj} & -\varrho_j\Upsilon_j & 0 & 0 & G_{dj}^t & \varrho_j A_{d_j}^t \mathcal{W}_j \\ \bullet & \bullet & -\varrho_j\mathcal{W}_j & 0 & 0 & 0 & 0 \\ \bullet & \bullet & \bullet & -\Xi_{nj} & 0 & 0 & \varrho_j \sum_{k=1}^{n_s} E_{kj}\mathcal{W}_j \\ \bullet & \bullet & \bullet & \bullet & -\gamma_j^2 I_j & \Phi_j^t & \varrho_j \Gamma_j^t \mathcal{W}_j \\ \bullet & \bullet & \bullet & \bullet & \bullet & -I_j & 0 \\ \bullet & \bullet & \bullet & \bullet & \bullet & \bullet & -\varrho_j\mathcal{W}_j \end{bmatrix} < 0, \quad (1.241)$$

where

$$\begin{aligned} \Xi_{oj} &= \mathcal{P}_j[A_j + \sum_{k=1}^{n_s} F_{kj}] + [A_j + \sum_{k=1}^{n_s} F_{kj}]^t \mathcal{P}_j^t + \Theta_j + \Theta_j^t + \mathcal{Q}_j + \sum_{k=1}^{n_s} \mathcal{Z}_{kj}, \\ \Xi_{mj} &= \Upsilon_j + \Upsilon_j^t + (1 - \mu_j)\mathcal{Q}_j, \quad \Xi_{aj} = \mathcal{P}_j A_{dj} - \Theta_j + \Upsilon_j^t, \\ \Xi_{nj} &= \sum_{k=1}^{n_s} (1 - \mu_{kj})\mathcal{Z}_{kj}, \quad \Xi_{cj} = \mathcal{P}_j \sum_{k=1}^{n_s} E_{kj}. \end{aligned} \quad (1.242)$$

More detailed results can be found in [53].

1.14. Conclusions and future work

In this chapter, we overview the *stability and stabilization of time-delay systems* with emphasis on the following topics:

- Systems with time-delays constitute a good compromise between very simple models with finite dimension and very complex models of PDEs. The behavior features and the structural characteristics of delay systems are particular enough to justify specific techniques,
- The main Lyapunov-based tools are the corner-stone in developing robust stability in combination with model transformations. Several extensions are anticipated when examining different forms of Lyapunov–Krasovskii functionals.
- In the robust control area, existing results can be generally subdivided into two classes:

C1. The first class consists in systems with input or output delays (mainly, $\mathcal{H}_\infty$ performance or predictor-like techniques).

C2. The second class in state delays (discrete or distributed). The intersection of the two classes is still to be addressed.

An outlook of future research investigations are:

- Many contemporary dynamical systems with aftereffect are still requiring further investigation: For example, this is the case of delay systems with strong nonlinearities, as well as time-varying or state-dependent delays. In the growing research area of networked control systems the impact of packet delays is worth investigating.
- An interesting research area can be found on resilient delay-dependent stability and stabilization. This certainly will further strengthen the design and applications of robust control research.
- There are classes of nonlinear dynamical systems with delays including jump systems, fuzzy systems, and switched systems inviting additional research efforts. To this end, the class of distributed/multiagent systems has ample of open research topics, including design of consensus patterns and estimations/control over graphs.
- A surge of interests has been recently arisen regarding Wirtinger-based integral inequality and augmented Lyapunov–Krasovskii functionals [79,80]. The ensuing results triggered some development in the time-delay system stability. Further discussions and assessments of these results and related issues suggest attractive research directions at least from the computational standpoint.
- The class of uncertain nonlinear networked systems with both multiple stochastic time-varying communication delays and multiple packet dropouts was addressed in [81] for filtering design and in [82] for reliable control. A promising research direction is extension of the role of delay patterns to alternative forms.

Acknowledgment

This research work is supported by the deanship of scientific research (DSR) at KFUPM through research project no. IN 161065.

References

[1] V.L. Kharitonov, Robust stability analysis of time delay systems: a survey, in: Proc. the Fourth IFAC Conference on System Structure and Control, Nantes, France, 1998, pp. 1–12.

[2] V.L. Kharitonov, D. Melchior-Aguliar, On delay-dependent stability conditions, Systems & Control Letters 40 (1) (2000) 71–76.

[3] J.K. Hale, S.M.V. Lunel, Introduction to Functional Differential Equations, Springer, New York, 1993.
[4] K. Gu, V.L. Kharitonov, J. Chen, Stability of Time-Delay Systems, Birkhäuser, Boston, 2003.
[5] J.P. Richard, Time-delay systems: an overview of some recent advances and open problems, Automatica 39 (2003) 1667–1694.
[6] M.S. Mahmoud, Robust Control and Filtering for Time-Delay Systems, Marcel Dekker Inc., New York, 2000.
[7] A. Halanay, Differential Equations: Stability, Oscillations, Time Lags, Academic Press, New York, 1966.
[8] E. Fridman, Introduction to Time-Delay Systems: Analysis and Control, Springer, 2014.
[9] M.S. Mahmoud, Switched Time-Delay Systems, Springer-Verlag, Boston, July 2010.
[10] E.F. Infante, W.B. Castelan, A Lyapunov functional for a matrix difference-differential equation, Journal of Differential Equations 29 (1978) 439–451.
[11] W. Huang, Generalization of Lyapunov theorem in a linear delay system, Journal of Mathematical Analysis and Applications 142 (1989) 83–94.
[12] J. Louisell, A stability analysis for a class of differential-delay equations having time-varying delay, in: S. Busenberg, M. Martelli (Eds.), Delay Differential Equations and Dynamical Systems, in: Lecture Notes in Mathematics, vol. 1475, Springer, Berlin, 1991, pp. 225–242.
[13] M.S. Mahmoud, M. Zribi, H_∞ controllers for time-delay systems using linear matrix inequalities, Journal of Optimization Theory and Applications 100 (1) (January 1999) 89–123.
[14] M.S. Mahmoud, N.F. Al-Muthairi, Design of robust controllers for time–delay systems, IEEE Transactions on Automatic Control AC-39 (5) (1994) 995–999.
[15] M.S. Mahmoud, N.F. Al-Muthairi, Quadratic stabilization of continuous-time systems with state delay and norm-bounded time-varying uncertainties, IEEE Transactions on Automatic Control AC-39 (10) (1994) 2135–2139.
[16] K. Gu, A generalized discretization scheme of Lyapunov functional in the stability problem of linear uncertain time-delay systems, International Journal of Robust and Nonlinear Control 9 (1999) 1–14.
[17] K. Gu, Discretization schemes for Lyapunov–Krasovskii functionals in time delay systems, Kybernetica 37 (4) (2001) 479–504.
[18] S. Boyd, L.E. Ghaoui, E. Feron, E. Balakrishnan, Linear Matrix Inequalities in System and Control Theory, SIAM, Philadelphia, PA, 1994.
[19] X.-J. Jing, D.-L. Tan, Y.-C. Wang, An LMI approach to stability of systems with severe time-delay, IEEE Transactions on Automatic Control 49 (7) (2004).
[20] M.S. Mahmoud, Resilient Control of Uncertain Dynamical Systems, Springer, Heidelberg, 2004.
[21] S. Xu, T. Chen, Robust H_∞ control for uncertain discrete-time systems with time-varying delays via exponential output feedback controllers, Systems & Control Letters 51 (2004) 171–183.
[22] S.K. Nguang, P. Shi, S. Ding, Delay dependent fault estimation for uncertain time delay nonlinear systems: an LMI approach, International Journal of Robust and Nonlinear Control 16 (18) (2006) 913–933.
[23] G.P. Liu, Y.Q. Xia, J. Chen, D. Rees, W.S. Hu, Networked predictive control of systems with random network delays in both forward and feedback channels, IEEE Transactions on Industrial Electronics 54 (3) (2007) 1282–1297.
[24] W.-A. Zhang, L. Yu, A robust control approach to stabilization of networked control systems with time-varying delays, Automatica 45 (2009) 2440–2445.

[25] X.M. Sun, G.P. Liu, W. Wang, D. Rees, Stability analysis for networked control systems based on average dwell time method, International Journal of Robust and Nonlinear Control 20 (15) (2010) 1774–1784.
[26] X.-G. Yu, An LMI approach to robust H_∞ filtering for uncertain systems with time-varying distributed delays, Journal of the Franklin Institute 345 (2008) 877–890.
[27] C. Peng, Y.-C. Tian, Delay-dependent robust H_∞ control for uncertain systems with time-varying delay, Journal of Information Science 179 (2009) 3187–3197.
[28] Y.Q. Xia, M.Y. Fu, H.J. Yang, G.P. Liu, Robust sliding-mode control for uncertain time-delay systems based on delta operator, IEEE Transactions on Industrial Electronics 56 (9) (2009) 3646–3655.
[29] M.F. Miranda, Valter J.S. Leite, Robust stabilization of polytopic discrete-time systems with time-varying state delay: a convex approach, Journal of the Franklin Institute 348 (2011) 568–588.
[30] F. Yang, Z. Wang, Y.S. Hung, M. Gani, Control for networked systems with random communication delays, IEEE Transactions on Automatic Control 51 (3) (2006) 511–517.
[31] W.-A. Zhang, L. Yu, Modelling and control of networked control systems with both network-induced delay and packet-dropout, Automatica 44 (2008) 3206–3210.
[32] H. Song, W.-A. Zhang, L. Yu, H_∞ filtering of network-based systems with communication constraints, IET Signal Processing 4 (1) (2010) 69–77.
[33] D. Wu, J. Wu, S. Chen, Stability of networked control systems with polytopic uncertainty and buffer constraint, IEEE Transactions on Automatic Control 55 (5) (2010) 1202–1208.
[34] P. Naghshtabrizi, J. Hespanha, A. Teel, Exponential stability of impulsive systems with application to uncertain sampled-data systems, Systems & Control Letters 57 (2008) 378–385.
[35] E. Fridman, A refined input delay approach to sampled-data control, Automatica 46 (2010) 421–427.
[36] L. Mirkin, Some remarks on the use of time-varying delay to model sample and hold circuits, IEEE Transactions on Automatic Control 52 (2007) 1109–1112.
[37] K. Liu, V. Suplin, E. Fridman, Stability of linear systems with general sawtooth delay, IMA Journal of Mathematical Control and Information 27 (2010) 419–436.
[38] M.S. Mahmoud, Control and Estimation Methods Over Communication Networks, Springer-Verlag, UK, 2014.
[39] P. Park, A delay-dependent stability criterion for systems with uncertain time-invariant delays, IEEE Transactions on Automatic Control 44 (1999) 876–877.
[40] Y.S. Moon, P. Park, W.H. Kwon, Y.S. Lee, Delay-dependent robust stabilisation of uncertain state-delayed systems, International Journal of Control 74 (2001) 1447–1455.
[41] V. Suplin, E. Fridman, U. Shaked, H_∞ control of linear uncertain time-delay systems – a projection approach, IEEE Transactions on Automatic Control 51 (2006) 680–685.
[42] K. Gu, An integral inequality in the stability problem of time-delay systems, in: Proc. the 39th IEEE Conference on Decision and Control, Sydney, Australia, 2000, pp. 2805–2810.
[43] E. Fridman, New Lyapunov–Krasovskii functionals for stability of linear retarded and neutral type systems, Systems & Control Letters 43 (2001) 309–319.
[44] E. Fridman, U. Shaked, An improved stabilization method for linear time-delay systems, IEEE Transactions on Automatic Control 47 (11) (2002) 1931–1937.
[45] E. Fridman, U. Shaked, Delay-dependent stability and H_∞ control: constant and time-varying delays, International Journal of Control 76 (1) (2003) 48–60.

[46] W.H. Chen, Z.H. Guan, X. Lu, Delay-dependent guaranteed cost control for uncertain discrete-time systems with delay, IEE Proceedings. Control Theory and Applications 150 (2003) 412–416.
[47] E.K. Boukas, Discrete-time systems with time-varying time delay: stability and stabilizability, Mathematical Problems in Engineering (2006) 42489.
[48] V.J. Leite, S. Tarbouriech, P.L.D. Peres, A convex approach for robust state-feedback control of discrete-time systems with state delay, in: Proc. American Control Conference, Boston, June 30–July 2, 2004, pp. 2870–2875.
[49] M.S. Mahmoud, New results on robust control design of discrete-time uncertain systems, IEE Proceedings. Control Theory and Applications 152 (2005) 453–459.
[50] Z. Wang, G. Wei, Gang Feng, Reliable H_∞ control for discrete-time piecewise linear systems with infinite distributed delays, Automatica 45 (2009) 2991–2994.
[51] Y. Liu, Z. Wang, J. Liang, X. Liu, Synchronization and state estimation for discrete-time complex networks with distributed delays, IEEE Transactions on Systems, Man and Cybernetics. Part B 38 (5) (2008) 1314–1325.
[52] V.B. Kolmanovskii, J.P. Richard, Stability of some linear systems with delay, IEEE Transactions on Automatic Control 44 (5) (1999) 984–989.
[53] M.S. Mahmoud, Decentralized stabilization of interconnected nonlinear systems with time-varying delays, IEEE Transactions on Automatic Control 54 (11) (2009) 2663–2668.
[54] M. Wu, Y. He, J.H. She, G.P. Liu, New delay-dependent stability criteria for robust stability of time-varying delay systems, Automatica 40 (8) (2004) 1435–1439.
[55] Y. He, Q.G. Wang, C. Lin, M. Wu, Augmented Lyapunov functional and delay-dependent stability criteria for neutral systems, International Journal of Robust and Nonlinear Control 15 (18) (2005) 923–933.
[56] Y. He, Q.G. Wang, L. Xie, C. Lin, Further improvement of free-weighting matrices technique for systems with time-varying delay, IEEE Transactions on Automatic Control 52 (2) (2007) 293–299.
[57] J. Sun, G.P. Liu, J. Chen, Delay-dependent stability and stabilization of neutral time-delay systems, International Journal of Robust and Nonlinear Control 19 (12) (2009) 1364–1375.
[58] J. Sun, G.P. Liu, On improved delay-dependent stability criteria for neutral time-delay systems, European Journal of Control 15 (6) (2009) 613–623.
[59] M.S. Mahmoud, A.Y. Al-Rayyah, Efficient parameterization to stability and feedback synthesis of linear time-delay systems, IET Control Theory & Applications 3 (8) (2009) 1107–1118.
[60] F. Gouaisbaut, D. Peaucelle, Delay-dependent robust stability of time delay systems, in: The 5th IFAC Symposium on Robust Control Design, Toulouse, France, July 2006, pp. 5–7.
[61] S. Xu, J. Lam, A survey of linear matrix inequality techniques in stability analysis of delay systems, International Journal of Systems Science 39 (12) (December 2008) 1095–1113.
[62] M.S. Mahmoud, S.Z. Selim, P. Shi, Global exponential stability criteria for neural networks with probabilistic delays, IET Control Theory & Applications 4 (11) (2011) 2405–2415.
[63] G.C. Walsh, H. Ye, L.G. Bushnell, Stability analysis of networked control systems, IEEE Transactions on Control Systems Technology 10 (3) (2002) 438–446.
[64] S. Yin, L. Yu, W.A. Zhang, A switched system approach to networked H_∞ filtering with packet losses, Circuits, Systems, and Signal Processing 30 (January 2011) 1341–1354.
[65] R.A. Gupta, M.Y. Chow, Networked control system: overview and research trends, IEEE Transactions on Industrial Electronics 57 (7) (2010).

[66] Y. Tipsuwan, M.Y. Chow, Control methodologies in networked control systems, Control Engineering Practice 11 (2003) 1099–1111.
[67] W. Zhang, M.S. Branicky, S.M. Phillips, Stability of networked control systems, IEEE Transactions on Control Systems Magazine 9 (2001) 84–99.
[68] S. Zampieri, Trends in networked control systems, in: Proc. the 17th World Congress, Seoul, Korea, 2008, pp. 2886–2894.
[69] L.A. Montestruque, P.J. Antsaklis, On the model-based control of networked systems, Automatica 39 (10) (2003) 1837–1843.
[70] P. Seiler, R. Sengupta, An $\mathcal{H}_\infty$ approach to networked control, IEEE Transactions on Automatic Control 50 (3) (2005) 356–364.
[71] R.A. Gupta, Mo-Yuen Chow, Networked control system: overview and research trends, IEEE Transactions on Industrial Electronics 57 (7) (2010).
[72] M.S. Mahmoud, H.N. Nounou, Y. Xia, Robust dissipative control for Internet-based switching systems, Journal of the Franklin Institute 347 (1) (2010) 154–172.
[73] P. Seiler, R. Sengupta, Analysis of communication losses in vehicle control problems, in: Proc. 2001 Amer. Contr. Conf., vol. 2, 2001, pp. 1491–1496.
[74] F.L. Lian, J. Moyne, D. Tilbury, Network design consideration for distributed control systems, IEEE Transactions on Control Systems Technology 10 (2) (2002).
[75] Y. Mikheev, V. Sobolev, E. Fridman, Asymptotic analysis of digital control systems, Automation and Remote Control 49 (1988) 1175–1180.
[76] E. Fridman, A. Seuret, J.-P. Richard, Robust sampled-data stabilization of linear systems: an input delay approach, Automatica 40 (2004) 1441–1446.
[77] K. Liu, E. Fridman, Wirtinger's inequality and Lyapunov-based sampled-data stabilization, Automatica 48 (2012) 102–108.
[78] K. Liu, E. Fridman, Networked-based stabilization via discontinuous Lyapunov functionals, International Journal of Robust and Nonlinear Control 22 (2012) 420–436.
[79] A. Seuret, F. Gouaisbaut, Wirtinger-based integral inequality: application to time-delay systems, Automatica 49 (2013) 2860–2866.
[80] O.M. Kwon, M.J. Park, Ju H. Park, S.M. Lee, Stability and stabilization of T-S fuzzy systems with time-varying delays via augmented Lyapunov-Krasovskii functionals, Information Sciences 372 (2016) 1–15.
[81] H. Dong, Z. Wang, H. Gao, Robust $\mathcal{H}_\infty$ filtering for a class of nonlinear networked systems with multiple stochastic communication delays and packet dropouts, IEEE Transactions on Signal Processing 58 (4) (2010) 1957–1966.
[82] Y. Liu, Z. Wang, W. Wang, Robust reliable control for discrete-time-delay systems with stochastic nonlinearities and multiplicative noises, Optimal Control Applications & Methods 32 (2011) 285–297.
[83] B.D.O. Anderson, J.B. Moore, Optimal Filtering, Prentice Hall, New York, 1979.
[84] M.S. Mahmoud, N.F. Al-Muthairi, S. Bingulac, Robust Kalman filtering for continuous time-lag systems, Systems & Control Letters 38 (4) (December 1999) 309–319.
[85] M.S. Mahmoud, Y. Xia, Applied Control Systems Design: State-Space Methods, Springer-Verlag, UK, 2012.
[86] M. Benosman, A survey of some recent results on nonlinear fault tolerant control, Mathematical Problems in Engineering 2010 (2010) 586169.
[87] Y. Zhang, J. Jiang, Issues on integration of fault diagnosis and reconfigurable control in active fault-tolerant control systems, in: Proc. the 6th IFAC Symposium of Fault Detection Supervision and Safety for Technical Processes, Beijing, China, 2006, pp. 1513–1524.
[88] M. Staroswiecki, H. Yang, B. Jiang, Progressive accommodation of aircraft actuator faults, in: Proc. the 6th IFAC Symposium of Fault Detection Supervision and Safety for Technical Processes, Beijing, China, 2006, pp. 877–882.
[89] J. Chen, R.J. Patton, Robust Model-Based Fault Diagnosis for Dynamic Systems, Kluwer Academic Publisher, Boston, 1999.

[90] M. Witczak, Modelling and Estimation Strategies for Fault Diagnosis of Nonlinear Systems: From Analytical to Soft Computing Approaches, Springer, London, 2007.
[91] S.K. Nguang, P. Zhang, S.X. Ding, Parity relation based fault estimation for nonlinear systems: an LMI approach, International Journal of Automation and Computing 4 (2007) 164–168.
[92] B. Halder, N. Sarkar, Robust fault detection of robotic manipulator, The International Journal of Robotics Research 26 (2007) 273–285.
[93] B. Halder, N. Sarkar, Robust nonlinear analytic redundancy for fault detection and isolation in mobile robot, International Journal of Automation and Computing 2 (2007) 177–182.
[94] Z. Mao, B. Jiang, P. Shi, Observer based fault-tolerant control for a class of nonlinear networked control systems, Journal of the Franklin Institute 347 (2010) 940–956.
[95] B. Jiang, F.N. Chowdhury, Fault estimation and accommodation for linear MIMO discrete-time systems, IEEE Transactions on Control Systems Technology 13 (3) (2005) 493–499.
[96] P.M. Frank, Analytical and qualitative model-based fault diagnosis-a survey, European Journal of Control 2 (1996) 6–28.
[97] R.S. Burns, Advanced Control Engineering, Butterworth-Heinemann, Jordan Hill, Oxford OX2 8DP, 2001.
[98] C. Bonivento, A. Isidori, L. Marconi, A. Paoli, Implicit fault-tolerant control: application to induction motors, Automatica 40 (3) (2004) 355–371.
[99] H. Lens, J. Adamy, Observer based controller design for linear systems with input constraints, in: Proc. the 17th World Congress, IFAC, Seoul, Korea, July 6–11, 2008, pp. 9916–9921.
[100] C. Bonivento, L. Gentili, A. Paoli, Internal model based fault tolerant control of a robot manipulator, in: Proc. the 43rd IEEE Conference on Decision and Control, vol. 5, 2004, pp. 5260–5265.
[101] M. Benosman, K.Y. Lum, Passive actuators' fault-tolerant control for affine nonlinear systems, in: Proc. 17th IFAC Word Congress, Seoul, Korea, 2008, pp. 14229–14234.
[102] G. Tao, S. Chen, S.M. Joshi, An adaptive actuator failure compensation controller using output feedback, IEEE Transactions on Automatic Control 47 (3) (2002) 506–511.
[103] N.E. Wu, Y. Zhang, K. Zhou, Detection, estimation, and accommodation of loss of control effectiveness, International Journal of Adaptive Control and Signal Processing 14 (7) (2000) 775–795.
[104] M. Du, R. Gandhi, P. Mhaskar, An integrated fault detection and isolation and safe-parking framework for networked process systems, Industrial & Engineering Chemistry Research 50 (2011) 5667–5679.
[105] P. Mhaskar, C. McFall, A. Gani, P.D. Christofides, J.F. Davis, Isolation and handling of actuator faults in nonlinear systems, Automatica 44 (2008) 53–62.
[106] T.C. Yang, Networked control system: a brief survey, IEE Proceedings. Control Theory and Applications 153 (4) (2006) 403–412.
[107] M.S. Mahmoud, Networked control systems analysis and design: an overview, The Arabian Journal for Science and Engineering 41 (3) (March 2016) 711–758.
[108] J.P. Hespanha, P. Naghshtabrizi, Yonggang Xu, A survey of recent results in networked control systems, Proceedings of the IEEE 95 (1) (2007) 138–162.
[109] P. Antsaklis, J. Baillieul, Special issue on technology of networked control systems, Proceedings of the IEEE 95 (1) (2007) 5–8.
[110] X. Zhang, T. Parisini, M.M. Polycarpou, Adaptive fault-tolerant control of nonlinear uncertain systems: an information-based diagnostic approach, IEEE Transactions on Automatic Control 49 (8) (2004) 1259–1274.

[111] R. Seliger, P.M. Frank, Robust component fault detection and isolation in nonlinear dynamic systems, in: Proc. IFAC Symposium on Fault Detection, Supervision and Safety of Technical Process, Baden-Baden, Germany, 1991, pp. 313–318.
[112] J.P. Gauthier, H. Hammouri, S. Othman, A simple observer for nonlinear systems: application to bio-reactors, IEEE Transactions on Automatic Control 37 (6) (1992) 875–880.
[113] R.M. Guerra, R. Garrido, A.O. Miron, The fault detection problem in nonlinear systems using residual generators, IMA Journal of Mathematical Control and Information 22 (2005) 119–136.
[114] H. Hammouri, M. Kinnaert, E.H.E. Yaagoubi, Observer-based approach to fault detection and isolation for nonlinear systems, IEEE Transactions on Automatic Control 44 (10) (1999) 1879–1884.
[115] G. Besancon, High-gain observation with disturbance attenuation and application to fault detection and isolation for nonlinear systems, Automatica 39 (2003) 1095–1102.
[116] X.G. Yan, C. Edwards, Sensor fault detection and isolation for nonlinear systems based on sliding mode observer, International Journal of Adaptive Control and Signal Processing 21 (8–9) (2007) 657–673.
[117] X.G. Yan, C. Edwards, Nonlinear robust fault reconstructive and estimation using a sliding mode observers, Automatica 43 (9) (2007) 1605–1614.
[118] S. Jeong, T. Takahashi, Wheeled inverted pendulum type assistant robot: inverted mobile, standing, and sitting motions, in: IEEE/RSJ International Conference on Intelligent Robots and Systems, 2007, pp. 1932–1937.
[119] E.K. Boukas, Z.K. Liu, Deterministic and Stochastic Time-Delay Systems, Birkhäuser, Boston, 2002.
[120] B.D.O. Anderson, J.B. Moore, Linear Optimal Control, Prentice-Hall, Inc., Englewood Cliffs, NJ, 1971.
[121] K. Zhou, J.C. Doyle, Essentials of Robust Control, Prentice-Hall, New Jersey, 1998.
[122] Y. Shi, B. Yu, Robust mixed H_2/H_∞ control of networked control systems with random time delays in both forward and backward communication links, Automatica 47 (2011) 754–760.
[123] D. Zong, D. Yang, L. Hou, Q. Wang, Robust finite-time H_∞ control for Markovian jump systems with partially known transition probabilities, Journal of the Franklin Institute 7 (2013) 1562–1578.
[124] L. Zhang, J. Lam, Necessary and sufficient conditions for analysis and synthesis of Markov jump linear systems with incomplete transition descriptions, IEEE Transactions on Automatic Control 55 (7) (July 2010).
[125] S. Chae, D. Huang, S.K. Nguanga, H_∞ control of a class of extended Markov jump linear systems, in: American Control Conference, San Francisco, CA, USA, July 2011.
[126] Y. Zhang, H. Yong, W. Min, Stabilization for Markovian jump systems with partial information on transition probability based on free-connection weighting matrices, Automatica 47 (1) (Jan 2011) 79–84.
[127] Y.B. Zhao, Y. Kang, D. Rees, Stochastic stabilization of packet-based networked control systems, International Journal of Innovative Computing, Information & Control 7 (5(A)) (May 2011).
[128] H. Zhang, Y. Shi, A.S. Mehr, Robust static output feedback control and remote PID design for networked motor systems, IEEE Transactions on Industrial Electronics 58 (12) (December 2011) 5396–5405.
[129] G. Wang, Q. Zhang, V. Sreeram, Partially mode-dependent H_∞ filtering for discrete-time Markovian jump systems with partly unknown transition probabilities, Signal Processing 90 (2010) 548–556.

[130] G. Wang, Q. Zhang, Robust control of uncertain singular stochastic systems with Markovian switching via proportional-derivative state feedback, IET Control Theory & Applications 6 (8) (2012) 1089–1096.
[131] G. Wang, Q. Zhang, C. Yang, Exponential H_∞ filtering for singular systems with Markovian jump parameters, International Journal of Robust and Nonlinear Control 23 (2013) 792–806.
[132] W.H. Chen, Disturbance observer based control for nonlinear systems, IEEE/ASME Transactions on Mechatronics 9 (4) (2004) 706–710.
[133] P. Shi, M. Mahmoud, S. Nguang, A. Ismail, Robust filtering for jumping systems with mode-dependent delays, Signal Processing 86 (2006) 140–152.
[134] X. Luan, P. Shi, F. Liu, Stabilization of networked control systems with random delays, IEEE Transactions on Industrial Electronics 58 (9) (2011) 4323–4330.
[135] E.P. Gatzke, E.S. Meadows, C. Wang, F.J. Doyle III, Model-based control of a four-tank system, Computers & Chemical Engineering 24 (2000) 1503–1509.
[136] X.F. Jiang, Q.L. Han, S. Liu, A. Xue, A new H_∞ stabilization criterion for networked control systems, IEEE Transactions on Automatic Control 53 (4) (2008) 1025–1032.
[137] F.W. Yang, Z.D. Wang, Y.S. Hung, M. Gani, H_∞ control for networked systems with random communication delays, IEEE Transactions on Automatic Control 51 (3) (2006) 511–518.
[138] H. Yang, Y. Xia, P. Shi, Stabilization of networked control systems with nonuniform random sampling periods, International Journal of Robust and Nonlinear Control 21 (5) (2011) 501–526.
[139] B. Jiang, Z. Mao, P. Shi, $\mathcal{H}_\infty$ filter design for a class of networked control systems via T-S fuzzy model approach, IEEE Transactions on Fuzzy Systems 18 (1) (2010) 201–208.
[140] D.F. Delchamps, Stabilizing a linear system with quantized state feedback, IEEE Transactions on Automatic Control 35 (1990) 916–924.
[141] M.S. Mahmoud, Decentralized Control and Filtering in Interconnected Dynamical Systems, CRC Press, New York, 2010.
[142] D. Srinivasagupta, H. Schattler, B. Joseph, Time-stamped model predictive control: an algorithm for control of processes with random delays, Computers & Chemical Engineering 28 (8) (2004) 1337–1346.
[143] S.H. Kim, P. Park, Networked-based robust $\mathcal{H}_\infty$ control design using multiple levels of network traffic, Automatica 45 (3) (2009) 764–770.
[144] M.S. Mahmoud, Y. Xia, Robust stability and stabilization of a class of nonlinear switched discrete-time systems with time-varying delays, Journal of Optimization Theory and Applications 143 (2009) 329–355.
[145] B. Shen, Z. Wang, Y.S. Hung, Distributed consensus $\mathcal{H}^\infty$ filtering in sensor networks with multiple missing measurements: the finite-horizon case, Automatica 46 (10) (2010) 1682–1688.
[146] G. Liu, Predictive controller design of networked systems with communication delays and data loss, IEEE Transactions on Circuits and Systems. II 57 (6) (2010) 481–485.
[147] R. Yang, P. Shi, G.P. Liu, H. Gao, Networked-based feedback control design for systems with mixed delays based on quantization and dropout compensation, Automatica 47 (2011) 2805–2809.
[148] R. Luck, A. Ray, An observer-based compensator for distributed delays, Automatica 26 (5) (1990) 903–908.
[149] Y. Xia, M. Fu, G.-P. Liu, Analysis and Synthesis of Networked Control Systems, Springer, 2011.
[150] Y. Xia, M. Fu, P. Shi, Analysis and Synthesis of Dynamical Systems With Time-Delays, Springer, 2009.
[151] Y. Xia, Li Li, G.-P. Liu, P. Shi, $\mathcal{H}_\infty$ predictive control of networked control systems int, Journal of Control 84 (6) (2011) 1080–1097.

[152] Y. Xia, G.-P. Liu, M. Fu, P. Shi, Predictive control of networked systems with random delay and data dropout?, IET Control Theory & Applications 3 (11) (2009) 1476–1486.

[153] E. Fridman, M. Dambrine, Control under quantization, saturation and delay: a LMI approach, Automatica 45 (10) (2009) 2258–2264.

[154] G. Grassi, D.A. Miller, Theory and experimental realization of observer-based discrete-time hyperchaos synchronization, IEEE Transactions on Circuits and Systems. I, Fundamental Theory and Applications 49 (3) (Mar. 2002) 373–378.

[155] G.I. Bara, A. Zemouche, M. Boutayeb, Observer synthesis for Lipschitz discrete-time systems, in: Proc. IEEE Int. Symp. Circuits Syst., Kobe, Japan, May 2005, pp. 3195–3198.

[156] T.L. Liao, N.S. Huang, An observer-based approach for chaotic synchronization with applications to secure communications, IEEE Transactions on Circuits and Systems. I, Fundamental Theory and Applications 46 (9) (Sept. 1999) 1144–1150.

[157] D. Zhao, S.K. Spurgeon, X. Yan, An adaptive finite time sliding mode observer, in: J.B. Clempner, W. Yu (Eds.), New Perspectives and Applications of Modern Control Theory, Springer International Publishing AG, 2018, pp. 523–538.

CHAPTER TWO

Stabilization techniques for large input-delay systems with uncertainty

Majdeddin Najafi
Research institute for Avionics, Isfahan University of Technology, Isfahan, Iran

Contents

2.1. Introduction

Most of the industrial systems, somehow, are containing lag or input delays [1]. These systems, such as oil and gas factories, piping systems, chemical processes, network based systems, reactors, rolling mills, aircrafts, helicopters and air-conditioning systems, are specifically prone to delays. For example, the interactions between systems often involve communications, which cause a large delay [2]. Moreover, inherent lags in the sensors and actuators usually cause input delay. The time delay in states, inputs, or outputs usually decreases the stability margin of systems, especially for the systems with uncertainties or disturbances. So, many researchers know compensation of delays as the main challenge in designing a proper controller [3].

Input delays are vivid examples of difficulties in the design and implementation of practical control systems, especially when systems are unstable

https://doi.org/10.1016/B978-0-12-820599-0.00007-0

or the control loop is closed with a long delay. The literature on the compensation of delays is quite rich [3–5]. The research has covered over few decades from the early methods of the Smith predictor [6] to newer techniques such as modified forms of the Smith predictor [7,8], chain predictor [9], PDE-based predictor methods [10–13], and neural networks and fuzzy predictors [14,15].

Several techniques explicitly deal with delays. In these methods, which are known as predictor-based methods, delays can be directly compensated. The advantage offered by these methods is that the value of delay is irrelevant and the compensation can be reached for any delay. Also, the controller can be designed independently of the delay, which means that the performance of the closed-loop system is not necessarily compromised. Most of the model-reduction methods (e.g., Smith predictor, modified and unified Smith predictor, finite spectrum, and Artstein reduction [6,7,16,17]) try to eliminate the delayed terms with an analytical model-based reduced order.

One problem with the model reduction methods of prediction is the sensitivity to system model parameters [5]. A minor deviation from nominal values can result in instability of the closed-loop system [18]. This has even been observed in simulations where a minute calculation error can significantly drift the results to instability. To avoid this condition, an alternative approach has been proposed for delay compensation based on asymptotic predictors [9]. Asymptotic predictors usually benefit observer-like structures such as the Luenberger observer. However, this approach is limited by the value of delay that can be compensated [9]. A delay partitioning technique is then suggested in sequential predictors to overcome long delay [9,19–21]. Consequently, a chain of asymptotic predictors may be used to enlarge the value of delay that can tolerate model uncertainties [22]. These latter methods are known as sequential subpredictors (SSPs) or more recently as chain predictors.

In this chapter, we utilize a set of sequential subpredictors to solve a stability problem for the systems with large input delay and uncertainties. As previously mentioned, the state is predicted over a large range of delay times. We show that the proposed set of predictors enables a common state-feedback controller to stabilize the closed-loop system with large values of input delay. Simulation results show a successful performance of the proposed method.

2.2. Problem statement

Consider the basic state-space description of linear input time-delay system

$$\begin{cases} \dot{x}(t) = Ax(t) + Bu(t-d), \\ x(t) = \phi(t) \quad \forall t \in [-d, 0], \end{cases} \tag{2.1}$$

where $x \in \mathbb{R}^n$, $u \in \mathbb{R}^m$, and $d > 0$ are constant input lags. As explained before, the lag in the input prevents the real-time control of the system. Also, in each moment, it is just possible to control the systems after d seconds. For example, a closed-loop system with common feedback control $Kx(t)$ causes a state-delayed system

$$\begin{cases} \dot{x}(t) = Ax(t) + BKx(t-d), \\ x(t) = \phi(t) \quad \forall t \in [-d, 0], \end{cases} \tag{2.2}$$

where just the coefficient of the delayed part of system (k) is designable. In this case, achieving appropriate behavior is extremely difficult.

For unstable open-loop systems, it is possible to show that there is d_m such that there exists no state feedback for $d > d_m$. A preliminary estimate of d_m for non-Hurwitz A can be obtained by $d_m \approx \frac{1}{\|\lambda_{\max}(A)\|}$ (this estimate can be studied using one the best methods to find d_m presented by Xu et al. [23]). Therefore, for more unstable systems, finding a stabilizer state-feedback is more bounded and will be heavily dependent on input-delay value.

2.3. Asymptotical predictors

To overcome this challenge, the delay in the input must be omitted and compensated, or the system state is predicted for d seconds. Based on the previous discussion, any model-based predictors as the Smith predictor (SP), modified SP, unified SP, finite spectrum, and Artstein reduction are very sensitive to model uncertainty or bit computational error. So, it is inevitable to use asymptotical state predictors instead of model-based predictors.

The main difference between asymptotical predictors (AS) and model-based predictors is that in AS the predicted state $x_p(t)$ is not exactly equal to $x(t+d)$. Instead, a dynamic model is suggested such that x_p asymptotically converges to $x(t+d)$:

$$x_p(t) \longrightarrow x(t+d). \tag{2.3}$$

So, there always exists a prediction error $\bar{e}_p$ between $x_p(t)$ and $x(t+d)$,

$$\bar{e}_p(t) = x_p(t) - x(t+d), \tag{2.4}$$

which must converge to zero. However, (2.4) is not causal. Therefore an equivalent prediction error can be defined by

$$e_p(t) = x_p(t-d) - x(t). \tag{2.5}$$

Now the main challenge is the dynamic model of the asymptotical predictor. A primary suggestion can be an observer-like predictor dynamic model. So, the predictor can be presented as

$$\dot{x}_p(t) = Ax_p(t) + Bu(t) + L(x_p(t-d) - x(t)), \tag{2.6}$$

where $x_p \in \mathbb{R}^n$ is the predicted state. In this case, if the prediction error (2.5) converges to zero, then $x_p(t)$ forecasts $x(t-d)$ for d seconds. For stability analysis of the prediction error, we can obtain the dynamic equation for e_p:

$$\dot{e}_p(t) = \dot{x}_p(t-d) - \dot{x}(t) = Ae_p(t) + Le_p(t-d), \tag{2.7}$$

where A and d are known, and L is an arbitrary matrix that must be chosen such that $e_p(t)$ asymptotically converges to zero.

2.3.1 Closed-loop system with asymptotical predictor

Like the observer-based closed-loop systems, the predictor-based closed-loop system can be defined by

$$\begin{cases} \dot{x}(t) = Ax(t) + Bu(t-d), \quad x(t) = \phi(t) \;\; \forall t \in [-d, 0], \\ \dot{x}_p(t) = Ax_p(t) + Bu(t) + L(x_p(t-d) - x(t)), \\ u(t) = K\bar{x}(t). \end{cases} \tag{2.8}$$

Using (2.5), this equation can be decoupled by the equivalent system

$$\begin{cases} \dot{x}(t) = Ax(t) + BKx(t) + e_p(t), \\ \dot{e}_p(t) = Ae_p(t) + Le_p(t-d). \end{cases} \tag{2.9}$$

It is clear that the delay in the input is compensated by a predictor if and only if e_p asymptotically converges to zero. In other words, a closed-loop system is stable if and only if $(A+BK)$ is Hurwitz and the prediction error is asymptotically stable. Therefore the main problem is stability analysis of the prediction error.

2.3.2 Stability analysis of prediction error

A sufficient condition for the stability of (2.7) can be given by the following proposition.

Proposition 2.1 ([24]). *Consider system (2.7). If $(A+L)$ is Hurwitz, then there exists a sufficiently small $d_m > 0$ such that (2.7) is asymptotically stable for all $d \in [0, d_m]$.*

Proposition 2.1 shows that for Hurwitz A, a proper L can be founded (at least $L = 0$) such that e_p asymptotically converges to zero for any d. However, for non-Hurwitz A, the delay is bounded by $(d \leq d_m)$, where $(A+L)$ is Hurwitz. As L is an arbitrary matrix, the maximum value of d_m depends on A. As discussed before, d_m can be approximated by $d_m \approx \frac{1}{\|\lambda_{\max}(A)\|}$. So, these kinds of predictors can forecast the system state just for d_m seconds. The way to overcome this problem will be described in the next sections.

Another challenge about the predictor is the method of obtaining the optimal L such that (2.7) is stable for maximum possible value of d_m. Based on [23] and [9], the next theorem explains an approach to obtain suboptimal value of L.

Theorem 2.1. *Consider the prediction error equation (2.7). $e_p(t)$ is asymptotically stable, and also x_p predicts x for $d \in (0, d_m]$ seconds if for a given d_m, there exist symmetric matrices $\bar{P} > 0$, $\bar{S} > 0$, $\bar{Q} > 0$ and matrices U, $\bar{Y}$, and $\bar{W}$ such that the following inequality holds:*

$$\begin{bmatrix} \Psi & U - \bar{Y} + \bar{W}^T & d_m \bar{P} A^T & -d_m \bar{Y} \\ * & -\bar{Q} - \bar{W} - \bar{W}^T & d_m U^T & -d_m \bar{W} \\ * & * & -d_m \bar{S} & 0 \\ * & * & * & -d_m \bar{P}\bar{S}^{-1}\bar{P} \end{bmatrix} < 0, \tag{2.10}$$

where $\Psi = (A\bar{P} + \bar{P}A^T + \bar{Y} + \bar{Y}^T + \bar{Q})$. Moreover, L is given by

$$L = U\bar{P}^{-1}. \tag{2.11}$$

Proof. [9] To show the stability of the prediction error (2.7), we define a Lyapunov functional candidate

$$V(e_p) = V_1(e_p) + V_2(e_p) + V_3(e_p), \tag{2.12}$$

where

$$V_1(e_p) = e_p(t)^T P e_p(t), \tag{2.13}$$

$$V_2(e_p) = \int_{-d}^{0}\int_{t+\beta}^{t} \dot{e}_p(\alpha)^T S \dot{e}_p(\alpha) d\alpha d\beta, \tag{2.14}$$

$$V_3(e_p) = \int_{t-d}^{t} e_p(\alpha)^T Q e_p(\alpha) d\alpha. \tag{2.15}$$

Similarly to the approach in [9], (2.12) can be written as

$$\begin{aligned}
\dot{V}_1(e_p) &= 2e_p(t)^T P(Ae_p(t) + Le_p(t-d)) \\
&= \frac{1}{d}\int_{t-d}^{t} [2e_p(t)^T(PA+Y)e_p(t) + 2e_p(t)^T(PL - Y + W^T)e_p(t-d) \\
&- 2e_p(t-d)^T We_p(t-d) - 2e_p(t)^T dY\dot{e}_p(\alpha) - 2e_p(t-d)^T dW\dot{e}_p(\alpha)]d\alpha, \\
\dot{V}_2(e_p) &= \int_{-d}^{0} [\dot{e}_p(t)^T S\dot{e}_p(t) - \dot{e}_p(t+\beta)^T S\dot{e}_p(t+\beta)]d\beta \\
&= \int_{t-d}^{t} [\dot{e}_p(t)^T S\dot{e}_p(t) - \dot{e}_p(\alpha)^T S\dot{e}_p(\alpha)]d\alpha, \\
&= \frac{1}{d}\int_{t-d}^{t} [e_p(t)^T dA^T SAe_p(t) + e_p(t-d)^T dL^T SLe_p(t-d) \\
&+ 2e_p(t)^T dA^T SLe_p(t-d) - \dot{e}_p(\alpha)^T S\dot{e}_p(\alpha)]d\alpha, \\
\dot{V}_3(e_p) &= e_p(t)^T Qe_p(t) - e_p(t-d)^T Qe_p(t-d) \\
&= \frac{1}{d}\int_{t-d}^{t} [e_p(t)^T Qe_p(t) - e_p(t-d)^T Qe_p(t-d)]d\alpha.
\end{aligned}$$

Finally,

$$\dot{V}(e_p) = \frac{1}{d}\int_{t-d}^{t} \xi(t,\alpha)^T \Lambda(d)\xi(t,\alpha)d\alpha, \tag{2.16}$$

where

$$\xi(t,\alpha) = [e_p(t)^T \;\; e_p(t-d)^T \;\; \dot{e}_p(\alpha)^T]^T. \tag{2.17}$$

By using the Schur complement it is possible to show that $\Lambda(d)$ is negative definite (i.e., $\dot{V} < 0$) if the following linear matrix inequality (LMI) holds:

$$\begin{bmatrix} \Theta + dA^TSA + dYS^{-1}Y^T & PL - Y + W^T + dA^TSL + dYS^{-1}W^T \\ * & -Q - W - W^T + dL^TSL + dWS^{-1}W^T \end{bmatrix} < 0 \tag{2.18}$$

or

$$\begin{bmatrix} \Theta & PL - Y + W^T \\ * & -Q - W - W^T \end{bmatrix} + d\begin{bmatrix} A^TS & -Y \\ L^TS & -W \end{bmatrix}\begin{bmatrix} S^{-1} & 0 \\ 0 & S^{-1} \end{bmatrix}\begin{bmatrix} SA & SL \\ -Y^T & -W^T \end{bmatrix} < 0, \tag{2.19}$$

where $\Theta = PA + A^TP + Y + Y^T + Q$. Since $S > 0$, (2.19) holds for all $d \in (0, d_m]$ if it holds for $d = d_m$. Applying again the Schur complement for $d = d_m$ yields

$$\begin{bmatrix} PA + A^TP + Y + Y^T + Q & PL - Y + W^T & d_mA^TS & -d_mY \\ * & -Q - W - W^T & d_mL^TS & -d_mW \\ * & * & -d_mS & 0 \\ * & * & * & -d_mS \end{bmatrix} < 0. \tag{2.20}$$

Now consider $\bar{P} = P^{-1}$, $\bar{S} = S^{-1}$, $\bar{Q} = P^{-1}QP^{-1}$, $\bar{Y} = P^{-1}YP^{-1}$, $\bar{W} = P^{-1}WP^{-1}$. By pre- and postmultiplying both sides of LMI (2.20) by $diag\{\bar{P}, \bar{P}, \bar{S}, \bar{P}\}$ it is possible to obtain (2.10) based on (2.20). Therefore (2.7) is asymptotically stable if (2.10) holds. □

Note that (2.10) is a nonlinear matrix inequality (NMI) due to the term $\bar{P}\bar{S}^{-1}\bar{P}$. There exist several methods to solve this kind of NMIs, which are studied in some papers. One of the famous techniques is reducing the original nonconvex problem to an LMI-based nonlinear minimization problem. The cone complementarity linearization (CCL) algorithm in [25] can be used to get a solution. This modified form of the algorithm is explained in the last section of this chapter for a similar problem.

Example 2.1. Consider the system $\dot{x}(t) = Ax(t) + Bu(t - d)$ with

$$A = \begin{bmatrix} 0.5 & 1 \\ 0 & -2 \end{bmatrix}, \quad B = \begin{bmatrix} 1 \\ 2 \end{bmatrix}, \quad d = 1.36.$$

Using Theorem 2.1 and MCCL method, we obtain

$$L = \begin{bmatrix} -0.4575 & -0.2050 \\ -0.1251 & 0.005 \end{bmatrix}.$$

By Theorem 2.1 there exists suitable L such that (2.10) can be feasible for maximum $d_m = 1.97$ (note that $\lambda_{\max}(A) = 0.5$).

This example shows us that the primary asymptotical predictor may be not suitable for unstable systems with long time delay. So we introduce the idea of using the chain predictor, which is explained in the next section.

2.4. Sequential subpredictors

The primary asymptotical predictor (2.6) may be nonuseful for some unstable systems for a long time-delay, as it was discussed in the previous section. Therefore we suggest a sequential structure of subpredictors, which can be used to stabilize systems with any d. The idea is based on splitting the time delay and using a chain of subpredictors to successively forecast the states for a large delay. The global predictor contains r coupled subpredictors; each of them predicts the system for $\bar{d}$ seconds, where

$$\bar{d} = \frac{d}{r}, \quad r \in \mathbb{Z}^+. \tag{2.21}$$

The SSP dynamics can be described by [9]

$$\begin{cases} \dot{\bar{x}}_1(t) = A\bar{x}_1(t) + L_1(\bar{x}_1(t-\bar{d}) - \bar{x}_2(t)) + Bu(t), \\ \vdots \\ \dot{\bar{x}}_i(t) = A\bar{x}_i(t) + L_i(\bar{x}_i(t-\bar{d}) - \bar{x}_{i+1}(t)) + Bu(t-(i-1)\bar{d}), \\ \vdots \\ \dot{\bar{x}}_{r-1}(t) = A\bar{x}_{r-1}(t) + L_{r-1}(\bar{x}_{r-1}(t-\bar{d}) - \bar{x}_r(t)) + Bu(t-(r-2)\bar{d}), \\ \dot{\bar{x}}_r(t) = A\bar{x}_r(t) + L_r(\bar{x}_r(t-\bar{d}) - x(t)) + Bu(t-(r-1)\bar{d}), \\ \text{-------------------------------} \\ x_p(t) = \bar{x}_1(t), \end{cases} \tag{2.22}$$

where $\bar{x}_i \in \mathbb{R}^n$, $i = 1, \cdots, r$. In this structure of SSP, $\bar{x}_i(t + (i-1)\bar{d})$ approaches $\bar{x}_{i+1}(t + i\bar{d})$ $(i = 1, \cdots, r-1)$, so that the $x_p(t) = \bar{x}_1(t)$ converges to

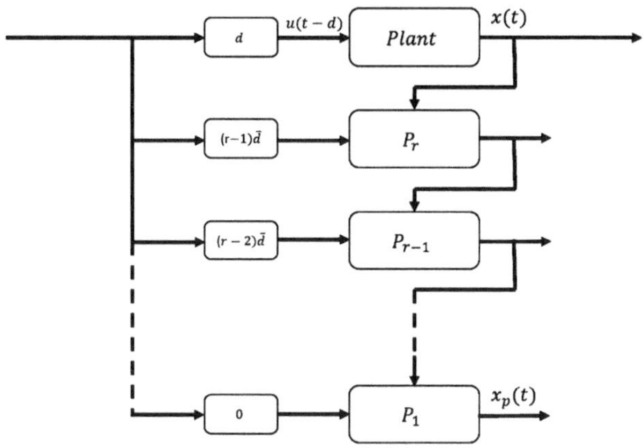

Figure 2.1 Sequential sub-predictor model.

the real state $x(t+r\bar{d})$ as all errors converge to zero. Fig. 2.1 introduces the model of SSP.

In the SSP the method of definition of the prediction errors plays a key role in the decoupling closed-loop system. One of these definitions can be assumed as follows:

$$\begin{cases} e_1(t) = \bar{x}_1(t-r\bar{d}) - \bar{x}_2(t-(r-1)\bar{d}), \\ \vdots \\ e_i(t) = \bar{x}_i(t-(r-i+1)\bar{d}) - \bar{x}_{i+1}(t-(r-i)\bar{d}), \\ \vdots \\ e_{r-1}(t) = \bar{x}_{r-1}(t-2\bar{d}) - \bar{x}_r(t-\bar{d}), \\ e_r(t) = \bar{x}_r(t-\bar{d}) - x(t), \\ \text{-----------------------------} \\ e_p(t) = e_1(t) + e_2(t) + \cdots + e_r(t). \end{cases} \tag{2.23}$$

Based on (2.23) and (2.22), the prediction error can be obtained as

$$\begin{cases} \dot{e}_1(t) = Ae_1(t) + L_1e_1(t-\bar{d}) - L_2e_2(t-\bar{d}), \\ \vdots \\ \dot{e}_i(t) = Ae_i(t) + L_ie_i(t-\bar{d}) - L_{i+1}e_{i+1}(t-\bar{d}), \\ \vdots \\ \dot{e}_{(r-1)}(t) = Ae_{(r-1)}(t) + L_{(r-1)}e_{(r-1)}(t-\bar{d}) - L_re_r(t-\bar{d}), \\ \dot{e}_r(t) = Ae_r(t) + L_re_r(t-\bar{d}). \end{cases} \tag{2.24}$$

The closed-loop system with $u(t) = Kx_p(t)$ also can be calculated based on (2.23) as

$$\bar{x}(t) = Ax(t) + BKx(t) + e_p(t). \tag{2.25}$$

It is clear that the closed-loop system is stable if $(A + BK)$ is Hurwitz and $e_p(t)$ asymptotically converges to zero. In view of the structure of prediction error (2.24), we can find that the stability of each error equation can be decoupled to each other from end to beginning. So it is possible to independently design each L_i.

On the other hand, Proposition 2.1 shows that there exists at least a small d_m such that error equation $\bar{e}_i(t) = Ae_i(t) + L_ie_i(t - \bar{d})$ is asymptotically stable with a suitable L_i (which can be obtained by Theorem 2.1). Therefore it is possible to choose r sufficiently big such that $\bar{d} \leq d_m$ (i.e., $d \leq rd_m$).

Remark 2.1. In view of definition (2.24), it is possible to obtain the prediction error system in the Laplace domain as

$$\begin{cases} E_i(s) = (-1)^{r-i+1}(sI - A - L_ie^{-\bar{d}s})L_{i+1}E_{i+1}(s)e^{-\bar{d}s}, & i = 1, \cdots, r-1, \\ E_r(s) = -(sI - A - L_re^{-\bar{d}s})\Phi(s), \end{cases} \tag{2.26}$$

where $\Phi(s)$ is the initial condition. So, the predictor dynamics is asymptotically stable if all $(A + L_ie^{-\bar{d}s})$ are Hurwitz for $i = 1, \cdots, r$. Therefore, based on Proposition 2.1, it is possible to choose R sufficiently large ($\bar{d}$ sufficiently small) such that there exist suitable L_i and K for which the closed loop is stable for any unstable system with a long time delay.

It is possible to synthesize all SSP parameters as presented in the next theorem.

Theorem 2.2. *Consider the prediction error equation* (2.24). $e_p(t)$ *is asymptotically stable, and also* x_p *predicts* x *for* $d \in (0, r\bar{d}]$ *seconds if for a given* $d_m \geq \bar{d}$, *there exist symmetric matrices* $\bar{P}_i > 0$ $(i = 1, \cdots, r)$, $\bar{S} > 0$, $\bar{Q} > 0$ *and matrices* U_j $(j = 1, \cdots, r)$, $\bar{Y}$, *and* $\bar{W}$ *such that inequality* (2.10) *holds with* A *replaced by* $\bar{A} = diag\{A, \cdots, A\}$, $\bar{P} = diag\{\bar{P}_1, \cdots, \bar{P}_R\}$,

$$U = \begin{bmatrix} U_1 & -U_2 & \ddots & \vdots \\ 0 & U_2 & \ddots & 0 \\ \vdots & \ddots & \ddots & -U_r \\ 0 & \cdots & \cdots & U_r \end{bmatrix}. \tag{2.27}$$

Table 2.1 Number of subpredictors (r) versus maximum delay (d).

r	Max delay ($d=r\bar{d}$)	$\bar{d}=d_m$
1	1.97	1.97
2	3.91	1.955
3	5.73	1.91

Moreover, L_i is given as

$$L_i = U_i \bar{P}_i^{-1}, \quad i = 1,\ 2,\ \cdots,\ r. \tag{2.28}$$

Proof. [9] Since (2.24) has the same structure as (2.7), replacing $e(t)$ with $e(t) = [e_1, e_2, \cdots, e_r]$ in the Lyapunov function (2.12), the rest of the proof can follow that of Theorem 2.1. □

Remark 2.2. In Theorem 2.2, R must be chosen sufficiently big to make inequality (2.10) feasible. Also, for simplicity, it is possible to consider $U_1 = U_2 = \cdots = U_r$ and $\bar{P}_1 = \bar{P}_2 = \cdots = \bar{P}_r$; however, this assumption reduces the degrees of freedom.

The following examples illustrate the ability of the proposed method to control and stabilize unstable systems.

Example 2.2. Consider the system in Example 2.1. Table 2.1 shows the maximum possible delay for the asymptotic SSP based on Theorem 2.2 (for maximum iteration number of 100 and delay step increment .001). This example shows that in the SSP method, the maximum time delay is directly proportional to r, whereas d_m is approximately independent from r.

Example 5.3 in [9] shows the relations between maximum possible delay, the instability of a system, and the number of subpredictors. The previous discussion shows the ability of SSP to stabilize unstable systems with long time delay. In the next section, we present a general design of a predictor-based controller.

2.4.1 General controller designing

In this section, we explain a general controller designing where just the output of the systems is measurable. Consider the input delay system

$$\begin{cases} \dot{x}(t) = Ax(t) + Bu(t-d), \\ y(t) = Cx(t) + Du(t-d), \\ x(t) = \phi(t) \ \ \forall t \in [-d, 0], \end{cases} \tag{2.29}$$

where $y \in \mathbb{R}^q$. Whiteout loss of generality, we assume that $C = [I_q, 0]$. If C has a different form, then there always exists a transformation matrix such that C converts to this form. In this case the predictor structure is changed as follows:

$$\begin{cases} \dot{\bar{x}}_1(t) = A\bar{x}_1(t) + L_1(\bar{x}_1(t-\bar{d}) - \bar{x}_2(t)) + Bu(t), \\ \vdots \\ \dot{\bar{x}}_i(t) = A\bar{x}_i(t) + L_i(\bar{x}_i(t-\bar{d}) - \bar{x}_{i+1}(t)) + Bu(t-(i-1)\bar{d}), \\ \vdots \\ \dot{\bar{x}}_{r-1}(t) = A\bar{x}_{r-1}(t) + L_{r-1}(\bar{x}_{r-1}(t-\bar{d}) - \bar{x}_r(t)) + Bu(t-(r-2)\bar{d}), \\ \dot{\bar{x}}_r(t) = A\bar{x}_r(t) + L_r(C\bar{x}_r(t-\bar{d}) + Du(t-d) - y(t)) + Bu(t-(r-1)\bar{d}), \\ \hline x_p(t) = \bar{x}_1(t). \end{cases} \tag{2.30}$$

Based on predictor error (2.23), we obtain the error equation dynamics:

$$\begin{cases} \dot{e}_1(t) = Ae_1(t) + L_1e_1(t-\bar{d}) - L_2e_2(t-\bar{d}), \\ \vdots \\ \dot{e}_i(t) = Ae_i(t) + L_ie_i(t-\bar{d}) - L_{i+1}e_{i+1}(t-\bar{d}), \\ \vdots \\ \dot{e}_{(r-1)}(t) = Ae_{(r-1)}(t) + L_{(r-1)}e_{(r-1)}(t-\bar{d}) - L_re_r(t-\bar{d}), \\ \dot{e}_r(t) = Ae_r(t) + L_rCe_r(t-\bar{d}). \end{cases} \tag{2.31}$$

The following theorem presents the design of SSP and state feedback control based on $y(t)$ to achieve the stability of a closed-loop system.

Theorem 2.3. *Consider the closed-loop system (2.29), (2.30) and state feedback $u(t) = Kx_p(t) = K\bar{x}_1(t)$. Assume that (A, B) is stabilizable. The closed-loop system is asymptotically stable, and also $\bar{x}_1$ predicts x for $d \in (0, r\bar{d}]$ seconds if for a given $d_m \geq \bar{d}$, there exist symmetric matrices $\bar{P}_i > 0$ $(i = 0, 1, \cdots, r+1)$, $\bar{S} > 0$, $\bar{Q} > 0$ and matrices F_1, U_j $(j = 1, \cdots, r+1)$, $\bar{Y}$, and $\bar{W}$ such that*

$$\begin{bmatrix} \Psi & U - \bar{Y} + \bar{W}^T & d_m(\bar{P}\bar{A}^T + F^T\bar{B}^T) & -d_m\bar{Y} \\ * & -\bar{Q} - \bar{W} - \bar{W}^T & d_mU^T & -d_m\bar{W} \\ * & * & -d_m\bar{S} & 0 \\ * & * & * & -d_m\bar{P}\bar{S}^{-1}\bar{P} \end{bmatrix} < 0, \tag{2.32}$$

where $\bar{A} = diag\{A, \cdots, A\}$, $\bar{P} = diag\{\bar{P}_0, \bar{P}_1, \cdots, \bar{P}_{r+1}\}$, $F = [F_1 \; 0 \; \cdots \; 0]$,

$$\bar{B} = \begin{bmatrix} B \\ 0 \\ \vdots \\ 0 \\ 0 \end{bmatrix}, U = \begin{bmatrix} 0 & U_1 & 0 & \cdots & 0 \\ 0 & U_1 & -U_2 & \ddots & \vdots \\ 0 & 0 & U_2 & \ddots & 0 \\ \vdots & \ddots & \ddots & \ddots & -\bar{U}_r \\ 0 & 0 & \cdots & \cdots & \bar{U}_r \end{bmatrix}, \tag{2.33}$$

and

$$\bar{U}_r = \begin{bmatrix} U_r & 0 \\ U_{r+1} & 0 \end{bmatrix}. \tag{2.34}$$

Moreover, L_i *and* K *are given as*

$$K = F_1 \bar{P}_0^{-1}, \; L_i = U_i \bar{P}_i^{-1}, \quad i = 1, \; 2, \; \cdots, \; r-1, L_r = \begin{bmatrix} U_r \\ U_{r+1} \end{bmatrix} \bar{P}_r^{-1}. \tag{2.35}$$

Note that the dimensions of $\bar{P}_r$, $\bar{P}_{r+1}$, U_r, *and* U_{r+1} *are* $q \times q$, $(n-q) \times (n-q)$, $q \times q$, *and* $(n-q) \times q$, *respectively.*

Proof. (Similar to [22]) Proposition 2.1 implies that the stabilizability of (A, B) is a necessary condition for the stability of the closed-loop system. By considering (2.23) we define the following set of new state variables:

$$\begin{bmatrix} z_1 = x + e_1 + e_2 + \cdots + e_r \\ z_2 = e_1 \\ \vdots \\ z_{r+1} = e_r \end{bmatrix}. \tag{2.36}$$

Then the closed-loop system can be rewritten as

$$\dot{z}_c(t) = \bar{A}_c z_c(t) + \bar{L} z_c(t - \bar{d}), \tag{2.37}$$

where

$$z_c = [z_1^T \; z_2^T \; \cdots \; z_{r+1}^T]^T,$$
$$z_c(t) = [0 \; 0 \; \cdots \; -\phi(t)^T]^T \; \forall t \in [-d, 0], \; \bar{A}_c = \bar{A} + \bar{B}\bar{K}, \text{ and}$$

$$\bar{K} = [K \; 0 \; \cdots \; 0], \quad \bar{L} = \begin{bmatrix} 0 & L_1 & 0 & \cdots & 0 \\ 0 & L_1 & -L_2 & \ddots & \vdots \\ 0 & 0 & L_2 & \ddots & 0 \\ \vdots & \ddots & \ddots & \ddots & -L_rC \\ 0 & 0 & \cdots & \cdots & L_rC \end{bmatrix} \tag{2.38}$$

The stability of the closed-loop system is equivalent to that of (2.37) according to (2.23) and (2.36). Also, $x_p = \bar{x}_1$ predicts x for $d = r\bar{d}$ seconds, provided that z_i $(i = 2, \cdots, r+1)$ asymptotically converge to zero because

$$e_t(t) = x_p(t - r\bar{d}) - x(t) = e_1(t) + \cdots + e_r(t) = x_2(t) + \cdots + x_{r+1}(t). \tag{2.39}$$

Since (2.37) has the same structure as (2.7), replacing $\bar{A}_c$ and $\bar{L}$ by (2.38), the rest of the proof can follow the proof of Theorem 2.1. □

The following example considers an unstable system with input time delay stabilized using SSP.

Example 2.3. Consider the following system with an unstable pole:

$$\begin{cases} \dot{x}(t) = Ax(t) + Bu(t - 1.6), \\ y(t) = Cx(t), \end{cases} \tag{2.40}$$

$$A = \begin{bmatrix} 2 & -2 \\ 2 & -3 \end{bmatrix}, \quad B = \begin{bmatrix} 0 \\ 2 \end{bmatrix}, \quad C = \begin{bmatrix} -1 & -1 \end{bmatrix},$$

$$x(t) = [1 \quad 1.7]^T \quad \forall t \in [-1.6, 0].$$

By using Theorem 2.3 (for $r = 3$, maximum iteration number of 80) we obtain $\bar{d} = 1.6/3 = 0.533$. Using (2.28), we calculate the closed-loop matrices L_i and K as

$$K = \begin{bmatrix} -13.3656 & -8.6013 \end{bmatrix},$$

$$L_1 = \begin{bmatrix} -2.1395 & 1.1728 \\ -1.0963 & 0.7532 \end{bmatrix}, \quad L_2 = \begin{bmatrix} -1.6804 & 0.8944 \\ -0.2305 & 0.2075 \end{bmatrix},$$

$$L_3 = \begin{bmatrix} -1.0648 & 0.6328 \\ 0.6033 & -.1058 \end{bmatrix}.$$

The initial condition $x(t)$, $t \in [-1.6, 0]$, is unknown for the predictor (i.e., $\bar{x}_i(0) = 0$). Fig. 2.2 presents the closed-loop output $y(t)$, the predictor

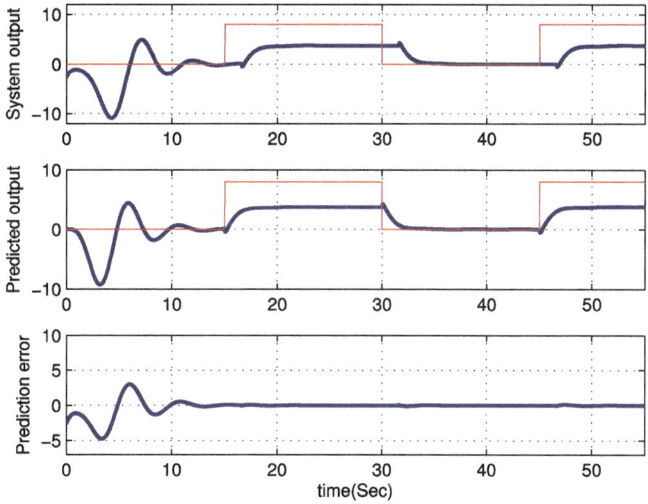

Figure 2.2 Pulse response of closed-loop system for Example 2.3.

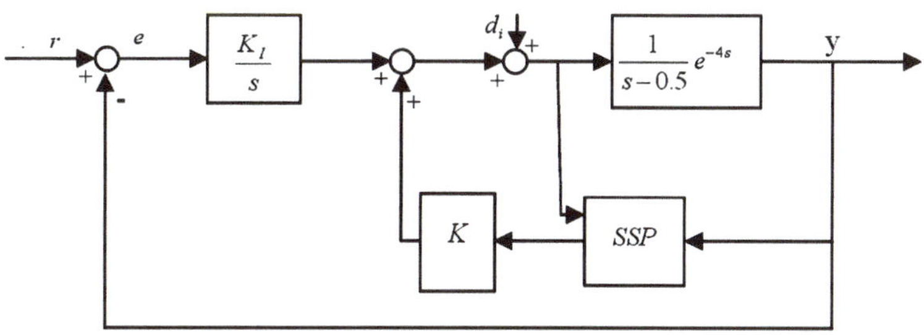

Figure 2.3 Closed-loop diagram of Example 2.4 [9].

output $\bar{y}(t) = C\bar{x}_1(t)$, and the prediction error $e_t(t) = C(\bar{x}_1(t - 1.6) - x(t))$ for a pulse input.

Example 2.4 ([9]). Consider the unstable input-delay plant $X(s) = U(s)e^{-4s}/(s - 0.5)$. In [26], it is shown that there exists no classical controller stabilizing this system. The proposed method is applied to make the closed-loop system stable.

To stabilize the system and enforce its steady-state error converge to zero, usually a state feedback is used together with an input integrator, as it is shown in Fig. 2.3. To stabilize the inner loop, we use Theorem 2.3 to design SSP matrices L_i and controller gain K. After stabilizing the inner loop, we can choose K_I sufficiently small to stabilize the closed-loop system. Using

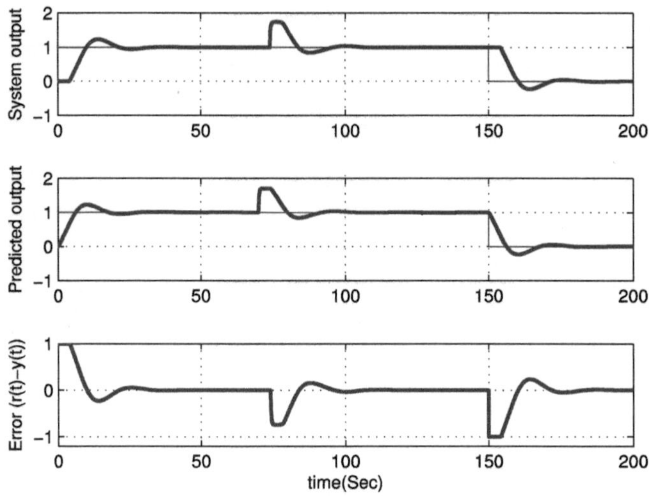

Figure 2.4 Pulse response of closed-loop system of Example 2.4 [9].

this method (for $r = 3$, $\bar{d} = 1.33$, the maximum iteration number of 40, and the delay step increment 0.1), we design the parameters $K = -6.224$, $L_1 = -0.6858$, $L_2 = -0.6807$, $L_3 = -0.6047$, and $K_I = 1$. Fig. 2.4 presents the response to a pulse input and a unit step disturbance after $t > 70$ s. The designed controller yields zero static tracking error and disturbance rejection as it is shown in Fig. 2.4c (defined by $e(t) = r(t) - y(t)$).

Another important benefit of applying SSP is the ability of stabilizing dead time systems with uncertainty, which is explained in the next section.

2.5. Robust controller synthesis using SSP

In this section, we design a robust controller for an uncertain input-delay system. Consider the following uncertain time-delay system:

$$\begin{cases} \dot{x}(t) = (A + \Delta A)x(t) + (B + \Delta B)u(t - d), \\ y(t) = Cx(t) + Du(t - d), \\ x(t) = \phi(t) \quad \forall t \in [-d, 0], \end{cases} \tag{2.41}$$

where ΔA and ΔB are two time-varying bounded matrices satisfying the following condition:

$$[\Delta A \ \Delta B] = DJ(t)[E \ E_B], \tag{2.42}$$

$$J(t)^T J(t) \leq I. \tag{2.43}$$

The following theorem presents a method to design a robust controller for system (2.41). The next lemma is necessary for proving the theorem.

Lemma 2.1 ([27]). *Let Ω, Γ, and Σ be matrices of appropriate dimensions with symmetric Ω. Then*

$$\Omega + \Gamma J(\tau)\Sigma + \Sigma^T J^T(\tau)\Gamma^T < 0 \tag{2.44}$$

for all $J(\tau)$ satisfying $J(\tau)^T J(\tau) \leq I$ if and only if there exists a scalar $\epsilon > 0$ such that

$$\Omega + \epsilon\Gamma\Gamma^T + \epsilon^{-1}\Sigma^T\Sigma < 0. \tag{2.45}$$

Theorem 2.4 ([22]). *Consider the closed-loop system (2.41) and (2.30), and let $u(t) = Kx_p(t) = K\bar{x}_1(t)$. Assume that (A, B) is stabilizable. The closed-loop system is asymptotically stable, and also $\bar{x}_1$ predicts x for $d \in (0, r\bar{d}]$ seconds if for a given $d_m \geq \bar{d}$, there exist symmetric matrices $\bar{P}_i > 0$ $(i = 0, 1, \cdots, r+1)$, $\bar{S} > 0$, and $\bar{Q} > 0$, matrices F_1, U_j $(j = 1, \cdots, r+1)$, $\bar{Y}$, and $\bar{W}$, and a scalar $\epsilon > 0$ such that the following inequality holds:*

$$\begin{bmatrix} \Psi & U - \bar{Y} + \bar{W}^T & d_m(\bar{P}\bar{A}^T + F^T\bar{B}^T + \epsilon\bar{D}\bar{D}^T) & \bar{P}\bar{E}^T + F^T E_B^T & -d_m\bar{Y} \\ * & -\bar{Q} - \bar{W} - \bar{W}^T & d_m U^T & 0 & -d_m\bar{W} \\ * & * & -d_m\bar{S} + d_m^2\epsilon\bar{D}\bar{D}^T & 0 & 0 \\ * & * & * & -\epsilon I & 0 \\ * & * & * & * & -d_m\bar{P}\bar{S}^{-1}\bar{P} \end{bmatrix} < 0, \tag{2.46}$$

where $\Psi = (\bar{A}\bar{P} + \bar{P}\bar{A}^T + \bar{Y} + \bar{Y}^T + \bar{Q} + \bar{B}F + F^T\bar{B}^T) + \epsilon\bar{D}\bar{D}^T$, and $\bar{A}$, $\bar{P}$, F, $\bar{B}$, and U appear in (2.33). Also, $\bar{D} = [0 \ \cdots \ 0 \ \bar{D}^T]^T$ and $\bar{E} = [-E \ E \ \cdots \ E \ E]$. Moreover, L_i and K are given as

$$K = F_1\bar{P}_0^{-1}, \ L_i = U_i\bar{P}_i^{-1}, \ i = 1, 2, \cdots, r-1, L_r = \begin{bmatrix} U_r \\ U_{r+1} \end{bmatrix} \bar{P}_r^{-1}. \tag{2.47}$$

Proof. Consider the set of state variables (2.36). Then the closed-loop system can be rewritten as

$$\dot{z}_c(t) = \bar{A}_c z_c(t) + \bar{L}z_c(t - \bar{d}), \tag{2.48}$$

where

$$\bar{A}_c = (\bar{A} + \Delta\bar{A}) + (\bar{B} + \Delta\bar{B})\bar{K}, \ \Delta\bar{A} = \begin{bmatrix} 0 & 0 & \cdots & 0 \\ \vdots & \ddots & \ddots & 0 \\ 0 & 0 & \ddots & 0 \\ -\Delta A & \Delta A & \cdots & \Delta A \end{bmatrix}, \quad (2.49)$$

$$\Delta\bar{B} = \begin{bmatrix} 0 \\ \vdots \\ 0 \\ -\Delta B \end{bmatrix}, \quad (2.50)$$

and z_c, $\bar{K}$, and $\bar{L}$ appear in (2.38). The stability of the closed-loop system and that of (2.48) are equivalent according to (2.23) and (2.36). (2.48) has the same structure as (2.37), and therefore the proof can follow that of Theorem 2.3, replacing $\bar{A}_c$ as defined in (2.49). So, it leads to following inequality:

$$\begin{bmatrix} \Psi & U - \bar{Y} + \bar{W}^T & d_m(\bar{P}(\bar{A} + \Delta\bar{A})^T + F^T(\bar{B} + \Delta\bar{B})^T) & -d_m\bar{Y} \\ * & -\bar{Q} - \bar{W} - \bar{W}^T & d_m U^T & -d_m\bar{W} \\ * & * & -d_m\bar{S} & 0 \\ * & * & * & -d_m\bar{P}\bar{S}^{-1}\bar{P} \end{bmatrix} < 0,$$

(2.51)

where $\Psi = ((\bar{A} + \Delta\bar{A})\bar{P} + \bar{P}(\bar{A} + \Delta\bar{A})^T + \bar{Y} + \bar{Y}^T + \bar{Q} + (\bar{B} + \Delta\bar{B})F + F^T(\bar{B} + \Delta\bar{B})^T)$. By using Lemma 2.1 and the Schur complement we can rewrite (2.51) as (2.46). □

Next examples illustrate the capability of SSP to stabilize unstable uncertain input delay with a large value of delay.

Example 2.5 ([9]). Consider the unstable system $G(s) = \frac{1}{s-1}e^{-s}$. In the MSP method, we can obtain the predictor compensator Z as $Z(s) = \frac{1}{s-1}e^{-1} - \frac{1}{s-1}e^{-s}$. C in MSP must stabilize the delay-free part of system (i.e., $\frac{1}{s-1}$). Let $C(s)$ be chosen as constant K_P. We obtain the MSP controller

$$\frac{C}{1 + CZ} = \frac{K_P(e^{-1} - e^{-s})(s-1)}{(s-1) + K_P(e^{-1} - e^{-s})}. \quad (2.52)$$

There is an unstable hidden mode in MSP. For example, the zero at $s = 1$ can lead to an unstable zero-pole cancellation. Therefore this method is very sensitive to modeling uncertainty and computational error. Fig. 2.5a presents the step response using MSP for the original system ($C(s) = 10$).

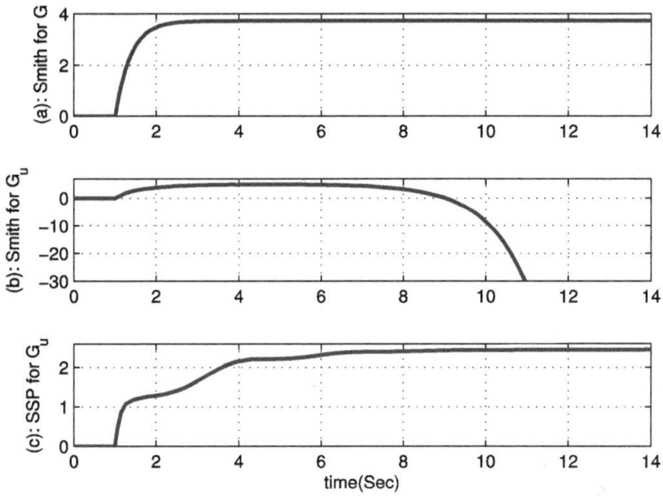

Figure 2.5 Step response for (a) MSP for exact system, (b) SP for uncertain system, and (c) SSP for uncertain system, Example 2.5 [9].

But this method is unstable for systems with any small uncertainty, for example, $G_u(s) = \frac{1}{s-(1+0.1)}e^{-s}$, which is shown in Fig. 2.5b. Fig. 2.5c illustrates the step response of the closed-loop system in the presence of uncertainty using SSP that is stable. Using Theorem 2.3, we calculate the parameters $L_1 = -2.662$, $L_2 = -2.3808$, $L_3 = -1.9022$, and $K = -10.078$ (for $r = 3$, iteration number 40 and delay increment 0.01).

Example 2.6 ([22]). Consider system (2.41) with

$$A = \begin{bmatrix} -0.5 \pm 0.1\Delta_1 & 1.5 \pm 0.3\Delta_2 & 1 \\ 0 & -1 \pm 0.4\Delta_3 & 0 \\ 0 & 0 & \pm 0.1\Delta_3 \end{bmatrix}, \quad B = \begin{bmatrix} 14 & 3.5 \\ 1 & 1 \\ -5 & -1 \end{bmatrix}, \text{ and}$$

$$C = \begin{bmatrix} 1 & -1 & 3 \end{bmatrix}.$$

The uncertainties are considered as $\|\Delta_i\| < 1$. Moreover, the transfer function between controllable input to measurable output can be described as

$$G(s) = \begin{bmatrix} \frac{0.2(s-6.5\pm1.9)}{(s-0.5\pm0.1)(s+1\pm0.4)} e^{-3s} \\ \frac{(1.2\pm0.3)(s-0.25\pm0.1)}{(s\pm0.1)(s-0.5\pm0.1)(s+1\pm0.4)} e^{-3s} \end{bmatrix}, \tag{2.53}$$

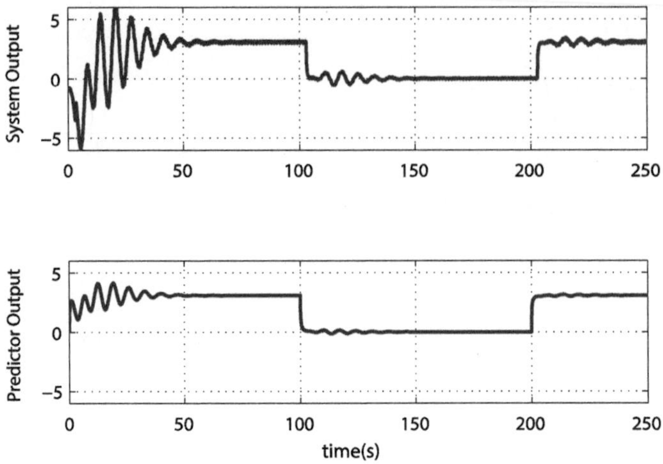

Figure 2.6 Pulse response of closed-loop system and predictor, Example 2.6 [22].

which is an unstable nonminimum phase system with large input delay. This uncertain system is unstable and has an integrator term and right half zeroes. Also, just $y(t)$ is measurable. The system can be robustly stabilized for delay $d = 3$ seconds by using the SSP method. The obtained parameters are

$$r = 3,\ L_1 = \begin{bmatrix} -1.579 & -1.4175 & -.797 \\ 0.197 & 0.025 & -0.297 \\ 0.302 & 0.064 & -0.125 \end{bmatrix},$$

$$L_2 = \begin{bmatrix} -1.522 & -1.474 & -0.821 \\ 0.223 & 0.088 & -0.2712 \\ 0.260 & 0.090 & -0.13 \end{bmatrix},\ L_3 = \begin{bmatrix} -0.979 \\ 0.016 \\ -0.0956 \end{bmatrix},$$

$$K_3 = \begin{bmatrix} 0.015 & 0.144 & 0.357 \\ -2.398 & -1.480 & -2.256 \end{bmatrix}.$$

For the initial condition $x(0) = [-1 \ \ 1 \ -0.5]$, Fig. 2.6 presents the pulse response of the closed-loop system and predictor output ($y_p(t) = Cx_p(t)$) for $\Delta_i = \sin(\omega_i t)$. Fig. 2.7 also presents $u(t)$ and the prediction error $y_p(t-3) - y(t)$, which converges to zero.

The next section explains the MCCL method to solve nonlinear matrix inequalities used in the theorems. Also, we present a sample code for MATLAB® to design SSP.

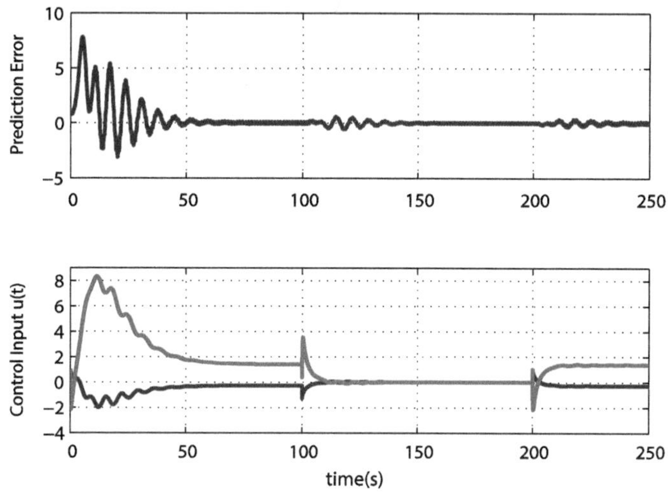

Figure 2.7 Summation of prediction errors and $u(t)$, Example 2.6 [22].

2.6. MCCL method

In this section, we describe a modified CCL algorithm to solve nonlinear matrix inequalities (2.10), (2.32), and other similar NMIs. Consider the following general inequality:

$$\begin{bmatrix} \Xi & \Pi \\ \Pi^T & -d_m \bar{P}\bar{S}^{-1}\bar{P} \end{bmatrix} < 0, \tag{2.54}$$

where Ξ and Π must be chosen based on (2.10), (2.32), and other similar NMIs. To solve (2.54) using the modified CCL, the Schur complement implies that (2.54) holds if

$$\begin{bmatrix} \Xi & \Pi \\ \Pi^T & -d_m \bar{T} \end{bmatrix} < 0, \tag{2.55}$$

where

$$-\bar{P}\bar{S}^{-1}\bar{P} < -\bar{T}, \tag{2.56}$$

which can be rewritten as

$$\bar{P}^{-1}\bar{S}\bar{P}^{-1} < \bar{T}^{-1}. \tag{2.57}$$

Taking $\bar{T}=T^{-1}$, $\bar{S}=S^{-1}$, and $\bar{P}=P^{-1}$, from the Schur complement (2.57) is equivalent to

$$\begin{bmatrix} T & P \\ * & S \end{bmatrix} > 0. \tag{2.58}$$

The existence of a solution for (2.55) and (2.58) is a sufficient condition for the feasibility of (2.54), which is imposes some degree of conservativeness. However, using this technique, the original nonconvex problem is casted into the following LMI-based nonlinear minimization problem:

Minimize $Tr(P\bar{P}+S\bar{S}+T\bar{T})$ subject to (2.55) and

$$\begin{bmatrix} T & P \\ * & S \end{bmatrix} > 0, \begin{bmatrix} \bar{P} & I \\ * & P \end{bmatrix} > 0, \begin{bmatrix} \bar{S} & I \\ * & S \end{bmatrix} > 0, \begin{bmatrix} \bar{T} & I \\ * & T \end{bmatrix} > 0, \tag{2.59}$$

where Tr denotes the trace operator. Then the modified CCL algorithm can be used to solve this optimization problem and find the maximum possible value of delay d_m by the following procedure.

Step 1: Solve (2.55) and (2.59) for sufficiently small initial value of d_{m0} and find a feasible set $\{P_0, \bar{P}_0, S_0, \bar{S}_0, T_0, \bar{T}_0, \bar{Q}_0, \bar{Y}_0, \bar{W}, U_0, F_0\}$ satisfying them. Set $i=0$ and $j=0$.

Step 2: Solve the LMI problems (2.55) and (2.59) for all variables: Minimize $Tr(P\bar{P}_i+\bar{P}P_i+S\bar{S}_i+\bar{S}S_i+T\bar{T}_i+\bar{T}T_i)$ subject to (2.55) and (2.59). Set $P_{i+1}=P$, $\bar{P}_{i+1}=\bar{P}$, $S_{i+1}=S$, $\bar{S}_{i+1}=\bar{S}$, $T_{i+1}=T$, and $\bar{T}_{i+1}=\bar{T}$.

Step 3: If (2.54) is satisfied and $d \le d_{m_i}$ ($d \le Rd_{m_i}$ for Theorems 2.1, 2.2 and similar ones), then the problem is already solved. If (2.54) is not satisfied within a specified number of iterations (j), then exit with no solution. If (2.54) is satisfied but $d > d_{m_i}$ ($d > Rd_{m_i}$ for Theorems 2.1, 2.2 and similar ones), then set $i=i+1$, $j=0$, increment d_{m_i}, and go to Step 2. Otherwise, set $i=i+1$, $j=j+1$ and go to Step 2.

The original CCL seeks for a new set of feasible parameters after each delay increment d_{m_i}, whereas in the proposed modified CCL the value of parameters are propagated to the next stage. This modification accelerates the algorithm since it updates the parameters based on the previous values rather than searching from the initial values. Moreover, if there exists no solution in this algorithm, it is possible to increase the number of subpredictors R and try again.

2.6.1 Matlab sample code

In this section, we present a MATLAB sample code to find the method of designing an SSP for input-delay systems. The first function is the main iterative MCCL loop to obtain a proper observer coefficient L and feedback gain K. For simplicity, the code is just presented for maximum $r=2$. Note that for greater r, the definition code for LMI matrices (like s P, S, ...) must be extended.

```
function [KK]=Mccl(a,b,dd,r,N)
%a=A matrix, b=B matrix, d= delay, r= number of predictors,
%N=iterative number
[i,j]=size(b);
for k=1 : r+1
    A(i*(k-1)+1:i*(k),i*(k-1)+1:i*(k))=a;
    B(i*(k-1)+1:i*(k),1:j)=0*b;
end
B(1:i,1:j)=b;
d=dd/r;
ii=i*(r+1);
I=eye(i*(r+1),i*(r+1));
I=[I,I,I,I,I,I];
iii=1;jjj=0;
while iii<N & jjj == 0
    iii=iii+1;
PR=k_L_santccl(a,b,dd,r,I);
P=PR(1:ii,1:ii);PP=PR(1:ii,ii+1:2*ii);
T=PR(1:ii,2*ii+1:3*ii);TT=PR(1:ii,3*ii+1:4*ii);
S=PR(1:ii,4*ii+1:5*ii);SS=PR(1:ii,5*ii+1:6*ii);
Q=PR(1:ii,6*ii+1:7*ii);Y=PR(1:ii,7*ii+1:8*ii);
W=PR(1:ii,8*ii+1:9*ii);U=PR(1:ii,9*ii+1:10*ii);
F=PR(1:ii,10*ii+1:10*ii+j)';
lmi=[A*P+P*A'+Y+Y'+Q+B*F+F'*B' U-Y+W' -d*Y d*(P*A'+F'*B');
    U'-Y'+W -W-W'-Q -d*W d*U';
    -d*Y' -d*W' -d*P*inv(S)*P 0*S;
    d*(A*P+B*F) d*U 0*S -d*S];
if max(real(eig(lmi)))<0
    jjj=1;
    %iii
    'ok'
    K=F*inv(P);
    L=U*inv(P);
    KK=[L,K'];
    eig([A+B*K+L]);
else
    iii
end
```

```
I=PR;
end
```

In the above code, in each iterative, LMIs are solved in next function:

```
function PR=k_L_santccl(a,b,d,r,PR0)
eps=.0;
[i,j]=size(b);
r=r+1;                    %for control designing
for k=1 : r
    A(i*(k-1)+1:i*(k),i*(k-1)+1:i*(k))=a;
    B(i*(k-1)+1:i*(k),1:j)=0*b;
end
B(1:i,1:j)=b;
ii=r*i;
d=d/(r-1);
I=eye(i,i);

setlmis([])
if r==2
    [P0,n,sp0] = lmivar(1,[i 1]);
    [P1,n,sp1] = lmivar(1,[i 1]);
    [P,n,sp] = lmivar(3,[sp0 0*I;0*I sp1]);
    [PP0,n,spp0] = lmivar(1,[i 1]);
    [PP1,n,spp1] = lmivar(1,[i 1]);
    [PP,n,spp] = lmivar(3,[spp0 0*I;0*I spp1]);
elseif r==3
    [P0,n,sp0] = lmivar(1,[i 1]);
    [P1,n,sp1] = lmivar(1,[i 1]);
    [P2,n,sp2] = lmivar(1,[i 1]);
    [P,n,sp] =
        lmivar(3,[sp0 0*I 0*I;0*I sp1 0*I;0*I 0*I sp2]);
    [PP0,n,spp0] = lmivar(1,[i 1]);
    [PP1,n,spp1] = lmivar(1,[i 1]);
    [PP2,n,spp2] = lmivar(1,[i 1]);
    [PP,n,spp] =
       lmivar(3,[spp0 0*I 0*I;0*I spp1 0*I;0*I 0*I spp2]);
end
[T,n,st] = lmivar(1,[ii 1]);
[TT,n,stt] = lmivar(1,[ii 1]);
[S,n,ss] = lmivar(1,[ii 1]);
[SS,n,sss] = lmivar(1,[ii 1]);
[Q,n,sq] = lmivar(1,[ii 1]);
[W,n,sw] = lmivar(2,[ii ii]);
[Y,n,sy] = lmivar(2,[ii ii]);
[F1,n,sf1] = lmivar(2,[j i]);
if r==2
    [F,n,sf] = lmivar(3,[sf1 ,zeros(j,i)]);
```

```
    [U1,n,su1] = lmivar(2,[i i]);
    [U,n,su] = lmivar(3,[0*I su1;0*I su1]);
elseif r==3
    [F,n,sf] = lmivar(3,[sf1, zeros(j,i), zeros(j,i)]);
    [U1,n,su1] = lmivar(2,[i i]);
    [U2,n,su2] = lmivar(2,[i i]);
    [U,n,su] =
      lmivar(3,[0*I su1 0*I;0*I su1 -su2;0*I 0*I su2]);
% P is positive
lmiterm([-1 1 1 P],1,1);lmiterm([ 1 1 1 0],eps);
% Q is positive
lmiterm([-2 1 1 Q],1,1);lmiterm([ 2 1 1 0],eps);
% S is positive
lmiterm([-3 1 1 S],1,1);lmiterm([ 3 1 1 0],eps);
% basic LMI
lmiterm([ 4 1 1 P],A,1,'s');lmiterm([ 4 1 1 F],B,1,'s');
lmiterm([ 4 1 1 Y],1,1,'s');lmiterm([ 4 1 1 Q],1,1);
lmiterm([ 4 1 2 U],1,1);lmiterm([ 4 1 2 Y],-1,1);
lmiterm([ 4 1 2 -W],1,1);lmiterm([ 4 1 3 P],1,d*(A)');
lmiterm([ 4 1 3 -F],1,d*B');lmiterm([ 4 1 4 Y],-d,1);
lmiterm([ 4 2 2 Q],-1,1);lmiterm([ 4 2 2 W],-1,1,'s');
lmiterm([ 4 2 3 -U],d,1);lmiterm([ 4 2 4 W],-d,1);
lmiterm([ 4 3 3 S],-d,1);lmiterm([ 4 4 4 T],-d,1);
lmiterm([-4 1 1 0],-eps);
%CCL LMIs
lmiterm([ -5 1 1 P],1,1);lmiterm([ -5 1 2 0],eye(ii));
lmiterm([ -5 2 2 PP],1,1);lmiterm([ 5 1 1 0],0);
lmiterm([ -6 1 1 T],1,1);lmiterm([ -6 1 2 0],eye(ii));
lmiterm([ -6 2 2 TT],1,1);lmiterm([ 6 1 1 0],0);
lmiterm([ -7 1 1 TT],1,1);lmiterm([ -7 1 2 PP],1,1);
lmiterm([ -7 2 2 SS],1,1);lmiterm([ 7 1 1 0],0);
lmiterm([ -8 1 1 S],1,1);lmiterm([ -8 1 2 0],eye(ii));
lmiterm([ -8 2 2 SS],1,1);lmiterm([ 8 1 1 0],0);
%LMI Creat
LMIs = getlmis;
cc=[];
P0=PR0(1:ii,1:ii);PP0=PR0(1:ii, ii+1:2*ii);
T0=PR0(1:ii,2*ii+1:3*ii);TT0=PR0(1:ii,3*ii+1:4*ii);
S0=PR0(1:ii,4*ii+1:5*ii);SS0=PR0(1:ii,5*ii+1:6*ii);
% set c optimization parameter subject to CCL method
for o=1 : r
    cc=coeffm(cc,PP0((o-1)*i+1:o*i,(o-1)*i+1:o*i),i);
end
for o=1 : r
    cc=coeffm(cc,P0((o-1)*i+1:o*i,(o-1)*i+1:o*i),i);
end
cc=coeffm(cc,TT0,ii);cc=coeffm(cc,T0,ii);
cc=coeffm(cc,SS0,ii);cc=coeffm(cc,S0,ii);
```

```
[jjj ,jj]=size(cc);
cc(jj+1:n)=0;
% LMI Solve
options = [1e-5,0,0,0,1];
[copt,xopt] = mincx(LMIs,cc,options);
xfeas=xopt';
P = dec2mat(LMIs,xfeas,P);
PP = dec2mat(LMIs,xfeas,PP);T = dec2mat(LMIs,xfeas,T);
TT =dec2mat(LMIs,xfeas,TT);S =dec2mat(LMIs,xfeas,S);
SS =dec2mat(LMIs,xfeas,SS);Q = dec2mat(LMIs,xfeas,Q);
W = dec2mat(LMIs,xfeas,W);Y = dec2mat(LMIs,xfeas,Y);
U = dec2mat(LMIs,xfeas,U);F = dec2mat(LMIs,xfeas,F);
PR=[P,PP,T,TT,S,SS,Q,Y,W,U,F'];
end
```

Also, the optimization parameter c must be chosen such as explained in MCCL. The next code explains the method of c definition by using *coeffm.m* code:

```
function cc=coeffm(cc,P0,i)
[jjj ,jj]=size(cc);
ir=1;ic=1;ff=1;
for ii=jj+1:jj+i*(i+1)/2
    if ir==ic
        ff=1;
    else
        ff=2;
    end
    cc(ii)=ff*P0(ir,ic);
    if ir==ic
        ic=ic+1;
        ir=1;
    else
        ir=ir+1;
    end
end
```

2.7. Conclusion and future works

In this chapter, we propose an asymptotic-predictor method based on a state observer that can stabilize all linear systems with any pure input delay. The stability of the predictor and the closed-loop system is proved using Lyapunov functions. Furthermore, to stabilize unstable systems with a long time delay, we suggest a chain predictor (called sequential subpredictors) by cascading a set of subpredictors to piecewisely forecast the states during a long delay. More importantly, a robust controller is designed for dead time systems with uncertainty using the SSP approach. We present simulation examples illustrating the significance of the predictor for stabilizing unstable systems with long time-delay. The SSP method is used to robust H_∞ control of input delay systems in [22]. Now the SSP is extended for nonlinear systems [28].

References

[1] E.W. Kamen, Industrial Controls and Manufacturing, Elsevier, 1999.

[2] C.-L. Lai, P.-L. Hsu, Design the remote control system with the time-delay estimator and the adaptive Smith predictor, IEEE Transactions on Industrial Informatics 6 (1) (2009) 73–80.

[3] I. Karafyllis, M. Krstic, Predictor Feedback for Delay Systems: Implementations and Approximations, Springer, 2017.

[4] M. Krstic, Delay Compensation for Nonlinear, Adaptive, and PDE Systems, Springer, 2009.

[5] Q.-C. Zhong, Robust Control of Time-Delay Systems, Springer Science & Business Media, 2006.

[6] O.J. Smith, A controller to overcome dead time, ISA Journal 6 (2) (1959) 28–33.

[7] K. Watanabe, M. Ito, A process-model control for linear systems with delay, IEEE Transactions on Automatic Control 26 (6) (1981) 1261–1269.

[8] Q.-C. Zhong, G. Weiss, A unified Smith predictor based on the spectral decomposition of the plant, International Journal of Control 77 (15) (2004) 1362–1371.

[9] M. Najafi, S. Hosseinnia, F. Sheikholeslam, M. Karimadini, Closed-loop control of dead time systems via sequential sub-predictors, International Journal of Control 86 (4) (2013) 599–609.

[10] M. Krstic, On compensating long actuator delays in nonlinear control, IEEE Transactions on Automatic Control 53 (7) (2008) 1684–1688.

[11] M. Krstic, Input delay compensation for forward complete and strict-feedforward nonlinear systems, IEEE Transactions on Automatic Control 55 (2) (2010) 287–303.

[12] N. Bekiaris-Liberis, M. Krstic, Robustness of nonlinear predictor feedback laws to time-and state-dependent delay perturbations, Automatica 49 (6) (2013) 1576–1590.

[13] I. Karafyllis, M. Krstic, T. Ahmed-Ali, F. Lamnabhi-Lagarrigue, Global stabilisation of nonlinear delay systems with a compact absorbing set, International Journal of Control 87 (5) (2014) 1010–1027.

[14] J.-Q. Huang, F.L. Lewis, Neural-network predictive control for nonlinear dynamic systems with time-delay, IEEE Transactions on Neural Networks 14 (2) (2003) 377–389.

[15] F.L. Lewis, J. Campos, R. Selmic, Neuro-Fuzzy Control of Industrial Systems With Actuator Nonlinearities, vol. 24, SIAM, 2002.

[16] Z. Artstein, Linear systems with delayed controls: a reduction, IEEE Transactions on Automatic Control 27 (4) (1982) 869–879.

[17] A. Olbrot, Stabilizability, detectability, and spectrum assignment for linear autonomous systems with general time delays, IEEE Transactions on Automatic Control 23 (5) (1978) 887–890.

[18] D. Wang, D. Zhou, Y. Jin, S.J. Qin, A strong tracking predictor for nonlinear processes with input time delay, Computers & Chemical Engineering 28 (12) (2004) 2523–2540.

[19] T. Ahmed-Ali, E. Cherrier, F. Lamnabhi-Lagarrigue, Cascade high gain predictors for a class of nonlinear systems, IEEE Transactions on Automatic Control 57 (1) (2012) 221–226.

[20] S. Battilotti, Nonlinear predictors for systems with bounded trajectories and delayed measurements, Automatica 59 (2015) 127–138.
[21] M. Farza, M. M'Saad, T. Menard, M.L. Fall, O. Gehan, E. Pigeon, Simple cascade observer for a class of nonlinear systems with long output delays, IEEE Transactions on Automatic Control 60 (12) (2015) 3338–3343.
[22] M. Najafi, F. Sheikholeslam, Q.-G. Wang, S. Hosseinnia, Robust hα control of single input-delay systems based on sequential sub-predictors, IET Control Theory & Applications 8 (13) (2014) 1175–1184.
[23] S. Xu, J. Lam, Improved delay-dependent stability criteria for time-delay systems, IEEE Transactions on Automatic Control 50 (3) (2005) 384–387.
[24] L. Dugard, E.I. Verriest, Stability and Control of Time-Delay Systems, vol. 228, Springer, 1998.
[25] L. El Ghaoui, F. Oustry, M. AitRami, A cone complementarity linearization algorithm for static output-feedback and related problems, IEEE Transactions on Automatic Control 42 (8) (1997) 1171–1176.
[26] S.C. Lee, Q.-G. Wang, et al., Stabilizing control for a class of delay unstable processes, ISA Transactions 49 (3) (2010) 318–325.
[27] I.R. Petersen, A stabilization algorithm for a class of uncertain linear systems, Systems & Control Letters 8 (4) (1987) 351–357.
[28] A. Vafaei, M.J. Yazdanpanah, A chain observer for nonlinear long constant delay systems: a matrix inequality approach, Automatica 65 (2016) 164–169.

CHAPTER THREE

Discrete-time adaptive control of time-delayed linear systems

Khalid Abidi[a] **and Hang Jian Soo**[b]

[a]Newcastle University (Singapore Campus), SIT Building @ Nanyang Polytechnic, Ang Mo Kio, Singapore

[b]Technology Centre for Offshore and Marine, Singapore (TCOMS), Prince George's Park, Singapore

Contents

3.1. Introduction

When uncertainty in the plant parameters cannot be ignored, it can be mitigated using either robust or adaptive control approaches. The decision to use one or the other must be based on understanding the nature of the parameter uncertainty. If the range of plant parameter variations can be characterized, then robust controllers can ensure that performance specifications are met. On the other hand, if these specifications are tight and parameter variations occur over a larger range than can be reasonably handled by a robust control law, then an adaptive controller may be a better choice. However, the designer must be mindful that the parameter adaptation process takes time, and frequently varying parameters can adversely affect the transient performance of the closed-loop system.

Various continuous-time model-reference adaptive controllers for linear plants with known input time delay and uncertain parameters have been

https://doi.org/10.1016/B978-0-12-820599-0.00008-2

proposed over the years, for example, by Ortega and Lozano [3], Niculescu and Annaswamy [4], and the adaptive posicast controller (APC) by Yildiz et al [5]. These controllers may all be considered model-reference adaptive versions of the Finite Spectrum Assignment (FSA) regulator introduced by Manitius and Olbrot [2]. FSA can be used to compensate plants with any combination of multiple discrete input delays and distributed input delays. In the simplest case of a single input delay τ,

$$\dot{\mathbf{x}}(t) = A\mathbf{x}(t) + B\mathbf{u}(t-\tau), \tag{3.1}$$

the control law is given by

$$\mathbf{u}(t) = F\mathbf{x}(t) + F\int_{-\tau}^{0} e^{-(\tau+\eta)A}B\mathbf{u}(t+\eta)d\eta. \tag{3.2}$$

This results in a closed-loop system with finite spectrum, which can be assigned (under some controllability conditions) by a suitable choice of F.

Whereas FSA was originally formulated in a state variable setting, in the above-mentioned adaptive controllers, the idea has been adapted for plants where only output feedback is available. In this setting, it is hard to design stable adaptation laws for plants with relative degree $n^* \geq 1$, because the resulting error dynamics is not strict positive real. The augmented error approach can be used to overcome this difficulty [8] and is employed in the controller by Ortega and Lozano [3], which enables it to handle plants of arbitrary relative degree. If such flexibility is not required, then a simpler controller for plants with $n^* \leq 2$ is given by Niculescu and Annaswamy [4].

A shortcoming of both adaptive controllers in [3,4] is that all plant poles must have multiplicity one. This constraint is removed in APC [5]; however, it can only be used on plants with $n^* \leq 2$. Nevertheless, an important advantage of APC is its clearly discernible observer–predictor structure, and it can be readily applied to both output feedback and state feedback cases. The APC control law for the latter is given by

$$\mathbf{u}(t) = \Theta_x^\top(t)\mathbf{x}(t) + \int_{-\tau}^{0} \Lambda_u(t,\eta)\mathbf{u}(t+\eta)d\eta + \Theta_r^\top(t)\mathbf{r}(t), \tag{3.3}$$

where $\Theta_x(t)$, $\Lambda_u(t,\eta)$, and $\Theta_r(t)$ are adaptive parameters [6].

The integral term present in continuous-time FSA-based control laws has posed a challenge to practical implementation. It was found that certain numerical approximations of this integral would lead to closed-loop instability, even if the time-steps used are very small [7]. These offending approximations result in frequency responses that are not strictly proper, with the consequence that the approximation error grows significantly at high frequencies, causing instability. For a detailed analysis of this issue and an improved approximation formula, the reader is advised to consult [13]. Therefore from a practical perspective it would be desirable to have on hand a *discrete-time* control law.

The presence of disturbances to the plant dynamics must be taken into consideration in adaptive controller design. This is not merely because it is generally desirable to analyze and minimize the effect of disturbances on control performance; an adaptation law that does not take into account disturbances can suffer from "parameter drift", where the adaptive parameters grow unbounded in the presence of a bounded or even vanishing disturbance [10]. This issue can be resolved by introducing a dead zone into the adaptation law. The idea is to "turn off" the parameter adaptation process when the measured estimation error falls below some level, since small errors are likely dominated by the disturbance rather than the true error.

In the adaptive control schemes discussed so far, the input time delay is assumed to be known. It is not as straightforward to deal with an unknown time delay, because even a linear plant is not linear in the time delay, and hence the delay cannot be estimated like the other uncertain parameters. One solution is the adaptive controller in [11], which models the time delay as a transport PDE, so it can be estimated explicitly. A different approach is truncated predictor feedback, which is a method of finding a robust control law that ensures stability without knowledge of the time delay, provided that the time delay lies within certain bounds [12].

In this chapter, we introduce to the reader a discrete-time version of the APC.[1] This is a model-reference adaptive controller that can be applied to linear systems with a known input delay and subject to matched unmeasurable disturbances to the plant state. It is assumed that the plant states are measurable.

The key ideas underpinning the discrete APC are summarized as follows:

1. A disturbance observer is used to estimate the unmeasurable disturbance. Although some estimation error is inevitable, it can be shown that this decreases with the length of the sampling interval, given reasonable assumptions about the disturbance.

[1] This work was originally presented in [1]. An earlier version was given in [9], in which disturbances were not considered, and delays were assumed to be integer multiples of the sampling interval.

2. The control law predicts the future state of the plant *taking into account the estimated disturbance* and compensates for it in order to achieve the desired reference model dynamics in the closed-loop system.
3. To cope with the plant uncertainty, the control law parameters are adjusted by adaptation laws. It will be shown that these parameters are simply those of a dynamics derived from the model tracking error dynamics. Hence the adaptation laws directly estimate the control law parameters. The adaptation laws employed are a modified recursive least-squares estimator.
4. A dead zone is built into the adaptation law, which turns off the adaptation process if the model tracking error falls below the maximum perturbation attributable to the disturbance. This prevents the disturbance from degrading the transient performance of the closed-loop system and ensures stability.

The discrete-time APC approach can also be extended to plants with constant but uncertain input delay, provided that an upper bound for the delay is known. This involves expressing the plant model in a form that does not explicitly depend on the delay, but only its upper bound. The control law is then obtained in a similar manner; however, the adaptation law now includes some parameters that must be manually tuned.

The rest of this chapter is organized as follows. The control problem is defined in Section 3.2. The discrete APC controller and stability proof are presented for the case of a known time delay in Section 3.3, and for the case of an uncertain time delay in Section 3.4. Finally, these results are illustrated with simulation examples in Section 3.5.

3.2. Problem definition

Consider the continuous-time system

$$\dot{\mathbf{x}}(t) = A\mathbf{x}(t) + B\left(\mathbf{u}(t-\tau) + \mathbf{f}(t)\right), \tag{3.4}$$

where $\mathbf{x} \in \Re^n$ is the vector of states, $A \in \Re^{n\times n}$ is a constant uncertain state matrix, $B \in \Re^{n\times m}$ is a constant uncertain input gain matrix, $\mathbf{u} \in \Re^m$ is the vector of the control inputs, $\tau \geq 0$ is a known input time-delay, and $\mathbf{f}(t)$ is a matched unmeasurable exogenous disturbance. For system (3.4), we make the following assumptions.

Assumption 3.1. There exists a known nominal gain matrix B_n such that $B = B_\mathrm{n}\Lambda$, where $\Lambda \in \Re^{m\times m}$ is a constant uncertain positive definite matrix.

Assumption 3.2. The disturbance $\mathbf{f}(t)$ is smooth and bounded.

Consider that the reference model reflecting the desired response characteristics

$$\dot{\mathbf{x}}_m(t) = A_m\mathbf{x}_m + B_m\mathbf{r}(t-\tau), \tag{3.5}$$

where $A_m \in \Re^{n\times n}$ is a constant Hurwitz matrix, $B_m \in \Re^{n\times q}$ is a constant matrix, and $\mathbf{r} \in \Re^q$ is the desired reference command. Note that the dynamics given in the reference model (3.5) is the best that can be achieved with any kind of control law in terms of handling the time delay, that is, the time delay cannot be eliminated from the closed-loop dynamics, but only its effect on the stability can be eliminated. The control problem is finding a bounded control input $\mathbf{u}$ such that $\lim_{t\to\infty} \|\mathbf{x}_m(t) - \mathbf{x}(t)\| = 0$ while keeping all the system signals bounded.

The reference model (3.5) is obtained by applying a nominal controller to the nominal system, which is the system given in (3.4) assuming no uncertainty in the matrices A and B. Note that a conventional state feedback or linear quadratic regulator (LQR) controller cannot be used to obtain (3.5). This is because the delay τ introduces infinitely many eigenvalues to the closed-loop system unless a nominal controller that eliminates the effect of the delay on the eigenvalues is used. Therefore a finite spectrum assignment controller [2] is used to design the nominal controller. This point is important also for the theoretical sound augmentation of the nominal controller with the adaptive controller [18].

3.3. Control law design and stability analysis

In this section, we present the discrete-time design of the APC. The approach is presented for high-order systems with known time-delay. An extension to uncertain input time delay will be presented in the next section.

3.3.1 Control law design

Consider system (3.4). If a constant $p \in Z^+$ is selected such that $pT \le \tau \le (p+1)T$, where the time delay τ is known, and T is the sampling interval, then the sampled-data form of (3.4) is obtained as

$$\mathbf{x}_{k+1} = \Phi\mathbf{x}_k + \Gamma_1\mathbf{u}_{k-p} + \Gamma_2\mathbf{u}_{k-p-1} + \mathbf{d}_k, \tag{3.6}$$

where the matrices $\Phi \in \Re^{n\times n}$, $\Gamma_1 \in \Re^{n\times m}$, and $\Gamma_2 \in \Re^{n\times m}$ are considered uncertain and are computed using the relations

$$\Phi = e^{AT},\ \Gamma_1 = \int_0^{(p+1)T-\tau} e^{A\sigma}\, d\sigma\, B,\ \text{and } \Gamma_2 = \int_{(p+1)T-\tau}^{T} e^{A\sigma}\, d\sigma\, B,$$

and the disturbance vector $\mathbf{d}_k$ is computed by the relation

$$\mathbf{d}_k = \int_0^T e^{A\sigma} B\mathbf{f}\big((k+1)T-\sigma\big)d\sigma.$$

Based on the assumptions on $\mathbf{f}(t)$, the following properties are defined for $\mathbf{d}_k$, [14], [15].

Property 3.1. The norm of the difference between two successive disturbance signals is at most of the order of the square of the sampling interval, that is, $\|\mathbf{d}_k - \mathbf{d}_{k-1}\| \leq \Delta$ for some $\Delta \in O(T^2)$.

Property 3.2. The relationship between $\mathbf{d}_k$ and $\mathbf{f}(t)$ is represented as

$$\mathbf{d}_k = \Gamma_{\mathrm{n}}\left(L_0\mathbf{f}_k + L_1\dot{\mathbf{f}}_k\right) + O(T^3) = \Gamma_{\mathrm{n}}\mathbf{w}_k + O(T^3),$$

where Γ_{n} is a known nominal input gain matrix, $(L_0, L_1) \in \Re^{m\times m}$ are some uncertain constant matrices, $\mathbf{f}_k = \mathbf{f}(kT)$, and $\mathbf{w}_k = L_0\mathbf{f}_k + L_1\dot{\mathbf{f}}_k$.

Remark 3.1. If the time delay τ is an integer multiple of the sampling interval, then p is selected such that $pT = \tau$, and, consequently,

$$\Gamma_1 = \int_0^T e^{A\sigma}\, d\sigma\, B,\ \text{and } \Gamma_2 = 0.$$

Remark 3.2. For typical implementations, sampling intervals are selected as $T < 1$, and therefore minimizing the influence of the disturbance to at most $O(T^2)$ is desirable.

Now consider the reference model (3.5) in the sampled-data form

$$\mathbf{x}_{\mathrm{m},k+1} = \Phi_{\mathrm{m}}\mathbf{x}_{\mathrm{m},k} + \Gamma_{\mathrm{m}}\mathbf{r}_{k-p}, \tag{3.7}$$

where Φ_{m} has eigenvalues inside the unit circle. The objective is designing a proper control law that will ensure that system (3.6) will track the reference model (3.7) and thereby achieve $\lim_{k\to\infty} \|\mathbf{x}_{\mathrm{m},k} - \mathbf{x}_k\| \leq \varepsilon$ for some constant ε.

To design a proper control law for system (3.6), assume that the system is without uncertainty and that there exists $\Theta \in \Re^{m\times n}$, a positive-definite $\Theta_{\gamma_1} \in \Re^{m\times m}$, and a positive-definite $\Theta_{\gamma_2} \in \Re^{m\times m}$ such that

$$\Phi - \Gamma_{\rm n}\Theta = \Phi_{\rm m},\ \Gamma_1 = \Gamma_{\rm n}\Theta_{\gamma_1},\ \text{and}\ \Gamma_2 = \Gamma_{\rm n}\Theta_{\gamma_2}. \tag{3.8}$$

Now consider the delay observer given as

$$\hat{\mathbf{d}}_k = \mathbf{d}_{k-1} = \mathbf{x}_k - \Phi\mathbf{x}_{k-1} - \Gamma_1 u_{k-p-1} - \Gamma_2 u_{k-p-2} \tag{3.9}$$

and select a matrix $D \in \Re^{m\times n}$ such that from Property 3.2 and (3.8) it follows that

$$\mathbf{w}_{k-1} = (D\Gamma_{\rm n})^{-1} D(\mathbf{x}_k - \Phi\mathbf{x}_{k-1}) - \Theta_{\gamma_1}\mathbf{u}_{k-p-1} - \Theta_{\gamma_2}\mathbf{u}_{k-p-2} + O\left(T^3\right), \tag{3.10}$$

where D is selected such that $D\Gamma_{\rm n}$ is nonsingular. Adding and subtracting the term $\mathbf{d}_{k-1}$ on the right-hand side of (3.6) and using Property 3.2, we obtain that

$$\begin{aligned}\mathbf{x}_{k+1} &= \Phi\mathbf{x}_k + \Gamma_1\mathbf{u}_{k-p} + \Gamma_2\mathbf{u}_{k-p-1} + \mathbf{d}_k - \mathbf{d}_{k-1} + \Gamma_{\rm n}\mathbf{w}_{k-1} + O\left(T^3\right)\\ &= \Phi\mathbf{x}_k + \Gamma_1\mathbf{u}_{k-p} + \Gamma_2\mathbf{u}_{k-p-1} + \Gamma_{\rm n}\mathbf{w}_{k-1} + \bar{\varsigma}_k,\end{aligned} \tag{3.11}$$

where $\|\bar{\varsigma}_k\| = \|\mathbf{d}_k - \mathbf{d}_{k-1}\| \leq \Delta \in O\left(T^2\right)$, and that, for a reasonable selection of sampling interval, $O\left(T^2\right) \approx O\left(T^2\right) + O\left(T^3\right)$.

Remark 3.3. Note that using the delay disturbance observer (3.10), the influence of the disturbance can increase to $O(T)$ at certain instances, for example, when the disturbance is suddenly applied or changes direction. However, overall influence is maintained at $O(T^2)$, as was demonstrated experimentally in [22].

Substituting (3.8) and (3.10) into (3.11), we obtain that

$$\begin{aligned}\mathbf{x}_{k+1} = \Phi_{\rm m}\mathbf{x}_k + \Gamma_{\rm n}\big(&\bar{\Theta}\mathbf{x}_k - \Phi_{\rm x}\mathbf{x}_{k-1} + \Theta_{\gamma_1}\mathbf{u}_{k-p} + \Theta_{\gamma_{21}}\mathbf{u}_{k-p-1}\\ &- \Theta_{\gamma_2}\mathbf{u}_{k-p-2}\big) + \bar{\varsigma}_k,\end{aligned} \tag{3.12}$$

where $\bar{\Theta} = \Theta + (D\Gamma_{\rm n})^{-1} D$, $\Phi_{\rm x} = (D\Gamma_{\rm n})^{-1} D\Phi$, and $\Theta_{\gamma_{21}} = \Theta_{\gamma_2} - \Theta_{\gamma_1}$. Performing successive substitutions on (3.12), we obtain that

$$\mathbf{x}_{k+p+1} = \Phi_{\rm m}\mathbf{x}_{k+p} + \Gamma_{\rm n}\left(\Theta_{\rm x}\bar{\chi}_k + \Theta_{\rm u}\bar{\xi}_k + \Theta_{\gamma_1}\mathbf{u}_k\right) + \bar{\delta}_{k+p}, \tag{3.13}$$

where the matrices $\Theta_x \in \Re^{m\times 2n}$ and $\Theta_u \in \Re^{m\times m(p+2)}$ contain the matrices $\bar{\Theta}$, Φ_x, Θ_{γ_1}, Θ_{γ_2}, and $\Theta_{\gamma_{21}}$. The vectors $\bar{\chi}_k \in \Re^{2n}$ and $\bar{\boldsymbol{\xi}}_k \in \Re^{m(p+2)}$ are defined as

$$\bar{\chi}_k = \begin{bmatrix} \mathbf{x}_k \\ \mathbf{x}_{k-1} \end{bmatrix} \text{ and } \bar{\boldsymbol{\xi}}_k^\top = \begin{bmatrix} \mathbf{u}_{k-1}^\top \\ \vdots \\ \mathbf{u}_{k-p-2}^\top \end{bmatrix}.$$

Furthermore, the disturbance term $\bar{\delta}_k \in \Re^n$ is given as

$$\bar{\delta}_{k+p} = \bar{\varsigma}_{k+p} + \left(\bar{\Theta}\Upsilon_0 - \Phi\right)\bar{\varsigma}_{k+p-1} + \cdots + \left(\bar{\Theta}\Upsilon_{p-2} - \Phi\Upsilon_{p-3}\right)\bar{\varsigma}_k, \tag{3.14}$$

where

$$\Upsilon_\eta = \bar{\Phi}^{\eta+1} + \sum_{i=0}^{\lfloor\frac{\eta-1}{2}\rfloor}\left[\frac{(-1)^{i+1}}{(i+1)!}\prod_{j=0}^{i}(\eta - 2i + j)\right]\bar{\Phi}^{\eta-2i-1}\Phi^{i+1} \tag{3.15}$$

with $\bar{\Phi} = \Phi + I$, and $\lfloor\cdot\rfloor$ is the floor function. Note that the number of terms on the right-hand side of (3.14) is finite and is equal to $p+1$. Furthermore, from the analysis in [14] it can be shown that the order of the norm of each term on the right-hand side of (3.14) is at most $O\left(T^2\right)$ and that therefore $\bar{\delta}_k$ is bounded. Finally, since $O\left(T^2\right)$ terms are summed $p+1$ times, the order of $\|\bar{\delta}_{k+p}\|$ is at most $O\left(pT+T\right)\cdot O(T) = O(\tau + T)\cdot O(T)$. Then the control law is selected as

$$\mathbf{u}_k = -\Theta_{\gamma_1}^{-1}\left(\Theta_x\bar{\chi}_k + \Theta_u\bar{\xi}_k - \Theta_r\mathbf{r}_k\right), \tag{3.16}$$

where $\Theta_r \in \Re^{m\times q}$ is selected such that $\Gamma_m = \Gamma_n\Theta_r$. If the tracking error is defined as $\mathbf{e}_k = \mathbf{x}_k - \mathbf{x}_{m,k}$, then the control law (3.16) leads to the closed-loop error dynamics given as

$$\mathbf{e}_{k+1} = \Phi_m\mathbf{e}_k + \bar{\delta}_k, \tag{3.17}$$

which is stable, and $\lim_{k\to\infty}\|\mathbf{e}_k\| \le \varepsilon \in O(T+\tau)$.

Proceeding now with uncertain Φ, Γ_1, and Γ_2, the parameters Θ_x, Θ_u, and Θ_{γ_1} become uncertain. The control law (3.16) is then modified to the form

$$\mathbf{u}_k = -\hat{\Theta}_{\gamma_1,k}^{-1}\left(\hat{\Theta}_{x,k}\bar{\chi}_k + \hat{\Theta}_{u,k}\bar{\boldsymbol{\xi}}_k - \Theta_r\mathbf{r}_k\right), \tag{3.18}$$

where $\hat{\Theta}_{\mathrm{x},k}$, $\hat{\Theta}_{\mathrm{u},k}$ and $\hat{\Theta}_{\gamma_1,k}$ are the estimates of Θ_{x}, Θ_{u} and Θ_{γ_1} respectively. To derive the adaptive law for $\hat{\Theta}_{\mathrm{x},k}$, $\hat{\Theta}_{\mathrm{u},k}$, and $\hat{\Theta}_{\gamma_1,k}$, it is necessary to derive the closed-loop system.

Consider p delayed sampling instants (3.13). Adding and subtracting the term $\Gamma_{\mathrm{n}}\hat{\Theta}_{\gamma_1,k-p}\mathbf{u}_{k-p}$, we obtain that

$$\begin{aligned}\mathbf{x}_{k+1} = \Phi_{\mathrm{m}}\mathbf{x}_k + \Gamma_{\mathrm{n}}\big(&\Theta_{\mathrm{x}}\bar{\chi}_{k-p} + \Theta_{\mathrm{u}}\bar{\xi}_{k-p} + \Theta_{\gamma_1}\mathbf{u}_{k-p} - \hat{\Theta}_{\gamma_1,k-p}\mathbf{u}_{k-p}\\ &+ \hat{\Theta}_{\gamma_1,k-p}\mathbf{u}_{k-p}\big) + \bar{\delta}_{k+p}.\end{aligned} \tag{3.19}$$

Substituting (3.18) into (3.19) and defining the estimation errors as $\tilde{\Theta}_{\mathrm{x},k} = \Theta_{\mathrm{x}} - \hat{\Theta}_{\mathrm{x},k}$, $\tilde{\Theta}_{\mathrm{u},k} = \Theta_{\mathrm{u}} - \hat{\Theta}_{\mathrm{u},k}$, and $\tilde{\Theta}_{\gamma_1,k} = \Theta_{\gamma_1} - \hat{\Theta}_{\gamma_1,k}$, we obtain the closed-loop error dynamics as

$$\mathbf{e}_{k+1} = \Phi_{\mathrm{m}}\mathbf{e}_k + \Gamma_{\mathrm{n}}\big(\tilde{\Theta}_{\mathrm{x},k-p}\bar{\chi}_{k-p} + \tilde{\Theta}_{\mathrm{u},k-p}\bar{\boldsymbol{\xi}}_{k-p} + \tilde{\Theta}_{\gamma_1,k-p}\mathbf{u}_{k-p}\big) + \bar{\delta}_k. \tag{3.20}$$

It is convenient to rewrite the error dynamics (3.20) in the augmented form

$$\mathbf{e}_{k+1} = \Phi_{\mathrm{m}}\mathbf{e}_k + \Gamma_{\mathrm{n}}\tilde{\Psi}_{k-p}^{\top}\bar{\boldsymbol{\zeta}}_{k-p} + \bar{\delta}_k, \tag{3.21}$$

where

$$\begin{aligned}\tilde{\Psi}_k^{\top} &= \left[\ \tilde{\Theta}_{\mathrm{x},k}\ \middle|\ \tilde{\Theta}_{\mathrm{u},k}\ \middle|\ \tilde{\Theta}_{\gamma_1,k}\ \right] \in \Re^{m\times(2n+m(p+3))} \text{ and}\\ \bar{\boldsymbol{\zeta}}_k^{\top} &= \left[\ \bar{\chi}_k^{\top}\ \middle|\ \bar{\boldsymbol{\xi}}_k^{\top}\ \middle|\ u_k^{\top}\ \right] \in \Re^{2n+m(p+3)}.\end{aligned}$$

To proceed with the formulation of the adaptive law, define

$$\mathbf{z}_{k+1} = D_{\gamma}^{\top}(\mathbf{e}_{k+1} - \Phi_{\mathrm{m}}\mathbf{e}_k) \tag{3.22}$$

such that from the substitution of (3.21) into (3.22) we obtain that

$$\mathbf{z}_{k+1} = \tilde{\Psi}_{k-p}^{\top}\bar{\boldsymbol{\zeta}}_{k-p} + \bar{\upsilon}_k, \tag{3.23}$$

where $\mathbf{z}_k \in \Re^m$, $D_{\gamma}^{\top} = (D\Gamma_{\mathrm{n}})^{-1}D^{\top}$, and $\bar{\upsilon}_k = D_{\gamma}^{\top}\bar{\delta}_k \in \Re^m$. The adaptive law must be formulated with the objective of minimizing $\mathbf{z}_{k+1}$ so that the tracking error would follow the dynamics $\mathbf{e}_{k+1} = \Phi_{\mathrm{m}}\mathbf{e}_k + \bar{\delta}_k$. Therefore the adaptive law is formulated as follows:

$$\hat{\Psi}_{k+1} = \begin{cases}\hat{\Psi}_{k-p} + \alpha_k\dfrac{\beta_k}{1 + \alpha_k\bar{\boldsymbol{\zeta}}_{k-p}^{\top}Q\bar{\boldsymbol{\zeta}}_{k-p}}Q\bar{\boldsymbol{\zeta}}_{k-p}\mathbf{z}_{k+1}^{\top} & \forall k \in [p,\infty),\\ \hat{\Psi}_0 & \forall k \in [0,p),\end{cases} \tag{3.24}$$

where $\alpha_k \in \Re$ is a positive coefficient used to ensure a nonsingular $\hat{\Theta}_{\gamma_1,k}$, and the matrix $Q \in \Re^{(2n+m(p+3))\times(2n+m(p+3))}$ is a positive-definite adaptive gain matrix. Considering that the disturbance term $\mathbf{f}(t)$ is bounded, there exists a bound on $\bar{\upsilon}_k$ such that the weighing coefficient β_k defined as

$$\beta_k = \begin{cases} 1 - \frac{\bar{\upsilon}_{\max}}{\|\mathbf{z}_{k+1}\|}, & \text{if } \|\mathbf{z}_{k+1}\| \geq \bar{\upsilon}_{\max}, \\ 0, & \text{if } \|\mathbf{z}_{k+1}\| < \bar{\upsilon}_{\max}, \end{cases} \tag{3.25}$$

where $\|\bar{\upsilon}_k\| \leq \bar{\upsilon}_{\max}$, ensures that the closed-loop system is robust to the influence of the uncertain term $\bar{\upsilon}_k$. Selection of the value of the coefficient α_k is outlined in the following lemma.

Lemma 3.1. *The matrix $\hat{\Theta}_{\gamma_1,k}$ is nonsingular if and only if*

$$\alpha_k^{-1} \neq \lambda\left[-\hat{\Theta}_{\gamma_1,k-p}^{-1} S \frac{\beta_k}{1+\alpha_k \bar{\boldsymbol{\xi}}_{k-p}^{\top} Q \bar{\boldsymbol{\xi}}_{k-p}} Q \bar{\boldsymbol{\xi}}_{k-p} \mathbf{z}_{k+1}^{\top} \right],$$

where $\lambda[\cdot]$ is the set of eigenvalues, and $S = [\mathbf{0} \cdots \mathbf{0}\ I] \in \Re^{m\times(n+m(p+3))}$.

Proof. From the adaptive law (3.24), the matrix $\hat{\Theta}_{\gamma_1,k+1}$ is given as

$$\hat{\Theta}_{\gamma_1,k+1} = \hat{\Theta}_{\gamma_1,k-p} + \alpha_k S \frac{\beta_k}{1+\alpha_k \bar{\boldsymbol{\xi}}_{k-p}^{\top} Q \bar{\boldsymbol{\xi}}_{k-p}} Q \bar{\boldsymbol{\xi}}_{k-p} \mathbf{z}_{k+1}^{\top}. \tag{3.26}$$

Consider now that the initial parameters are selected such that $\hat{\Theta}_{\gamma_1,0}$ is nonsingular. Then, for a time step of $k = p$, the adaptive law (3.26) is written as

$$\begin{aligned} \hat{\Theta}_{\gamma_1,p+1} &= \hat{\Theta}_{\gamma_1,0} + \alpha_p S \frac{\beta_p}{1+\alpha_k \bar{\boldsymbol{\xi}}_0^{\top} Q \bar{\boldsymbol{\xi}}_0} Q \bar{\boldsymbol{\xi}}_0 \mathbf{z}_{p+1}^{\top} \\ &= \alpha_p \hat{\Theta}_{\gamma_1,0} \left[\frac{1}{\alpha_p} I + \hat{\Theta}_{\gamma_1,0}^{-1} S \frac{\beta_p}{1+\alpha_p \bar{\boldsymbol{\xi}}_0^{\top} Q \bar{\boldsymbol{\xi}}_0} Q \bar{\boldsymbol{\xi}}_0 \mathbf{z}_{p+1}^{\top} \right]. \end{aligned} \tag{3.27}$$

From (3.27), the matrix $\hat{\Theta}_{\gamma_1,p+1}$ is nonsingular if the condition

$$\alpha_p^{-1} \neq \lambda\left[-\hat{\Theta}_{\gamma_1,0}^{-1} S \frac{\beta_p}{1+\alpha_p \bar{\boldsymbol{\xi}}_0^{\top} Q \bar{\boldsymbol{\xi}}_0} Q \bar{\boldsymbol{\xi}}_0 \mathbf{z}_{p+1}^{\top} \right] \tag{3.28}$$

is satisfied. The result (3.28) is then generalized as

$$\alpha_k^{-1} \neq \lambda\left[-\hat{\Theta}_{\gamma_1,k-p}^{-1} S \frac{\beta_k}{1+\alpha_k \bar{\boldsymbol{\xi}}_{k-p}^{\top} Q \bar{\boldsymbol{\xi}}_{k-p}} Q \bar{\boldsymbol{\xi}}_{k-p} \mathbf{z}_{k+1}^{\top} \right]. \tag{3.29}$$

This completes the proof. □

Remark 3.4. The coefficient $\alpha_k > 0$ can be selected from a range of predefined values to ensure that $\hat{\Theta}_{\gamma_1,k}$ is nonsingular.

3.3.2 Stability analysis

In this section, we present the stability analysis of the system under control law (3.18) and adaptive law (3.24).

Theorem 3.1. *System* (3.6) *and the adaptive law* (3.24) *results in a closed-loop system with bounded* $\tilde{\Psi}_k$ *and* $\lim_{k\to\infty} \|\mathbf{e}_k\| \leq \varepsilon \in O(\tau + T)$.

Proof. Let

$$\mathbf{z}_k^{\top} = \begin{bmatrix} z_{1,k} \mid z_{2,k} \mid \cdots \mid z_{m,k} \end{bmatrix}^{\top} \text{ and } \tilde{\Psi}_k^{\top} = \begin{bmatrix} \tilde{\boldsymbol{\psi}}_{1,k} \mid \tilde{\boldsymbol{\psi}}_{2,k} \mid \cdots \mid \tilde{\boldsymbol{\psi}}_{m,k} \end{bmatrix}^{\top},$$

where $\tilde{\boldsymbol{\psi}}_{j,k} \in \Re^{(n+m(p+3))\times 1}, j = 1, \ldots, m$. Now, consider the positive function

$$V_k = \sum_{j=1}^{m} \left(\sum_{i=0}^{p} \tilde{\boldsymbol{\psi}}_{j,k-i}^{\top} Q^{-1} \tilde{\boldsymbol{\psi}}_{j,k-i} \right). \tag{3.30}$$

The forward difference of (3.30) is given by

$$\begin{aligned} \Delta V_k &= V_{k+1} - V_k \\ &= \sum_{j=1}^{m} \left[\tilde{\boldsymbol{\psi}}_{j,k+1}^{\top} Q^{-1} \tilde{\boldsymbol{\psi}}_{j,k+1} - \tilde{\boldsymbol{\psi}}_{j,k-p}^{\top} Q^{-1} \tilde{\boldsymbol{\psi}}_{j,k-p} \right]. \end{aligned} \tag{3.31}$$

Consider the adaptive law (3.24). Subtracting both sides from $\boldsymbol{\psi}_j$, we obtain that

$$\boldsymbol{\psi}_j - \hat{\boldsymbol{\psi}}_{j,k+1} = \boldsymbol{\psi}_j - \hat{\boldsymbol{\psi}}_{j,k-p} - \frac{\alpha_k \beta_k}{1 + \alpha_k \bar{\boldsymbol{\zeta}}_{k-p}^{\top} Q \bar{\boldsymbol{\zeta}}_{k-p}} Q \bar{\boldsymbol{\zeta}}_{k-p} z_{j,k+1}. \tag{3.32}$$

Defining $\tilde{\boldsymbol{\psi}}_{j,k} = \boldsymbol{\psi}_j - \hat{\boldsymbol{\psi}}_{j,k}$, we simplify (3.32) to the form

$$\tilde{\boldsymbol{\psi}}_{j,k+1} = \tilde{\boldsymbol{\psi}}_{j,k-p} - \frac{\alpha_k \beta_k}{1 + \alpha_k \bar{\boldsymbol{\zeta}}_{k-p}^{\top} Q \bar{\boldsymbol{\zeta}}_{k-p}} Q \bar{\boldsymbol{\zeta}}_{k-p} z_{j,k+1}. \tag{3.33}$$

Substituting (3.33) into (3.31), we obtain that

$$\Delta V_k = \sum_{j=1}^{m} \left[\frac{-2\alpha_k \beta_k \tilde{\boldsymbol{\psi}}_{j,k-p}^{\top} \bar{\boldsymbol{\zeta}}_{k-p} z_{j,k+1}}{1 + \alpha_k \bar{\boldsymbol{\zeta}}_{j,k-p}^{\top} Q \bar{\boldsymbol{\zeta}}_{k-p}} + \frac{\alpha_k^2 \beta_k^2 \bar{\boldsymbol{\zeta}}_{k-p}^{\top} Q \bar{\boldsymbol{\zeta}}_{k-p} z_{j,k+1}^2}{\left(1 + \alpha_k \bar{\boldsymbol{\zeta}}_{k-p}^{\top} Q \bar{\boldsymbol{\zeta}}_{k-p}\right)^2} \right]. \tag{3.34}$$

Using the fact that

$$\frac{\alpha_k \bar{\xi}_{k-p}^{\top} Q \bar{\xi}_{k-p}}{1+\alpha_k \bar{\xi}_{k-p}^{\top} Q \bar{\xi}_{k-p}} < 1, \tag{3.35}$$

it is possible to reduce (3.34) to the form

$$\begin{aligned} \Delta V_k &\leq \sum_{j=1}^{m} \left[\frac{-2\alpha_k \beta_k \tilde{\boldsymbol{\psi}}_{j,k-p}^{\top} \bar{\xi}_{k-p} z_{j,k+1}}{1+\alpha_k \bar{\xi}_{k-p}^{\top} Q \bar{\xi}_{k-p}} + \frac{\alpha_k \beta_k^2 z_{j,k+1}^2}{1+\alpha_k \bar{\xi}_{k-p}^{\top} Q \bar{\xi}_{k-p}} \right] \\ &\leq -\frac{2\alpha_k \beta_k \bar{\xi}_{k-p}^{\top} \tilde{\Psi}_{k-p} \mathbf{z}_{k+1}}{1+\alpha_k \bar{\xi}_{k-p}^{\top} Q \bar{\xi}_{k-p}} + \frac{\alpha_k \beta_k^2 \mathbf{z}_{k+1}^{\top} \mathbf{z}_{k+1}}{1+\alpha_k \bar{\xi}_{k-p}^{\top} Q \bar{\xi}_{k-p}}. \end{aligned} \tag{3.36}$$

Furthermore, from (3.25) we obtain that

$$\beta_k^2 \|\mathbf{z}_{k+1}\|^2 = \beta_k \|\mathbf{z}_{k+1}\|^2 - \bar{\upsilon}_{\max} \cdot \|\mathbf{z}_{k+1}\|. \tag{3.37}$$

Thus, substituting (3.23) and (3.37) into (3.36), we obtain that

$$\Delta V_k \leq -\frac{\alpha_k \beta_k^2 \mathbf{z}_{k+1}^{\top} \mathbf{z}_{k+1}}{1+\alpha_k \bar{\xi}_{k-p}^{\top} Q \bar{\xi}_{k-p}}, \tag{3.38}$$

which implies that V_k is nonincreasing, and thus $\tilde{\Psi}_k$ is bounded. Note that for any $k \in [k_0, \infty)$,

$$V_{k+1} = V_{k_0} + \sum_{i=0}^{k-k_0} \Delta V_{k_0+i}. \tag{3.39}$$

Substituting (3.38) into (3.39), we obtain that

$$\lim_{k\to\infty} V_{k+1} \leq \max_{k_0 \in [0,p)} V_{k_0} - \lim_{k\to\infty} \sum_{i=0}^{k-k_0} \frac{\alpha_k \beta_k^2 \mathbf{z}_{k_0+i+1}^{\top} \mathbf{z}_{k_0+i+1}}{1+\alpha_k \bar{\xi}_{k_0+i-p}^{\top} Q \bar{\xi}_{k_0+i-p}}. \tag{3.40}$$

Suppose that V_{k+1} is nonnegative and V_{k_0} is finite in the interval $[0, p)$. Then, according to the convergence theorem of the sum of series,

$$\lim_{k\to\infty} \frac{\alpha_k \beta_k^2 \mathbf{z}_{k+1}^{\top} \mathbf{z}_{k+1}}{1+\alpha_k \bar{\xi}_{k-p}^{\top} Q \bar{\xi}_{k-p}} = 0. \tag{3.41}$$

To guarantee that $\lim_{k\to\infty}\beta_k\|\mathbf{z}_{k+1}\|=0$, we must guarantee that $\|\bar{\boldsymbol{\zeta}}_k\|\leq \mu_0+\mu_1\max_{i\in[0,k+1]}\|\mathbf{z}_i\|$. By the definition of $\mathbf{z}_k$ we have

$$\mathbf{e}_{k+1}=\Phi_{\mathrm{m}}\mathbf{e}_k+\mathbf{z}_{k+1}. \tag{3.42}$$

Since Φ_{m} has eigenvalues inside the unit circle, $\mathbf{e}_k=\mathbf{x}_{\mathrm{m},k}-\mathbf{x}_k$, and $\mathbf{x}_{\mathrm{m},k}$ is bounded, there exist constants c_0 and c_1 such that [19]

$$\|\mathbf{x}_{k+1}\|\leq c_0+c_1\|\mathbf{z}_{k+1}\|. \tag{3.43}$$

Since the system has stable open-loop zeros, the control input is bounded as

$$\|\mathbf{u}_{k-p}\|\leq\kappa_0+\kappa_1\max_{i\in[0,k+1]}\|\mathbf{x}_i\| \tag{3.44}$$

for some constants κ_0 and κ_1. Looking at the signal growth rates, $\bar{\boldsymbol{\zeta}}_k$ is a vector containing $\mathbf{x}_k$ and $\mathbf{u}_k$. Then we have

$$\|\bar{\boldsymbol{\zeta}}_{k-p}\|\leq c_0^0+c_1^0\|\mathbf{z}_k\|\leq c_0^0+c_1^0\max_{i\in[0,k+1]}\|\mathbf{z}_i\|, \tag{3.45}$$

and therefore, using the key technical lemma [16], [19], we have $\lim_{k\to\infty}\beta_k\|\mathbf{z}_{k+1}\|=0$. Further, using the definition of β_k given by (3.37), z_{k+1} will converge to a bound of $|\bar{v}_k|\in O(\tau+T)\cdot O(T)$, and from (3.42) as $k\to\infty$, the following stable error dynamics is achieved:

$$\mathbf{e}_{k+1}=\Phi_{\mathrm{m}}\mathbf{e}_k+\bar{v}_k. \tag{3.46}$$

Finally, using the results in [9], $\lim_{k\to\infty}\|\mathbf{e}_k\|\leq\varepsilon\in O(\tau+T)$ is guaranteed. □

3.4. Control law design for uncertain time delay

Consider system (3.4) with an uncertain input time-delay τ such that its sampled-data representation is given as

$$\mathbf{x}_{k+1}=\Phi\mathbf{x}_k+\Gamma_1\mathbf{u}_{k-\ell}+\Gamma_2\mathbf{u}_{k-\ell-1}+\mathbf{d}_k, \tag{3.47}$$

where the delay $\ell\in Z$ is uncertain and given as $\ell T\leq\tau\leq(\ell+1)T$. The uncertain time delay τ is assumed to have an upper-bound for $\tau\leq\tau_{\mathrm{p}}$ and that the upper-bound can be set such that it is an integer multiple of the sampling interval, that is, $\tau_{\mathrm{p}}=pT$ for some known p. Similar to the known time-delay problem, the controller design is preceded by the reformulation of the system dynamics (3.47) such that the influence of the disturbance $\mathbf{d}_k$ is minimized. This is accomplished by using the delay observer (3.10) which in this case is given by

$$\mathbf{w}_{k-1} = (D\Gamma_{\mathrm{n}})^{-1} D(\mathbf{x}_k - \mathbf{\Phi}\mathbf{x}_{k-1}) - \Theta_{\gamma_1}\mathbf{u}_{k-\ell-1} - \Theta_{\gamma_2}\mathbf{u}_{k-\ell-2} + O\big(T^3\big). \tag{3.48}$$

Substituting (3.8) and (3.48) into (3.47), we obtain that

$$\begin{aligned}\mathbf{x}_{k+1} = \mathbf{\Phi}_{\mathrm{m}}\mathbf{x}_k + \Gamma_{\mathrm{n}}\big(&\bar{\Theta}\mathbf{x}_k - \mathbf{\Phi}_{\mathrm{x}}\mathbf{x}_{k-1} + \Theta_{\gamma_1}\mathbf{u}_{k-\ell} + \Theta_{\gamma_{21}}\mathbf{u}_{k-\ell-1}\\ &- \Theta_{\gamma_2}\mathbf{u}_{k-\ell-2}\big) + \bar{\varsigma}_k.\end{aligned} \tag{3.49}$$

Performing successive substitutions on (3.49), we obtain that

$$\mathbf{x}_{k+p+1} = \mathbf{\Phi}_{\mathrm{m}}\mathbf{x}_{k+p} + \Gamma_{\mathrm{n}}\big(\Theta_{\mathrm{x}}\bar{\chi}_k + \Theta_{\mathrm{u}}\bar{\xi}_{k+p-\ell} + \Theta_{\gamma_1}\mathbf{u}_{k+p-\ell}\big) + \bar{\delta}_{k+p}, \tag{3.50}$$

where the matrices Θ_{x}, Θ_{u} are defined similarly to those in (3.13), and $\bar{\boldsymbol{\xi}}_{k+p-\ell}^{\top} = \left[\begin{array}{c|c|c} \mathbf{u}_{k+p-\ell-1}^{\top} & \cdots & \mathbf{u}_{k-\ell-2}^{\top} \end{array}\right] \in \Re^{m(p+2)}$.

Remark 3.5. In the case where the time delay is known, the successive substitution of (3.49) is done in such a way as to have no time delay between the control input and the terms on the right-hand side of (3.49). However, when the time delay is uncertain, the successive substitution will be done based on the upper bound of the delay p, and as a consequence some elements of the control history $\bar{\xi}_k$ will be leading, if $\ell < p$, the terms that are to be compensated by the control input.

To design a proper control law, system (3.50) will be revised so that the influence of the uncertain delay term ℓ on the system description is eliminated. Let $\Theta_{\mathrm{u}} = \left[\begin{array}{c|c|c} \Omega_{\ell+1} & \cdots & \Omega_{\ell+p+2} \end{array}\right] \in \Re^{m\times m(p+2)}$ and revise (3.50) so that

$$\begin{aligned}\mathbf{x}_{k+p+1} = \mathbf{\Phi}_{\mathrm{m}}\mathbf{x}_{k+p} + \Gamma_{\mathrm{n}}\big(&\Theta_{\mathrm{x}}\bar{\chi}_k + \Omega_{\ell}\mathbf{u}_{k+p-\ell} + \Omega_{\ell+1}\mathbf{u}_{k+p-\ell-1}\\ &+ \cdots + \Omega_p\mathbf{u}_k + \cdots + \Omega_{\ell+p+2}\mathbf{u}_{k-\ell-2}\big) + \bar{\delta}_{k+p},\end{aligned} \tag{3.51}$$

where $\Omega_\ell = \Theta_{\gamma_1}$ for convenience. Let the matrices $\Omega_1, \ldots, \Omega_{\ell-1}$ and $\Omega_{\ell+p+3}$, $\ldots$, Ω_{2p+2} be null matrices such that (3.51) can be written as

$$\mathbf{x}_{k+p+1} = \mathbf{\Phi}_{\mathrm{m}}\mathbf{x}_{k+p} + \Gamma_{\mathrm{n}}\Big(\Theta_{\mathrm{x}}\bar{\chi}_k + \Omega_1\mathbf{u}_{k+p-1} + \cdots + \Omega_{\ell-1}\mathbf{u}_{k+p-\ell+1}$$

$$+\,\Omega_\ell \mathbf{u}_{k+p-\ell} + \Omega_{\ell+1}\mathbf{u}_{k+p-\ell-1} + \cdots + \Omega_p \mathbf{u}_k + \cdots + \Omega_{\ell+p+2}\mathbf{u}_{k-\ell-2}$$
$$+\,\Omega_{\ell+p+3}\mathbf{u}_{k-\ell-3} + \cdots + \Omega_{2p+2}\mathbf{u}_{k-p-2}\Big) + \bar{\delta}_{k+p}. \tag{3.52}$$

Now define the augmented matrices $\bar{\Omega}_1 = \left[\begin{array}{c|c|c} \Omega_1 & \cdots & \Omega_{p-1} \end{array}\right] \in \Re^{m\times m(p-1)}$ and $\bar{\Omega}_2 = \left[\begin{array}{c|c|c} \Omega_{p+1} & \cdots & \Omega_{2p+2} \end{array}\right] \in \Re^{m\times m(p+2)}$ such that (3.52) can be simplified as

$$\mathbf{x}_{k+p+1} = \Phi_{\mathrm{m}}\mathbf{x}_{k+p} + \Gamma_{\mathrm{n}}\Big(\Theta_{\mathrm{x}}\bar{\chi}_k + \bar{\Omega}_1\bar{\mathbf{u}}_{k+p} + \Omega_p\mathbf{u}_k + \bar{\Omega}_2\bar{\xi}_k\Big) + \bar{\delta}_{k+p}, \tag{3.53}$$

where $\bar{\mathbf{u}}_k^\top = \left[\begin{array}{c|c|c} \mathbf{u}_{k-1}^\top & \cdots & \mathbf{u}_{k-p+1}^\top \end{array}\right] \in \Re^{m(p-1)}$. Then the control law can be selected as

$$\mathbf{u}_k = -\hat{\Omega}_{p,k}^{-1}\left(\hat{\Theta}_{\mathrm{x},k}\bar{\chi}_k + \hat{\bar{\Omega}}_{2,k}\bar{\xi}_k - \Theta_{\mathrm{r}}\mathbf{r}_k\right), \tag{3.54}$$

where $\hat{\Omega}_{p,k}$, $\hat{\Theta}_{\mathrm{x},k}$, and $\hat{\bar{\Omega}}_{2,k}$ are the estimates of Ω_p, Θ_{x}, and $\bar{\Omega}_2$, respectively.

To select that adaptive law for the parameters $\hat{\Omega}_{p,k}$, $\hat{\Theta}_{\mathrm{x},k}$, and $\hat{\bar{\Omega}}_{2,k}$, substitute (3.54) into p delayed sampling instants (3.53) and subtract the result from the reference model (3.7); then we obtain the error dynamics

$$\mathbf{e}_{k+1} = \Phi_{\mathrm{m}}\mathbf{e}_k + \Gamma_{\mathrm{n}}\tilde{\Psi}_{k-p}^\top\bar{\zeta}_{k-p} + \Gamma_{\mathrm{n}}\bar{\Omega}_1^\top\bar{\mathbf{u}}_k + \bar{\delta}_k, \tag{3.55}$$

where

$$\tilde{\Psi}_k^\top = \left[\begin{array}{c;{2pt/2pt}c;{2pt/2pt}c} \tilde{\Theta}_{\mathrm{x},k} & \tilde{\bar{\Omega}}_{2,k} & \tilde{\Omega}_{p,k} \end{array}\right] \in \Re^{m\times(2n+m(p+3))} \text{ and}$$
$$\bar{\boldsymbol{\zeta}}_k^\top = \left[\begin{array}{c;{2pt/2pt}c;{2pt/2pt}c} \bar{\chi}_k^\top & \bar{\boldsymbol{\xi}}_k^\top & \mathbf{u}_k^\top \end{array}\right] \in \Re^{2n+m(p+3)}.$$

Note that the error dynamics (3.55) is similar to (3.21) with the only difference being the additional term $\Gamma_{\mathrm{n}}\bar{\Omega}_1^\top\bar{\mathbf{u}}_k$, which exists due to the uncertainty in the delay. If the delay ℓ is known and $\ell = p$, then $\bar{\Omega}_1$ is the null matrix.

Remark 3.6. Note that the number of terms in $\bar{\Omega}_1\bar{\mathbf{u}}_{k+p}$ increases as the delay upper bound p is set further from the actual delay ℓ. The number of terms also increases if smaller values of the sampling interval are selected. An increase in the number of terms in $\bar{\Omega}_1\bar{\mathbf{u}}_{k+p}$ may lead to poorer performance of the controller. Although it is desirable to use smaller sampling intervals to reduce the effect of the disturbances, care must be taken so as not to degrade the performance of the controller due to the increase of the terms in $\bar{\Omega}_1\bar{\mathbf{u}}_{k+p}$.

Proceeding further, using $\mathbf{z}_{k+1} = (D^\top \Gamma_n)^{-1} D^\top (\mathbf{e}_{k+1} - \Phi_m \mathbf{e}_k)$, we obtain

$$\mathbf{z}_{k+1} = \tilde{\Psi}_{k-p}^\top \bar{\boldsymbol{\xi}}_{k-p} + \bar{\Omega}_1^\top \bar{\mathbf{u}}_k + \bar{\upsilon}_k, \tag{3.56}$$

where $\mathbf{z}_{k+1} \in \Re^m$. The adaptive law will be formulated to be robust to the term $\bar{\Omega}_1^\top \bar{\mathbf{u}}_k$. Based on (3.56), following an approach similar to that in [21], we propose the adaptive law

$$\hat{\Psi}_{k+1} = \begin{cases} \hat{\Psi}_{k-p} + \alpha_k \frac{\beta_k}{\vartheta_k} Q \bar{\boldsymbol{\xi}}_{k-p} \mathbf{z}_{k+1}^\top & \forall k \in [p, \infty), \\ \hat{\Psi}_0 & \forall k \in [0, p), \end{cases} \tag{3.57}$$

where the scalar function $\vartheta_k = 1 + \alpha_k \bar{\boldsymbol{\xi}}_{k-p}^\top Q \bar{\boldsymbol{\xi}}_{k-p} + \alpha_k \gamma_c \lambda_c^2 \|\bar{\mathbf{u}}_k\|^2$, the matrix Q is a constant positive definite adaptive gain matrix of dimension $2n + m(p+3)$, γ_c and λ_c are positive tuning constants, β_k is a positive weighing coefficient, and $\alpha_k > 0$ is a coefficient used to ensure a nonsingular $\hat{\Omega}_{p,k}$.

Remark 3.7. Note that similarly to the previous cases, α_k must be selected such that α_k^{-1} is not an eigenvalue of $-\hat{\Omega}_{p,k-p}^{-1} S \frac{\beta_k}{\vartheta_k} Q \bar{\boldsymbol{\xi}}_{k-p} \mathbf{z}_{k+1}^\top$, where $S = [\mathbf{0} \ \cdots \ \mathbf{0} \ I] \in \Re^{m \times (n + m(p+3))}$.

Consider the constant uncertainty $\bar{\Omega}_1$ and assume that $\|\bar{\Omega}_1\| = \lambda_c \rho$, where ρ is an uncertain positive constant. It is easy to see that $\|\bar{\Omega}_1^\top \bar{\mathbf{u}}_k\| \le \lambda_c \rho \|\bar{\mathbf{u}}_k\|$. *Further, the weighing coefficient β_k can be defined as*

$$\beta_k = \begin{cases} 1 - \frac{\lambda_c \hat{\rho}_k \|\bar{\mathbf{u}}_k\| + \bar{\upsilon}_{\max}}{\|\mathbf{z}_{k+1}\|}, & \text{if } \|\mathbf{z}_{k+1}\| \ge \lambda_c \hat{\rho}_k \|\bar{\mathbf{u}}_k\| + \bar{\upsilon}_{\max}, \\ 0, & \text{if } \|\mathbf{z}_{k+1}\| < \lambda_c \hat{\rho}_k \|\bar{\mathbf{u}}_k\| + \bar{\upsilon}_{\max}, \end{cases} \tag{3.58}$$

where $\|\bar{\upsilon}_k\| \le \bar{\upsilon}_{\max}$, $\hat{\rho}_k$ is the estimate of ρ, and λ_c can be chosen as any constant satisfying $0 < \lambda_c < \lambda_{c,\max}$ with $\lambda_{c,\max}$ defined later. The estimation law for ρ can be given as

$$\hat{\rho}_{k+1} = \hat{\rho}_k + \alpha_k \frac{\beta_k \lambda_c \gamma_c \|\bar{\mathbf{u}}_k\| \cdot \|\mathbf{z}_{k+1}\|}{\vartheta_k}. \tag{3.59}$$

Note that if $\|\mathbf{z}_{k+1}\| \ge \lambda_c \hat{\rho}_k \|\bar{\mathbf{u}}_k\| + \bar{\upsilon}_{\max}$, then from (3.58) we obtain that

$$\beta_k^2 \mathbf{z}_{k+1}^\top \mathbf{z}_{k+1} = \beta_k \mathbf{z}_{k+1}^\top \mathbf{z}_{k+1} - \beta_k \big(\lambda_c \hat{\rho}_k \|\bar{\mathbf{u}}_k\| + \bar{\upsilon}_{\max}\big) \cdot \|\mathbf{z}_{k+1}\|. \tag{3.60}$$

The validity of this adaption law is verified by the following theorem.

Theorem 3.2. *Under the adaptive law* (3.57) *and the closed-loop dynamics* (3.56), *the tracking error* $\mathbf{e}_k$ *is bounded.*

Proof. Consider the positive functions

$$V_k = \sum_{j=1}^{m} \left(\sum_{i=k-p}^{k} \tilde{\boldsymbol{\psi}}_{j,i}^{\top} Q^{-1} \tilde{\boldsymbol{\psi}}_{j,i} \right) + \frac{1}{\gamma_c} \tilde{\rho}_k^2, \tag{3.61}$$

where $\tilde{\Psi}_k^{\top} = \left[\; \tilde{\boldsymbol{\psi}}_{1,k} \mid \tilde{\boldsymbol{\psi}}_{2,k} \mid \cdots \mid \tilde{\boldsymbol{\psi}}_{m,k} \; \right]^{\top}$. The difference of V_k with respect to a single time step is

$$\begin{aligned} \Delta V_k &= V_{k+1} - V_k \\ &= \sum_{j=1}^{m} \left(\tilde{\boldsymbol{\psi}}_{j,k+1}^{\top} Q^{-1} \tilde{\boldsymbol{\psi}}_{j,k+1} - \tilde{\boldsymbol{\psi}}_{j,k-p}^{\top} Q^{-1} \tilde{\boldsymbol{\psi}}_{j,k-p} \right) \\ &\quad + \frac{1}{\gamma_c} \left(\tilde{\rho}_{k+1}^2 - \tilde{\rho}_k^2 \right). \end{aligned} \tag{3.62}$$

Defining $\tilde{\rho}_k = \rho - \hat{\rho}_k$, it is possible to obtain

$$\tilde{\rho}_{k+1} = \tilde{\rho}_k - \alpha_k \frac{\beta_k \lambda_c \gamma_c \|\bar{\mathbf{u}}_k\| \cdot \|\mathbf{z}_{k+1}\|}{\vartheta_k}. \tag{3.63}$$

Substituting (3.57) and (3.63) into (3.62) and simplifying further, we obtain that

$$\begin{aligned} \Delta V_k &= \sum_{j=1}^{m} \left(\tilde{\boldsymbol{\psi}}_{j,k-p} - \alpha_k \frac{\beta_k}{\vartheta_k} Q \bar{\boldsymbol{\zeta}}_{k-p} z_{j,k+1} \right)^{\top} Q^{-1} \left(\tilde{\boldsymbol{\psi}}_{j,k-p} \right. \\ &\quad \left. - \alpha_k \frac{\beta_k}{\vartheta_k} Q \bar{\boldsymbol{\zeta}}_{k-p} z_{j,k+1} \right) - \sum_{j=1}^{m} \tilde{\boldsymbol{\psi}}_{j,k-p}^{\top} Q^{-1} \tilde{\boldsymbol{\psi}}_{j,k-p} \\ &\quad + \frac{1}{\gamma_c} \left(\tilde{\rho}_k - \frac{\alpha_k \beta_k \lambda_c \gamma_c \|\bar{\mathbf{u}}_k\| \|\mathbf{z}_{k+1}\|}{\vartheta_k} \right)^2 - \frac{1}{\gamma_c} \tilde{\rho}_k^2 \\ &\le \sum_{j=1}^{m} \left(\frac{\alpha_k^2 \beta_k^2}{\vartheta_k^2} \bar{\boldsymbol{\zeta}}_{k-p}^{\top} Q \bar{\boldsymbol{\zeta}}_{k-p} z_{j,k+1}^2 - 2 \frac{\alpha_k \beta_k}{\vartheta_k} \tilde{\boldsymbol{\psi}}_{j,k-p}^{\top} \bar{\boldsymbol{\zeta}}_{k-p} z_{j,k+1} \right) \\ &\quad + \frac{\alpha_k^2 \beta_k^2}{\vartheta_k^2} \gamma_c \lambda_c^2 \|\bar{\mathbf{u}}_k\|^2 \mathbf{z}_{k+1}^{\top} \mathbf{z}_{k+1} - 2 \frac{\alpha_k \beta_k}{\vartheta_k} \lambda_c \tilde{\rho}_k \|\bar{\mathbf{u}}_k\| \|\mathbf{z}_{k+1}\| \\ &\le -2 \frac{\alpha_k \beta_k}{\vartheta_k} \tilde{\Psi}_{k-p} \bar{\boldsymbol{\zeta}}_{k-p}^{\top} \mathbf{z}_{k+1} + 2 \frac{\alpha_k \beta_k}{\vartheta_k} \lambda_c \tilde{\rho}_k \|\bar{\mathbf{u}}_k\| \|\mathbf{z}_{k+1}\| \end{aligned}$$

$$
\begin{aligned}
&-2\frac{\alpha_k\beta_k}{\vartheta_k}\lambda_c\tilde{\rho}_k\|\bar{\mathbf{u}}_k\|\,\|\mathbf{z}_{k+1}\| \\
&+\frac{\alpha_k^2\beta_k^2}{\vartheta_k^2}\left(\bar{\boldsymbol{\xi}}_{k-p}^\top Q\bar{\boldsymbol{\xi}}_{k-p}+\gamma_c\lambda_c^2\|\bar{\mathbf{u}}_k\|^2\right)\mathbf{z}_{k+1}^\top\mathbf{z}_{k+1} \\
\le&-\frac{\alpha_k\beta_k}{\vartheta_k}\left[2\tilde{\Psi}_{k-p}\bar{\boldsymbol{\xi}}_{k-p}^\top-\frac{\alpha_k\beta_k}{\vartheta_k}\left(\bar{\boldsymbol{\xi}}_{k-p}^\top Q\bar{\boldsymbol{\xi}}_{k-p}+\gamma_c\lambda_c^2\|\bar{\mathbf{u}}_k\|^2\right)\mathbf{z}_{k+1}^\top\right]\mathbf{z}_{k+1}.
\end{aligned}
\tag{3.64}
$$

Using (3.60) and the fact that $\frac{\alpha_k\bar{\boldsymbol{\xi}}_{k-p}^\top Q\bar{\boldsymbol{\xi}}_{k-p}+\alpha_k\gamma_c\lambda_c^2\|\bar{\mathbf{u}}_k\|^2}{\vartheta_k}<1$, we obtain that

$$\Delta V_k < -\alpha_k\frac{\beta_k^2}{\vartheta_k}\mathbf{z}_{k+1}^\top\mathbf{z}_{k+1}. \tag{3.65}$$

Following the same steps that have led to (3.41) in Theorem 3.1, we conclude that

$$\lim_{k\to\infty}\alpha_k\frac{\beta_k^2}{\vartheta_k}\mathbf{z}_{k+1}^\top\mathbf{z}_{k+1}=0. \tag{3.66}$$

Estimate (3.65) shows that $\tilde{\Theta}_{x,k}$, $\tilde{\Omega}_{p,k}$, $\tilde{\Omega}_{2,k}$, and $\tilde{\rho}_k$ are bounded because V_k is nonincreasing. Using arguments similar to those in Theorem 3.1, it is possible to establish that $\|\bar{\boldsymbol{\xi}}_k\|\le c_1^1+c_2^1\|\mathbf{z}_{k+1}\|$ for some positive constants c_1^1 and c_2^1, thus establishing the condition required by the key technical lemma, which guarantees that $\lim_{k\to\infty}\beta_k\|\mathbf{z}_{k+1}\|=0$. Furthermore, there must exist a positive constant ε such that $\max_{i\in[0,k]}\{\beta_i\|\mathbf{z}_{i+1}\|\}\le\varepsilon$. Then according to the definition of β_k in (3.58),

$$\max_{i\in[0,k]}\|\mathbf{z}_{i+1}\|\le\varepsilon+\lambda_c\bar{\rho}\max_{i\in[0,k]}\|\bar{\mathbf{u}}_i\|\le\varepsilon+\lambda_c\bar{\rho}\max_{i\in[0,k]}\|\bar{\boldsymbol{\xi}}_i\|, \tag{3.67}$$

where $\max_{i\in[0,k]}\hat{\rho}_i\le\bar{\rho}$. Following the analysis in Theorem 3.1 and the bound on $\bar{\boldsymbol{\xi}}_k$, we find the maximum tracking error bound:

$$\max_{i\in[0,k]}\|\mathbf{z}_{i+1}\|\le\varepsilon+\lambda_c\bar{\rho}\max_{i\in[0,k]}\{c_1^1+c_2^1\|\mathbf{z}_i\|\}, \tag{3.68}$$

which results in

$$\|\mathbf{z}_{k+k_0}\|\le\left(\lambda_c\bar{\rho}c_2^1\right)^k\|\mathbf{z}_{k_0}\|+\left(\varepsilon+\lambda_c\bar{\rho}d_1\right)\sum_{i=0}^{k-1}\left(\lambda_c\bar{\rho}c_2^1\right)^i, \tag{3.69}$$

implying that to guarantee the boundedness of $\|\mathbf{z}_k\|$, we must take $|\lambda_c|<\lambda_{c,\max}<\frac{1}{\bar{\rho}c_2^1}$.

From (3.69) *the steady-state bound of* $\|\mathbf{z}_k\|$ *is given as* $\left(\varepsilon+\lambda_c\bar{\rho}d_1\right)\times\sum_{i=0}^{k-1}\left(\lambda_c\bar{\rho}c_2^1\right)^i$, *and since,* $\mathbf{e}_{k+1}=\Phi_m\mathbf{x}_k+\mathbf{z}_{k+1}$, *the bound of* $\|\mathbf{e}_k\|$ *can be similarly evaluated.* □

3.5. Simulation example

To illustrate the advantages of the new approach, we use the flight control example in [9], where extensive comparisons between the discrete-time APC, continuous-time APC, and MRAC are presented, and it was shown that the discrete-time APC has superior performance. In this example, we compare the continuous-time APC and discrete-time APC in [9] to the new approach presented in this chapter.

Now consider the longitudinal dynamics of a four-engine jet transport aircraft [20]. The aircraft flies straight and at a level flight of 40,000 ft with a velocity of 600 ft/sec. Under these conditions, the nominal short period dynamics is given by

$$\begin{bmatrix}\dot{x}_1(t)\\ \dot{x}_2(t)\end{bmatrix}=\begin{bmatrix}-0.323 & 1\\ -1.169 & -0.480\end{bmatrix}\begin{bmatrix}x_1(t)\\ x_2(t)\end{bmatrix}+\begin{bmatrix}-0.018\\ -1.379\end{bmatrix}\left(u(t-\tau)+f(t)\right), \tag{3.70}$$

where x_1 is the angle of attack in radians, x_2 is the pitch rate in radians per second, u is the elevator deflection also in radians, and $f(t)=0.025\left(1+\sin\left(0.5t-\frac{\pi}{2}\right)\right)$ is a persisting exogenous disturbance. The time-delay value used in the simulation is given as $\tau=0.21$ s. The eigenvalues are $-0.4017\pm1.0785i$, giving a nominal short period natural frequency of $\omega_n=1.1423$ rad/s and a nominal damping ratio of $\zeta_n=0.3517$.

To obtain a challenging scenario, we introduced control effectiveness uncertainty resulting in a 30% decrease in elevator effectiveness. In addition, by adding further uncertainty to the state matrix, proximity of the open loop poles to the imaginary axis was halved, and the damping ratio was reduced by 48%. The resulting uncertain system becomes

$$\begin{bmatrix}\dot{x}_1(t)\\ \dot{x}_2(t)\end{bmatrix}=\begin{bmatrix}-0.323 & 1.005\\ -1.176 & -0.077\end{bmatrix}\begin{bmatrix}x_1(t)\\ x_2(t)\end{bmatrix}+\begin{bmatrix}-0.009\\ -0.689\end{bmatrix}\left(u(t-\tau)+f(t)\right). \tag{3.71}$$

To implement the controller (3.16), the sampled-data reference model needs to be computed. To do this, the nominal system (3.70) will be stabilized using the LQR method ignoring the time-delay, and the result is sampled at $T = 0.02$ s, resulting in the sampled-data reference model

$$\begin{bmatrix} x^{\mathrm{m}}_{1,k+1} \\ x^{\mathrm{m}}_{2,k+1} \end{bmatrix} = \begin{bmatrix} 0.9924 & 0.0179 \\ -0.0622 & 0.9078 \end{bmatrix} \begin{bmatrix} x^{\mathrm{m}}_{1,k} \\ x^{\mathrm{m}}_{2,k} \end{bmatrix} + \begin{bmatrix} 0.0021 \\ 0.0905 \end{bmatrix} r_{k-p}. \tag{3.72}$$

3.5.1 New approach vs continuous-time APC

Referring to [17], the continuous-time gains are designed as $\Psi_{\mathrm{x}} = \mathrm{diag}(3.7, 8.3) \times 10^3$ and $\Psi_{\mathrm{r}} = 8.4 \times 10^3$ (refer to [17] for a more detailed description of the gains Ψ_{x} and Ψ_{r}). The gains used in the integral approximation are tuned to get the best performance. As for the new approach, adaptive parameters are initialized with the nominal plant values, and the adaptive gain is selected as $Q = 30I$. The Matlab® code for the proposed approach simulation is shown in Listing 3.1. The performance of the two controllers in the absence of disturbance is shown in Fig. 3.1. The performance of the continuous-time APC is oscillatory at certain intervals, whereas the discrete-time APC has a short oscillatory period and then dampens quickly by the time the second pulse reference comes into effect.

```
%nominal plant parameters
A_n = [-0.323 1;-1.169 -0.480];
B_n = [-0.018;-1.379];
C = [1 0];
n = 2; %system order
tau = 0.21; %input-delay
%reference model parameter
A_m = [-0.3505 0.9441;-3.2754 -4.8036];
B_m = [0.0614;4.7465];
%actual plant parameters
A = [-0.323 1.005;-1.176 -0.077];
B = [-0.009;-0.689];
%Sampling interval
T = 0.02;
%sampled-data nominal plant parameters
Phi_n_Aug = expm([A_n B_n;0 0 0]*T);
Phi_n = Phi_n_Aug(1:n,1:n);
Ga_n = Phi_n_Aug(1:n,n+1);
p = round(tau/T); %delay in time-steps
%sampled-data actual plant parameters
Phi_Aug = expm([A B;0 0 0]*T);
Phi = Phi_Aug(1:n,1:n);
Ga = Phi_Aug(1:n,n+1);
%sampled-data reference model
Phi_m_Aug = expm([A_m B_m;0 0 0]*T);
Phi_m = Phi_m_Aug(1:n,1:n);
Ga_m = Phi_m_Aug(1:n,n+1);
eig_m = eig(Phi_m);
%simulation settings
t_final = 100;
k_final = t_final/T; %simulation end in time-steps
%signals initial values
```

```
x(:,1) = [0 0]';
x_m(:,1) = [0 0]';
x_0 = [0 0]';
v_0 = 0;
x_est(:,1) = [0 0]';
zeta(:,1) = [x(:,1)' 0]';
%parameters initial values
Psi_est(:,:,1) = [Phi_n Ga_n]';
Phi_est(:,:,1) = Psi_est(1:n,1:n,1)';
Ga_est(:,:,1) = Psi_est(n+1,1:n,1)';
%controller and adaptive gain values
Theta(:,:,1) = acker(Phi_est(:,:,1),Ga_est(:,:,1),eig_m);
K_r(1) = Ga_m(2)/Ga_est(2,1);
Q = 30*eye(n+1); %adaptive gain
%other simulation parameters
delta = 5e-6; %error bound
m = 0; %counter value
sum_c = zeros(n,1);

for k = 1:k_final
    w(k) = 0.025*(1 + sin(0.5*(k-1)*T - pi/2)); %sampled-data disturbance
    sum_c_0 = zeros(n,1);
    for j = 1:m  %control input history term for the future state calculation
             sum_c = sum_c_0 + (Phi_est(:,:,k)^(j-1))*Ga_est(:,k)*u(k-j);
             sum_c_0 = sum_c;
    end
    x_f_est(:,k) = (Phi_est(:,:,k)^p)*x(:,k) + sum_c; %future state calculation
    for i = 1:k_final + 1  %generating the pulse train reference signal r(i)
             if i <= 1/T
                  r(i) = 0;
            end
            if i <= 26/T
                 if i > 1/T
                      r(i) = 5*pi/180;
                 end
            end
            if i <= 51/T
                 if i > 26/T
                      r(i) = 6.3*pi/180;
                 end
            end
            if i <= 76/T
                 if i > 51/T
                      r(i) = 4.3*pi/180;
                end
            end
            if i <= 101/T
                 if i > 76/T
                      r(i) = 5.3*pi/180;
                 end
            end
    end
    w_est(k) = inv(C*Ga_est(:,k))*(C* x(:,k) - C*Phi_est(:,:,k)*x_0) - v_0; %delay disturbance estimation
    u(k) = -Theta(:,:,k)*x_f_est(:,k) - w_est(k) + K_r(k)*r(k); %control law

    if k <= p   %delay in the control input
         v(k) = 0;
         r_d(k) = 0;
    else
         v(k) = u(k - p);
         r_d(k) = r(k-p);
         r_d(k+1) = r(k-p+1);
    end
```

```
    v_0 = v(k);
    x_0 =  x(:,k);
    x(:,k+1) = Phi*x(:,k) + Ga*v(k) + Ga*w(k);   %plant dynamics
    x_m(:,k+1) = Phi_m*x_m(:,k) + Ga_m*r_d(k);   %reference plant dynamics
    zeta(:,k) = [x(:,k)' v(k)]';  %signal vector
    x_est(:,k+1) = Psi_est(:,:,k)'*zeta(:,k);  %estimator plant dynamics
    x_error(:,k+1) = x(:,k+1) - x_est(:,k+1);
    if norm(x_error(:,k+1)) > delta    %dead-zone
            beta(k) = 1 - delta/norm(x_error(:,k+1));
    else
            beta(k) = 0;
    end
    Psi_est(:,:,k+1) = Psi_est(:,:,k) + beta(k)*Q*zeta(:,k)*x_error(:,k+1)'/(1 + zeta(:,k)'*Q*zeta(:,k)); %adaptive law
    Phi_est(:,:,k+1) = Psi_est(1:n,1:n,k+1)';       %state matrix derived from the adaptive parameter
    Ga_est(:,:,k+1) = Psi_est(n+1,1:n,k+1)';      %gain matrix derived from the adaptive parameter
    Theta(:,:,k+1) = acker(Phi_est(:,:,k+1),Ga_est(:,:,k+1),eig_m);  %state feedback gain computed from the adaptive state and gain matrices
    K_r(k+1) = Ga_m(2)/Ga_est(2,k+1);
    t(k) = (k-1)*T;     %time calculation
    t(k+1) = (k)*T;
    if m < p                %counter for the future state calculation
           m = m +1;
    end
end
plot(t,x_m(1,:)*180/pi,t,x(1,:)*180/pi,'LineWidth',1)
legend('Reference','New Approach');
xlabel('t [sec]')
ylabel('x_1')
```

Listing 3.1: Proposed approach.

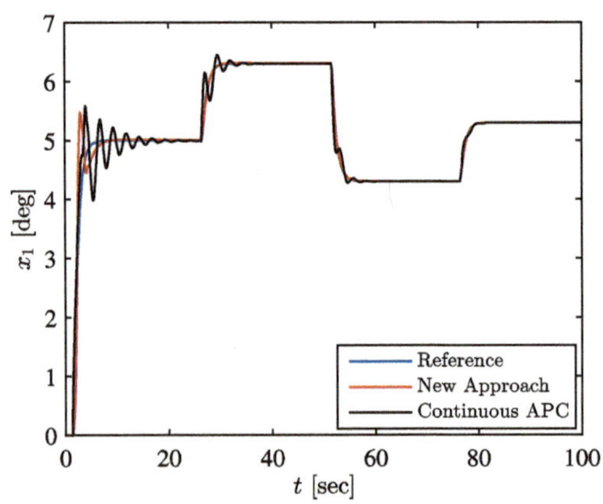

Figure 3.1 Tracking performance of the new approach vs continuous-time APC for $\tau = 0.21$ and no disturbance.

3.5.2 New approach vs discrete-time APC

The adaptive gains for the discrete-time APC [9] are selected as $Q = \text{diag}(Q_x, Q_u, Q_\gamma)$, where $Q_x = \text{diag}(35, 28, 30, 22)$, $Q_u = 8I_{p+2\times p+2}$, and $Q_\gamma = 0.12$. The performance of the new approach is shown in Fig. 3.2

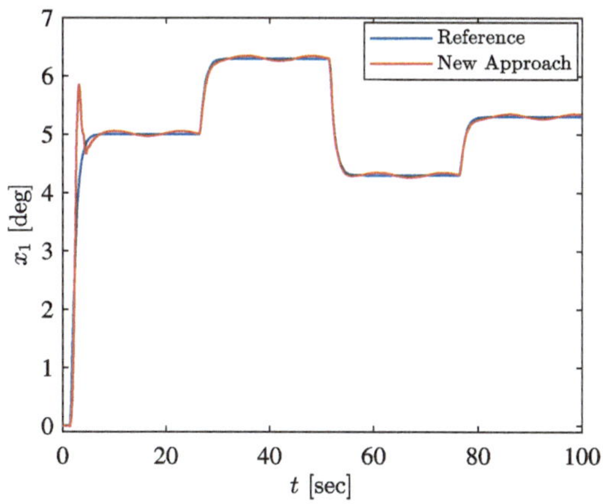

Figure 3.2 Tracking performance of the new approach with $\tau = 0.21$ s.

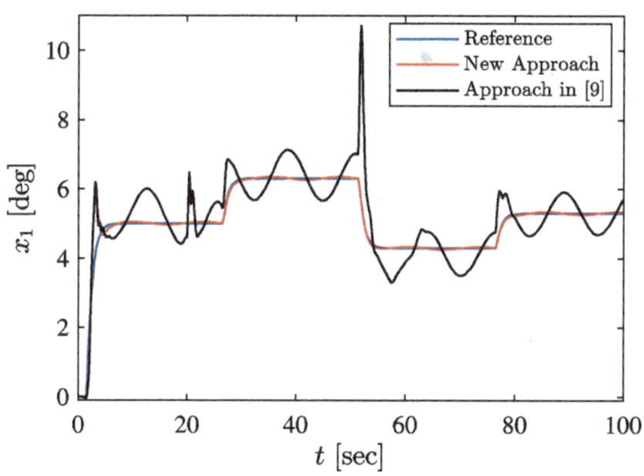

Figure 3.3 Tracking performance of the new approach vs the approach in [9].

and compared to the approach of [9] in Fig. 3.3. As it can be seen, the new approach is capable of attenuating the influence of the disturbance as opposed to the approach of [9], which is incapable of attenuating the disturbance at all. This confirms that the new approach is capable of dealing with uncertain time-delay systems subjected to exogenous disturbances.

3.5.3 Uncertain time delay

Consider system (3.70) where the time delay is assumed to be $\tau_p = 0.5$ s, whereas the actual time-delay is $\tau = 0.41$ s. For the approach in [9], the gains are selected as $\lambda_c = 0.015$, $\gamma_c = 100$, and $Q = \text{diag}(300, 60I_{p\times p}, 150)$. As for the adaptive law (3.57), the gains are selected as $\lambda_c = 0.01$, $\gamma_c = 112$,

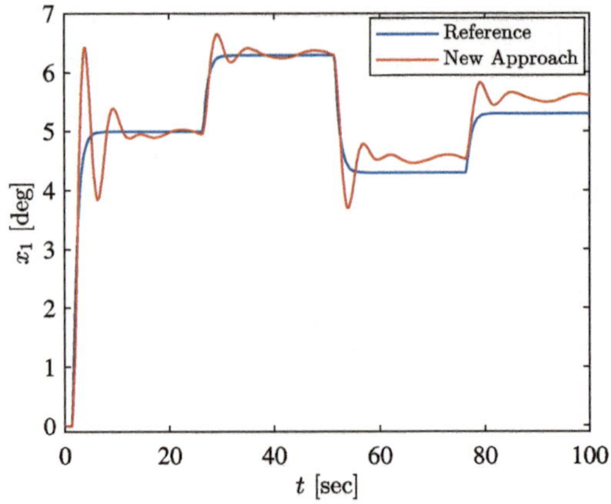

Figure 3.4 Tracking performance of the new approach with $\tau = 0.41$ s and $\tau_p = 0.5$ s.

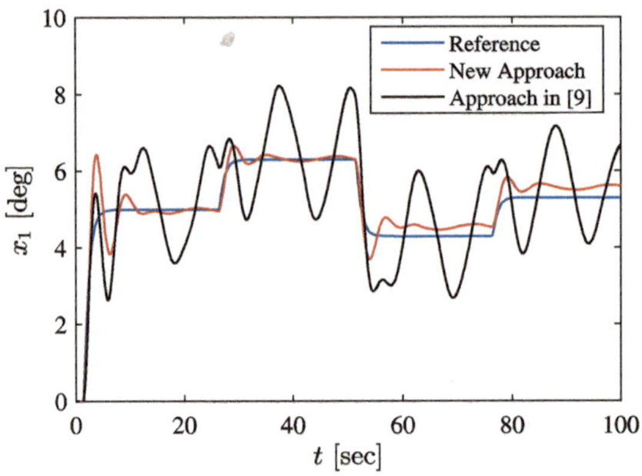

Figure 3.5 Tracking performance of the new approach vs the approach in [9].

and $Q = \text{diag}(320, 70I_{p+2\times p+2}, 140)$. The system is simulated under these conditions, and the results can be seen in Figs. 3.4 and 3.5. The results show that the system converges within a reasonable error bound around the desired trajectory, as opposed to the approach of [9], which is incapable of attenuating the persisting disturbance.

3.6. Conclusion

In this work, we developed a discrete version of the adaptive posicast controller for sampled-data systems in the presence of disturbances and input time delays that are possibly noninteger multiples of the sampling interval. We introduced a disturbance observer to the controller design, the utilization of which minimizes the effect of the disturbance on the closed-loop system performance. In addition, we presented an extension of the method for the case of unknown input delays. We also presented simulation studies that verify the performance of the method.

References

[1] K. Abidi, Y. Yildiz, A. Annaswamy, Control of uncertain sampled-data systems: an adaptive posicast control approach, IEEE Transactions on Automatic Control 62 (5) (2017) 2597–2602.
[2] A. Manitius, A. Olbrot, Finite spectrum assignment problem for systems with delays, IEEE Transactions on Automatic Control 24 (4) (1979) 541–552.
[3] R. Ortega, R. Lozano, Globally stable adaptive controller for systems with delay, International Journal of Control 47 (1) (1988) 17–23.
[4] S. Niculescu, A. Annaswamy, An adaptive Smith-controller for time-delay systems with relative degree $n^* \leq 2$, Systems & Control Letters 49 (5) (2003) 347–358.
[5] Y. Yildiz, et al., Adaptive posicast controller for time-delay systems with relative degree $n* \leq 2$, Automatica 46 (2) (2010) 279–289.
[6] K. Abidi, J.-X. Xu, Advanced Discrete-Time Control: Designs and Applications, Springer, 2015.
[7] V. Van Assche, M. Dambrine, J. Lafay, J. Richard, Some problems arising in the implementation of distributed-delay control laws, in: Proceedings of the 38th IEEE Conference on Decision and Control, vol. 5, Phoenix, AZ, USA, 1999, pp. 4668–4672.
[8] K. Narendra, A. Annaswamy, Stable Adaptive Systems, Dover Publications, 2005.
[9] K. Abidi, Y. Yildiz, On the discrete adaptive posicast controller, in: 12th IFAC Workshop on Time-Delay Systems, Ann Arbor, MI, 2015.
[10] P.A. Ioannou, J. Sun, Robust Adaptive Control, Dover Publications, 2012.
[11] M. Krstic, Delay Compensation for Nonlinear, Adaptive, and PDE Systems, Birkhäuser, 2009.
[12] Y. Wei, Z. Lin, Stability criteria of linear systems with multiple input delays under truncated predictor feedback, Systems & Control Letters 111 (2018) 9–17.
[13] Q. Zhong, On distributed delay in linear control laws-part I: discrete-delay implementations, IEEE Transactions on Automatic Control 49 (11) (2004) 2074–2080.
[14] K. Abidi, X.-J. Xu, On the discrete-time integral sliding-mode control, IEEE Transactions on Automatic Control 52 (4) (2007) 709–715.
[15] T. Nguyen, W.-C. Su, Z. Gajic, Output feedback sliding mode control for sampled-data systems, IEEE Transactions on Automatic Control 55 (7) (2010) 1684–1689.
[16] G. Goodwin, P. Ramadge, P. Caines, Discrete-time multivariable adaptive control, IEEE Transactions on Automatic Control 25 (3) (1980) 449–456.
[17] Y. Yildiz, A. Annaswamy, D. Yanakiev, I. Kolmanovsky, Adaptive idle speed control for internal combustion engines, in: Proceedings of the American Control Conference, New York City, NY, 2007, pp. 3700–3705.
[18] Y. Yildiz, Adaptive control for time-delay systems applied to flight control, in: Proceedings of the AIAA Guidance, Navigation, and Control Conference, AIAA Paper 2010-7576, Keystone, CO, 2010.
[19] K.P.V. Kokotovic, Foundations of Adaptive Control, Springer-Verlag, New York, 1991.
[20] J.H. Blakelock, Automatic Control of Aircraft and Missiles, 2nd edition, John Wiley & Sons, 1991.

[21] K. Abidi, A robust discrete-time adaptive control approach for systems with almost periodic time-varying parameters, International Journal of Robust and Nonlinear Control 24 (1) (2014) 166–178.
[22] K. Abidi, J.-X. Xu, J.-H. She, A discrete-time terminal sliding-mode control approach applied to a motion control problem, IEEE Transactions on Industrial Electronics 56 (9) (2009) 3619–3627.

CHAPTER FOUR

On a quasioptimal spectrum assignment for delayed systems

Libor Pekař
Faculty of Applied Informatics, TBU in Zlín, Zlín, Czechia

Contents

4.1. Introduction

Time-delay systems (TDSs) represent a family of processes, mechanisms, organisms, structures, and so on that are difficult to be modeled or analyzed and, in many cases, their controller design can be arduous [9,39,66,73,86]. Not only input/output but also internal (or state) delays are considered in these systems. The main reason for those unpleasant properties lies in their infinite-dimensional nature due to the transcendental

Control Strategy for Time-Delay Systems
https://doi.org/10.1016/B978-0-12-820599-0.00009-4

behavior of the characteristic equation [13,55]. It is caused by the existence of exponential terms in the characteristic equation. Moreover, another issue with TDSs is that there is a lack of well-applicable and technically acceptable control design methods and techniques for TDS.

Regarding the control design of linear time-invariant (LTI) TDSs, some approaches result in controller laws giving rise to a finite feedback spectrum. Usually, in these cases an unconventional and complicated controller structure is obtained. One possibility is to suggest branched control system structures driven by the aim to compensate for input/output delays [48] such that the characteristic polynomial (instead of a quasipolynomial) is reached. Within this family, the Smith predictor (and its numerous modifications) represents the most famous dead-time compensator [1,22,76]. However, the Smith predictor is primarily not intended to compensate for internal delays – in such cases, the feedback characteristic equation might, unfortunately, be transcendental.

Thus another possibility to get a finite spectrum is using the finite spectrum assignment controller paradigm [40,84]. In an ideal case, this concept allows assigning a finite number of desired closed-loop poles arbitrarily. It is based on the prediction of states or outputs over the delay period, which is characterized by a control law that includes a distributed delay element. The delay distribution is expressed by a convolution integral; see the referenced sources for further details.

On the other hand, the use of standard proportional-integral-derivative (PID) controllers yields an infinite-dimensional feedback control system. Indeed, there is no term in the controller law that may cancel exponential (delay) terms of the controlled plant. Moreover, many PID controller design procedures and tuning rules for TDSs lead to poor control performance [68]. Because of that, numerous advanced design and tuning methods have been developed. For instance, we can use optimization techniques to get the best control performance in a particular sense [77], avoid unacceptable values in the controller parameter space by the determination of stabilizing regions or guaranteed spectral abscissa [17,72] or even place the dominant (i.e., the rightmost) subset of system poles to desired loci [14,37,43,92]. However, there exist specific limitations [29] that ought to be taken into account here.

An algebraic approach for TDSs using the ring of stable and proper quasipolynomial meromorphic functions (R_{QM}) was eventually proposed by the author in [60]. The ring definition covers either retarded or neutral delayed systems, as well as those with lumped or distributed delays. It reflects the fact that the Laplace transform and the z-transform operators are mutually dependent from the functional point of view. Moreover, it is not limited to commensurate delays, does not require any rational approximation, and is simple enough to be used for stabilization and control issues in practice. Note that it was proved, among others, that the ring constitutes a commutative ring yet neither an integral domain nor a prime ideal. It can be deduced by examples that the controller design procedure in the ring via the Bézout identity followed by the Youla–Kučera parameterization [16,33,80] may yield a finite-dimensional control loop for stable controlled plants; however, only an infinite-dimensional control system is reached when the plant is unstable [51,61].

Therefore we have to cope with an infinite number of poles in an unstable case. Moreover, the same scenario occurs when using a forced PID controller, even if the algebraic procedure is used. But there is only a finite number of adjustable controller parameters – therefore a researcher or an engineer has to handle only a part of the spectrum. Usually, the rightmost (i.e., the dominant) subset of system poles is taken into account. Note that some authors consider the dominant part in a different way, as the effect of poles loci to the overall system dynamics is not straightforward for TDSs [69].

Hence the author of this chapter extended the famous quasicontinuous shifting algorithm (QCSA) developed in [42,44] in two ways. Namely, (first) he attempted to optimize the global loci of the whole spectrum by minimizing the spectral abscissa and (second) to place a finite subset of the dominant part of the spectrum to the desired position while using the spectral abscissa minimization for the remaining spectrum.

Regarding the former problem, the author's pioneering work [61] was aimed at the transition of the pole placement problem into the minimization of the spectral abscissa, which was initially solved via the QCSA followed by a comparative utilization of three iterative optimization algorithms there. Namely, a direct-search Nelder–Mead algorithm (NMA) [47], a gradient-based extended gradient sampling algorithm (EGSA) [6,78,81], and a bioinspired self-organizing migration algorithm (SOMA) [74,89] were used. An unstable model of a roller skater on the controller swaying bow (RSCSB) [93] was applied to numerically verify the proposed theoretical results. Later, revised and extended results were published in [51,58]. The latter problem was initially solved by the author in [50]. On the contrary to most of the similar other methods proposed hitherto [32,45], not only closed-loop poles but also its zeros are handled, which was a crucial contribution. The zero-pole-assessment task can be viewed as a model-matching problem where the desired model can be determined by the analysis of the dynamical properties of a finite-dimensional system. Further improvements and extensions of the primordial idea and preliminary results toward additional substrategies and assumption of TDSs with delays of the neutral type were published in [51,52,59]. It is worth noting that spectral shaping of even neutral-delay systems has a long tradition in control systems research [7]. However, it was found that results presented in the above-introduced publications suffer from some gaps and imperfections. In particular, there are two significant problems to be assumed. First, we have to be careful when selecting the number and multiplicity of desired zeros and poles due to the feasibility of the matching model and root-shifting solvability. Second, the preliminary works ignored the so-called parity interlacing property (PIP) [19], which decides whether a complex conjugate pair of poles can move to the left over a real zero. Therefore the eventual improved, revised, extended, and more advanced version of the zero-pole placement shifting-based algorithm has been published recently in [56].

In this chapter, we adopt an algebraic approach via the R_{QM} ring and the use of standard PI(D) controllers are adopted to meet both the above-introduced goals. That is, first, we perform a spectral abscissa optimization by applying the QCSA followed by the use of numerical optimization algorithms. Second, we acquaint the reader with the simultaneous placing of the dominant (herein, the rightmost) poles and zeros of the eventual delayed infinite-dimensional feedback control system. This procedure is done by matching the actual roots loci with the desired ones. The selection of desired zeros and poles loci is based on a simple finite-dimensional model and time responses of which are known. Meanwhile, the remaining parts of the spectra are shifted to the left to minimize both the spectral abscissae. Although retarded systems are primarily considered, the reader is provided with some notes on the use of the proposed techniques to, more tricky, neutral-type LTI TDSs. Theoretical ideas and algorithms are supported by simulation (numerical) examples that indicate the advantages and drawbacks of the proposed methods. Potential future directions are elucidated as well.

To highlight and recall the main contribution of this chapter, the point-to-point list of goals, used tools and proposed techniques follow:

1. Successive placement of the finite dominant subset of characteristic roots and system zeros for retarded- and neutral-type LTI TDSs to prescribed desired loci. In contrast to other known procedures, zeros are simultaneously considered as well.

2. The use of the R_{QM} ring and elegant algebraic parlance to design stabilizing controllers' structures with the ability of their further parameterization. The ring covers commensurate, incommensurate, retarded-type, neutral-type, lumped, and distributed delays.
3. The use of a quasicontinuous shifting technique followed by a numerical optimization of the selected cost function reflecting the closeness of the obtained dominant spectra to the desired ones and spectral abscissae of remaining spectra. On the contrary to other techniques, initially placed loci may change during the optimization. The PIP property is also taken into account.
4. Advanced mathematics is not required; therefore the whole proposed procedure is effortlessly applicable in practice.

The standard notation is used throughout the chapter. $\mathbb{C}$, $\mathbb{R}$, and $\mathbb{N}$ denote, respectively, the sets of complex, real, and natural numbers (with zero). $\mathbf{Re}\,(s)$ and $\mathbf{Im}\,(s)$ mean the real and imaginary parts of $s \in \mathbb{C}$, respectively. $\mathbb{R}_+^n$ is the n-dimensional Euclidean space of positive real-valued vectors. The superscript "T" denotes the vector or matrix transpose. The element in the ith row and jth column of a matrix or a vector $\mathbf{M}$ is denoted by M_{ij}.

The rest of the chapter is organized as follows. Section 4.2 provides the reader with essential preliminaries about LTI TDSs. Namely, we give a model of LTI TDSs in the input–output space governed by the transfer function (matrix). Then spectral properties of the system and essential spectrum and their abscissae are introduced. Definitions of exponential and strong stability and some of their issues follow. In Section 4.3, we present the background and the definition of the R_{QM} ring and its application to controller structure design for a simple feedback control system. The designed control law ensures control feedback stability (in the sense of the R_{QM} ring definition), asymptotic reference tracking, and load disturbance rejection. Our spectral abscissa minimization procedure is introduced in Section 4.4. After a concise problem statement, the well-established root shifting QCSA algorithm is reiterated. Specific issues of a neutral-type LTI TDS, such as the nonsmooth behavior and an estimation of the abscissa of its essential spectrum, are then discussed. The eventually proposed quasioptimization algorithm follows. A simulation example incorporating a controlled retarded model of a skater on the swaying bow is given as well. In the example the simple feedback control design via the R_{QM} ring is used. The main contribution of this chapter can be found in Section 4.5. First, the problem of the assignment of the dominant finite subsets of both spectra of system poles and zeros is stated. As desired root loci are based on the selection of a suitable matching model, the reader is acquainted with the analysis of dynamics of the reference 2nd-order finite-dimensional model. The introduction of PIP and its various scenarios (supported by a simple explanatory example) that may become during the root shifting toward desired loci follow. The core of this section consists in the proposition of the original zero-pole placement algorithm based on the simultaneous shifting to the desired root loci and the minimization of a given cost function. Two numerical examples conclude the section. Namely, first, feedback poles and zeros are adjusted when controlling the roller skater movement. Several settings of cost function parameter values and numerical optimization algorithms are compared within this example. The necessity of the primordial use of the QCSA is demonstrated. Second, a fictive plant controlled by using a branched feedback loop yielding a neutral-type control system is tuned by the proposed algorithm. Possible future research is concisely discussed in Section 4.6. Especially, an advanced cost function selection or the transition from the system spectrum to its pseudospectrum to get better robustness are discussed. Finally, the whole chapter is concluded in Section 4.7, providing the point-by-point findings.

4.2. Preliminaries

To specify the family of LTI TDSs, we introduce a concise description of their models. The eventual input–output model is given to the reader via the Laplace transform from the initial state-space model. Further, we provide selected basic properties of the eigenvalue spectrum of a representative of the specified family of systems. Poles loci and their dependence on system parameters are stressed. Elementary remarks on exponential stability are added as well.

4.2.1 LTI TDS input–output model

We denote by $\mathbf{x}(t) \in \mathbb{R}^n$, $\mathbf{u}(t) \in \mathbb{R}^m$, $\mathbf{y}(t) \in \mathbb{R}^l$ the vectors of state and input and output variables, respectively. The function $\dot{\mathbf{x}}(t)$ expresses the time derivative of $\mathbf{x}(t)$. Real values $0 = \vartheta_{\cdot,0} < \vartheta_{\cdot,1} < \cdots < \vartheta_{\cdot,n.} = L$ denote *lumped* delays, and $0 \le \tau \le L$ means *distributed* ones. Then the following functional (delay) differential equation and the algebraic equation formulate a general continuous-time LTI-TDS with constant delays [27,55]

$$\dot{\mathbf{x}}(t) - \sum_{i=1}^{n_H} \mathbf{H}_i \dot{\mathbf{x}}(t - \vartheta_{H,i}) - \int_0^L \mathbf{H}_d(\tau)\dot{\mathbf{x}}(t-\tau)\mathrm{d}\tau$$

$$= \sum_{i=0}^{n_A} \mathbf{A}_i \mathbf{x}(t - \vartheta_{A,i}) + \sum_{i=0}^{n_B} \mathbf{B}_i \mathbf{u}(t - \vartheta_{B,i})$$
$$+ \int_0^L [\mathbf{A}_d(\tau)\mathbf{x}(t-\tau) + \mathbf{B}_d(\tau)\mathbf{u}(t-\tau)]\mathrm{d}\tau$$
$$\mathbf{y}(t) = \sum_{i=0}^{n_C} \mathbf{C}_i \mathbf{x}\left(t - \vartheta_{C,i}\right) + \int_0^L \mathbf{C}_d(\tau)\,\mathbf{x}(t-\tau)\,\mathrm{d}\tau, \tag{4.1}$$

where $\mathbf{A}_i$, $\mathbf{A}_d(\tau)$, $\mathbf{B}_i$, $\mathbf{B}_d(\tau)$, $\mathbf{C}$, $\mathbf{C}_d(\tau)$, $\mathbf{H}$, and $\mathbf{H}_d(\tau)$ are real-valued matrices of compatible dimensions.

The system is said to be of *neutral* type (NTDS) if $\mathbf{H}_d(\tau) \neq \mathbf{0}$ or $\mathbf{H}_i \neq \mathbf{0}$ for some i. Otherwise, the system is of *retarded* type (RTDS).

The Laplace transform of (4.1) has the following form:

$$\left[\mathbf{I} - \underbrace{\left(\sum_{i=1}^{n_H} \mathbf{H}_i \exp(-s\vartheta_{H,i}) + \int_0^L \mathbf{H}_d(\tau)\exp(-s\tau)\mathrm{d}\tau\right)}_{H(s)}\right]\mathbf{x}(s)$$
$$= \underbrace{\left(\sum_{i=0}^{n_A} \mathbf{A}_i \exp(-s\vartheta_{A,i}) + \int_0^L \mathbf{A}_d(\tau)\exp(-s\tau)\mathrm{d}\tau\right)}_{A(s)}\mathbf{x}(s)$$
$$+ \underbrace{\left(\sum_{i=0}^{n_B} \mathbf{B}_i \exp\left(-s\vartheta_{B,i}\right) + \int_0^L \mathbf{B}_d(\tau)\exp(-s\tau)\,\mathrm{d}\tau\right)}_{B(s)}\mathbf{u}(s),$$
$$\mathbf{y}(s) = \underbrace{\left(\sum_{i=0}^{n_C} \mathbf{C}_i \exp\left(-s\vartheta_{C,i}\right) + \int_0^L \mathbf{C}_d(\tau)\exp(-s\tau)\,\mathrm{d}\tau\right)}_{C(s)}\mathbf{x}(s), \tag{4.2}$$

where s is the complex-valued Laplace transform variable, and $\mathbf{I}$ stands for the unit matrix. By using the notation from (4.2) the transfer function matrix of (4.1) reads

$$\mathbf{G}(s) = \mathbf{C}(s)\left[s\left(\mathbf{I} - \mathbf{H}(s)\right) - \mathbf{A}(s)\right]^{-1}\mathbf{B}(s). \tag{4.3}$$

The *characteristic function* (a *quasipolynomial* or an *entire function*) is

$$\Delta(s) = \det\left(s\left(\mathbf{I} - \mathbf{H}(s)\right) - \mathbf{A}(s)\right). \tag{4.4}$$

Note that a quasipolynomial $q(s)$ has the general form

$$q(s) = s^n + \sum_{i=0}^{n} \sum_{j=0}^{h_i} q_{ij} s^i \exp(-\tau_{ij} s), \tag{4.5}$$

where $\tau_{i0} = 0$ and $\tau_{ij} \in \mathbb{R}_+$ (i.e., $j \neq 0$) represent delays, and $q_{n0} = 0$, $q_{ij} \in \mathbb{R}$. If $\Delta(s)$ is an entire function, then it has the form of a quasipolynomial fraction where the denominator roots are algebraically canceled [12,31,55].

4.2.2 LTI TDS spectrum and its properties

The *characteristic values* (eigenvalues, *poles*) of the system are elements of the *system spectrum*

$$\Sigma_P := \{s : \Delta(s) = 0\}. \tag{4.6}$$

By comparing (4.4) and (4.6) it is clear that system poles agree with eigenvalues of the matrix $s\mathbf{H}(s) + \mathbf{A}(s)$. Moreover, the *essential spectrum* $\Sigma_{P,ess}$ can be defined for NTDSs as the set of all eigenvalues of the matrix $\mathbf{H}(s)$ [44]. Let us introduce $\Delta_{num}(s) = s^n + \sum_{i=0}^{n} \sum_{j=0}^{h_i} d_{ij} s^i \exp(-\tau_{ij} s)$, the (monic) numerator of $\Delta(s)$ (i.e., if there are no distributed delays, we simply have $\Delta_{num}(s) = \Delta(s)$). Then the exponential polynomial following the highest s-power of $\Delta_{num}(s)$ reads $d_n(s) = 1 + \sum_{j=1}^{h_n} d_{nj} \exp(-\tau_{nj} s)$. It is called the *associated exponential polynomial*. Then the essential spectrum can be also expressed also as

$$\Sigma_{P,ess} := \{s : d_n(s) = 0\}. \tag{4.7}$$

Properties of $\Sigma_{P,ess}$ differ from those of Σ_P and have a great impact on system dynamics and stability. We first provide the reader with some basic facts about $\Sigma_{P,ess}$ [3,27,43,44,53,55].

Proposition 4.1. *For $\Sigma_{P,ess}$, we have:*

1. *The elements $s_k \in \Sigma_{P,ess}$ constitute chains parallel to the imaginary axis with $|\mathrm{Re}\, s_k| < \infty$ and $\mathrm{Im}\, s_k \to \infty$ as $|s_k| \to \infty$.*
2. *Let $\alpha_{P,ess} := \sup\{\mathrm{Re}(s) : s \in \Sigma_{P,ess}\}$. The value of $\alpha_{P,ess}$ is not continuous with respect to delays, but it behaves continuously with respect to nondelay parameters.*
3. *Define $\bar{\alpha}_{P,ess} := \sup \lim_{\varepsilon \to 0} \{\alpha_{P,ess}(\boldsymbol{\tau} + \delta\boldsymbol{\tau}) : \forall \|\delta\boldsymbol{\tau}\| < \varepsilon, \varepsilon > 0\}$, where $\boldsymbol{\tau}$ is the vector of all delays. Then the estimation of the safe upper bound $\bar{c}_{P,ess}$ on*

$\bar{\alpha}_{P,ess}$ can be given by

$$\bar{c}_{P,ess} = \arg_{c\in\mathbb{R}} \left(\sum_{j=1}^{h_n} \left| d_{nj} \right| \exp\left(-c\tau_{nj}\right) = 1 \right); \tag{4.8}$$

see [83].

It is worth noting that the value of $\bar{c}_{P,ess}$ is independent of particular delays, whereas $\alpha_{P,ess}$ is a function of $\boldsymbol{\tau}$.

For selected properties of Σ_P and their relation to $\Sigma_{P,ess}$, see [25,42,43].

Proposition 4.2. *For Σ_P, we have:*

1. *For $\Delta_{num}(s)$ being a quasipolynomial (rather than polynomial), $|\Sigma_P| = \infty$.*
2. *For RTDSs, all $s_k \in \Sigma_P$ are isolated, and there are finitely many poles in the half-plane $\mathrm{Re}\, s > \beta$ for any finite $0 > \beta \in \mathbb{R}$; however, infinitely many ones rest in the half-plane $\mathrm{Re}\, s < \beta$.*
3. *For NTDSs, the system poles $s_k \in \Sigma_P$ constitute strips, and $s_k \to s_{k,ess} \in \Sigma_{P,ess}$ as $|s_k| \to \infty$ (i.e., Σ_P asymptotically converges to $\Sigma_{P,ess}$ at high frequencies). It implies that there can lie infinitely many poles in the subset $\mathbb{C}_{\beta_1 \le \mathrm{Re}\, s \le \beta_2} \subset \mathbb{C}$ for some finite $\beta_1, \beta_2 \in \mathbb{R}$. In particular, infinitely many system characteristic roots may be located in the half-plane $\mathrm{Re}\,(s) > 0$.*
4. *For NTDSs, assume that the sequence $|s_{k-1}| < |s_k|$, $s_k \in \Sigma_{P,sub} \subset \Sigma_P$. Then $\lim_{k\to\infty} \mathrm{Im}\, s_k = \infty$ for $\lim_{k\to\infty} s_k = \alpha_{P,ess}$. Only a finite number of poles are located in the half-plane with $\mathrm{Re}\, s > \bar{c}_{P,ess} \ge \bar{\alpha}_{P,ess}$, and they are isolated.*
5. *The isolated single system poles behave continuously and smoothly with respect to $\boldsymbol{\tau}$ and nondelay parameters.*
6. *Let $\alpha_P := \sup\{\mathrm{Re}\,(s) : s \in \Sigma_P\}$ express the spectral abscissa (of system poles). It can be nonsmooth, nonconvex, or even non-Lipschitz, especially for NTDSs; see details in* [78].

Exponential stability of LTI TDSs is closely related to system poles loci (unlike some other types of stability) [3,23].

Proposition 4.3 ([25,27,43]). *An RTDS is exponentially stable if and only if $\alpha_P < 0$. An NTDS is exponentially stable if and only if there exists $\varepsilon > 0$ such that $\alpha_P < -\varepsilon$.*

Due to items (2) and (3) of Proposition 4.1, we have to be careful when analyzing whether the system spectrum remains in the left half-plane. Hence the specific feature of TDSs is the so-called *strong stability* [43,44].

Definition 4.1. System (4.1) is said to be strongly stable if $\bar{\alpha}_{P,ess} < 0$.

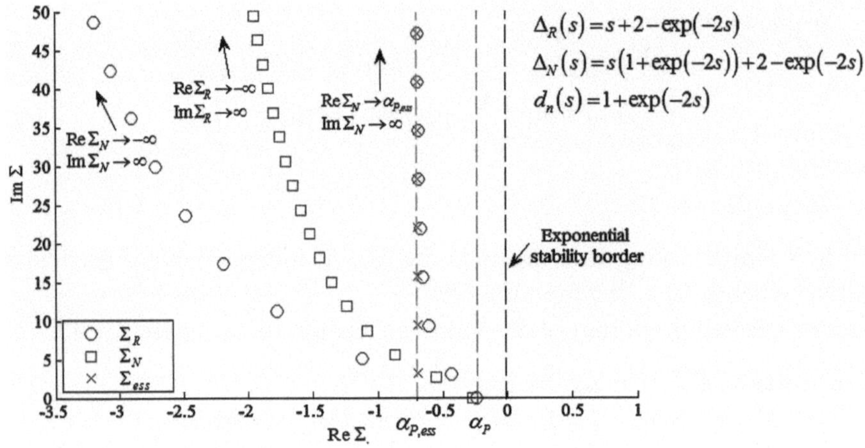

Figure 4.1 LTI TDS spectrum and stability.

Proposition 4.4. *An NTDS is strongly stable if and only if*

$$1 < \sum_{j=1}^{h_n} \left| d_{nj} \right| . \tag{4.9}$$

Some spectral features introduced in Propositions 4.1 and 4.2 are displayed in Fig. 4.1. In the figure, Σ_R expresses the spectrum corresponding to the characteristic quasipolynomial $\Delta_R(s)$ of the selected RTDS, whereas Σ_N, $\Delta_N(s)$ are those for the chosen NTDS, where $d_n(s)$ is the corresponding associated exponential polynomial.

4.3. The use of R_{QM} ring to controller design

The determination of a parameterized controller structure usually represents an important step when designing the eventual control law. Among numerous possible approaches, methods, and techniques, the family of algebraic approaches plays an important role. As introduced in [54], first attempts to use algebraic tools for TDSs were made in the 1970s [30,46]. Later a subset of these methods started to utilize the input/output fractional representation [16,33,80] extending the standard transfer function (matrix) description toward TDS [13,21]. The R_{QM} ring [60] is one of such tools that is hand in hand with the well-established algebraic tools and procedures (such as the linear Diophantine equation or the Youla–Kučera parameteri-

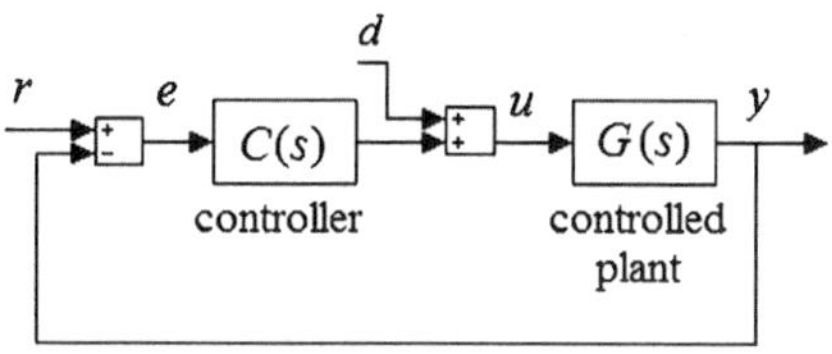

Figure 4.2 Simple negative control feedback loop.

zation), giving rise to a complex approach. The author has recently utilized it when proposing a robust control of technological processes [54].

In this section, we introduce the ring definition together with some necessary notions. Then we give a sketch of the control system synthesis for a simple negative feedback loop to the reader. We restrict the further presentation to single-input single-output controlled systems.

4.3.1 R_{QM} ring definition

Definition 4.2. The R_{QM} set constitutes the algebraic ring the elements $T(s)$ of which satisfy:

1. $T(s) \in H_\infty$;
2. $T(s)$ is formally stable;
3. $T(s) = \frac{q_n(s)}{q_d(s)}$, $q_{n0}(s), q_d(s)$ are quasipolynomials with $q_n(s) = q_{n0}(s)\,\mathrm{e}^{-\tau s}$, $\tau \geq 0$.

Definition 4.3. System (4.1) is formally stable if $\alpha_{P,ess} < 0$.

Note that, by comparison of Definitions 4.1 and 4.3, formal and strong stability are very close to each other. It is worth noting that strong stability cannot be included in Definition 4.2 due to required ring properties; see details in [38,60]. Moreover, the inclusion of $T(s)$ in H_∞ yields its properness [49].

4.3.2 Simple control feedback-loop synthesis

We concisely introduce the controller structure design procedure and utilized tools using the R_{QM} ring for the standard negative control feedback loop; see Fig. 4.2.

In the scheme, r stands for the reference, e means the control error, u is the control action affected by the load disturbance d, and y represents the controlled output. Consider the plant governed by the (single-variable) transfer function expressed by the Bézout coprime factorization [12,75]

$$G(s) = \frac{B(s)}{A(s)},$$
$$A(s), B(s) \in R_{QM}. \tag{4.10}$$

The feedback control system is stable in the R_{QM} sense if and only if the following Bézout identity holds:

$$A(s)\,Q_0(s) + B(s)\,P_0(s) = 1, \tag{4.11}$$

where $C(s) = Q_0(s)/P_0(s)$ is the coprime transfer function of the controller with $Q_0(s), P_0(s) \neq 0 \in R_{QM}$ being a particular solution of (4.11). The Bézout identity can effectively be solved by using the extended (generalized) Euclidean algorithm through the determination of the greatest common divisor of $A(s)$, $B(s)$, that is, $GCD(A, B)$; see [60] for more detail. Then all controllers satisfying (4.11) can be found by the parameterization

$$Q(s) = Q_0(s) \pm \frac{A(s)T(s)}{GCD(A, B)},$$
$$P(s) = P_0(s) \mp \frac{B(s)\,T(s)}{GCD(A, B)} \neq 0, \tag{4.12}$$

where $T(s) \in R_{QM}$ is arbitrary [33].

In particular, asymptotic reference tracking and load disturbance rejection are simultaneously reached in $T(s)$ in (4.12) if

$$\frac{A(s)\,P(s)}{F_r(s)} \in R_{QM}, \quad \frac{B(s)\,P(s)}{F_d(s)} \in R_{QM}, \tag{4.13}$$

where $F_r(s), F_d(s) \in R_{QM}$ are given by coprime factorizations of the Laplace forms of r and d,

$$R(s) = \frac{H_r(s)}{F_r(s)}, \quad D(s) = \frac{H_d(s)}{F_d(s)}. \tag{4.14}$$

For simplicity, we further take the following two assumptions thorough the rest of the chapter:

1. There are no distributed delays in the controlled system, that is, $\Delta_{num}(s) = \Delta(s)$;
2. There are no common zeros of $A(s)$ and $B(s)$.

4.4. Spectral abscissa minimization

In this section, we formulate and analyze the task of the spectral abscissa minimization. The spectral abscissa α_P has been defined in item (6) of Proposition 4.2. It can be considered either as a value characterizing the current set of system characteristic values or as the function $\alpha_P(\mathbf{p})$ expressing the dependence on system parameters $\mathbf{p}$:

$$\begin{aligned} \alpha_P &: \mathbb{R}^r \mapsto \mathbb{R} \\ \mathbf{p} &\mapsto \alpha_P(\mathbf{p}) \end{aligned} \tag{4.15}$$

Particularly, it is interesting the case where the vector $\mathbf{p}$ means adjustable controller parameters or system delays and $\alpha_P(\mathbf{p})$ is the abscissa of the control feedback spectrum. It is clear from Proposition 4.3 that $\alpha_P(\cdot)$ is closely related to exponential stability; thus we can attempt to reach the minimum of this function:

$$\mathbf{p}_{opt} = \arg\min_{\mathbf{p}} \alpha_P(\mathbf{p}), \tag{4.16}$$

that is, the "maximal" exponential stability [72]. However, special attention has to be paid to NTDSs due to strong stability (see Definition 4.1 and Proposition 4.4).

Therefore we further give some ideas and strategies on how to solve (4.16), including the specific features of NTDSs. It is worth noting that the abscissa $\alpha_Z(\cdot)$ of the set of all system zeros $\Sigma_Z := \{s : G(s) = 0\}$ can be defined and handled analogously.

4.4.1 QCSA

First of all, we have to introduce the well-established QCSA algorithm [42] for RTDSs since it is the initial step of our procedure. It is based on the iterative shifting of the right-most poles to the left by small changes of $\mathbf{p}$. The algorithm was extended to NTDSs [44]; nevertheless, we separately discuss this modification in Section 4.4.2.

Algorithm 4.1 (QCSA).
Input: Closed-loop characteristic quasipolynomial $\Delta_{num}(s, \mathbf{p})$, initially desired poles loci, optimization problem (4.16).

1. Assign the number $n_P \leq r$ of feedback poles s_i, $i = 1, 2, \ldots, n_P$, directly to the desired initial positions (see details. e.g., in [94]), which yields the current parameter setting $\mathbf{p}_0$. (Or simply set $\mathbf{p}_0$ arbitrarily.)

2. Unexpected poles may emerge right from the desired ones, that is, they are dominant (rightmost). Therefore set algorithm termination parameters and the initial number of moved (shifted, controlled) poles to $m := 1$ ($m := 2$ if the rightmost pair is complex-conjugate).
3. Compute the number of m rightmost poles, for example, by using the DDE-BIFTOOL [18] or the quasipolynomial rootfinder (QPmR) [82, 83] Matlab® packages.
4. Compute the sensitivity of m rightmost poles with respect to changes in $\mathbf{p}$ by means of the sensitivity matrix $\mathbf{S} \in \mathbb{C}^{m \times r}$ with the elements

$$S_{ij} = -\frac{\left[\frac{\partial \Delta_{num}(s,\mathbf{p})}{\partial p_j}\right]_{\substack{s=s_i \\ \mathbf{p}=\mathbf{p}_0}}}{\left[\frac{\partial \Delta_{num}(s,\mathbf{p})}{\partial s}\right]_{\substack{s=s_i \\ \mathbf{p}=\mathbf{p}_0}}}. \tag{4.17}$$

5. Select desired poles displacement $\Delta\boldsymbol{\delta} = [\delta_1, \delta_2, \ldots, \delta_m]^T$ with $\operatorname{Re}\delta_i < 0$ in the vicinity of the currently shifted poles loci and calculate the estimation of the change of parameters $\Delta\mathbf{p}$ via

$$\Delta\mathbf{p} = \mathbf{S}^+ \Delta\boldsymbol{\sigma}, \tag{4.18}$$

where the superscript + means the Moore–Penrose (pseudo)inverse [62]. Then, the set of parameters is updated to $\mathbf{p} := \mathbf{p}_0 + \Delta\mathbf{p}$.
6. If necessary, the value of m is increased (until $m = n_P$). If the available degrees of freedom in the controller do not allow to further reduce $\alpha_P(\mathbf{p})$, then terminate the algorithm; otherwise, set $\mathbf{p}_0 := \mathbf{p}$ and go to step (3).

Output: The eventual parameter values $\mathbf{p}_0$ and the corresponding spectrum (of system poles) Σ_P with its abscissa $\alpha_P(\mathbf{p}_0)$.

A sketch of a Matlab code example for the direct pole placements and a simple quasicontinuous shifting (without the increment of m) follows:

```
syms s C0 C1;
C_sym=[C0; C1]; % Parameters to set
tau=0.33; % I/O (transmission) delay
% Characteristic quasipolynomial - 1st order unstable system + I/O delay
% + PI controller in the feedback
M=s^2+(2*C1*exp(-tau*s)-1)*s+2*C0*exp(-tau*s);
% Derivatives with respect to "s"
dMds=diff(M,s);
ddMds=diff(dMds,s);
dddMds=diff(ddMds,s);
```

```
% Derivatives with respect to "C0, C1"
dMdC=[diff(M,C0); diff(M,C1)];

C=[1; 1]; % Initial setting [C0, C1]
M1=subs(M,C_sym,C);
% ----- Initial poles determination -----
Region = [-5 1 0 5];
GridDensity = 0.05;
Error_QPMR = 10^(-6);
P=qpmr(M1, Region, GridDensity, Error_QPMR, 1); % QPmR
% ----- Direct pole placement -----
dM1dC=subs(dMdC,C_sym,C);
M1=subs(M,C_sym,C);
LeftSide(1,:)=real(subs(dM1dC,s,PrescribedPoles));
LeftSide(2,:)=imag(subs(dM1dC,s,PrescribedPoles));
RightSide(1)=real(subs(M1,s,PrescribedPoles));
RightSide(2)=imag(subs(M1,s,PrescribedPoles));
C=C-pinv(LeftSide)*RightSide.';
C=eval(C);
M1=subs(M,C_sym,C);
P=qpmr(M1, Region, GridDensity, Error_QPMR, 1); % Sorted poles
% ---- Quasi-continuous pole shifting  ----
Counter=0; % Iteration counter
Step=0.001; % Pole shift step
Stop=0;
Stop_error=10^(-6); % Parameters increment limit
dM1dC=subs(dMdC,C_sym,C);
dM1ds=subs(dMds,C_sym,C);
while (~Stop)
    Counter=Counter+1;
    Shift_vector=(PrescribedPoles-P(1))/norm(PrescribedPoles-P(1))*Step;
    Shift(1)=real(Shift_vector);
    % Sensitivity matrix computation
    for n=1:1:length(dMdC)
        SensitivityMatrix(1,n)=real([(subs(-dM1dC(n)/dM1ds,s,P(1)))]);
    end
    if imag(Shift_vector)>=1.5*Error_QPMR % Non-zero shift in imaginary axis
        Shift(2)=imag(Shift_vector);
        for n=1:1:length(dMdC)
            SensitivityMatrix(2,n)=imag([(subs(-dM1dC(n)/dM1ds,s,P(1)))]);
        end
    end
    DC=real(pinv(SensitivityMatrix)*Shift.');
    P_list(Counter,1:3)=P.'; % List of poles 3 dominant poles (history)
    if size(C)~=size(DC)
        DC=DC';
    end
    C=C+DC; % Parameters update
```

```
    if sum(abs(DC))<Stop_error
        Stop=1;
    end
    M1=subs(M,C_sym,C);
    dM1dC=subs(dMdC,C_sym,C);
    dM1ds=subs(dMds,C_sym,C);
    P=qpmr(M1, Region, GridDensity, Error_QPMR, 0);
end
```

Let us make some notes to Algorithm 4.1. Regarding steps (4) and (5), if all complex conjugates for s_i are included in $\mathbf{S}$ and $\Delta\boldsymbol{\sigma}$, then (4.18) gives $\Delta\mathbf{p} \in \mathbb{R}^r$. The task of the spectral abscissa minimization naturally requires $\Delta\boldsymbol{\sigma} \in \mathbb{R}^m$; then (4.18) is adjusted to

$$\Delta\mathbf{p} = (\operatorname{Re}\mathbf{S})^+ \Delta\boldsymbol{\sigma}. \tag{4.19}$$

We suppose that the sensitivity matrix has the full rank. The used Moore–Penrose (pseudo)inverse minimizes the norm $\|\mathbf{p}\|_2 = \sum_{i=1}^{r} p_i^2$, which is beneficial since it prevents the parameters from too high values [35].

The value of m is usually increased in step (6) when there exists a complex conjugate pair of controlled poles that becomes a pair of single real poles and vice versa, or there are more rightmost poles with almost identical real parts. In these cases the shifting procedure does not result in a better value of $\alpha_P(\cdot)$. This feature is closely related to item (6) of Proposition 4.2.

4.4.2 Neutral delay case

Due to Proposition 4.1, it is necessary to be aware of the existence of the essential spectrum $\Sigma_{P,ess}$ for an NTDS. More specifically, the value of the abscissa $\bar{\alpha}_{P,ess}$ has to be checked because of the sensitivity of $\alpha_{P,ess}$ to infinitesimal delay variations. Its qualified estimation is given, for example, by the a safe upper bound $\bar{c}_{P,ess}$ on $\bar{\alpha}_{P,ess}$ (4.8), and the verification of strong stability is done via condition (4.9) of Proposition 4.4. Hence the spectral abscissa optimization can be abandoned as soon as the associated exponential polynomial $d_n(s) = 1 + \sum_{j=1}^{h_n} d_{nj} \exp(-\tau_{nj}s)$ (i.e., the term following the highest s-power in $\Delta_{num}(s)$, see Section 4.2.2) does not include any adjustable parameter and (4.9) is not satisfied.

If $d_n(s, \mathbf{p})$ for some $\mathbf{p} \neq \varnothing$, then we have to check (4.9) in every single step of Algorithm 4.1. As the eventual parameters do not satisfy (4.9), the last $\mathbf{p}$ for which the string stability condition (4.9) holds must be eventually taken.

Let us now recall a computational estimation of $\alpha_{P,ess}$ [49,65].

Proposition 4.5. *Let $\tilde{d}_n(s) = 1 + \sum_{j=1}^{\tilde{h}_n} \tilde{d}_{nj} \exp(-j\tau_0 s)$ be an approximation of $d_n(s)$ where $\tilde{d}_{nj} \in \mathbb{R}$ and $\tau_0 \in \mathbb{R}_+$ is a base delay. Using the substitution $q = \exp(-\tau_0 s)$, we introduce the approximate associate polynomial*

$$\tilde{d}_n(q) = 1 + \sum_{j=1}^{\tilde{h}_n} \tilde{d}_{nj} q^j. \tag{4.20}$$

Then all $s_i \in \Sigma_{P,ess}$ are calculated from

$$s_i = -\frac{\ln(q_i)}{\tau_0}. \tag{4.21}$$

By (4.21) there exist chains of poles of neutral type asymptotically approaching the vertical lines

$$\operatorname{Re} s = -\frac{\ln(|q_i|)}{\tau_0} \tag{4.22}$$

for each root q_i of $\tilde{d}_n(q)$. This implies that

$$\alpha_{P,ess} = \max\left\{-\frac{\ln(|q_i|)}{\tau_0} : \tilde{d}_n(q_i) = 0\right\}. \tag{4.23}$$

Note that (4.21) means that whenever s_i is a zero of $\tilde{d}_n(s)$, then every

$$s_k = s_i \pm \mathrm{j}k\frac{2\pi}{\tau_0}, \; k \in \mathbb{N}, \tag{4.24}$$

is also the zero of $\tilde{d}_n(s)$, which provides an estimation of complete $\Sigma_{P,ess}$.

Some authors [44] suggested to avoid dealing with poles left from $\bar{\alpha}_{P,ess}$ (expressed by its estimation $\bar{c}_{P,ess}$) since the knowledge of $\bar{c}_{P,ess}$ may prevent spending much control action to poles with smaller real parts, which are mostly useless when strong stabilization and arguable for the spectral abscissa minimizing. This idea, however, does not apply when $d_n(s, \mathbf{p})$ can be adjusted by suitable values of $\mathbf{p}$. It can be observed that isolated poles may lie either to the right (see Fig. 4.1) or to the left (see Fig. 4.3) from $\alpha_{P,ess}$.

In the latter case, it was observed that it is desirable to control also isolated poles left from this vertical line with a sufficiently small modulus [51]. Both the spectral abscissa and the abscissa of the essential spectrum can be adjusted by handling these characteristic roots.

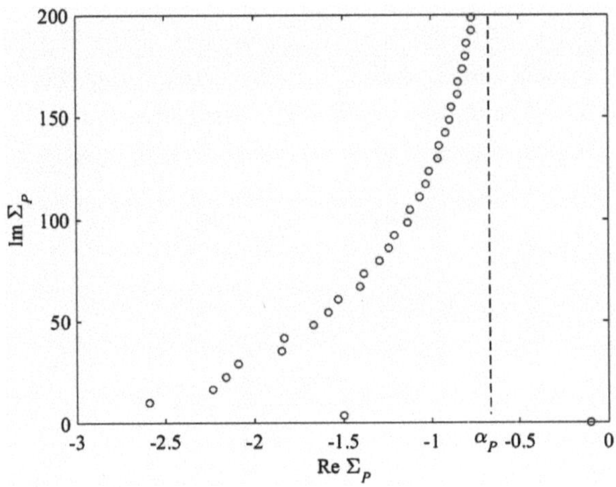

Figure 4.3 Isolated poles left from $\bar{c}_{P,ess}$ ($\alpha_{P,ess}$).

4.4.3 Proposed optimization procedure

In our algorithm, we use the QCSA combined with the awareness of strong stability and an additional numerical (iterative) optimization algorithm (direct-search, bioinspired, swarm, genetic, etc.).

Algorithm 4.2 (Spectral abscissa minimization).

Input: Closed-loop characteristic quasipolynomial $\Delta_{num}(s, \mathbf{p})$, initially desired poles loci, optimization problem (4.16), selected numerical optimization algorithm.

1. If $\Delta_{num}(s, \mathbf{p})$ with $\mathbf{p} = \varnothing$ and (4.9) does not hold, abandon the algorithm.
2. Apply Algorithm 4.1 (i.e., the QCSA) until $m = r$. Take only poles right from $\alpha_{P,ess}$ (4.23) or those with a small modulus. In every single step, check (4.9): if the condition becomes unsatisfied, then save $\tilde{\mathbf{p}} := \mathbf{p}_0$. This step gives rise to the current parameter values $\mathbf{p}_0$.
3. Use the selected iterative algorithm starting with $\mathbf{p}_0$ resulting from (2). Again, check (4.9) as in (2).
4. If the eventual values of $\mathbf{p}$ meet (4.9), then set $\mathbf{p}_{opt} := \mathbf{p}$; else, set $\mathbf{p}_{opt} := \tilde{\mathbf{p}}$.

Output: The optimal parameter values $\mathbf{p}_{opt}$, the corresponding spectrum $\Sigma_{P,opt}$, and the essential spectrum $\Sigma_{P,ess,opt}$ with their abscissae $\alpha_{P,opt}$ and $\alpha_{P,ess,opt}$, respectively.

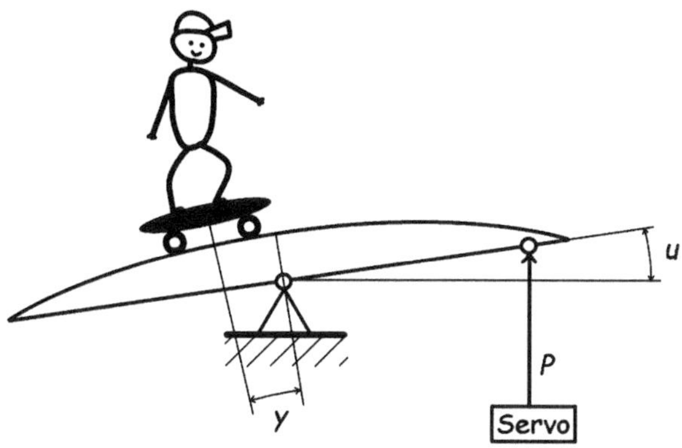

Figure 4.4 A sketch of a roller skater on the remotely controlled swaying bow.

Note that the use of the QCSA according to step (2) of Algorithm 4.2 prior to the deployment of an iterative algorithm as in step (3) is beneficial. It was shown by examples that when step (2) is omitted, the found minimum of $\alpha_{P,opt}$ is worse than in the case of the a priori utilization of the QCSA.

A simulation example follows [52,58].

4.4.4 Numerical example

Assume the following mathematical model [93] of a roller skater on the remotely controlled swaying bow (see Fig. 4.4) [57]:

$$y^{(4)}(t) + a\ddot{y}(t-\vartheta) = bu(t-\vartheta-\tau) \Leftrightarrow G(s) = \frac{Y(s)}{U(s)} = \frac{b\exp(-(\tau+\vartheta)s)}{s^2(s^2+a\exp(-\vartheta s))}, \tag{4.25}$$

where $u(t)$ and $y(t)$ mean the slope angle of a bow caused by the external force $P(t)$ and the deviation of the skater from the bow axis of symmetry, respectively. The skater manipulates the servo by sending signals into its electronics. The delays τ and ϑ in model (4.25) express the skater and servo latencies, respectively. Let the nominal values be $b=0.2$, $a=-1$, $\tau=0.3$ s, $\vartheta=0.1$ s.

Let us first sketch the controller design in the R_{QM} ring (see Section 4.3.1). Consider the control system as in Fig. 4.2. If $A(s), B(s) \in R_{QM}$ in (4.10) are set as

$$B(s) = \frac{0.2\exp(-0.4s)}{(s+\lambda)^4},$$

$$A(s) = \frac{s^2\left(s^2 - \exp(-0.1s)\right)}{(s+\lambda)^4}, \lambda > 0, \tag{4.26}$$

then a particular solution of (4.11) yields, for example,

$$\begin{aligned} Q_0(s) &= \frac{\left(q_3 s^3 + q_2 s^2 + q_1 s + q_0\right)(s+\lambda)^4}{\Delta_{P,qp}(s)}, \\ P_0(s) &= \frac{\left(s^3 + p_2 s^2 + p_1 s + p_0\right)(s+\lambda)^4}{\Delta_{P,qp}(s)}, \\ \Delta_{P,qp}(s) &= s^2\left(s^2 - \exp(-0.1s)\right)\left(s^3 + p_2 s^2 + p_1 s + p_0\right) \\ &\quad + 0.2\exp(-0.4s)\left(q_3 s^3 + q_2 s^2 + q_1 s + q_0\right), \end{aligned} \tag{4.27}$$

where $\mathbf{p} = [q_3, q_2, q_1, q_0, p_2, p_1, p_0]^T \in \mathbb{R}^7$ are tunable parameters of controller governed by the transfer function $C(s) = Q_0(s)/P_0(s)$.

Apparently, stabilizing solution (4.27) meets the requirements (4.13) of asymptotic reference tracking and disturbance rejection for stepwise and linearwise external inputs since their Laplace forms (4.14) read, respectively,

$$\frac{H_.(s)}{F_.(s)} = \frac{\frac{h_0}{s+\lambda}}{\frac{s}{s+\lambda}}, \quad \frac{H_.(s)}{F_.(s)} = \frac{\frac{h_0}{s+\lambda}}{\frac{s^2}{s+\lambda}}. \tag{4.28}$$

In other words, all $F_.(s)$ always divide $A(s)$. However, if (for instance) the reference and input disturbance signals are quadraticwise, conditions (4.13) do not hold. We can prove that if

$$T(s) = -\frac{1}{0.2}\frac{p_0\lambda^4(s+\lambda)^4}{\Delta_{P,qp}(s)} \tag{4.29}$$

is taken in (4.12) (note that $GCD(A, B) = 1$), then the eventual form of $P(s)$ satisfies (4.13) for

$$\frac{H_.(s)}{F_.(s)} = \frac{\frac{h_0}{s+\lambda}}{\frac{s^3}{s+\lambda}}. \tag{4.30}$$

Then the reference-to-output transfer function reads

$$\begin{aligned} G_{RY}(s) &= \frac{1}{(s+\lambda)^4}\frac{g_N(s)}{\Delta_{P,qp}(s)}\exp(-0.4s), \\ g_N(s) &= 0.2\left(\left(q_3 s^3 + q_2 s^2 + q_1 s + q_0\right)(s+\lambda)^4 + p_0\lambda^4 s^2\left(s^2 - \exp(-0.1s)\right)\right), \end{aligned} \tag{4.31}$$

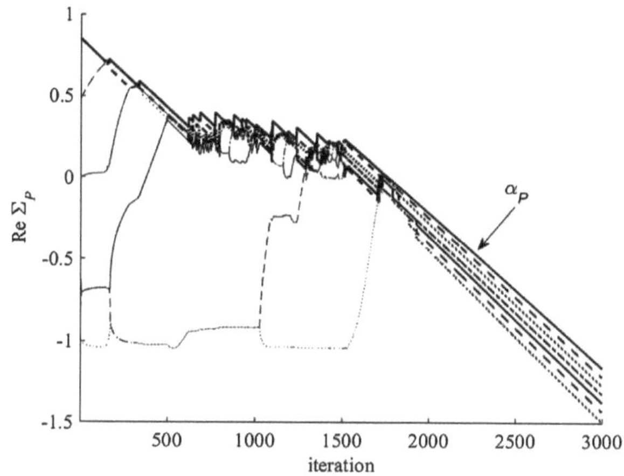

Figure 4.5 Iterative evolution of α_P (**p**) and real parts of the dominant roots of $\Delta_{P,qp}(s)$ by using the QCSA for iteration steps 1 to 3000.

that is, the characteristic quasipolynomial has the form

$$\Delta_P(s) = \Delta_{P,num}(s) = (s+\lambda)^4 \Delta_{P,qp}(s). \tag{4.32}$$

It is worth noting that (4.31) hold also for controller (4.27). The factor $(s+\lambda)^4$ in (4.32) has a quadruple real pole. To cancel its influence, λ must be much higher than the maximum absolute value of roots of the quasipolynomial $\Delta_{P,qp}(s)$.

Now let us provide the reader with the spectral abscissa minimization following Algorithm 4.2. Starting from the trial parameters setting $\mathbf{p}_0 = [1, 1, 1, 1, 1, 1, 1]^T$, giving rise to the unstable control feedback system with $\alpha_P(\mathbf{p}_0) = 0.8492$, the use of the QCSA results in a stabilizing setting

$$\mathbf{p}_{3305} = [106523133, 26247749, 5617613, 469418.2, 640264.2, \\ 10560107, 8222650]^T \tag{4.33}$$

with $\alpha_P(\mathbf{p}_{3305}) = -1.4454$ (i.e., after the number of 3305 iteration steps). The evolution of $\alpha_P(\mathbf{p})$ and real parts of other dominant poles up to the iteration step of 3000 is roughly displayed in Fig. 4.5, and iterations form 3001 to 3520 are focused in Fig. 4.6.

Note that the eventual values $\mathbf{p}_{3520}$ are not further considered since they are too high.

Set again $\mathbf{p}_0 := \mathbf{p}_{3305}$. Let the well-known NMA [47] be used for further spectral abscissa minimization. The comparison of the evolution of resulting

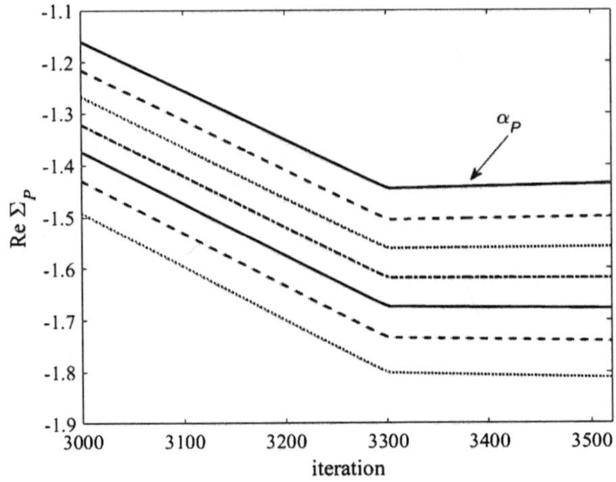

Figure 4.6 A detailed view on the evolution of $\alpha_P(\mathbf{p})$ and real parts of the dominant roots of $\Delta_{P,qp}(s)$ by using the QCSA for iteration steps 3001 to 3520.

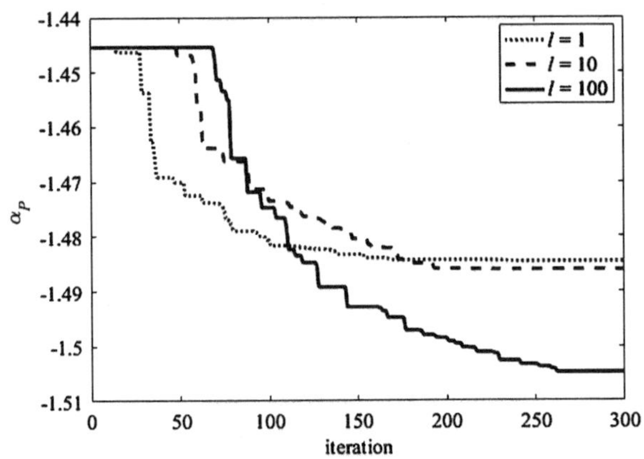

Figure 4.7 Iterative evolution of $\alpha_P(\mathbf{p})$ for $\Delta_{P,qp}(s)$ by using the NMA for several initial simplex edge lengths.

abscissae for three different initial simplex edge lengths ($l = 1, 10, 100$) is displayed in Fig. 4.7. Apparently, the largest initial simplex yields the best performance. However, the obtained parameter values remain very close to (4.33).

Besides the direct-search NMA, the bioinspired (swarm) optimization SOMA [89] algorithm and the extended gradient-based EGSA algorithm [6] for even nonconvex problems have been benchmarked. In Fig. 4.8, we give a comparison of the course of $\alpha_P(\mathbf{p})$ during iteration steps for these

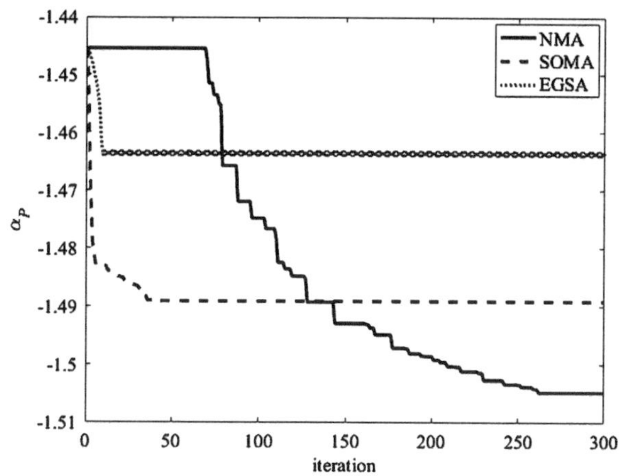

Figure 4.8 A comparison of the evolution of $\alpha_P(\mathbf{p})$ for $\Delta_{P,qp}(s)$ by using the NMA, SOMA, and EGSA.

three algorithms (for standard versions of the algorithms and their selected suitable settings). Notice that the use of NMA gives the best result within this benchmark.

4.5. Quasioptimal zeros-poles assignment algorithm

In this section, we further extend the concept of the spectral abscissa minimization so that not only the real part of the dominant characteristic root but its general position in the complex plane is attempted to be reached. Moreover, system zeros are considered as well. The presented concept has recently been published in [56], where two substrategies have been proposed. In this chapter, we present a less complicated substrategy, which, moreover, gives better numeric results. Regarding the remaining substrategy, we refer the reader to [56] for more detail. We herein extend it to NTDSs, which is supported by a simple simulation example.

4.5.1 Problem statement

Let us first provide the reader with motivation and the problem statement. Recall that the subject of the study of this chapter is an infinite spectrum of control feedback characteristic values, resulting from a control design (e.g., using the R_{QM} ring). It is worth noting that zeros in LTI TDSs have the same impact on the system dynamics as for finite-dimensional systems (it implies, e.g., that zeros located right from α_P may cause a control response overshoot even for aperiodic control systems). Hence it is desirable to adjust the spectrum Σ_Z of zeros as well, which enables us to adjust the feedback control dynamics more comprehensively. The goal is to place a selected subset of the rightmost poles and zeros to desired loci and, simultaneously, to minimize both abscissae of the remaining spectra.

A sketch of the algorithm framework is as follows. The initial step consists in the direct zero/pole placement (analogously to step (1) of Algorithm 4.1). Whenever the placed roots are not dominant, a subset of the rightmost roots is quasicontinuously shifted to the prescribed loci (via the QCSA of an iterative optimization algorithm), whereas the rests of both spectra are attempted to be pushed to the left. The optimization procedure should take into account the distance of dominant adjusted roots from the prescribed loci and abscissae of rests of the spectra. However, the task is rather tricky because the controller parameters are generally simultaneously included in the numerator and denominator of the closed-loop transfer function. In addition, the use of habitual optimization techniques may go wrong due to the properties introduced in item (6) of Proposition 4.2.

Denote by $\Delta_Z(s)$ the numerator (quasi)polynomial of $G_{RY}(s) = Y(s)/R(s)$. Assume, moreover, that there is no s such that it is a common root of $\Delta_P(s)$ and $\Delta_Z(s)$, that is, $\Sigma_{ZP} := \{s : \Delta_Z(s) = \Delta_P(s) = 0\} = \varnothing$. The matching model transfer function $G_{RY,m}(s) = \Delta_{Z,m}(s)/\Delta_{P,m}(s)$ gives rise to the desired zero and pole loci ζ_i, $i = 1, 2, \ldots, n_Z = \deg \Delta_{Z,m}(s)$ and σ_i, $i = 1, 2, \ldots, n_P = \deg \Delta_{P,m}(s)$, respectively. The values n_Z, n_P include possible root multiplicities. Let selectable (tunable) parameters in the feedback system numerator and denominator, respectively, be elements of vectors $\mathbf{p}_Z \in \mathbb{R}^{r_Z}$ and $\mathbf{p}_P \in \mathbb{R}^{r_P}$, $r_P \geq n_P$. The numbers of currently shifting zeros and poles are m_Z, m_P, respectively.

Assuming that the sensitivity matrix $\mathbf{S} \in \mathbb{C}^{(m_Z+m_P)\times(r_Z+r_P)}$ has the full rank, the natural requirement is that

$$r_Z \geq n_Z, \quad r_P \geq n_P \tag{4.34}$$

to get a solvable model matching problem. In addition, some tunable parameters must serve to the minimization of abscissae of the remaining spectra, not to the approaching of the dominant roots to the desired ones, that is, a stricter condition is to be met:

$$0 < n_P < r_P, \quad 0 \leq n_Z < r_Z. \tag{4.35}$$

Condition (4.35) is separately applied to the numerator and denominator. Moreover, there must be taken into consideration the overall number of prescribed roots and free parameters, $n = n_Z + n_P$ and $r = r_Z + r_P$, respectively, under the restriction

$$n < r. \tag{4.36}$$

Regarding the matching model itself, it must be strictly proper (i.e., strongly feasible):

$$n_Z < n_P. \tag{4.37}$$

Conditions (4.34)–(4.37) must be considered when selecting $G_{RY,m}(s)$. The particular desired zero/pole loci can be chosen based on the required model dynamics. A concise analysis of simple-model dynamics is elaborated is Section 4.5.2.

4.5.2 Approximate reference dynamics analysis

The matching model can be chosen as the second-order undelayed stable periodic or aperiodic system governed by the transfer function

$$G_{RY,m}(s) = \frac{\Delta_{Z,m}(s)}{\Delta_{P,m}(s)} = \frac{f_1 s + f_0}{s^2 + d_1 s + d_0} = k\frac{s - \zeta}{(s - \sigma_1)(s - \sigma_2)} \tag{4.38}$$

with

$$\begin{aligned} &\operatorname{Re}\sigma_1 < 0,\ \operatorname{Re}\sigma_2 < 0,\ \operatorname{Re}\zeta < 0, \\ &\frac{f_0}{d_0} = -\frac{k\zeta}{\sigma_1\sigma_2} = 1 \Rightarrow k = -\frac{\sigma_1\sigma_2}{\zeta}, \end{aligned} \tag{4.39}$$

all the coefficients being real. Consider, for instance, a desired periodic control response (see Fig. 4.9), that is, $\sigma_1 = \bar{\sigma}_2 = \alpha + j\omega$, $\sigma_1\sigma_2 = |\sigma|^2$, where bar means the complex conjugate.

It can be calculated that the matching model step response and impulse response read, respectively,

$$\begin{aligned} h_m(t) &= \frac{k}{|\sigma|^2}\left[\exp(\alpha t)\left(\zeta\cos(\omega t) - \frac{\zeta\alpha - |\sigma|^2}{\omega}\sin(\omega t)\right) - \zeta\right] \\ &= 1 - \exp(\alpha t)\left(\cos(\omega t) + \frac{|\sigma|^2 - \zeta\alpha}{\zeta\omega}\sin(\omega t)\right), \\ g_m(t) &= \dot{h}_m(t) = k\exp(\alpha t)\left(\cos(\omega t) + \frac{\alpha - \zeta}{\omega}\sin(\omega t)\right). \end{aligned} \tag{4.40}$$

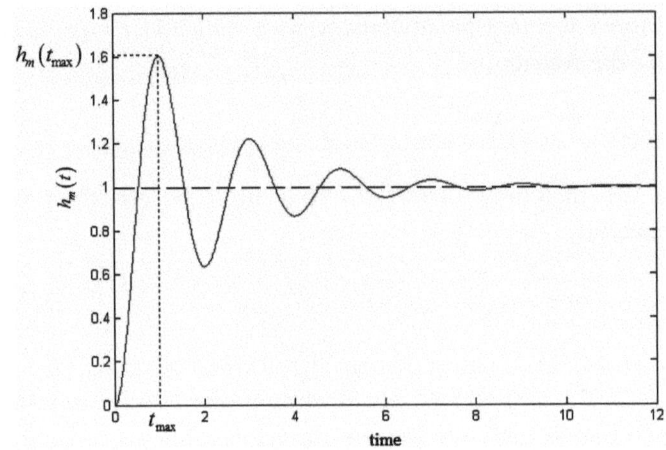

Figure 4.9 Matching model step response with the maximum overshoot and the overshoot time.

Note that conditions (4.39) are applied in (4.40). Let the model zero/pole loci be calculated from the desired *maximum overshoot* and the *overshoot time*, as indicated in Fig. 4.9. The overshoot time is defined as

$$\begin{aligned} t_{\max} &:= \arg\max_t (h_m(t) - 1) \\ &= \min\left\{t > 0 : g_m(t) = 0\right\}, \end{aligned} \tag{4.41}$$

and then the maximum overshoot is

$$\Delta h_{m,\max} := h_m(t_{\max}) - 1. \tag{4.42}$$

From (4.41) we can find the solution

$$t_{\max} = \min_{>0}\left(\frac{1}{\omega}\arccos\left(\pm\frac{|\alpha-\zeta|}{\sqrt{(\alpha-\zeta)^2+\omega^2}}\right)\right) = \min_{>0}\left(\frac{1}{\omega}\arctan\left(\frac{\omega}{\zeta-\alpha}\right)\right). \tag{4.43}$$

In addition, when the following relative damping values are introduced,

$$\xi_Z = -\frac{\zeta}{\omega},\ \xi_P = -\frac{\alpha}{\omega},\ \xi_\Delta = \frac{\zeta-\alpha}{\omega}, \tag{4.44}$$

(4.42) and (4.43) become

$$\Delta h_{m,\max} = \exp\left(-\xi_\alpha \bar{t}_{\max}\right)\left(\cos\left(\bar{t}_{\max}\right) - \frac{1}{\xi_z}\left(1+\xi_\alpha^2 - \xi_\alpha\xi_z\right)\sin\left(\bar{t}_{\max}\right)\right),$$

$$\bar{t}_{\max} := \omega t_{\max} = \min_{>0}\left(\arccos\left(\pm\frac{|\xi_\Delta|}{\sqrt{\xi_\Delta^2+1}}\right)\right) = \min_{>0}\left(\arctan\left(\frac{1}{\xi_\Delta}\right)\right), \tag{4.45}$$

respectively. Hence the computation of the desired (model) zero/pole loci starts with the selection of $\Delta h_{m,\max}$, $t_{\max}$ (or $\bar{t}_{\max}$), and one of the parameters α, ζ, ω, ξ_Z, ξ_P, ξ_Δ. Then the remaining parameters are computed either from (4.42) and (4.43) or from (4.45) by taking (4.44) into account. As these computations use mostly the inverse of a nonlinear function, we may utilize precalculated charts displaying, for example, the relations $\Delta h_{m,\max}(\xi_Z, \xi_P)$ and $\bar{t}_{\max}(\xi_Z, \xi_P)$.

4.5.3 PIP

The PIP problem [19] has to be also solved when designing ζ_i, σ_i. Namely, the matching model ought to be reset when PIP is not reached, or it cannot be further met. In some cases, we can solve it by increasing m_P. The definition of PIP and four possible scenarios for spectrum optimization follow.

Definition 4.4. The system satisfies PIP if there always exists an even number of poles (counting multiplicities) lying between any two real zeros and right to the rightmost real zero.

The given definition can be rephrased in such a way that the shifted poles can be combined to form a complex conjugate pair that can jump over the blocking real zero. If PIP does not hold, then the fixed zero blocks the shifted poles of an odd number to move over the blocking zero to the left. In such a case, additional (controlled) poles must be introduced so that there exists an even number of shifted poles (including their multiplicity) between any two consecutive blocking real zeros. Nevertheless, the increase of the shifted poles cannot be made in some cases, for instance, when there are no free degrees of freedom (i.e., a sufficient number of adjustable controller parameters).

Generally, the following four *scenarios* during the quasicontinuous shifting can appear [56]:

1. The PIP is satisfied.
2. The PIP is not satisfied, but the zero loci making PIP unsatisfied can be changed by using some (given) parameters from $\mathbf{p}_Z$. Moreover, the desired zero loci ζ_i may eventually satisfy PIP with respect to current pole loci s_i.

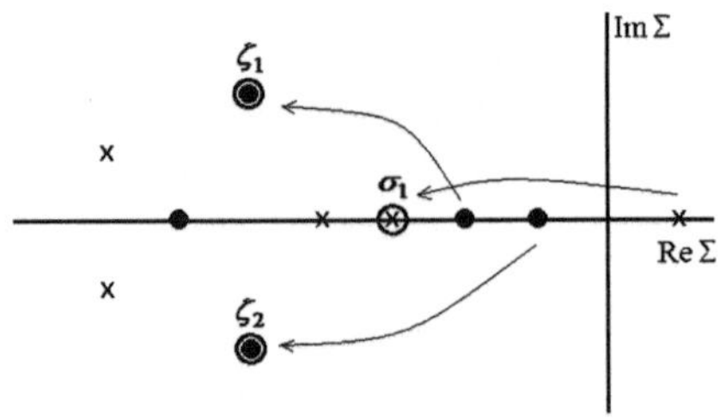

Figure 4.10 An example of scenario (2), where zero and pole loci are indicated by full circles (•) and crosses (×), respectively.

3. The PIP is not satisfied, but the zero loci making PIP unsatisfied can be changed by using some (given) parameters from $\mathbf{p}_Z$. However, the desired zero loci cannot reach PIP for the actual dominant poles.
4. The PIP is not satisfied, and unpleasant zero loci cannot be further adjusted via given parameters from $\mathbf{p}_Z$.

Whereas items (1) and (2) do not pose a problem, the last two items require reset some parameters. The following example attempts to shed light on the above-given scenarios [56].

Example. Assume that a hypothetical control system transfer function $G_{RY}(s)$ has five adjustable numerator parameters and four denominator parameters, whereas three of them are included in both the quasipolynomials, that is,

$$G_{RY}(s) = \frac{\Delta_Z(s, \mathbf{p}_Z)}{\Delta_P(s, \mathbf{p}_P)} = \frac{\Delta_Z(s, p_1, p_2, p_3, p_4)}{\Delta_P(s, p_3, p_4, p_5)}. \tag{4.46}$$

Let the rightmost subsets of the spectra be $\Sigma_Z = \{-1, -2, -6, \ldots\}$ and $\Sigma_P = \{1, -4, -7 \pm \mathrm{j}, \ldots\}$. This means that PIP is not satisfied. Select the following desired dominant root loci: $\zeta_{1,2} = -5 \pm 2\mathrm{j}$ and $\sigma_1 = -3$, see Fig. 4.10. The rightmost zero is to be shifted by using p_5, whereas the position of the dominant pair of zeros should be adjusted via $p_{1,2}$. Hence the controlled zeros can eventually satisfy PIP. This situation agrees with item (2) of the above-introduced scenarios.

4.5.4 ZPPSM algorithm

We finally provide the reader with the selected substrategy of the so-called zero-pole-placement shifting-based controller tuning method (ZPPSM), originally introduced in [56]. Once the algorithm is presented, the reader is acquainted with numerical examples, which go beyond the mentioned original research, especially, regarding the inclusion of NTDSs.

Algorithm 4.3 (ZPPSM).

Input: $G_{RY}(s)$, $\mathbf{p}_Z$, $\mathbf{p}_P$ (i.e., $r = r_Z + r_P$ are also known), objective (cost) function $\Phi(\mathbf{p})$ reflecting the distance of actual dominant root loci from the prescribed positions and spectral abscissae of the rests of both the spectra.

1. Set $n_P := r_P - 1$ or select $n_P \in \mathbb{N}$, $n_P < r_P$, as high as desired.
2. If there exists a number $n_Z \in \mathbb{N}$ satisfying

$$0 \leq n_Z < \min(n_P, r - n_P, r_Z), \tag{4.47}$$

then fix n_Z and go to step (4). Otherwise, set

$$n_P := n_P - 1. \tag{4.48}$$

3. If $r_P > n_P \geq \max\{r - n_P, r_Z\}$, go to step (2); otherwise, terminate the algorithm (a feasible and reasonable $G_{RY,m}(s)$ does not exist).
4. Choose model $G_{RY,m}(s)$ of a stable finite-dimensional system with the unit static gain, the numerator degree n_Z, the denominator degree n_P. Select its zero and pole loci, that is, ζ_i and σ_i, respectively.
5. Verify that the zeros and poles can be assigned independently, that is,

$$\begin{aligned}\operatorname{rank}\left(\left[\frac{\partial}{\partial p_{Z,j}}\Delta_Z(\zeta_i, \mathbf{p}_Z)\right]_{\substack{i=1,2,\ldots,n_Z\\ j=1,2,\ldots,r_Z}}\right) &= n_Z,\\ \operatorname{rank}\left(\left[\frac{\partial}{\partial p_{P,j}}\Delta_P(\sigma_i, \mathbf{p}_P)\right]_{\substack{i=1,2,\ldots,n_P\\ j=1,2,\ldots,r_P}}\right) &= n_P.\end{aligned} \tag{4.49}$$

If (4.49) does not hold, then return to step (4) and reset the desired model (zeros/poles selection). Otherwise, place roots into the prescribed loci by using the numbers n_Z and n_P of coefficients from the parameter sets $\mathbf{p}_Z$ and $\mathbf{p}_P$, respectively [94]. The numbers of $r_Z - n_Z$ and $r_P - n_P$ remaining coefficients are set arbitrarily.
6. Examine the PIP scenarios for the initially obtained spectra Σ_Z, Σ_P. If item (3) of the list of scenarios hold, go back to step (4) and reset σ_i and ζ_i. Else, if item (4) of the PIP scenarios is satisfied, the return to step (5) and select different parameters to perform the initial direct zeros/poles placement. Else, if item (1) or (2) of the scenarios holds, then go to step (7).

7. If all the initially set poles are dominant, then initialize the counter of currently shifted poles as

$$m_{P,mut} := n_P, \quad m_{P,opt} := 1, \quad m_P := m_{P,mut} + m_{P,opt} = n_P + 1. \tag{4.50a}$$

Otherwise, set

$$m_{P,mut} := n_P, \quad m_{P,opt} := 0, \quad m_P := m_{P,mut} + m_{P,opt} = n_P. \tag{4.50b}$$

The analogous setting is done for zeros as well, that is, set

$$m_{Z,mut} := n_Z, \quad m_{Z,opt} := 1, \quad m_Z := m_{Z,mut} + m_{Z,opt} = n_Z + 1 \tag{4.51a}$$

whenever they are dominant; else, set

$$m_{Z,mut} := n_Z, \quad m_{Z,opt} := 0, \quad m_Z := m_{Z,mut} + m_{Z,opt} = n_Z + 1. \tag{4.51b}$$

8. If

$$m_P < r_P \text{ and } m_Z < r_Z \text{ and } m_P + m_Z < r, \tag{4.52}$$

then go to step (9); else, go to (10).

9. If item (4) from the PIP scenarios holds for the fixed current m_Z, then set

$$m_{Z,opt} := m_{Z,opt} + 1 \quad \rightarrow \quad m_Z := m_Z + 1 \tag{4.53}$$

and go back to step (8).
Else, if item (3) from the set of PIP scenarios is satisfied, then go to step (12). Otherwise, go to (11).

10. If items (3) or (4) from the PIP scenarios hold, then go to step (12); otherwise, go to (11).

11. Apply the QCSA (see Section 4.4.1) for the numbers $m_{Z,mut}$ and $m_{P,mut}$ of mutual zeros and poles, respectively, toward ζ_i and σ_i by means, respectively, of $\mathbf{p}_Z$ and $\mathbf{p}_P$. Simultaneously, the numbers $m_{Z,opt}$ and $m_{P,opt}$ of less dominant zeros and poles, respectively, are forced to move left along the real axis. The incrementation of $m_{Z,opt}$ (4.53) or $m_{P,opt}$ (or both) is made during the run of the QCSA if necessary (see step (6) of Algorithm 4.1) and if (4.52) holds. After the incrementation, go to step (8). When the run of the QCSA is finished, go to step (12).

12. If all $m_{Z,mut}$ zeros and $m_{P,mut}$ poles are dominant, then the algorithm is finished. Else, minimize $\Phi(\mathbf{p})$ applying the initial settings resulting from step (11).

Output: The optimal parameter values $\mathbf{p}_{opt}$ and the corresponding spectra $\Sigma_{Z,opt}$, $\Sigma_{P,opt}$.

It is worth stressing that Algorithm 4.3 is useless for an NTDS whenever $\Delta_P(s, \mathbf{p}_P)$ incorporates $\mathbf{p}_P = \varnothing$ and (4.9) is not satisfied, that is, strong stability cannot be reached for NTDSs; see Algorithm 4.2 analogously. However, it might be reasonable to consider some roots left from $\bar{\alpha}_{P,ess}$ (estimated by $\bar{c}_{P,ess}$ (4.8)) since only part of the dominant spectrum serves to the stabilization or the minimization of α_P. Note that we can sufficiently use Proposition 4.5 to get an estimation of $\alpha_{P,ess}$.

The algorithm represents a substrategy of the ZPPSM in which zero and pole loci are simultaneously adjusted by using all tunable parameters. It is useful whenever $\mathbf{p}_P \subseteq \mathbf{p}_Z$ or $\mathbf{p}_P \supseteq \mathbf{p}_Z$. A flowchart of the main result of this chapter is displayed in Fig. 4.11.

In [45], another independently proposed technique of sequential pole placement was published, which may resemble the ZPPSM. In contrast to the herein presented algorithm, there are some essential differences. We list some of them:

1. Our algorithm is composed in the input–output space governed by meromorphic Laplace transfer functions, whereas the state space in the time domain is used in [45].
2. The ZPPSM comprises feedback poles and zeros (which yields a richer impact to the dynamics), whereas the referenced research deals solely with system poles.
3. Positions of the dominant roots are changing during the run of Algorithm 4.3. Contrariwise, poles remain in the desired loci, and the unrestrained remainder of the spectrum is moved to the left.

An interesting and difficult task is to select ζ_i and σ_i such that they are dominant for the given set $\mathbf{p}$ of adjustable parameters. If such matching roots are chosen, then Algorithm 4.3 reduces to Algorithm 4.2, that is, the minimization of both the abscissae. Although some interesting and promising results have recently been published [4,8,41,91], they are still quite far from being widely, effectively, and effortlessly used in practice.

4.5.5 Numerical examples

The following two simulation examples concisely demonstrate the use of Algorithm 4.3 to an RTDS and an NTDS.

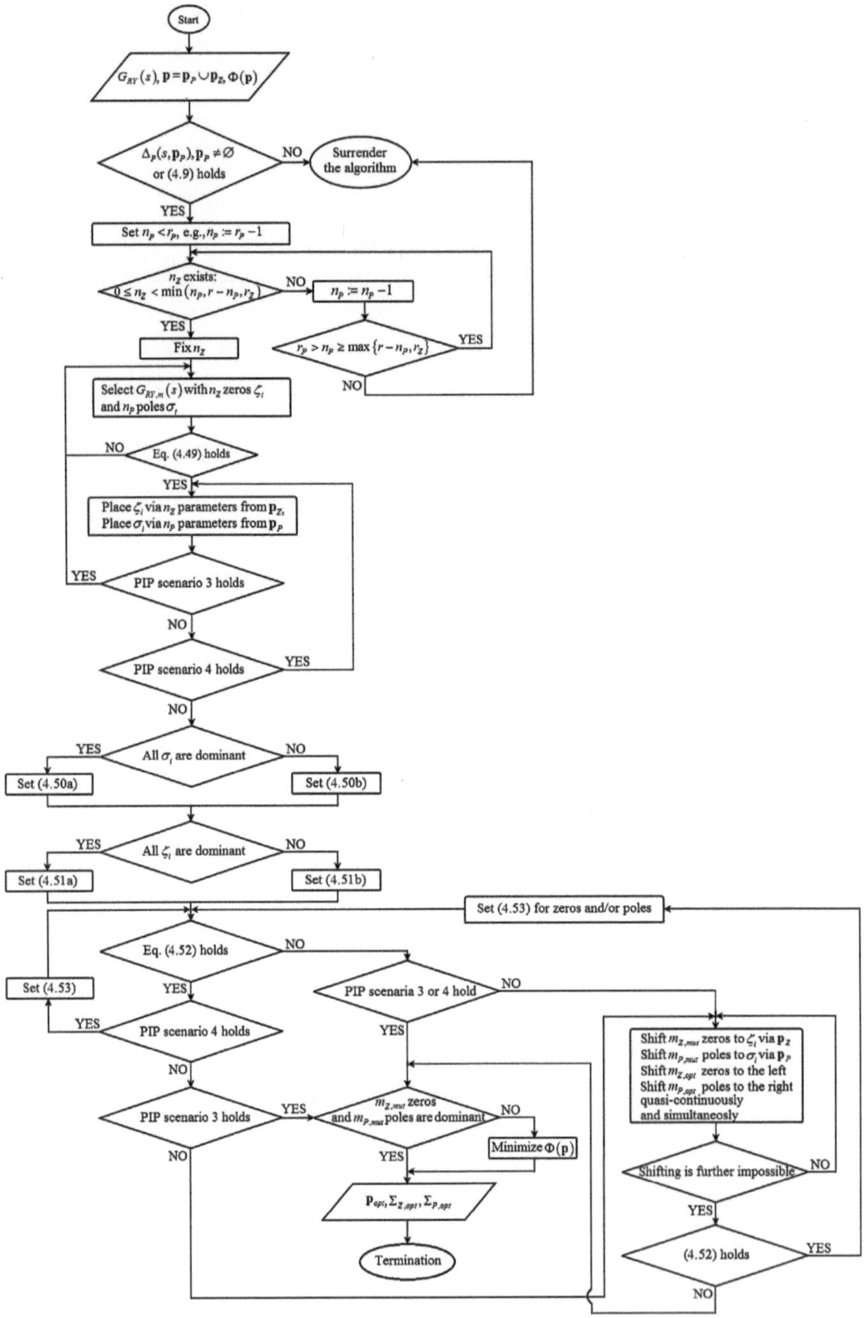

Figure 4.11 A flowchart of Algorithm 4.3.

Example 4.1. Consider model (4.25) of a roller skater on the remotely controlled swaying bow with nominal parameters as in Section 4.4.4. Let the controller designed yield the control system transfer function (4.31). Thus we have tunable parameters $\mathbf{p} = [q_3, q_2, q_1, q_0, p_2, p_1, p_0]^T \in \mathbb{R}^7$, where $\mathbf{p}_P = \mathbf{p}$, $\mathbf{p}_Z = [q_3, q_2, q_1, q_0, p_0]^T \in \mathbb{R}^5$, that is, $r = r_P = 7$, $r_Z = 5$, whereas $m \in \mathbb{R}_+$ stands for a selectable yet fixed parameter giving rise to the real root of the polynomial factor in $\Delta_P(s)$. To defuse the effect of the polynomial factor in (4.32), we have to set $m \gg \min\{\mathrm{Re}\,\sigma_i\}$, where σ_i are the selected model poles.

We construct the objective function as

$$\Phi(\mathbf{p}) = \sum_{i=1}^{n_P} |\sigma_i - s_i| + \sum_{i=1}^{n_Z} \beta_i |\zeta_i - z_i| + \lambda_1 \alpha_{P,r}(\mathbf{p}) + \lambda_2 \alpha_{Z,r}(\mathbf{p}), \qquad (4.54)$$

where $\beta_i, \lambda_1, \lambda_2 > 0$ are weight parameters, σ_i, ζ_i are selected desired pole and zero loci, respectively, and $\alpha_{P,r}(\mathbf{p})$, $\alpha_{Z,r}(\mathbf{p})$ denote, respectively, the spectral abscissae of the remainders of Σ_P, Σ_Z.

Now follow the steps of Algorithm 4.3. According to step (1), set $n_P := 2 < 7 = r_P$ initially. It can be proved that (4.47) is satisfied, that is, $0 \leq n_Z < 2$. Thus set $n_Z := 1$. It implies that the matching model is governed by the transfer function (4.38), which must satisfy (4.39) (see step (4)) of the algorithm. Consider a complex conjugated pair of model poles. In accordance with (4.41)–(4.44), do the following desired setting: $\Delta h_{m,\max} = 0.5$, $t_{\max} = 10$ s, $\xi_P = 0.5$. The graphical interpretation of formulae (4.45) for selected ranges of ξ_P, ξ_Z including this setting is displayed in Figs. 4.12 and 4.13.

We can deduce from Fig. 4.12 that $\xi_Z \approx 0.9$ for $\Delta h_{m,\max} = 0.5$, $\xi_P = 0.5$. Then Fig. 4.13 indicates that $\bar{t}_{\max} \approx 2$ for $\xi_P = 0.5$, $\xi_Z \approx 0.9$. These data imply $\sigma_{1,2} \approx -0.1 \pm 0.2\mathrm{j}$, $\zeta \approx -0.18$ from (4.44). Set $\lambda = 6$ to defuse the effect of the polynomial factor in (4.32).

According to step (5), we can verify that (4.49) holds and the initial direct zero/pole placement yields the coefficients and the spectra as follows:

$$\begin{aligned} \mathbf{p}_0 &= [1.1014, 0.9852, -0.0113, 0.0171, 1.113, 0.7, 0.2411]^T, \\ \Sigma_{P,0} &= \{0.8959, 0.1445, -0.1 \pm 0.2\mathrm{j}, -0.5201 \pm 0.5029\mathrm{j}, -1.0114, \ldots\}, \\ \Sigma_{Z,0} &= \left\{-0.1373 \pm 0.1536\mathrm{j}, -0.18, -1.0822, -2.3507 \pm 0.8423\mathrm{j}, \ldots\right\}. \end{aligned} \qquad (4.55)$$

Regarding PIP, we can see that Definition 4.4 is satisfied for the rightmost parts (4.55) of both the spectra; therefore go to step (7). Note that although

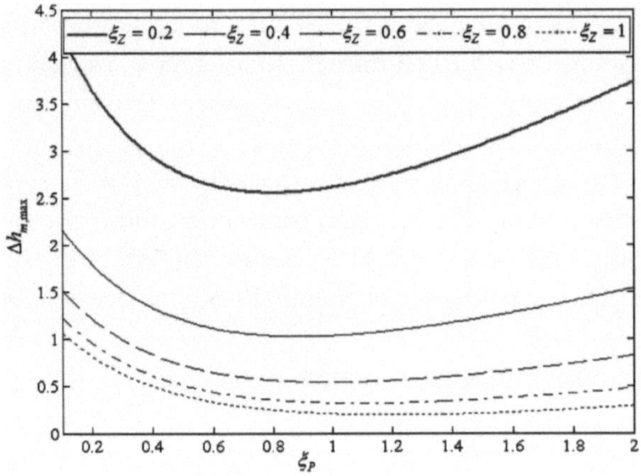

Figure 4.12 Maximum overshoot $\Delta h_{m,\max}$ for the selected range of ξ_P, ξ_Z according to (4.45).

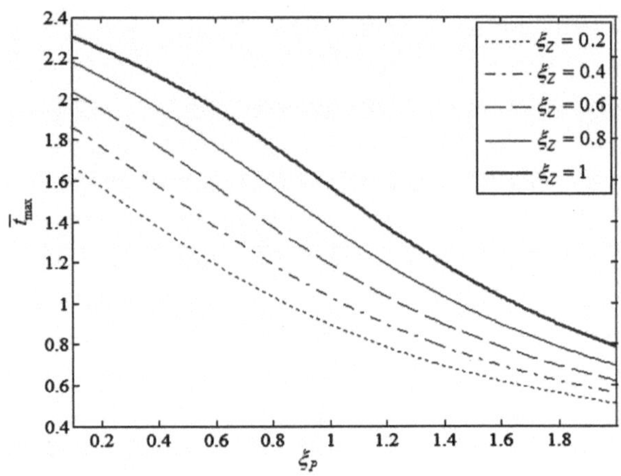

Figure 4.13 Normalized overshoot time $\bar{t}_{\max}$ for the selected range of ξ_P, ξ_Z according to (4.45).

the spectra contain the desired roots, they are not dominant. Moreover, the feedback control system is unstable. It implies that settings (4.50b) and (4.51b) are applied, that is, $m_{P,mut} = 2$, $m_{P,opt} = 0$, $m_{Z,mut} = 1$, $m_{Z,opt} = 0$. Now go to steps (9)–(11) of Algorithm 4.3 to perform the QCSA (setting (4.53) is not applied). The situation is displayed in Fig. 4.14, where only a half-plane is given for simplicity, and the desired loci are indicated by circles.

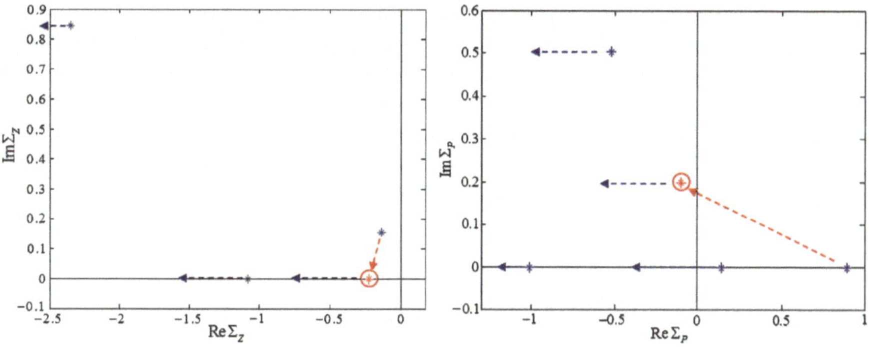

Figure 4.14 Initial zero (a) and pole (b) loci (4.55) given by the direct placement.

The run of the QCSA attempts to keep the rightmost zero and pole loci as close to the prescribed ones as possible. Simultaneously, the subdominant roots are shifted to the left (analogously to the method described in Section 4.4). It can, however, be observed that the distance $|\sigma_1 - s_1|$ is changing in cycles [56]. The eventual parameters and dominant subsets of the spectra after the number of 20636 iteration steps read

$$\begin{aligned}
\mathbf{p}_{20636} &= [5051.79, 9734.95, 1046.11, 78.957, 32.3117, 1.7838,\\
&\quad 954.866]^T,\\
\Sigma_{P,20636} &= \{-0.0945 \pm 1.1788\mathrm{j}, -0.117 \pm 5.0206\mathrm{j}, -0.1192 \pm 0.0715\mathrm{j},\\
&\quad -8.0986, \ldots\},\\
\Sigma_{Z,20636} &= \{-0.1792, -0.2191 \pm 0.1312\mathrm{j}, -0.759, -2.7833 \pm 8.2924\mathrm{j}, \ldots\}.
\end{aligned} \tag{4.56}$$

Obviously, the desired zero locus is almost reached; however, the rightmost pair of poles is still quite far from the prescribed pair given by the matching model. Hence go to step (12) of Algorithm 4.3 and let us choose the following various trial settings in (4.54):

$$\beta_1 = 0.5,\ \lambda_1 = 0.05,\ \lambda_2 = 0.05, \tag{4.57a}$$

$$\beta_1 = 0.7,\ \lambda_1 = 0.01,\ \lambda_2 = 0.01, \tag{4.57b}$$

$$\beta_1 = 0.7,\ \lambda_1 = 0.1,\ \lambda_2 = 0.02, \tag{4.57c}$$

$$\beta_1 = 1,\ \lambda_1 = 0.01,\ \lambda_2 = 0.01. \tag{4.57d}$$

Note that we have decided to optimize (4.54) with (4.57a) by the NMA with two distinct initial rectangular simplex edge lengths [47]:

$$\boldsymbol{\Delta}_1 = [500, 500, 100, 10, 10, 1, 100]^T, \tag{4.58a}$$
$$\boldsymbol{\Delta}_2 = [50, 50, 50, 50, 50, 50, 50]^T. \tag{4.58b}$$

Settings (4.57b) and (4.57c) are combined only with (4.58a), and the NMA is applied here again. Contrariwise, the SOMA [89] is used for (4.57d). The optimization control parameters are set as follows: The initial population radius $\rho = 10$, population size $N = 15$, step length $l = 3$, the discrete step $\Delta l = 0.21$, and the perturbation parameter $prt = 0.6$. The standard "all to one" strategy is used.

Setting (4.57a) suppresses the influence of zeros and makes the closeness of dominant and desired poles better than the other settings. Regarding the comparison of (4.58a) and (4.58b), the use of $\boldsymbol{\Delta}_1$ leads to a better $\alpha_{P,r}(\mathbf{p})$ suppression, whereas $\boldsymbol{\Delta}_2$ provides an expressive closeness of dominant and prescribed poles and a better improvement of $\alpha_{Z,r}(\mathbf{p})$. Compared to (4.57b)–(4.57d), the advantage of setting (4.57a) is that $\alpha_{Z,r}(\mathbf{p}) \not\approx \alpha_Z(\mathbf{p})$, which makes the dominancy of ζ stronger.

Regarding (4.57b) and (4.57c), the dominant zero is located at the almost exact desired position in both cases, yet $\alpha_{Z,r}(\mathbf{p}) \approx \operatorname{Re}\zeta$, that is, the dominancy is "weak". Moreover, the pair of eventual dominant poles is quite far from the prescribed locus, where the distance is more or less the same for both settings, whereas higher values of weights λ_1, λ_2 result in a better abscissa declination.

The use of the SOMA for the combination of (4.57d) with (4.58b) gives an improvement of $\alpha_{Z,r}(\mathbf{p})$ compared to the preceding setting. Contrariwise, we get $\alpha_{P,r}(\mathbf{p}) \approx \operatorname{Re}\sigma$, and the desired pole position is still far from the dominant pair.

Overall, unfortunately, none of the obtained results is satisfactory for control tasks. Control responses suffer from much higher overshoots compared to that given by the matching model. It is caused mainly by the relatively long distances of the eventual dominant and the desired poles loci. This result demonstrates that it is not possible to match an arbitrary desired model for LTI TDSs.

The combination of settings (4.57c) with (4.58a) for the NMA yields

$$\begin{aligned}
\mathbf{p}_{opt} &= [6122.69, 9116.67, 1235.11, 116.0975, 32.4834, -0.428,\\
&\quad 829.829]^T,\\
\Sigma_{P,opt} &= \{-0.1501 \pm 0.0933\mathrm{j}, -0.2848 \pm 4.163\mathrm{j}, -0.5436 \pm 1.4021\mathrm{j}\},\\
\Sigma_{Z,opt} &= \{-0.18, -0.18 \pm 0.2126\mathrm{j}, -0.9267, -3.024 \pm 7.5097\mathrm{j}\},
\end{aligned} \tag{4.59}$$

which gives the best integral absolute error criterion. For instance, the use of SOMA results in

$$\begin{aligned}\mathbf{p}_{opt} &= [5235.169, 9829.219, 1060.87, 78.2405, 30.9684, 1.763,\\ &\quad 947.517]^T,\\ \Sigma_{P,opt} &= \{-0.1158 \pm 0.0674\mathrm{j}, -0.1161 \pm 5.1163\mathrm{j}, -0.1211 \pm 1.2103\mathrm{j}\},\\ \Sigma_{Z,opt} &= \{-0.1801, -0.2247 \pm 0.1032\mathrm{j}, -0.7607, -2.817 \pm 8.1939\mathrm{j}\}.\end{aligned} \tag{4.60}$$

Finally, let us demonstrate the appropriateness of the prior utilization of the QCSA as in step (11) of Algorithm 4.3 rather than the direct deployment of an advanced iterative optimization algorithm. If this step is omitted when using the NMA for the setting (4.57c) and (4.58a), then the following spectra are obtained:

$$\begin{aligned}\Sigma_{P,opt} &= \{0.6551 \pm 0.0668\mathrm{j}, -0.0896 \pm 0.0586\mathrm{j}, -0.5547 \pm 0.448\mathrm{j},\\ &\quad -1.0762, \ldots\},\\ \Sigma_{Z,opt} &= \{-0.173, -0.173 \pm 0.0687\mathrm{j}, -1.0267, -3.223 \pm 6.9096\mathrm{j}, \ldots\}.\end{aligned} \tag{4.61}$$

Apparently, solution (4.61) converging to a local minimum gives an unstable feedback system, which is definitely unacceptable.

Example 4.2. This example is aimed at a demonstration of the use of an ad hoc modification of Algorithm 4.3 to a neutral control feedback system [52]. Special circumstances of NTDSs are stressed.

Let a hypothetical controlled plant be governed by the transfer function

$$G(s) = \frac{s+4}{s+1-2\exp(-0.4s)}\exp(-s). \tag{4.62}$$

Consider the control system with inner and outer controllers governed by transfer functions $G_i(s)$, $G_o(s)$, respectively; see Fig. 4.15.

Suppose that the following setting is made:

$$G_i(s) = \frac{q_2 s^2 + q_1 s + q_0}{s^2 + p_1 s + p_0}, \quad G_o(s) = \frac{r_2 s^2 + r_1 s + q_0 - p_0/4}{s^2 + p_1 s + p_0}, \tag{4.63}$$

where $\mathbf{p}_Z = [p_0, q_0, r_2, r_1]^T$, $\mathbf{p}_P = [p_1, p_0, q_1, q_0, q_2]^T$, and $\mathbf{p} = [p_1, p_0, q_1, q_0, q_2, r_1, r_2]^T$. It can be verified that a stepwise reference signal is asymptotically tracked and a disturbance variable from the same family of functions is

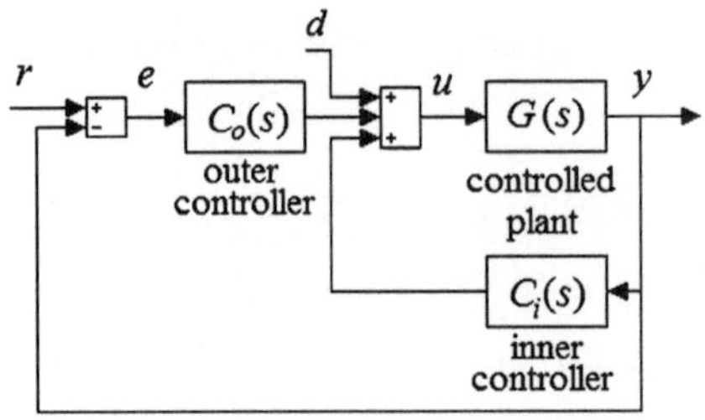

Figure 4.15 Control feedback loop with inner and outer controllers.

asymptotically rejected. Note that (4.63) can also be obtained via the algebraic approach in the R_{MQ} ring. The reference-to-output transfer function with (4.62) and (4.63) reads

$$G_{RY}(s) = \frac{(s+4)\left(r_2 s^2 + r_1 s + q_0 - p_0/4\right)\exp(-s)}{(s+4)\exp(-s)\left(q_2 s^2 + q_1 s + q_0\right) + (s+1-2\exp(-0.4s))\left(s^2 + p_1 s + p_0\right)}. \tag{4.64}$$

Notice that although $G(s)$ is of a retarded type, $G_{RY}(s)$ is an NTDS due to the nonstrict feasibility of $G(s)$. The controlled plant is unstable with $\alpha_P = 0.5836$.

The associated exponential polynomial has the simple form

$$d_n(s) = 1 + q_2 \exp(-s). \tag{4.65}$$

According to Proposition 4.4, the control system is strongly stable if and only if

$$\left|q_2\right| < 1. \tag{4.66}$$

By Proposition 4.5, $\tilde{d}_n(s) = d_n(s)$ and $\tilde{d}_n(q) = 1 + q_2 q$, where $q = \exp(-s)$. Then the essential spectrum is

$$s_i = -\ln q_i \pm \mathrm{j}2k\pi,\ k \in \mathbb{N},\ q_i = -\frac{1}{q_2}; \tag{4.67}$$

see (4.21) and (4.24) with the spectral abscissa

$$\alpha_{P,ess} = -\ln\left|\frac{1}{q_2}\right| \tag{4.68}$$

according to (4.23). Therefore for both, the exponential and strong stability of (4.64), (4.66) must hold, which must be checked during the poles shifting and the optimization of the spectral abscissae; see step (2) of Algorithm 4.2.

Let us now select the identical matching model as in the preceding example, that is, $\sigma_{1,2} \approx -0.1 \pm 0.2\mathrm{j}$, $\zeta \approx -0.18$ for (4.38). It is worth noting that the vertical strip of neutral poles approaching $\Sigma_{P,ess}$ is located left from the prescribed pair $\sigma_{1,2}$ if

$$-\ln\left|\frac{1}{q_2}\right| < -0.1 \Rightarrow |q_2| < 0.948. \tag{4.69}$$

Since the numerator Δ_Z of (4.64) is a polynomial, its roots (i.e., feedback zeros) can be placed directly and exactly via the parameters r_1, r_2 not included in $\mathbf{p}_P$. Let the second root z_2 of Δ_Z be far left from $\zeta = -0.18$, for instance, $z_2 = -5$ [52]. Hence the only adjustable parameters are $\mathbf{p}_P$; therefore the problem is reduced to (solely) pole placement.

The initial direct pole assignment gives

$$\begin{aligned}\mathbf{p}_{P,0} &= [0.821, 0.4329, -0.0722, -0.0445, 0.7664]^T,\\ \Sigma_{P,0} &= \{0.539 \pm 2.8461\mathrm{j}, -0.1 \pm 0.2\mathrm{j}, -0.2186 \pm 21.8116\mathrm{j},\\ &\quad -0.2409 \pm 15.5849\mathrm{j}, -0.2418 \pm 9.0022\mathrm{j}, \ldots\}\end{aligned} \tag{4.70}$$

which is an interesting result. Namely, the control system is strongly stable due to (4.66); however, it does not meet exponential stability since $\alpha_P(\mathbf{p}_{P,0}) = 0.539 \geq 0$. In addition, the pair of poles located at prescribed positions is not dominant. The QCSA is applied by setting $m_{P,mut} = 2$, $m_{P,opt} = 0$. Eventually, two rightmost poles are shifted toward the desired complex pair, and three other subdominant poles are pushed to the left. After 19400 iterations, we get

$$\begin{aligned}\mathbf{p}_{P,19400} &= [1.2097, 0.5984, -0.043, 0.186, -0.013]^T,\\ \Sigma_{P,19400} &= \{-0.0999 \pm 0.8432\mathrm{j}, -4.0358 \pm 2.3484\mathrm{j}, -4.0392 \pm 18.5881\mathrm{j},\\ &\quad -4.3022 \pm 2.6475\mathrm{j}, -4.4912 \pm 24.7169\mathrm{j}, \ldots\}\end{aligned} \tag{4.71}$$

The system is both exponentially and strongly stable with $\alpha_{P,ess}(\mathbf{p}_{P,19400}) = -4.343$; see (4.68).

Let us set the objective function to be minimized simply as

$$\Phi(\mathbf{p}_P) = |\sigma_1 - s_1| + 0.05\alpha_{P,r}(\mathbf{p}_P). \tag{4.72}$$

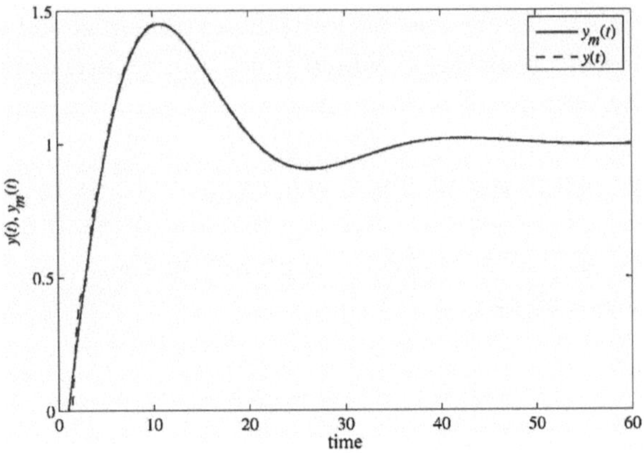

Figure 4.16 Matching model step response $y_r(t)$ versus the eventual control system step response $y(t)$.

The use of the NMA (after 158 iteration steps) yields

$$\begin{aligned}\mathbf{p}_{P,opt} &= [3.4051, 1.302, -0.9145, -0.2446, 0.1729]^T,\\ \Sigma_{P,opt} &= \{-0.1 \pm 0.2j, -0.5884 \pm 3.8128j, -1.6076 \pm 16.082j,\\ &\quad -1.6302 \pm 22.1068j, -1.6509 \pm 9.6930j, \ldots\}\end{aligned} \tag{4.73}$$

which is a desirable result. The direct zero placement using (4.73) gives $[r_1, r_2] = [-3.2815, -0.6335]$. Thus the eventual controller parameters are

$$\mathbf{p} = [3.4051, 1.302, -0.9145, -0.2446, 0.1729, -0.6335, -3.2815]^T. \tag{4.74}$$

Figs. 4.16 and 4.17 display the graphical comparison of controlled output step responses $y_m(t)$, $y(t)$ of the matching model and of the control system with controllers (4.63) tuned by (4.74), respectively [52]. The latter figure provides the reader with a more detailed view on the responses near the origin due to a very good agreement between responses displayed in the former figure. The error is caused by the existence of an infinite number of higher-frequency poles left from the desired dominant pair.

4.6. Future research possibilities

In this closing section, we discuss some possible future research directions and potential open problems. First, we is highlight the task of selection

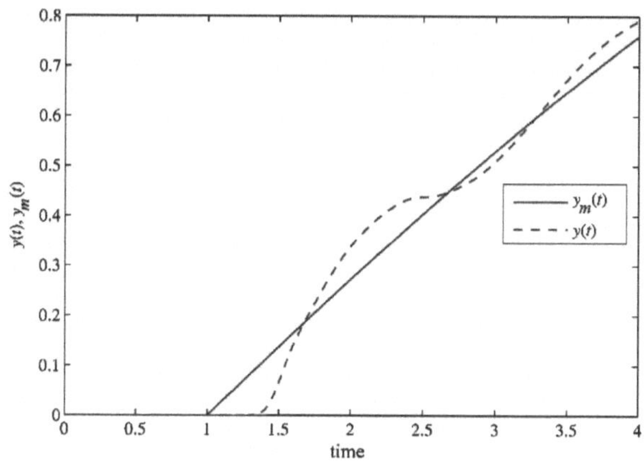

Figure 4.17 A detailed view of Fig. 4.16.

of the objective function $\Phi(\mathbf{p})$, which plays an important role in the attempt to improve the ZPPSM. Examples presented in Section 4.5.5 clearly indicate how this task can be tricky, especially, when $\alpha_{\cdot,r}(\mathbf{p})$ has a negative value.

Hence let us concentrate on the suppression of $\alpha_{\cdot,r}(\mathbf{p})$ and on strong stability, that is, when (4.9) holds after the minimization is finished.

Introduce

$$\Phi(\mathbf{p}) = \sum_{i=1}^{np} |\sigma_i - s_i| + \sum_{i=1}^{nz} \beta_i |\zeta_i - z_i| + \Pi(\mathbf{p}) + \Gamma(\mathbf{p}) \tag{4.75}$$

instead of (4.54), where $\Pi(\mathbf{p})$ and $\Gamma(\mathbf{p})$ are the *penalty* and *barrier* functions, respectively. Consider the conditions

$$\Psi(\mathbf{p}) \leq \mathbf{0} \tag{4.76}$$

on $\mathbf{p}$ to be satisfied. The penalty function behaves as

$$\Pi(\mathbf{p}) \begin{cases} = 0 & \text{if (4.76) holds,} \\ > 0 & \text{else.} \end{cases} \tag{4.77}$$

We have the following statement for penalty functions [20].

Proposition 4.6. *Consider the constrained optimization problem*

$$\min f(\mathbf{p}) \in \mathbb{R}, \ \mathbf{p} \in \mathbb{R}^n$$

$$
\begin{aligned}
&s.t.\\
&g_i(\mathbf{p}) \le 0,\ i = 1, \ldots, n_g,\\
&h_j(\mathbf{p}) = 0,\ j = 1, \ldots, n_h,
\end{aligned}
\tag{4.78}
$$

where $f(\mathbf{p})$, $g_i(\mathbf{p})$, *and* $h_j(\mathbf{p})$ *are continuous functions, and* $f(\mathbf{p})$ *is bounded from below.*

Let $\{\lambda_k\}_{k=1}^{\infty}$ *be an increasing sequence with* $\lim_{k\to\infty}\lambda_k = \infty$, $\lambda_k \in \mathbb{R}_+$, *and define*

$$
\begin{aligned}
&\Phi(\mathbf{p}) := f(\mathbf{p}) + \lambda_k \left(\sum_{i=1}^{n_g} \left(\max\left(0, g_i(\mathbf{p})\right)\right)^l + \sum_{j=1}^{n_h} \left|h_l(\mathbf{p})\right|^l \right),\\
&l \in \mathbb{N} \ge 1.
\end{aligned}
\tag{4.79}
$$

Then

$$
\begin{aligned}
&\lim_{k\to\infty} \mathbf{p}_{\min,k} = \mathbf{p}_{\min},\\
&\lim_{k\to\infty} \Phi\left(\mathbf{p}_{\min,k}, \alpha_k\right) = f\left(\mathbf{p}_{\min}\right),\ \text{and}\\
&\lim_{k\to\infty} \left(\sum_{i=1}^{n_g} \left(\max\left(0, g_i(\mathbf{p}_{\min,k})\right)\right)^l + \sum_{j=1}^{n_h} \left|h_j(\mathbf{p}_{\min,k})\right|^l \right) = 0
\end{aligned}
\tag{4.80}
$$

for the sequence of solutions

$$
\left\{\mathbf{p}_{\min,k}\right\}_{k=1}^{\infty} := \left\{ \arg\min_{\mathbf{p}} \Phi\left(\mathbf{p}, \lambda_k\right) \right\}_{k=1}^{\infty},
\tag{4.81}
$$

where $\mathbf{p}_{\min}$ *is the solution of problem* (4.78).

For the barrier function, we have

$$
\Gamma(\mathbf{p}) \begin{cases} < \infty & \text{if (4.76) holds,} \\ \to \infty & \text{else.} \end{cases}
\tag{4.82}
$$

Penalty functions can be used to set the desired values for particular spectral abscissae. Let

$$
\overline{\alpha_r} < \max_k \operatorname{Re} \sigma_k
\tag{4.83}
$$

be the desired eventual spectral abscissa of the rest of the particular spectra excluding the set of desired root (zero/pole) loci σ_k, $k = 1, 2, \ldots$. In the light of Proposition 4.6, we can set

$$\Pi(\mathbf{p}) = \lambda_1 \max\left(0, \alpha_{P,r}(\mathbf{p}) - \overline{\alpha_{P,r}}\right)^l + \lambda_2 \max\left(0, \alpha_{Z,r}(\mathbf{p}) - \overline{\alpha_{Z,r}}\right)^l, \quad (4.84)$$

where λ_1, λ_2 are either entries of increasing sequences $\{\lambda_{i,k}\}_{k=1}^{\infty}$, $i = 1, 2$, or at least sufficiently large real numbers. Then

$$\alpha_{\cdot,r}(\mathbf{p}) - \overline{\alpha_{\cdot,r}} \to 0 \Leftrightarrow \alpha_{\cdot,r}(\mathbf{p}) \to \overline{\alpha_{\cdot,r}} \quad (4.85)$$

as $\lambda_i \to \infty$; see (4.80). Note that (4.84) can also be used to Algorithm 4.2 where $\Phi(\mathbf{p}) = \Pi(\mathbf{p})$ and only poles are considered.

The concept of penalty functions can be utilized to meet strong stability for NTDSs as well. If, for instance

$$\Pi(\mathbf{p}) = \lambda \max\left(0, 1 - \varepsilon - \sum_{j=1}^{h_n} \left|d_{nj}(p)\right|\right)^2, \quad 0 < \varepsilon \ll 1, \quad (4.86)$$

then

$$\lim_{\lambda\to\infty}\left(1 - \varepsilon - \sum_{j=1}^{h_n} \left|d_{nj}(\mathbf{p})\right|\right) = 0 \Leftrightarrow \sum_{j=1}^{h_n} \left|d_{nj}(\mathbf{p})\right| \to 1 - \varepsilon. \quad (4.87)$$

Both penalty functions (4.84) and (4.86) can be included in (4.76).

Regarding barrier functions, they can serve to maintain strong stability. Whenever the optimization procedure (see step (12) of Algorithm 4.3) starts with a strongly stable parameter setting, the system remains strongly stable if

$$\Gamma(\mathbf{p}) = -\log\left(1 - \varepsilon - \sum_{j=1}^{h_n} \left|a_{nj}(\mathbf{p})\right|\right). \quad (4.88)$$

In other words, whenever the value of $\sum_{j=1}^{h_n} \left|a_{nj}(\mathbf{p})\right|$ approaches $1 - \varepsilon$ from below, the value of $\Gamma(\mathbf{p})$ becomes large, which means that the optimization procedure forces to decrease it.

The selection of a suitable minimization method represents another substantial question. There exist plenty of possible modern user-friendly metaheuristic algorithms with promising applicability to engineering problems [34]. To name just a few, the genetic algorithm [85] and the differential evolution [15] as representatives of evolutionary optimization methods [90], the particle swarm optimization algorithm [70] working with a population of solution candidates that follow leaders (best current solutions), the bee algorithm imitating behavior of honey bee colonies [64], the ant colony optimization algorithm [10] inspired by actions in an ant colony and position-following pheromone-based communication of ant, the cuckoo search approach [11] simulating the holoparasitism of some cuckoo species and behavior of host birds, or elephant herding optimization algorithm based on the clan updating and separation behavior of elephants [36] as representatives of a wide family of multiagent, swarm-intelligence, and swarm-optimization methods inspired by nature [63,87] ought to be mentioned. As introduced in Proposition 4.2, the spectral abscissa, however, can be non-Lipschitz, which disqualifies many habitual gradient-based methods.

However, metaheuristic algorithms do not guarantee that the globally optimal solution is found [2]. Therefore a sufficiently accurate initial solution estimation has to be set. A quasicontinuous shifting procedure might be a good choice. To adjust the control parameters of a particular optimization method that is used to the given cost function, we may utilize a metaoptimization tuning procedure [28,67]. Finding suitable behavioral parameters for an optimizer via a metaoptimizer can lead to getting over a suboptimal or locally optimal solution. A run of the algorithm can also be improved by its learning, for example, via deep neural networks [71] or based on the convergence velocity [36].

For this reason, this unpleasant feature opens the question of a fast and accurate spectrum computation and the demarcation of a suitable region in the complex plane. This is the most time-consumptive computational operation of the presented version of the ZPPSM. Whereas the position of the rightmost pair of roots of neutral NTDSs can be estimated by (4.8) or (4.23), there is almost no suggestion for retarded systems. Nevertheless, some scholars have attempted to overcome this problem. For instance, the updated (second) version of the QPmR [82] utilizes asymptotic properties of the chains of quasipolynomial zeros, the boundary of which can also be determined for further use of argument principle methods [8].

By considering other computational aspects special attention must be paid to particular shifting strategy techniques for the ZPPSM. In the current version the rightmost actual roots approach the desired loci regardless of the type of root. A multiple root is considered as a nest of single roots. For instance, if the aim is shifting a triple root toward a single desired locus, it is considered as a triplet of distinct single roots. Moreover, whenever the task is shifting a real root to a complex conjugate pair, the desired pair is considered as a single point of attraction. However, it might not be a problem, because the type of roots eventually coincides when shifting. These two aspects can be revised in the next versions of the algorithm to improve its convergence and speed.

The speed of an algorithm can also be enhanced by the utilization of modern advanced hardware and software tools, such as distributed computations on graphical cards, for example, by using of the open computing language (OpenCL) or the compute unified device architecture (CUDA). However, such techniques usable to cope with many engineering problems [88] require a deeper analysis of the ZPPSM to get an effective and well-applicable programming code.

The selection of a suitable matching model constitutes another problem of Algorithm 4.3. The presented idea of the desired maximal step response overshoot, the overshoot time, and the relative dumping is only one of possibilities to cope with that. In addition, there are different views of the zero/pole dominancy; for instance, the distance of a root from the complex plane origin can be taken as the measure rather than the position of a root in the real axis, or the idea of the "weights" of the impulse response modes [69] can be adopted.

Last but not least, another possible open task is to consider the *pseudospectral* abscissa [5,43] or the *smoothed spectral* abscissa [24,79] instead of the spectral abscissa (see Proposition 4.2). Let us concisely introduced these concepts.

Consider system (4.1) governed by (4.3) with the notation introduced therein. Let $\mathbf{H}(s) = \mathbf{0}$ for simplicity.

Definition 4.5. The pseudospectral abscissa is defined as

$$\alpha_{P,\varepsilon} := \sup\left\{\operatorname{Re}(s) : s \in \Sigma_{P,\varepsilon}\right\}, \tag{4.89}$$

where

$$\begin{aligned}\Sigma_{P,\varepsilon} &:= \left\{s \in \mathbb{C} : \det(s\mathbf{I} - (\mathbf{A}(s) + \mathbf{X})) = 0, \text{ for some } \mathbf{X} \in \mathbb{C}^{n \times n} \text{with } \|\mathbf{X}\|_2 \leq \varepsilon\right\} \\ &= \left\{s \in \mathbb{C} : \left\|\frac{1}{s\mathbf{I} - \mathbf{A}(s)}\right\|_2 \geq \frac{1}{\varepsilon}\right\}\end{aligned} \tag{4.90}$$

is the so-called pseudospectrum.

The value of $\alpha_{P,\varepsilon}$ represents the worst-case stability measure of the system under uncertainties bounded by ε. It can also be written [5] as

$$\begin{aligned}\alpha_{P,\varepsilon} &:= \sup\left\{\lambda \in \mathbb{R} : \left\| \frac{1}{(\lambda + j\omega)\mathbf{I} - \mathbf{A}(s)} \right\|_2 \geq \frac{1}{\varepsilon} \text{ for some } \omega \in \mathbb{R}\right\} \\ &= \sup\left\{\lambda \in \mathbb{R} : \left\| \frac{1}{s\mathbf{I} - (\mathbf{A}(s) - \lambda\mathbf{I})} \right\|_\infty \geq \frac{1}{\varepsilon}\right\}. \end{aligned} \quad (4.91)$$

The task of optimization of $\alpha_{P,\varepsilon}(\mathbf{p})$ is a trade-off between the minimization of $\alpha_P(\mathbf{p})$ and the H_∞ norm in (4.91). It is worth noting that $\alpha_{P,\varepsilon}(\mathbf{p})$ is a nonsmooth function with respect to $\mathbf{p}$. On the contrary, the smoothed spectral abscissa is a smooth function of $\mathbf{p}$ [79].

Definition 4.6. The smoothed spectral abscissa is defined as

$$\begin{aligned}\tilde{\alpha}_{P,\varepsilon} &:= \sup\left\{\lambda \in \mathbb{R} : \left\| \frac{1}{s\mathbf{I} - (\mathbf{A}(s) - \lambda\mathbf{I})} \right\|_2 \geq \frac{1}{\varepsilon}\right\} \\ &= \lambda \in \mathbb{R} : \left\| \frac{1}{s\mathbf{I} - (\mathbf{A}(s) - \lambda\mathbf{I})} \right\|_\infty = \frac{1}{\varepsilon}. \end{aligned} \quad (4.92)$$

Analogously to $\alpha_{P,\varepsilon}(\mathbf{p})$, the smoothed abscissa provides a compromise between the minimization of $\alpha_P(\mathbf{p})$ and the H_2 norm included in (4.92).

Both concepts have been used to design robust controllers [24,26]; however, there is still a research gap in their combination with pole and pole placement.

4.7. Conclusions

It is always difficult to tackle optimal or suboptimal control design and tuning for time-delay systems, even linear and time-invariant ones. Overview and summary of the spectral abscissa optimization and a model-matching-based suboptimal zero/pole placement are presented in this chapter. A list of key findings follows:

1. The controller design approach for input/output TDS models in a special ring of stable quasipolynomial meromorphic function is given. The design is applicable to linear systems with a wide class of delay types.
2. We have recalled the known quasicontinuous shifting procedure applied to the spectral abscissa minimization and the desired-root-loci placement. The main novelty of the latter algorithm lies in the simultaneous assignment of zeros and poles.

3. Real zeros can block poles when shifting to the left in the complex plane in some cases; therefore, a verification of the so-called parity interlacing property is included as well.
4. As root loci are adjusted to satisfy the desired loci, a time-domain response analysis of a finite-dimensional matching model is presented.
5. Specific issues of rather tricky neutral-type delays are concisely discussed.
6. The reader is also acquainted with numerical (simulation) examples to clarify the presented ideas and algorithms. However, these examples prove that the matching model (i.e., the desired root loci) cannot be chosen arbitrarily.
7. Finally, the reader is provided with some possible research directions.

Acknowledgments

The author would like to express his gratitude to the European Regional Development Fund and to the Ministry of Education, Youth and Sports for financial support of this work under the National Sustainability Programme project No. LO1303 (MSMT-7778/2014).

References

[1] J.M. Araújo, T.L.M. Santos, Control of a class of second-order linear vibrating systems with time-delay: Smith predictor approach, Mechanical Systems and Signal Processing 108 (2018) 173–187.
[2] C. Blum, A. Roli, Metaheuristics in combinatorial optimization: overview and conceptual comparison, ACM Computing Surveys 35 (3) (2003) 268–308.
[3] C. Bonnet, A.R. Fioravanti, J.R. Partington, Stability of neutral systems with commensurate delays and poles asymptotic to the imaginary axis, SIAM Journal on Control and Optimization 49 (2) (2011) 498–516.
[4] I. Boussaada, S.I. Niculescu, A. El Ati, R. Perez-Ramos, K.L. Trabelsi, Multiplicity-induced dominancy in parametric second-order delay differential equations: analysis and application in control design, ESAIM. Control, Optimisation and Calculus of Variations 26 (2020) 57.
[5] J.V. Burke, A.S. Lewis, M.L. Overton, Optimization and pseudospectra, with applications to robust stability, SIAM Journal on Mathematical Analysis and Applications 25 (1) (2003) 80–104.
[6] J.V. Burke, A.S. Lewis, M.L. Overton, A robust gradient sampling algorithm for nonsmooth, nonconvex optimization, SIAM Journal on Optimization 15 (3) (2005) 751–779.
[7] C.I. Byrnes, M.W. Spong, T.J. Tarn, A several complex variables approach to feedback stabilization of neutral delay-differential systems, Mathematical Systems Theory 17 (1) (1984) 97–133.
[8] T. Cai, H. Zhang, Novel properties of spectrum for stability test of linear delay differential systems, IEEE Access 7 (2019) 116424–116429.
[9] A. Callendar, D.R. Hartee, A. Porter, Time-lag in a control system, Philosophical Transactions of the Royal Society A: Mathematical, Physical and Engineering Sciences 235 (756) (1936) 415–444.

[10] B. Chandra Mohan, R. Baskaran, A survey: ant colony optimization based recent research and implementation on several engineering domain, Expert Systems with Applications 39 (4) (2012) 4618–4627.
[11] H. Chiroma, T. Herawan, I. Fister, S. Abdulkareem, L. Shuib, M.F. Hamza, Y. Saadi, A. Abubakar, Bio-inspired computation: recent development on the modifications of the cuckoo search algorithm, Applied Soft Computing 61 (2017).
[12] R.F. Curtain, K. Morris, Transfer functions of distributed parameters systems, Automatica 45 (5) (2009) 1101–1116.
[13] R.F. Curtain, H. Zwart, An Introduction to Infinite-Dimensional Linear Systems Theory, Springer-Verlag, New York, 1995.
[14] S. Das, K. Halder, A. Gupta, Delay handling method in dominant pole placement based PID controller design, IEEE Transactions on Industrial Informatics 16 (2) (2020) 980–991.
[15] S. Das, S.S. Mullick, P.N. Suganthan, Recent advances in differential evolution – an updated survey, Swarm and Evolutionary Computation 27 (2016) 1–30.
[16] C.A. Desoer, R.W. Liu, J. Murray, R. Seaks, Feedback system design: the fractional representation approach to analysis and synthesis, IEEE Transactions on Automatic Control 25 (3) (1980) 399–412.
[17] S. Elmadssia, K. Saadaoui, M. Benrejeb, PI controller design for time delay systems using an extension of the Hermite–Biehler theorem, Journal of Industrial Engineering 2013 (2013) 813037.
[18] K. Engelborghs, T. Luzyanina, G. Samaey, DDE-BIFTOOL V. 2.00: a Matlab package for bifurcation analysis of delay differential equations, Report TW 330, K.U. Leuven, Belgium, 2001.
[19] H.E. Erol, A. İftar, Decentralized controller design by continuous pole placement for incommensurate-time-delay systems, IFAC-PapersOnLine 48 (12) (2015) 462–467.
[20] R. Fletcher, Practical Methods of Optimization, Wiley, New York, 1987.
[21] C. Foias, Hitay Özbay, A. Tannenbaum, M. Thoma, Robust Control of Infinite Dimensional Systems: Frequency Domain Methods, Springer, 1996.
[22] P. García, P. Albertos, A new dead-time compensator to control stable and integrating processes with long dead-time, Automatica 44 (4) (2008) 1062–1071.
[23] R. Gerov, Z. Jovanović, G. Stojanović, Stability analysis of linear, time-invariant systems with time delay, in: 2019 14th International Conference on Advanced Technologies, Systems and Services in Telecommunications (TELSIKS), Nis, Serbia, 2019, pp. 404–407.
[24] M.A. Gomez, W. Michiels, Characterization and optimization of the smoothed spectral abscissa for time-delay systems, International Journal of Robust and Nonlinear Control 29 (13) (2019) 4402–4418.
[25] K. Gu, V.L. Kharitonov, J. Chen, Stability of Time-Delay Systems, Birkhäuser, Basel, Switzerland, 2003.
[26] S. Gumussoy, W. Michiels, A predictor-corrector type algorithm for the pseudospectral abscissa computation of time-delay systems, Automatica 46 (2010) 657–664.
[27] J.K. Hale, S.V. Lunel, Introduction to Functional Differential Equations, Springer, New York, 1993.
[28] C. Huang, Y. Li, X. Yao, A survey of automatic parameter tuning methods for metaheuristics, IEEE Transactions on Evolutionary Computation 24 (2) (2020) 201–216.
[29] M. Huba, D. Vrancíć, T. Huba, Evaluating performance limits in FOTD plant control, IFAC-PapersOnLine 49 (6) (2016) 218–225.
[30] E.W. Kamen, On the algebraic theory of systems defined by convolution operations, Mathematical Systems Theory 9 (1) (1975) 57–74.
[31] A.S. Kammer, N. Olgac, Non-conservative stability assessment of LTI dynamics with distributed delay using CTCR paradigm, in: 2015 American Control Conference, Palmer House Hilton, Chicago, IL, USA, 2015, pp. 4597–4602.

[32] S.S. Kandala, S. Chakraborty, T.K. Uchida, C.P. Vyasarayani, Hybrid method-of-receptances and optimization-based technique for pole placement in time-delayed systems, International Journal of Dynamics and Control 49 (2019).
[33] V. Kučera, Diophantine equations in control – a survey, Automatica 29 (6) (1993) 1361–1375.
[34] K.Y. Lee, Z. Vale, Applications of Modern Heuristic Optimization Methods in Power and Energy Systems, IEEE Press, 2020.
[35] A.S. Lewis, M.L. Overton, Nonsmooth optimization via quasi-Newton methods, Mathematical Programming 141 (1) (2013) 135–163.
[36] W. Li, G.-G. Wang, A.H. Alavi, Learning-based elephant herding optimization algorithm for solving numerical optimization problems, Knowledge-Based Systems 195 (2020) 105675.
[37] J. Liu, H. Wang, Y. Zhang, New result on PID controller design of LTI systems via dominant eigenvalue assignment, Automatica 62 (2015) 93–97.
[38] J.J. Loiseau, M. Cardelli, X. Dusser, Neutral-type time-delay systems that are not formally stable are not BIBO stabilizable, IMA Journal of Mathematical Control and Information 19 (1–2) (2002) 217–227.
[39] J.J. Loiseau, W. Michiels, S.-I. Niculescu, R. Sipahi, Topics in Time Delay Systems. Analysis, Algorithms and Control, Springer-Verlag, New York, 2009.
[40] A.Z. Manitius, A.W. Olbrot, Finite spectrum assignment problem for systems with delays, IEEE Transactions on Automatic Control 24 (4) (1979) 541–553.
[41] G. Mazanti, I. Boussaada, S.-I. Niculescu, On qualitative properties of single-delay linear retarded differential equations: characteristic roots of maximal multiplicity are necessarily dominant, preprint, arXiv:2002.06128, 2020.
[42] W. Michiels, K. Engelborghs, P. Vansevant, D. Roose, Continuous pole placement for delay equations, Automatica 38 (5) (2002) 747–761.
[43] W. Michiels, S.-I. Niculescu, Stability, Control, and Computation for Time-Delay Systems: An Eigenvalue-Based Approach, 2nd ed., SIAM, 2014.
[44] W. Michiels, T. Vyhlídal, An eigenvalue based approach for the stabilization of linear time-delay systems of neutral type, Automatica 41 (6) (2009) 991–998.
[45] W. Michiels, T. Vyhlídal, P. Zítek, Control design for time-delay systems based on quasi-direct pole placement, Journal of Process Control 20 (3) (2010) 337–343.
[46] A.S. Morse, Ring models for delay-differential systems, Automatica 12 (5) (1976) 529–531.
[47] J.A. Nelder, R. Mead, A simplex method for function minimization, Computer Journal 7 (4) (1965) 308–313.
[48] J. Normey-Rico, E. Camacho, Dead-time compensators: a survey, Control Engineering Practice 16 (4) (2008) 407–428.
[49] J.R. Partington, C. Bonnet, H_∞ and BIBO stabilization of delay systems of neutral type, Systems & Control Letters 52 (3–4) (2004) 283–288.
[50] L. Pekař, On a controller parameterization for infinite-dimensional feedback systems based on the desired overshoot, WSEAS Transactions on Systems 12 (6) (2013) 325–335.
[51] L. Pekař, Spectrum shaping based control strategy for unstable TDS, WSEAS Transactions on Systems 14 (2015) 203–214.
[52] L. Pekař, Extension of the pole-placement shifting based tuning algorithm to neutral delay systems: a case study, in: Intelligent Systems in Cybernetics and Automation Theory: Proceedings of the 4th Computer Science On-Line Conference 2015 (CSOC 2015), 2015, pp. 91–100.
[53] L. Pekař, On the numerical determination of stability regions in the delay space via dominant pole estimation, in: Q. Gao, H.R. Karimi (Eds.), Stability, Control and Application of Time-Delay Systems, Butterworth-Heinemann (Elsevier), 2019, pp. 1–22.

[54] L. Pekař, Controller design and tuning of a delayed heat exchanger plant in a robust sense, in: L. Pekař (Ed.), Advanced Analytic and Control Techniques for Thermal Systems With Heat Exchangers, Elsevier, 2020, pp. 263–284.
[55] L. Pekař, Q. Gao, Spectrum analysis of LTI continuous-time systems with constant delays: a literature overview of some recent results, IEEE Access 6 (2018) 35457–35491.
[56] L. Pekař, R. Matušů, A suboptimal shifting based zero-pole placement method for systems with delays, International Journal of Control, Automation, and Systems 16 (2) (2018) 594–608.
[57] L. Pekař, R. Matušů, R. Prokop, Gridding discretization-based multiple stability switching delay search algorithm: the movement of a human being on a controlled swaying bow, PLoS ONE 12 (6) (2017) e0178950.
[58] L. Pekař, R. Prokop, Algebraic optimal control in RMS ring: a case study, International Journal of Mathematics and Computers in Simulation 7 (1) (2013) 59–68.
[59] L. Pekař, R. Prokop, Dominant spectrum assignment for neutral time delay systems: a study case, in: Proceedings of the 30th European Conference on Modelling and Simulation 2016 (ECMS 2016), Regensburg, Germany, 2016, pp. 300–306.
[60] L. Pekař, R. Prokop, The revision and extension of the RMS ring for time delay systems, Bulletin of the Polish Academy of Sciences. Technical Sciences 65 (3) (2017) 341–350.
[61] L. Pekař, R. Prokop, P. Navrátil, Spectral abscissa minimization when algebraic control of unstable LTI-TDS, IFAC Proceedings Volumes 45 (3) (2012) 382–387.
[62] R. Penrose, On best approximate solutions of linear matrix equations, Mathematical Proceedings of the Cambridge Philosophical Society 52 (1) (1956) 17–19.
[63] D.T. Pham, M. Castellani, Benchmarking and comparison of nature-inspired population-based continuous optimisation algorithms, Soft Computing 18 (2014) 871–903.
[64] D.T. Pham, M. Castellani, A comparative study of the bees algorithm as a tool for function optimization, Cogent Engineering 2 (1) (2015) 1091540.
[65] R. Rabah, G.M. Skylar, A.V. Rezounenko, Stability analysis of neutral type systems in Hilbert space, Journal of Differential Equations 214 (2005) 391–428.
[66] J.-P. Richard, Time-delay systems: an overview of some recent advances and open problems, Automatica 39 (10) (2003) 1667–1694.
[67] B.M. Rudolf, Parameter Tuning for Numerical Optimization Algorithms, M.S. thesis, Department of Computer Science and Engineering, Czech Technical University in Prague, Prague, Czech Republic, 2017.
[68] M.L. Ruz, J. Garrido, F. Vazques, F. Morilla, Interactive tuning tool of proportional-integral controllers for first order plus time delay processes, Symmetry 10 (11) (2018) 569.
[69] M. Saadvandi, K. Meerbergen, W. Desmet, Parametric dominant pole algorithm for parametric model order reduction, Journal of Computational and Applied Mathematics 295 (2014) 259–280.
[70] S. Sengupta, S. Basak, R.A. Peters II, Particle swarm optimization: a survey of historical and recent developments with hybridization perspectives, Machine Learning and Knowledge Extraction 1 (2019) 157–191.
[71] A. Shrestha, A. Mahmood, Review of deep learning algorithms and architectures, IEEE Access 7 (2019) 53040–53065.
[72] R. Sipahi, S.-I. Niculescu, C.T. Abdallah, W. Michiels, K. Gu, Stability and stabilization of systems with time delay, IEEE Control Systems Magazine 31 (1) (2011) 38–65.
[73] R. Sipahi, T. Vyhlídal, S.-I. Niculescu, P. Pepe, Time Delay Systems: Methods, Applications and New Trends, Springer-Verlag, New York, 2012.
[74] L. Skanderová, T. Fabián, I. Zelinka, Self-adapting self-organizing migrating algorithm, Swarm and Evolutionary Computation 51 (2019) 100593.

[75] M.C. Smith, On stabilization and the existence of coprime factorization, IEEE Transactions on Automatic Control 34 (9) (1989) 1005–1007.
[76] O.J.M. Smith, Closer control of loops with dead time, Chemical Engineering Progress 53 (5) (1959) 217–219.
[77] S. Srivastava, A. Misra, S.K. Thakur, V.S. Pandit, An optimal PID controller via LQR for standard second order plus time delay systems, ISA Transactions 60 (1) (2016) 244–253.
[78] J. Vanbiervliet, K. Verheyden, W. Michiels, S. Vandewalle, A nonsmooth optimization approach for the stabilization of time-delay systems, ESAIM. Control, Optimisation and Calculus of Variations 14 (3) (2008) 478–493.
[79] J. Vanbiervliet, B. Vandereycken, W. Michiels, S. Vandewalle, M. Diehl, The smoothed spectral abscissa for robust stability optimization, SIAM Journal on Optimization 20 (1) (2009) 156–171.
[80] M. Vidyasagar, Control System Synthesis: A Factorization Approach, MIT Press, Cambridge, MA, 1985.
[81] T. Vyhlídal, W. Michiels, P. McGahan, Synthesis of a strongly stable state-derivative controller for a time delay system using constrained nonsmooth optimization, IMA Journal of Mathematical Control and Information 27 (4) (2010) 437–455.
[82] T. Vyhlídal, P. Zítek, Mapping based algorithm for large-scale computation of quasi-polynomial zeros, IEEE Transactions on Automatic Control 54 (1) (2009) 171–177.
[83] T. Vyhlídal, P. Zítek, QPmR - quasi-polynomial root-finder: algorithm update and examples, in: T. Vyhlídal, J.-F. Lafay, R. Sipahi (Eds.), Delay Systems: From Theory to Numerics and Applications, Springer, New York, 2014.
[84] K. Watanabe, M. Ito, M. Kaneko, T. Ouchi, Finite spectrum assignment problem for systems with delay in state variables, IEEE Transactions on Automatic Control 28 (4) (1983) 506–508.
[85] D. Whitley, A genetic algorithm tutorial, Statistics and Computing 4 (1994) 65–85.
[86] Y. Xia, M. Fu, P. Shi, Analysis and Synthesis of Dynamical Systems With Time-Delays, Springer-Verlag, Berlin, Heidelberg, 2009.
[87] X.-S. Yang, Nature-Inspired Metaheuristic Algorithms, second edition, Luniver Press, 2010.
[88] D.A. Yuen, L. Wang, X. Chi, L. Johnsson, W. Ge, Y. Shi, GPU Solutions to Multi-Scale Problems in Science and Engineering, Lecture Notes in Earth System Sciences, Springer, 2013.
[89] I. Zelinka, SOMA-self organizing migrating algorithm, in: G.C. Onwobolu, B.V. Babu (Eds.), New Optimization Techniques in Engineering, Springer, Berlin, 2004, pp. 167–217.
[90] I. Zelinka, A survey on evolutionary algorithms dynamics and its complexity – mutual relations, past, present and future, Swarm and Evolutionary Computation 25 (2015) 2–14.
[91] P. Zítek, J. Fišer, T. Vyhlídal, Dimensional analysis approach to dominant three-pole placement in delayed PID control loops, Journal of Process Control 23 (8) (2013) 1063–1074.
[92] P. Zítek, J. Fišer, T. Vyhlídal, Dynamic similarity approach to control system design: delayed PID control loop, International Journal of Control 92 (3) (2019) 329–338.
[93] P. Zítek, V. Kučera, T. Vyhlídal, Meromorphic observer-based pole assignment in time delay systems, Kybernetika 44 (5) (2008) 633–648.
[94] P. Zítek, T. Vyhlídal, Dominant eigenvalue placement for time delay systems, in: Proceedings of the 5th Portuguese Conference on Automation and Control (Controlo 2002), Aveiro, Portugal, 2002, pp. 605–610.

CHAPTER FIVE

Fuzzy model predictive control of discrete nonlinear systems with time-varying delay

Long Teng[a], Youyi Wang[b], and Wenjian Cai[b]
[a]Department of Materials and Production, Aalborg University, Aalborg, Denmark
[b]School of Electrical and Electronic Engineering, Nanyang Technological University, Singapore

Contents

5.1. Introduction

The time delay phenomenon is very common in industrial control systems. The delays can be induced by either the input or the output due to the system structure, measurement, and communication. Therefore the control of time delay is essential in order to improve the control performance. Many contributions have been achieved from different application scenarios, including robots [1,2], drones [3], intelligent vehicles [4], micro-grids [5], 5G communication [6], and so on.

From the perspective of control systems theory, the time delay control problem has received significant research results during the last decade. The stability of the control system could be deteriorated if the time delay effect is not appropriately considered. The most recent results of investiga-

Control Strategy for Time-Delay Systems
https://doi.org/10.1016/B978-0-12-820599-0.00010-0

tions of time delay in different control systems include continuous-time systems [7], discrete-time systems [8], switched systems [9], hybrid systems [10], stochastic systems [11], and networked control systems [12].

Generally, applications of Lyapunov theory to time-delay systems can be divided into two categories: the Lyapunov–Krasovskii functional (LKF) and the Lyapunov–Razumikhin function (LRF). The former method provides sufficient and necessary conditions for stability. However, the control synthesis is usually very complicated. LKF has been applied to different control systems [13–15]. In addition, applications of the Lyapunov–Razumikhin function have attracted many research interests in recent years. Stability analysis of discrete time-delay systems based on LRF are investigated by Liu and Marquez [16], who discussed different types of stability. More recently, Gielen et al. [17–19] have achieved many results on linear difference equations via LRF. Impulsive control is investigated for nonlinear systems with distributed time delay in [9]. Adaptive control is applied to stochastic nonlinear time-delay systems with perturbations in [11].

On the other hand, model predictive control has been more and more investigated and applied to industrial control systems due to the more powerful computing hardware and communication technology nowadays [20–23]. The model predictive control (MPC) is an advanced control method that makes use of the system model and optimizes future control actions in finite-time horizon. In particular, MPC can efficiently handle control constraints. As such, several results have been achieved on MPC of time-delay systems. MPC of delay difference equations is investigated in [18]. The robust model predictive control (RMPC) of time-delay systems with model uncertainties is achieved in [24]. The distributed model predictive control of multiagent systems with communication delays is presented in [25]. The stochastic model predictive control (SMPC) of driving systems with delays is discussed in [26].

For the control of nonlinear systems, the Takagi–Sugeno (T-S) fuzzy systems have been adopted extensively to approximate nonlinear systems. As a result, the traditional control methods for linear systems can be applied. Recent literature on time delay control of T-S fuzzy systems can be seen in [27,28]. In particular, the fuzzy model predictive control (FMPC) based on LRF is investigated for discrete nonlinear systems in [29,30], where multiple constant delays are involved. FMPC based on LKF is investigated for electrical vehicles with unknown constant delays in [31].

In this work, we investigate MPC of discrete-time nonlinear systems with unknown time-varying delay via the T-S fuzzy modeling and control approach. The LRF is employed so that new stabilization conditions are provided. The contributions of this work can be summarized as follows: (1) A novel FMPC approach based on LRF is investigated for discrete non-linear systems with time-varying delay. (2) In particular, the performance of the proposed approach is verified by different simulation scenarios and comparisons against the other method.

The rest of this chapter is structured as follows. The control problem is formulated in Section 5.2. The MPC approach is provided in Section 5.3. The simulation results are provided in Section 5.4. Finally, some conclusions and future directions are provided in Section 5.5.

Notations: By $\mathbb{R}$, $\mathbb{R}_+$, $\mathbb{Z}$, $\mathbb{Z}_+$, and $\mathbb{Z}_{[m,n]}$ we denote the set of real numbers, the set of nonnegative reals, the set of integers, the set of nonnegative integers, and the set of integers in the interval $[m, n]$, respectively. By $\|x\|$ we denote the Euclidean norm of a vector $x \in \mathbb{R}^n$. $\mathbf{x}_{[c1,c2]} := \{x(l)\}_{l\in\mathbb{Z}_{[c1,c2]}}$ with $c1, c2 \in \mathbb{Z}$, denotes a sequence ordered monotonically with respect to the index $l \in \mathbb{Z}_{[c1,c2]}$. $x \in \Omega_{a1}\backslash\Omega_{a2}$ means that $x \in \Omega_{a1}$ and $x \notin \Omega_{a2}$. A function $\alpha : \mathbb{R}_+ \to \mathbb{R}_+$ is said to belong to class $\mathcal{K}$ if it is continuous, strictly increasing, and $\alpha(0) = 0$. $\alpha \in \mathcal{K}_\infty$ if $\alpha \in \mathcal{K}$ and $\lim_{s\to\infty} a(s) = \infty$. The function $\beta : \mathbb{R}_+ \times \mathbb{R}_+ \to \mathbb{R}_+$ is said to belong to class $\mathcal{KL}$ if $\beta(\cdot, s) \in \mathcal{K}$ for each fixed $s \in \mathbb{R}_+$, $\beta(r, \cdot)$ is strictly deceasing for each fixed $r \in \mathbb{R}_+$, and $\lim_{s\to\infty} \beta(r, s) = 0$. The symbol $*$ represents a symmetric structure in matrix inequalities.

5.2. Problem formulation

5.2.1 System model

Consider the following discrete-time nonlinear system with time-varying state delay:

$$x(k+1) \in F(x(k), x_d(k), u(k)), \quad k \in \mathbb{Z}_+, \tag{5.1}$$

where $x(k), x_d(k) \in \mathbb{X}$ and $u(k) \in \mathbb{U}$ with $\mathbb{X} \subseteq \mathbb{R}^n$ and $\mathbb{U} \subseteq \mathbb{R}^m$, represent the system current and delayed states and inputs, respectively; $x_d(k)$ is defined as

$$x_d(k) := x(k + d(k)), \quad d(k) \in \mathbb{Z}_{[-h,-1]}, \tag{5.2}$$

where $d(k)$ denotes the number of delay, h denotes the upper bound of delay, and the minimal delay is set to 1.

Next, for time-delay systems, we introduce some definitions of positive invariance.

Definition 5.1 (Positively invariant set, D-invariance). For LRF conditions, considering system (5.1), a set Ω is called a positively invariant set for the closed-loop system corresponding to the control law π and for $d(k) \in \mathbb{Z}_{[-h,-1]}$ if for all $x \in \Omega$ and $x_d \in \Omega$, we have $x^+ \in \Omega$, where and in what follows, x, x_d, and x^+ are used to denote, at arbitrary time instant, the current state, delayed state, and state in the next instant, respectively.

Definition 5.2 (Lyapunov–Razumikhin function (LRF)). For LRF conditions, suppose that there exists a positive definite function $V(x)$ satisfying the following conditions:

i. There exist $\alpha_1, \alpha_2 \in \mathcal{K}_\infty$ satisfying

$$\alpha_1(\|x\|) \leq V(x) \leq \alpha_2(\|x\|) \tag{5.3}$$

for all $(x, x_d, x_1) \in \mathbb{R}^n \times \mathbb{R}^n \times F(x_0, x_d)$;

ii. For any initial condition,

$$V(x^+) \leq \bar{V}(x) \tag{5.4}$$

where $\bar{V}(x) = \max\{V(x), V(x_d)\}$. Then $V(x)$ is called an LRF for $x^+ \in F(x, x_d)$.

5.2.2 T-S fuzzy time-delay systems

Based on the aforementioned system model, nonlinear systems with time-varying delay can be represented by T-S fuzzy time-delay systems.

Plant Rule l:

IF: z_1 is F_1^l, …, z_v is F_v^l

THEN:

$$x^+ = A_l x + A_{dl} x_d + B_l u, \tag{5.5}$$

where $z := [z_1, \ldots, z_v]$ are premise variables, F_1^l, …, F_v^l are fuzzy sets. $l \in \mathbb{Z}_{[1,L]}$, and L is the number of fuzzy rules. The T-S fuzzy system (5.5) can be written as

$$x^+ = A_\mu x + A_{d\mu} x_d + B_\mu u, \tag{5.6}$$

where $A_\mu := \sum_{l=1}^{L} \mu_l(z)A_l$, $A_{d\mu} := \sum_{l=1}^{L} \mu_l(z)A_{dl}$, $B_\mu := \sum_{l=1}^{L} \mu_l(z)B_l$, and $\mu_l(z)$ is a normalized membership function.

Control law for aforementioned T-S fuzzy system is given as follows.

Control Rule l**:**
IF: z_1 is F_1^l, ..., z_v is F_v^l
THEN:

$$u = K_l x.$$

By using the fuzzy inference method with a singleton fuzzifier, product inference, and center average defuzzifiers the final control output can be obtained as follows:

$$u = K_\mu x, \tag{5.7}$$

where $K_\mu = \sum_{l=1}^{L} \mu_l(z)K_l$.

5.3. MPC approach

MPC requires solving an optimization problem over a finite horizon at each sampling instant. Then in the next instant the optimization problem is repeated based on the updated state information over a shifted horizon.

In this section, we investigate the MPC approach of systems with time-varying delay. We first introduce the online control scheme about MPC for the aforementioned T-S fuzzy time-delay systems. Then we provide conditions of terminal constraint set. Finally, we summarize the whole control algorithm and provide stability and feasibility analysis.

5.3.1 MPC scheme

For T-S fuzzy system (5.6), the prediction model for online approach is given by

$$x(k+s+1|k) = A_\mu x(k+s|k) + A_{d\mu} x_d(k+s|k) + B_\mu u(k+s|k). \tag{5.8}$$

MPC optimizes a cost function under some constraints. Usually, a finite-horizon cost function is adopted:

$$J(k) = \sum_{s=0}^{N-1} \ell(k+s|k) + V_T(x(k+N|k)), \tag{5.9}$$

where N is the prediction length, $\ell(k+s|k)$ is the stage cost at the predicted time instant, and $V_T(x(k+N|k))$ is the terminal cost with V_T being a positive definite function. The stage cost is chosen as

$$\ell(k+s|k) = x^T(k+s|k)Qx(k+s|k) + u^T(k+s|k)Ru(k+s|k), \tag{5.10}$$

where Q and R are positive matrices. In addition, the asymptotic convergence to an invariant set is considered [32,33]. As such, a positively invariant set is employed as the terminal constraint set that guarantees the system state to enter the terminal set at the end of prediction. The online optimization problem can be summarized as

$$\begin{cases} \min_{u(k+s|k)} \quad J(k), \\ u(k+s|k) \in \mathbb{U}, \\ x(k+N|k) \in \Omega_T, \end{cases}$$

where $s \in \mathbb{Z}_{[1,N-1]}$, and $\Omega_T \subseteq \mathbb{X}$ represents the terminal constraint set. The disturbance and control input satisfy the condition

$$u \in \mathbb{U} := \{u| \, |u_t| \leq u_{t,\max}\},$$

where η is a known constant, u_t is the tth element of the input vector with $t \in \mathbb{Z}_{[1,m]}$.

5.3.2 Terminal constraint set for online approach

Definition 5.3. The terminal constraint set Ω_T should satisfy two conditions.

Firstly, it should be a positively invariant set. Let $\Omega \subseteq \mathbb{X}$ be a positive invariant set described in Definition 5.1, and thus $x, x_d, \in \Omega$, and π is the corresponding control law. Define

$$\Omega := \left\{\{x, x_d\} | \max\{x^T P_\mu x, {x_d}^T P_\mu x_d\} \leq \xi\right\} \tag{5.11}$$

and the corresponding control law

$$\pi(x, x_d) := \left\{K_\mu x\right\}, \tag{5.12}$$

where $P_\mu = \sum_{l=1}^{L} \mu_l(x)P_l$ and $K_\mu = \sum_{l=1}^{L} \mu_l(x)K_l$.

Secondly, there exist $\alpha_3, \alpha_4 \in \mathcal{K}_\infty$ and a positive definite function $V(x) := x^T P_\mu x$ with $P_\mu = \sum_{l=1}^{L} \mu_l(x)P_l$ such that for all $x \in \Omega_T$,

$$\alpha_3(\|x\|) \leq V(x) \leq \alpha_4(\|x\|), \tag{5.13}$$

$$V(x^+) - \bar{V}(x) < -x^T Qx - u^T Ru, \tag{5.14}$$

where $\bar{V}(x) = \max\{x^T P_\mu x, x_d{}^T P_\mu x_d\}$. Definition 5.3 is based on the results in [34]. In the following theorem, we provide a condition that ensures the requirement of Definition 5.3.

Theorem 5.1. *Consider T-S fuzzy system (5.6). Suppose that the following matrix inequalities are feasible:*

$$\begin{bmatrix} \gamma\Xi & * & * & * & * \\ 0 & \gamma_d\Xi & * & * & * \\ \Psi & A_{da}G_b & -X_l & * & * \\ QG_b & 0 & 0 & -\xi Q & * \\ RY_b & 0 & 0 & 0 & -\xi R \end{bmatrix} < 0, \tag{5.15}$$

$$\begin{bmatrix} Z & * \\ Y_b{}^T & G_b + G_b{}^T - X_b \end{bmatrix} \geq 0, \; Z_{tt} \leq u_{t,\max}^2, \; t \in \mathbb{Z}_{[0,m]}, \tag{5.16}$$

where $\Xi = X_a - G_b - G_b{}^T$, $\Psi = A_a G_b + B_a Y_b$. $a, b, l \in \mathbb{Z}_{[1,L]}$, Z_{tt} *is the tth diagonal element of a matrix* Z, *and* γ *and* γ_d *are preset positive numbers that satisfy* $\gamma + \gamma_d = 1$. *Then* Ω_T *is a terminal constraint set regarding the terminal cost function* $\bar{V}(x)$.

Proof. Firstly, (5.11) can be easily obtained by resorting to the eigenvalues of the positive definite matrices P_l, $l \in \mathbb{Z}_{[1,L]}$. Let $\rho_{\min} := \min\{\rho_{\min}(P_l) | l \in \mathbb{Z}_{[1,L]}\}$, and $\rho_{\max} := \max\{\rho_{\max}(P_l) | l \in \mathbb{Z}_{[1,L]}\}$. Then $\rho_{\min}\|x\|^2 \leq V(x) \leq \rho_{\max}\|x\|^2$. Thus requirement (5.13) is satisfied.

Then the main effort is focused on the satisfaction of optimization problem (5.14). Resorting to the dilation lemma [35], we have

$$-G_b{}^T X_a{}^{-1} G_b \leq X_a - G_b{}^T - G_b. \tag{5.17}$$

Then multiplying $\mathrm{diag}\left\{G_b{}^{-T}, G_b{}^{-T}, I, I, I\right\}$ and its transpose from both sides of (5.15), respectively, yields that

$$\begin{bmatrix} -\gamma X_a{}^{-1} & * & * & * & * \\ 0 & -\gamma_d X_a{}^{-1} & * & * & * \\ \Upsilon & A_{da} & -X_l & * & * \\ Q & 0 & 0 & -\xi Q & * \\ RK_b & 0 & 0 & 0 & -\xi R \end{bmatrix} < 0, \tag{5.18}$$

where $\Upsilon = A_a + B_a K_b$. Inequality (5.18) is equivalent to

$$\begin{bmatrix} -\gamma\xi X_a{}^{-1} & * & * & * & * \\ 0 & -\gamma_d \xi X_a{}^{-1} & * & * & * \\ \Upsilon & A_{da} & -\frac{X_l}{\xi} & * & * \\ Q & 0 & 0 & -Q & * \\ RK_b & 0 & 0 & 0 & -R \end{bmatrix} < 0. \tag{5.19}$$

Substituting X_a^{-1} with P_a/ξ to (5.19), we rewrite (5.19) as

$$\begin{bmatrix} Q + K_\mu{}^T R K_\mu - \gamma P_\mu & * & * \\ 0 & -\gamma_d P_\mu & * \\ A_\mu + B_\mu K_\mu & A_{d\mu} & -P_l{}^{-1} \end{bmatrix} < 0 \tag{5.20}$$

Applying the Schur complement to (5.20), multiplying $\left[x^T, x_d{}^T\right]$ and its transpose from both sides of the resulting matrix inequality, respectively, and noticing that $P_\mu^+ = \sum_{l=1}^{L} \mu_l(x^+) P_l$, we get

$$x^{+T} P_\mu^+ x^+ + x^T Q x + x^T K_\mu{}^T R K_\mu x - (\gamma x^T P_\mu x + \gamma_d x_d{}^T P_\mu x_d) < 0. \tag{5.21}$$

With $\gamma + \gamma_d = 1$, it is easy to get $\gamma x^T P_\mu x + \gamma_d x_d{}^T P_\mu x_d \leq \bar{V}(x)$, and thus

$$V(x^+) - \bar{V}(x) < -x^T Q x - u^T R u. \tag{5.22}$$

Inequalities (5.16) guarantee the input constraint, and the proof can be seen in [36]. The proof is complete. □

5.3.3 MPC algorithm

Based on the previous results, in this subsection, we discuss the whole algorithm to complete the online MPC controller design and then analyze feasibility and stability.

The terminal constraint set Ω_T should satisfy the requirement of positively invariant set according to (5.11), which indicates that

$$\bar{V}(x) \leq \xi, \tag{5.23}$$

and thus $x(k), x_d(k) \in \Omega_T$. We consider the following problem:

$$\min \quad \xi, \text{ subjected to } \bar{V}(x) \leq \xi.$$

A sufficient condition for $\bar{V}(x) \leq \xi$ is given as

$$\begin{bmatrix} 1 & x^T(k+j) \\ x(k+j) & X_l \end{bmatrix} \geq 0, \; j \in \mathbb{Z}_{[-h,0]}, \; l \in \mathbb{Z}_{[1,L]}. \tag{5.24}$$

Remark 5.1. For clarity of exposition, at time instant k, it can be seen that the terminal constraint set only involves $x(k)$ and $x_d(k)$ according to (5.1), that is, $x(k), x_d(k) \in \Omega_k$. However, all possibly delayed states are employed in (5.24), which indicates that $x(k+j) \in \tilde{\Omega}_k$, $j \in \mathbb{Z}_{[-h,0]}$. Thus $\tilde{\Omega}_k$, rather than Ω_k, represents the terminal constraint set Ω_T computed at instant k, and it is easy to see that $\Omega_k \subseteq \tilde{\Omega}_k$. Although this choice may be less optimal because it may lead to a larger terminal constraint set, it is essential to guarantee the positive invariance of terminal constraint set as the delay is time-varying. It will be utilized further for the subsequent feasibility analysis.

Remark 5.2. (5.24) indicates that all possibly delayed states that are available at each time instant are involved. Moreover, the upper bound of delay is assumed to be known. As such, we need not know the exact delay in the controller design.

In summary, the MPC algorithm is given as follows.

MPC Algorithm.

Step 1. Obtain the system all (possibly delayed) states $x(k+j)$, $j \in \mathbb{Z}_{[-h,0]}$;

Step 2. Solve

$$\min_{\xi, X_a, G_b, Y_b, Z, \lambda} \quad \xi, \tag{5.25}$$

subject to (5.15), (5.16), and (5.24). Go to Step 3;

Step 3. Obtain the control input $u(k) = K_\mu x(k)$ with $K_\mu = \sum_{l=1}^{L} \mu_l K_l$ and $K_l = Y_l G_l^{-1}$. Then implement the control input to system (5.6). Move the time instant from k to $k+1$ and go to Step 1.

Remark 5.3. Consider the optimization problem (5.25), where (5.24) implies that the number of LMIs will increase with the number of delay. However, the other conditions, that is, (5.15) and (5.16), will remain unchanged with the delay. Therefore the computational burden of the online MPC algorithm will slightly increase with the number of delay.

Theorem 5.2. *For system (5.6), if the constrained optimization problem has a solution at time 0, then it will always be solvable. Furthermore, recursive feasibility can be achieved.*

Proof. Without loss of generality, assume that (5.25) is satisfied at time k according to the description in Remark 5.1, and thus $x(k+j) \in \tilde{\Omega}_k$, $j \in \mathbb{Z}_{[-h,0]}$, with $\tilde{\Omega}_k$ being the terminal constraint set at time k. Since $x(k) \in \Omega_k$ and $x_d(k) \in \Omega_k$ as stated in Remark 5.1, we have $x(k+1) \in \Omega_k$. Considering $\Omega_k \subseteq \tilde{\Omega}_k$, we have $x(k+1+j) \in \tilde{\Omega}_k$, $j \in \mathbb{Z}_{[-h,0]}$. Then we can conclude that (5.25) is solvable at time $k+1$. Moreover, the solution obtained at time k is feasible at time $k+1$. Applying the same process, we observe that the optimization problem is feasible at all times. □

Theorem 5.3. *If the constrained optimization problem is feasible at the initial time 0, then system (5.6) is stable.*

Proof. Consider the LRF $V(x)$ at time k and suppose that it gets its maximum value at the delayed states $x(k+\theta)$, $\theta \in \{d(k), 0\}$, with $d(k)$ defined in (5.2), that is, $\bar{V}^*(x(k)) = x^T(k+\theta)P_\mu{}^*(k)x(k+\theta)$, $P_\mu{}^*(k) = \sum_{l=1}^{L} \mu_l P_l{}^*(k)$, where $P_l{}^*(k)$ is an optimal value of $P_l(k)$ at time k. In addition, as $V^*(x(k)) = x^T(k)P_\mu{}^*(k)x(k)$, we can easily get

$$\rho_{\min}^* \left\| x(k) \right\|^2 \leq V^*(x(k)) \leq \rho_{\max}^* \left\| x(k) \right\|^2 \tag{5.26}$$

with

$$\begin{aligned} \rho_{\min}^* &:= \min\left\{ \rho_{\min}(P_l{}^*(k)) | l \in \mathbb{Z}_{[1,L]}, k \in \mathbb{Z}_+ \right\}, \\ \rho_{\max}^* &:= \max\left\{ \rho_{\max}(P_l{}^*(k)) | l \in \mathbb{Z}_{[1,L]}, k \in \mathbb{Z}_+ \right\}, \end{aligned}$$

where $\rho_{\max}(\cdot)$ and $\rho_{\min}(\cdot)$ represent the maximal and minimal eigenvalues, respectively.

Moreover, (5.14) implies

$$V_k{}^*(x(k+1)) - \bar{V}^*(x(k)) < -x(k)^T Q x(k) - u^T(k) R u(k), \tag{5.27}$$

where $V_k{}^*(x(k+1)) = x(k+1)^T P_\mu{}^*(k) x(k+1)$. Thus we have

$$V_k{}^*(x(k+1)) - \bar{V}^*(x(k)) < -x(k)^T Q x(k). \tag{5.28}$$

Because of the optimality of (5.25) at time $k+1$, we have

$$V_{k+1}^*(x(k+1)) \leq V_k^*(x(k+1)), \tag{5.29}$$

where $V^*_{k+1}(x(k+1))$ denotes the optimal $V(x)$ at time $k+1$. Then we have

$$V^*_{k+1}(x(k+1)) - \bar{V}^*(x(k)) < -x(k)^T Qx(k). \tag{5.30}$$

According to Definition 5.3, from (5.26) and (5.30) it follows that $V(x)$ is a Lyapunov function. Hence the closed-loop system is stable. □

5.4. Example: time-delay CSTR system

In this section, we provide a simulation on a time-delay continuous stirred-tank reactor (CSTR) system to verify the effectiveness of the proposed two algorithms [37]. In [38] the CSTR system is represented by discrete T-S fuzzy systems, and the sampling time is selected as $T_s = 0.2$ minutes. However, only constant time delay is investigated in [38]. In contrast, a more natural condition is given in this work that the time delay is allowed to be time-varying and unknown. Only an upper bound of delay is required, which is selected as $h = 10$.

There are three nominal states, that is, $x_{s1} = [0.1440; 0.8862]$, $x_{s2} = [0.4472; 2.7520]$, and $x_{s3} = [0.7646; 4.7052]$ when $u = 0$. Then the discrete T-S fuzzy time-delay systems can be formulated as follows.

Rule 1: If $x_2(k)$ is 0.8862, then

$$x(k+1) = A_1 x(k) + A_{d1} x_d(k) + B_1 u(k).$$

Rule 2: If $x_2(k)$ is 2.7520, then

$$x(k+1) = A_2 x(k) + A_{d2} x_d(k) + B_2 u(k).$$

Rule 3: If $x_2(k)$ is 4.7052, then

$$x(k+1) = A_3 x(k) + A_{d3} x_d(k) + B_3 u(k),$$

where

$$A_1 = \begin{bmatrix} 0.7500 & 0.0119 \\ -0.2238 & 0.8262 \end{bmatrix}, \quad B_1 = \begin{bmatrix} 0.0004 \\ 0.0546 \end{bmatrix},$$

$$A_2 = \begin{bmatrix} 0.6203 & 0.0762 \\ -1.2337 & 1.3265 \end{bmatrix}, \quad B_2 = \begin{bmatrix} 0.0023 \\ 0.0698 \end{bmatrix},$$

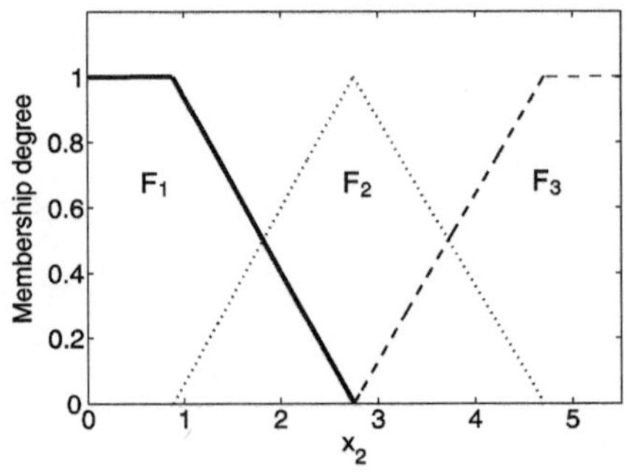

Figure 5.1 Fuzzy membership functions.

$$A_3 = \begin{bmatrix} 0.3068 & 0.0442 \\ -3.6621 & 1.0765 \end{bmatrix}, \quad B_3 = \begin{bmatrix} 0.0015 \\ 0.0634 \end{bmatrix},$$

$$A_{d1} = \begin{bmatrix} 0.0435 & 0.0003 \\ -0.0061 & 0.0455 \end{bmatrix},$$

$$A_{d2} = \begin{bmatrix} 0.0403 & 0.0019 \\ -0.0312 & 0.0581 \end{bmatrix},$$

$$A_{d3} = \begin{bmatrix} 0.0310 & 0.0013 \\ -0.1037 & 0.0528 \end{bmatrix}.$$

The membership functions of the T-S fuzzy time-delay systems (see Fig. 5.1) are given as follows:

$$\begin{cases} F_1 = \begin{cases} 1 & \text{if } x_2 \le 0.8862, \\ \frac{2.7520 - x_2}{2.7520 - 0.8862} & \text{if } 0.8862 < x_2 \le 2.7520, \\ 0 & \text{if } x_2 > 2.7520, \end{cases} \\ F_2 = \begin{cases} 1 - F_1 & \text{if } 0.8862 < x_2 \le 2.7520, \\ 1 - F_3 & \text{if } 2.7520 < x_2 \le 4.7052, \end{cases} \\ F_3 = \begin{cases} 0 & \text{if } x_2 \le 2.7520, \\ \frac{2.7520 - x_2}{4.7052 - 2.7520} & \text{if } 2.7520 < x_2 \le 4.7052, \\ 1 & \text{if } x_2 > 4.7052. \end{cases} \end{cases}$$

The weighting matrices Q and R are selected as diag$\{1e-6, 1e-9\}$ and 0.001, respectively. The input constraint is selected by: $\|u\| \le 10$. Considering $\gamma + \gamma_d = 1$, let $\gamma = 0.8$ and $\gamma_d = 0.2$.

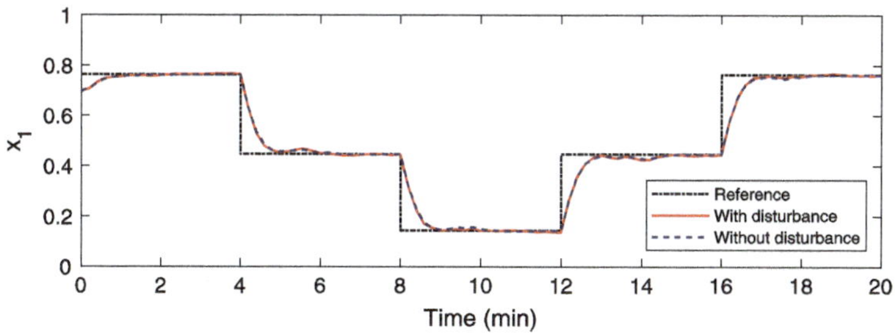

Figure 5.2 Trajectories of state x_1.

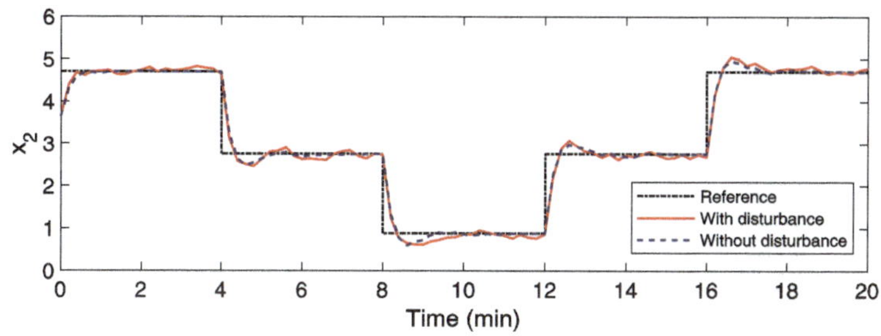

Figure 5.3 Trajectories of state x_2.

Different experiment scenarios and comparisons against other approaches are considered to evaluate the performance of the proposed FMPC approach.

Case 1: Application of the proposed method to the CSTR system without and with disturbances.

The simulation results of the proposed FMPC algorithm without/with disturbances are shown in Figs. 5.2–5.6, where the initial (delayed) states are set as $x(i)_{i\in\mathbb{Z}_{[-10,0]}} = [0.6972; 3.652]$. The piecewise reference signal sequentially changes from [0.7646; 4.7052] to [0.4472; 2.7520], [0.1440; 0.8862], [0.4472; 2.7520], and [0.7646; 4.7052]. Figs. 5.2 and 5.3 demonstrate that the closed-loop state response can generally track the piecewise reference well via the proposed MPC algorithm in both without/with disturbances circumstances, whereas from Fig. 5.3 we can see that the tracking performance is better in the case of without disturbances. Fig. 5.4 demonstrates the input constraint, from which we can see that the input constraint is satisfied. The number of unknown time-varying delay, which is randomly generated from MATLAB®, is displayed in Fig. 5.5. We can see that large

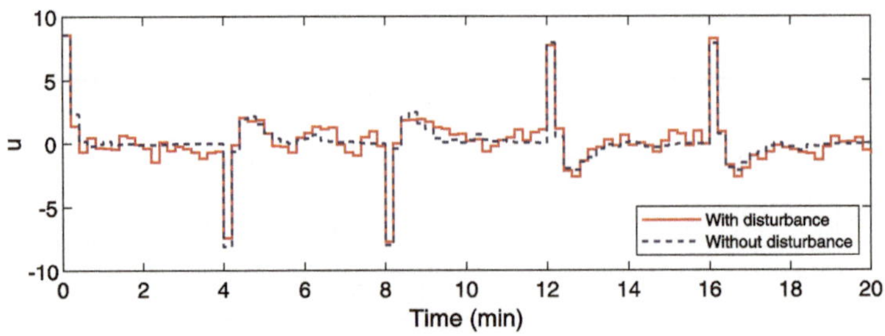

Figure 5.4 Trajectories of input *u*.

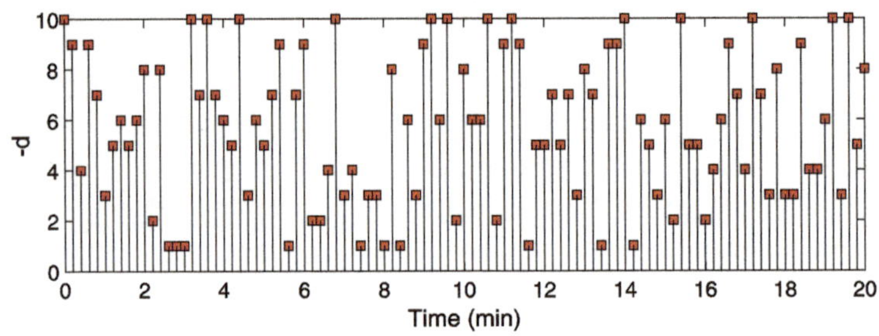

Figure 5.5 The number of time-varying delays.

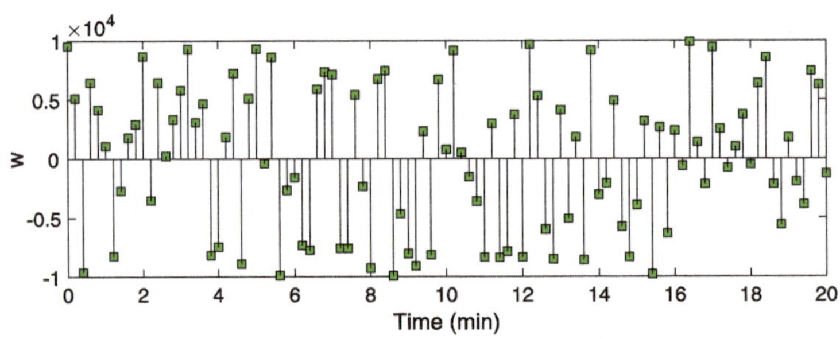

Figure 5.6 Disturbances.

time delays are involved as we can see that the upper bound of delay is set as 10. In addition, random disturbances are induced with the bound of disturbance set as $\|w\| \leq 10000$.

Case 2: Comparison of the proposed method and the FMPC approach for nondelayed systems in [39].

The simulation results are shown in Figs. 5.7–5.10. The trajectory tracking is demonstrated in Figs. 5.7 and 5.8, from which we can see that the

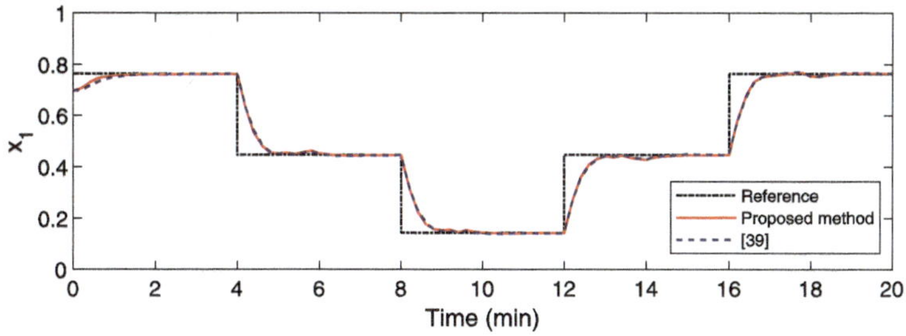

Figure 5.7 Trajectories of state x_1.

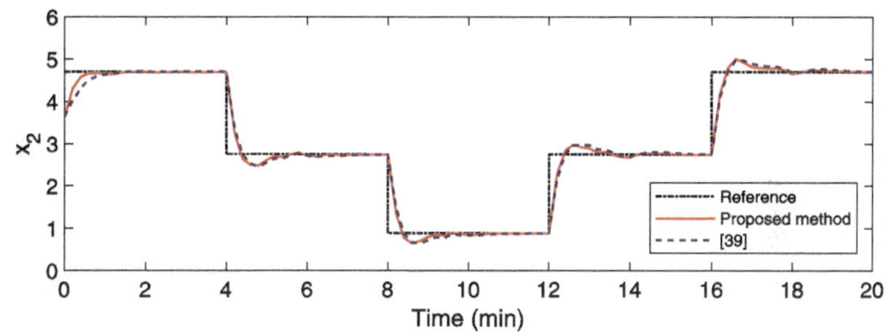

Figure 5.8 Trajectories of state x_2.

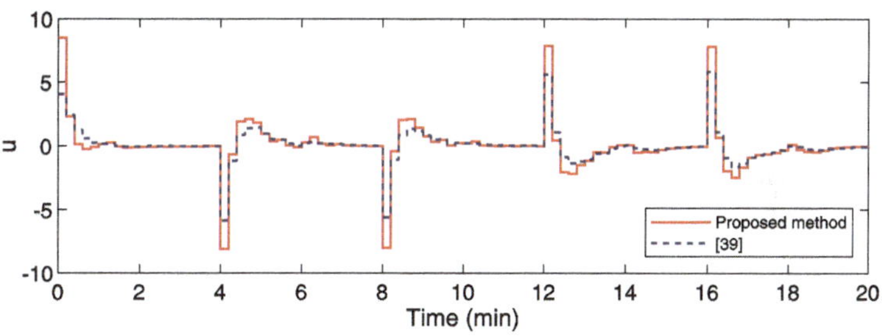

Figure 5.9 Trajectories of input u.

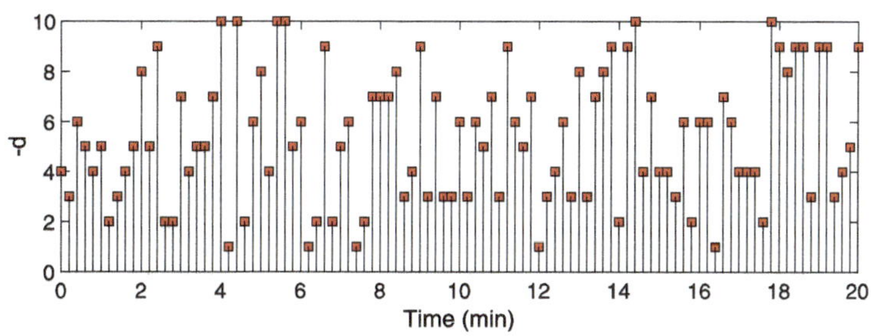

Figure 5.10 The number of time-varying delays.

proposed method achieves better performance compared with the FMPC approach for nondelayed systems in [39]. The input constraint is shown in Fig. 5.9, where we can see that the input constraint is satisfied for both control algorithms. Moreover, the number of unknown time-varying delay is demonstrated in Fig. 5.10.

5.5. Discussions

The performance of the proposed FMPC for time-delay systems is evaluated by simulation in different scenarios and comparisons with an existing approach, respectively. First, to verify the robustness of the proposed method against disturbances, we considered case 1 that CSTR system with persistent and unknown disturbances involved in comparison against that without any external disturbances. The simulation results show that tracking of reference signals is achieved even with a large number of external disturbances, as shown in Fig. 5.6. However, from Figs. 5.2 and 5.3 we observed that the tracking performance is not as good as the system without disturbances. Comparing Figs. 5.2 and 5.3, we see that the disturbances have stronger influence on x_2 than on x_1 due to the disturbances directly added to x_2 rather than to x_1. From Fig. 5.4 we can see that the control input with disturbances is generally stronger than that without disturbances, which demonstrates the effectiveness of the proposed controller in response to external disturbances.

Second, comparison against a recent FMPC method is discussed to further verify the tracking performance of the proposed method. The difference between the proposed method and FMPC in [39] is that the time delay effect is taken into account in the proposed control method. As such, it performs better in dealing with time delays that are frequently observed in practical systems. Figs. 5.7 and 5.8 demonstrate that the proposed control method achieves better transient performance in trajectory tracking.

5.6. Conclusions

In this work, we investigated FMPC of discrete nonlinear systems with unknown time-vary state delay. The proposed method adopts the Lyapunov–Razumikhin function for stabilization of time-delay control systems. The control synthesis and computation are simple due to the utilization of the Lyapunov–Razumikhin function. The control diagram of FMPC based on the Lyapunov–Razumikhin function is developed. The

MPC algorithm is formulated as solving a minimization problem of a cost function. The terminal constraint set based on the Razumikhin approach is analyzed. In the MPC algorithm, the input constraint is considered. The simulation results demonstrate the effectiveness of the proposed MPC algorithm when applied to a CSTR system.

The simulations are implemented with MATLAB, and all matrix inequalities are solved with the YALMIP toolbox [40] and the PENBMI toolbox [41]. In particular, for the interested readers, the MATLAB simulation files for the proposed approach can be accessed via https://drive.google.com/file/d/1uKAN2Z2J8Us9sauzQBSYE6qZmCSk6ack/view?usp=sharing.

References

[1] M. Bagheri, M. Krstić, P. Naseradinmousavi, Analytical and experimental predictor-based time delay control of Baxter robot, in: ASME 2018 Dynamic Systems and Control Conference, 2018.

[2] Y. Kali, M. Saad, K. Benjelloun, Optimal super-twisting algorithm with time delay estimation for robot manipulators based on feedback linearization, Robotics and Autonomous Systems 108 (2018) 87–99.

[3] J. Lee, D. Shin, H. Ryu, D. Lee, D.H. Shim, Fault tolerant adaptive control using time delay control scheme under motor faults of octocopter, in: 7th Inter. Conf. on Syst. and Contr. (ICSC), 2018, pp. 123–128.

[4] J. Guo, Y. Luo, C. Hu, C. Tao, K. Li, Robust combined lane keeping and direct yaw moment control for intelligent electric vehicles with time delay, International Journal of Automotive Technology 20 (2) (2019) 289–296.

[5] M.H. Khooban, N. Vafamand, T. Dragicevic, M.M. Mardani, R. Heydari, Modeling and hil real-time simulation for the secondary lfc in time-delay shipboard microgrids, in: International Conference on Electrical Systems for Aircraft, Railway, Ship Propulsion and Road Vehicles & International Transportation Electrification Conference (ESARS-ITEC), 2018, pp. 1–5.

[6] R.A. Delgado, K. Lau, R.H. Middleton, T. Wigren, Networked delay control for 5G wireless machine-type communications using multiconnectivity, IEEE Transactions on Control Systems Technology 27 (4) (2019) 1510–1525.

[7] M.M. Nicotra, T.W. Nguyen, E. Garone, I.V. Kolmanovsky, Explicit reference governor for the constrained control of linear time-delay systems, IEEE Transactions on Automatic Control 64 (7) (2019) 2883–2889.

[8] M.A. Regaieg, M. Kchaou, A. El Hajjaji, M. Chaabane, Robust H_∞ guaranteed cost control for discrete-time switched singular systems with time-varying delay, Optimal Control Applications & Methods 40 (1) (2019) 119–140.

[9] X. Liu, K. Zhang, Stabilization of nonlinear time-delay systems: distributed-delay dependent impulsive control, Systems & Control Letters 120 (2018) 17–22.

[10] Y. Jin, Y. Zhang, Y. Jing, J. Fu, An average dwell-time method for fault-tolerant control of switched time-delay systems and its application, IEEE Transactions on Industrial Informatics 66 (4) (2019) 3139–3147.

[11] H. Min, S. Xu, B. Zhang, Q. Ma, Globally adaptive control for stochastic nonlinear time-delay systems with perturbations and its application, Automatica 102 (2019) 105–110.

[12] H. Gao, T. Chen, New results on stability of discrete-time systems with time-varying state delay, IEEE Transactions on Automatic Control 52 (2) (2007) 328–334.
[13] C.E. De Souza, D. Coutinho, Robust stability and control of uncertain linear discrete-time periodic systems with time-delay, Automatica 50 (2) (2014) 431–441.
[14] H. Zhang, Y. Shi, M. Liu, H_{∞} step tracking control for networked discrete-time nonlinear systems with integral and predictive actions, IEEE Transactions on Industrial Informatics 9 (1) (2013) 337–345.
[15] V.N. Phat, K. Ratchagit, Stability and stabilization of switched linear discrete-time systems with interval time-varying delay, Nonlinear Analysis: Hybrid Systems 5 (4) (2011) 605–612.
[16] B. Liu, H.J. Marquez, Razumikhin-type stability theorems for discrete delay systems, Automatica 43 (7) (2007) 1219–1225.
[17] R.H. Gielen, A.R. Teel, M. Lazar, Tractable Razumikhin-type conditions for input-to-state stability analysis of delay difference inclusions, Automatica 49 (2) (2013) 619–625.
[18] R.H. Gielen, M. Lazar, Stabilisation of linear delay difference inclusions via time-varying control Lyapunov functions, IET Control Theory & Applications 6 (12) (2012) 1958–1964.
[19] R.H. Gielen, M. Lazar, Stabilization of polytopic delay difference inclusions via the Razumikhin approach, Automatica 47 (12) (2011) 2562–2570.
[20] J. Rodriguez, M.P. Kazmierkowski, J.R. Espinoza, P. Zanchetta, H. Abu-Rub, H.A. Young, C.A. Rojas, State of the art of finite control set model predictive control in power electronics, IEEE Transactions on Industrial Informatics 9 (2) (2013) 1003–1016.
[21] T.P.d. Nascimento, G.F. Basso, C.E.T. Dórea, L.M.G. Gonçalves, Perception-driven motion control based on stochastic nonlinear model predictive controllers, IEEE/ASME Transactions on Mechatronics 24 (4) (2019) 1751–1762.
[22] M. Moness, A.M. Moustafa, Real-time switched model predictive control for a cyber-physical wind turbine emulator, IEEE Transactions on Industrial Informatics 16 (6) (2020) 3807–3817.
[23] J. Kabzan, L. Hewing, A. Liniger, M.N. Zeilinger, Learning-based model predictive control for autonomous racing, IEEE Robotics and Automation Letters 4 (4) (2019) 3363–3370.
[24] B.D.O. Capron, M.T. Uchiyama, D. Odloak, Linear matrix inequality-based robust model predictive control for time-delayed systems, IET Control Theory & Applications 6 (1) (2012) 37–50.
[25] H. Li, Y. Shi, Distributed model predictive control of constrained nonlinear systems with communication delays, Systems & Control Letters 62 (10) (2013) 819–826.
[26] A. Katriniok, S. Kojchev, E. Lefeber, H. Nijmeijer, A stochastic model predictive control approach for driver-aided intersection crossing with uncertain driver time delay, in: 18th European Control Conference (ECC), 2019, pp. 243–249.
[27] L. Wang, H.-K. Lam, New stability criterion for continuous-time Takagi–Sugeno fuzzy systems with time-varying delay, IEEE Transactions on Cybernetics 49 (4) (2019) 1551–1556.
[28] J. Zhang, F. Zhu, H.R. Karimi, F. Wang, Observer-based sliding mode control for T–S fuzzy descriptor systems with time delay, IEEE Transactions on Fuzzy Systems 27 (10) (2019) 2009–2023.
[29] L. Teng, Y. Wang, W. Cai, H. Li, Robust model predictive control of discrete nonlinear systems with time delays and disturbances via T–S fuzzy approach, Journal of Process Control 53 (2017) 70–79.
[30] L. Teng, Y. Wang, W. Cai, H. Li, Efficient robust fuzzy model predictive control of discrete nonlinear time-delay systems via Razumikhin approach, IEEE Transactions on Fuzzy Systems 27 (2) (2019) 262–272.

[31] M.H. Khooban, N. Vafamand, T. Niknam, T. Dragicevic, F. Blaabjerg, Model-predictive control based on Takagi–Sugeno fuzzy model for electrical vehicles delayed model, IET Electric Power Applications 11 (5) (2017) 918–934.
[32] D.Q. Mayne, M.M. Seron, S.V. Raković, Robust model predictive control of constrained linear systems with bounded disturbances, Automatica 41 (2) (2005) 219–224.
[33] R. Ghaemi, J. Sun, I.V. Kolmanovsky, Robust control of constrained linear systems with bounded disturbances, IEEE Transactions on Automatic Control 57 (10) (2012) 2683–2688.
[34] B. Liu, D.J. Hill, Input-to-state stability for discrete time-delay systems via the Razumikhin technique, Systems & Control Letters 58 (8) (2009) 567–575.
[35] J.C. Geromel, P.D. Peres, J. Bernussou, On a convex parameter space method for linear control design of uncertain systems, SIAM Journal on Control and Optimization 29 (2) (1991) 381–402.
[36] N. Wada, K. Saito, M. Saeki, Model predictive control for linear parameter varying systems using parameter dependent Lyapunov function, IEEE Transactions on Circuits and Systems. II, Express Briefs 53 (12) (2006) 1446–1450.
[37] B. Lehman, J. Bentsman, S.V. Lunel, E.I. Verriest, Vibrational control of nonlinear time lag systems with bounded delay: averaging theory, stabilizability, and transient behavior, IEEE Transactions on Automatic Control 39 (5) (1994) 898–912.
[38] Y.-Y. Cao, P.M. Frank, Analysis and synthesis of nonlinear time-delay systems via fuzzy control approach, IEEE Transactions on Fuzzy Systems 8 (2) (2000) 200–211.
[39] W. Yang, G. Feng, T. Zhang, Robust model predictive control for discrete-time Takagi–Sugeno fuzzy systems with structured uncertainties and persistent disturbances, IEEE Transactions on Fuzzy Systems 22 (5) (2014) 1213–1228.
[40] J. Löfberg, YALMIP: a toolbox for modeling and optimization in MATLAB, in: IEEE International Symposium on Computer Aided Control Systems Design, 2004.
[41] M. Kocvara, M. Stingl, PENBMI users guide, http://www.penopt.com, 2006. (Accessed 4 February 2016).

CHAPTER SIX

Observability and observer design of time-delay linear systems with unknown inputs

Francisco Javier Bejarano[a] **and Gang Zheng**[b]
[a]SEPI, ESIME Ticomán, Instituto Politécnico Nacional, Mexico City, Mexico
[b]Defrost, INRIA - Lille Nord Europe, Lille, France

Contents

6.1. Introduction

The problem of estimating the state variables of a system is one of the main problems studied in control theory. The observability problem essentially consists in finding the conditions under which the trajectories of the state vector of the system can be reconstructed by means of the knowledge of the system output. Such a problem has been successfully solved for several sorts of dynamic systems, including linear systems, nonlinear systems, systems with delays, systems with partial differential equations (see, e.g., [22,32,37,45,54]). A generalization of the problem has been carried out by considering that the system under study is affected by unknown inputs. In such a case, for linear systems, observability and detectability

Control Strategy for Time-Delay Systems
https://doi.org/10.1016/B978-0-12-820599-0.00011-2

conditions were found by Molinari [46] and Hautus [31]. In [6], for linear singular systems, necessary and sufficient observability conditions were obtained. Unknown input observers were designed for linear systems in several papers: [11,17,33,34,42,61,63]. For the nonlinear case, in [30], conditions were given under which there exists an observer allowing for the state estimation.

Time-delay systems are widely used to model many applications, ranging from chemical and biological process to sampled data effects; see, for example, [48] and [56]. Many results have been published to treat this kind of systems for different aspects, such as controllability [22], stability [23], tracking [20], observability [65], identifiability [4,64], and disturbance decoupling [47]. The analysis of observation for time-delay systems can be dated back to the 1980s; see, for example, [44,51,55].

Different definitions of observability have been proposed, such as strong observability, spectral observability, and weak observability. The observability problem has been studied from different perspectives using, for instance, the functional analytic approach [10] and also using an algebraic method [12,22]. Different observers have been proposed. For instance, Fattouh et al. [21] designed an unknown input observer for linear systems with commensurate delays in state, input, and output variables, whereas the output was not affected by the unknown inputs. Inspired by the technique of output injection in [41], Hou et al. [35] solved this problem by transforming the studied system into a higher-dimensional observer canonical form with delayed output injection. Darouach [15,16] designed the unknown input observer for systems involving only one delay in the state and no delays in the input and output. Other observers for some classes of time-delay systems can be found in [14,24,57]. Nonetheless, the majority of the existing works on observability analysis are focused on time-delay systems whose outputs are not affected by unknown inputs. Moreover, usually, those methods admit no more than one delay. Up to our knowledge, not always it is possible to reduce the delay degree. There exist some methods to eliminate (or reduce the degree of) the delay, such as [43], [26], and [25]. The conditions to do the reduction of delay degree via a bicausal transformation with the same dimension can be found in [25].

On the other hand, representing dynamic systems over rings have been found to be useful since several decades ago [12,38,59]. A brief history and survey for linear systems over rings can be found in [38]. Recently, the use of rings and modules has been extended to nonlinear systems also; see, for example, [62] and [29]. Ohtsuka [50] studied some algebraic structures of nonlinear systems over rings obtained by immersion. Kotta and Tõnso [40] addressed the realization of discrete-time nonlinear input–output equations by using a noncommutative polynomial ring. The global observability of polynomial nonlinear systems is tackled by Kawano and Ohtsuka [39] using the localization of a polynomial ring. A particular class of dynamic systems that can be represented over a ring is that of systems with commensurate delays, where the delays can be grouped together, which allows for considering the coefficients of the mathematical model within a polynomial ring in case of linear systems (see [14]) and within a noncommutative ring in case of nonlinear systems [3]. There are some other systems that can be represented as linear systems over a principal ideal domain, for example, 2-D systems [18,53], linear systems with commensurate delays [7], systems over the integers [19], and parameterized systems [36]. It is also worth mentioning that some computational algebra tools to work with linear systems over rings have been given in [52].

Here we present recently obtained results concerning the observability of a general class of commensurate time-delay systems with unknown inputs. We *present sufficient conditions ensuring the observability of the system* (a precise definition is given further), which also allows for recovering the state variables of the system by two different ways. First, we show that the *state variables can be reconstructed by using a formula that expresses the state variables as a function of the actual and past values of the output and some its derivatives.* We also *design of a Luenberger-like observer for carrying out the state estimation.* It turns out that the conditions used for the presented observer design are necessary and sufficient for the case of a time-invariant linear system with unknown inputs (without delays). That is why we run to risk of saying that these required conditions are relaxed for designing a linear observer. The method used for the observer design was inspired by Hou et al. [35], who designed an observer for time-delay systems *without* unknown inputs. The main idea is to immerse the system into a one of higher dimension, allowing for putting all time-delays into an output injection. A pseudocode for checking the observability conditions and for the observer design is given in Appendix 6.A.

We use the following notation. $\mathbb{R}$ is the field of real numbers, $\mathbb{R}_{\neq 0}$ is the set of nonzero real numbers, and $\mathbb{N}$ is the set of positive integers. By $\mathbb{R}[\delta]$ we mean the polynomial ring over the field $\mathbb{R}$ whose indeterminate is δ, and $\mathbb{R}^n[\delta]$ is the $\mathbb{R}[\delta]$-module whose elements are the n-dimensional vectors whose entries are polynomials. By $\mathbb{R}^{q\times s}[\delta]$ we denote the set of $q \times s$ matrices whose entries are in $\mathbb{R}[\delta]$. For $f(\delta)$, a polynomial of $\mathbb{R}[\delta]$, $\deg f(\delta)$

is the degree of $f(\delta)$. For a matrix $M(\delta)$, $\deg M(\delta)$ (the degree of $M(\delta)$) is defined as the maximum degree of all the entries $m_{ij}(\delta)$ of $M(\delta)$. We use $\operatorname{rank}(M(\delta))$ to denote the rank of the matrix $M(\delta)$ over the ring $\mathbb{R}[\delta]$. The acronym for greatest common divisor is *gcd*.

6.2. Observability notions

We consider systems whose dynamics are governed by equations of the form

$$\dot{x}(t) = \sum_{i=0}^{k_a} A_i x(t - ih) + \sum_{i=0}^{k_b} B_i w(t - ih), \tag{6.1a}$$

$$y(t) = \sum_{i=0}^{k_c} C_i x(t - ih) + \sum_{i=0}^{k_d} D_i w(t - ih), \tag{6.1b}$$

where $x(\cdot) \in \mathbb{R}^n$, $y(\cdot) \in \mathbb{R}^p$, and $w(\cdot) \in \mathbb{R}^m$, and the initial condition $\varphi(t)$ is a piecewise continuous function $\varphi(t) : [-kh, 0] \to \mathbb{R}^n$ ($k = \max\{k_a, k_b, k_c, k_d\}$); thereby $x(t) = \varphi(t)$ on $[-kh, 0]$. All matrices A_i, B_i, C_i, and D_i appearing in (6.1) have their elements in $\mathbb{R}$.

By $x(t; \varphi, w)$ we mean the solution of the delay differential equation (6.1a) with the initial condition equal to φ and the input vector equal to w. Analogously, we define $y(t; \varphi, w) := C(\delta)x(t; \varphi, w) + D(\delta)w(t)$, that is, the system output of (6.1b) when $x(t) = x(t; \varphi, w)$.

Practically, we look for conditions allowing estimation of $x(t)$. To tackle the problem in a more formal way, we use the following observability definitions.

Definition 6.1 (Unknown input observability). System (6.1) is called *unknown input observable* (**UIO**) on the interval $[t_1, t_2]$ if there exist t_1' and t_2' ($t_1' < t_2'$) such that, for all inputs w and every initial condition φ,

$$y(t; \varphi, w) = 0 \text{ for all } t \in [t_1', t_2'] \text{ implies } x(t; \varphi, w) = 0 \text{ for } t \in [t_1, t_2].$$

Definition 6.2 (Backward UIO). System (6.1) is said to be *backward UIO* (**BUIO**) on $[t_1, t_2]$ if for each $\tau \in [t_1, t_2]$, there exist $t_1' < t_2' \le \tau$ such that, for all inputs w and every initial condition φ,

$$y(t; \varphi, w) = 0 \text{ for all } t \in [t_1', t_2'] \text{ implies } x(\tau; \varphi, w) = 0.$$

Remark 6.1. These definitions are essentially formulated following the observability definitions given in [37] for linear systems. Basically, UIO considers the case where the state vector can be reconstructed using past, actual, and future values of the system output. As for BUIO, it is related to the case where only actual and past values of the system output are needed for the actual state reconstruction. For the observer design, we will need the sufficient conditions obtained for BUIO.

Obviously, BUIO implies UIO. However, it should be noted that a system could be UIO (depending of the future values of the system output), and this does not exclude that the system is BUIO. That is, $x(t)$ may be reconstructed using future values of the system output, but it may be done also by using only past and actual values of the system output. For instance, the system

$$\dot{x}_1 = x_2, \quad \dot{x}_2 = x_1 + \delta x_2; \quad y_1 = \delta x_1, \quad y_2 = x_2$$

is UIO on $\left[t_1, t_1 + h\right]$, since, $y(t) = 0$ on $\left[\tau, t_1 + 2h\right]$ implies $x(\tau) = 0$. However, it is also BUIO on $\left[t_1, t_1 + h\right]$ $(t_1 \geq h)$ since, for each $\tau \in \left[t_1, t_1 + h\right]$, $y(t) = 0$ on $\left[t_1 - h, \tau\right]$ implies $x(\tau) = 0$.

By using the delay operator (backward time-shift operator) $\delta : x(t) \rightarrow x\left(t - h\right)$ system (6.1) may be represented in the following compact form:

$$\begin{aligned} \dot{x}(t) &= A(\delta)\, x(t) + B(\delta)\, w(t), \\ y(t) &= C(\delta)\, x(t) + D(\delta)\, w(t), \end{aligned} \tag{6.2}$$

where $A(\delta)$, $B(\delta)$, $C(\delta)$, and $D(\delta)$ are matrices over the polynomial ring $\mathbb{R}[\delta]$, defined as $A(\delta) := \sum_{i=0}^{k_a} A_i \delta^i$, $B(\delta) := \sum_{i=0}^{k_b} B_i \delta^i$, $C(\delta) := \sum_{i=0}^{k_c} C_i \delta^i$, and $D(\delta) := \sum_{i=0}^{k_d} D_i \delta^i$.

Now, for instance, the matrix $A(\delta)$ can be seen also as a linear transformation from the module $\mathbb{R}^n[\delta]$ to $\mathbb{R}^n[\delta]$ and likewise for the other matrices $B(\delta)$, $C(\delta)$, and $D(\delta)$.

6.3. Some concepts related to the polynomial ring $\mathbb{R}[\delta]$

Here we briefly review some concepts that are essential in the study carried out in this chapter. It deals mainly with invertibility properties of matrices over a polynomial ring and the concept of Noetherian modules.

A ring is like a field, but not all nonzero elements have an inverse. The polynomial ring $\mathbb{R}[\delta]$ is the set of all polynomials with indeterminate δ and real coefficients.

A module over a ring is like a vector space, but instead of a field as the set of scalars, we have a ring. So $\mathbb{R}^n[\delta]$ is the module over the polynomial ring, and its elements are n-dimensional vectors whose elements are polynomials.

By $\mathbb{R}^{n\times m}[\delta]$ we mean the set of matrices with n rows and m columns whose elements are within the polynomial ring $\mathbb{R}[\delta]$.

We recall the definition of invariant factors of a matrix over the polynomial ring. For any matrix $T(\delta) \in \mathbb{R}^{n\times m}[\delta]$, there exist two invertible matrices $P_1(\delta)$ and $P_2(\delta)$ (over $\mathbb{R}[\delta]$) that reduce $T(\delta)$ to its Smith normal form, that is,

$$P_1(\delta)\,T(\delta)\,P_2(\delta) = \begin{bmatrix} \operatorname{diag}(\psi_1(\delta)\cdots\psi_r(\delta)) & 0 \\ 0 & 0 \end{bmatrix}, \tag{6.3}$$

where $r = \operatorname{rank} T(\delta)$. Here the rank is calculated over the polynomial ring $\mathbb{R}[\delta]$.

Definition 6.3. The nonzero terms $\{\psi_i\}$ in the Smith normal form of $T(\delta)$ are called the invariant factors of $T(\delta)$. They satisfy $\psi_i|\psi_{i+1}$ and $d_i = d_{i-1}\psi_i$ ($d_0 = 1$), where each d_i is the greatest common divisor (gcd) of all $i \times i$ minors of $T(\delta)$. The $\{d_i\}$ are called the determinant divisors of $T(\delta)$.

For simplicity, for any matrix $T(\delta)$ with $\operatorname{rank} T(\delta) = \rho_T$, by $\operatorname{Inv}_S[T(\delta)] = \{\psi_i(\delta)\}_{1\le i\le r_T}$ we denote the set of invariant factors of the matrix $T(\delta)$.

Fact 6.1. *Given two matrices* $P(\delta)$ *and* $T(\delta) = \begin{bmatrix} P(\delta) \\ R(\delta) \end{bmatrix}$, *the identity* $\operatorname{Inv}_S P(\delta) = \operatorname{Inv}_S T(\delta)$ *is true if and only if* $R(\delta) = W(\delta)P(\delta)$ *for some matrix* $W(\delta)$ *(see, e.g., Lemma 3 in [7]).*

Definition 6.4. A matrix $M(\delta) \in \mathbb{R}^{n\times m}[\delta]$ is called *left unimodular* if there exists a matrix $[M(\delta)]_L^{-1} \in \mathbb{R}^{m\times n}[\delta]$ such that $[M(\delta)]_L^{-1} M(\delta) = I$. We call $[M(\delta)]_L^{-1}$ as a left inverse of $M(\delta)$. A *right unimodular* matrix is defined in an analogous way. Thus an invertible square matrix $M(\delta) \in \mathbb{R}^{n\times n}[\delta]$ is also called unimodular.

Fact 6.2. *A matrix* $M(\delta) \in \mathbb{R}^{n\times m}[\delta]$ *is left unimodular if and only if it has* m *constant (of zero degree) invariant factors. A matrix* $M(\delta) \in \mathbb{R}^{n\times m}[\delta]$ *is right unimodular if and only if it has* n *constant invariant zeros. Thus a square matrix* $M(\delta) \in \mathbb{R}^{n\times n}[\delta]$ *is unimodular if and only if it has* n *constant invariant zeros if and only if its determinant is a nonzero constant polynomial.*

Also, if a matrix $M(\delta) \in \mathbb{R}^{n\times m}[\delta]$ *is premultiplied or postmultiplied by a unimodular matrix, then the invariant factors of* $M(\delta)$ *do not change.*

Definition 6.5. Given an ascending chain of submodules

$$\mathcal{M}_1 \subseteq \mathcal{M}_2 \subseteq \cdots \subseteq \mathcal{M}_i \subseteq \cdots$$

of a module $\mathcal{M}$, we say that it is an *ascending stationary chain* if, for some k, $\mathcal{M}_{k+j} = \mathcal{M}_k$ for all $j \geq 1$. A *descending stationary chain* is defined in an analogous way by using $\supseteq$ instead of $\subseteq$ in the above chain.

Fact 6.3. *The* $\mathbb{R}[\delta]$*-module* $\mathbb{R}^n[\delta]$ *is a finitely generated (in fact, a free) Noetherian module, that is, every ascending chain of submodules of* $\mathbb{R}^n[\delta]$ *is stationary (see, e.g., [49]).*

For the reader interested in matrices over polynomial matrices, matrices over commutative rings, and module theory, we recommended the following references: [1,2,13,27,28].

6.4. Observability matrix

An easy way of checking the observability of a linear time-invariant system (without delays) is by using the so-called observability matrix. Likewise, we can generate a matrix in an iterative way that allows us to give sufficient conditions for the observability of system (6.2) and thus of system (6.1).

Algorithm to generate matrix $N_k(\delta)$**.**

Step 0 Define $\Delta_0 := 0_{1\times n}$, $G_0(\delta) := C(\delta)$, $F_0(\delta) := D(\delta)$.

Step 1 Define $S_0(\delta) \triangleq \begin{bmatrix} \Delta_0(\delta)B(\delta) \\ F_0(\delta) \end{bmatrix}$. Then Δ_1 is defined implicitly by the formula

$$\left[\begin{array}{c|c} F_1(\delta) & G_1(\delta) \\ \hline 0 & \Delta_1(\delta) \end{array}\right] := T_0(\delta)\left[\begin{array}{c|c} \Delta_0(\delta)B(\delta) & \Delta_0(\delta)A(\delta) \\ \hline F_0(\delta) & G_0(\delta) \end{array}\right], \tag{6.4}$$

where $T_0(\delta)$ is a unimodular matrix that reduces the matrix $S_0(\delta)$ into the form $\begin{bmatrix} F_1(\delta) \\ 0 \end{bmatrix}$, where $F_1(\delta)$ has full row rank equal to the rank of $S_0(\delta)$, that is, $\operatorname{rank}(F_1(\delta)) = \operatorname{rank}(S_0(\delta))$.

Thus in general we have the following:

Step k For $k \geq 1$, define $S_{k-1}(\delta) \triangleq \begin{bmatrix} \Delta_{k-1}(\delta) B(\delta) \\ F_{k-1}(\delta) \end{bmatrix}$. Then $\Delta_k(\delta)$ is defined implicitly by the formula

$$\left[\begin{array}{c|c} F_k(\delta) & G_k(\delta) \\ \hline 0 & \Delta_k(\delta) \end{array}\right] := T_{k-1}(\delta) \left[\begin{array}{c|c} \Delta_{k-1}(\delta) B(\delta) & \Delta_{k-1}(\delta) A(\delta) \\ F_{k-1}(\delta) & G_{k-1}(\delta) \end{array}\right], \tag{6.5}$$

where $T_{k-1}(\delta)$ is a unimodular matrix over $\mathbb{R}[\delta]$ that reduces the matrix S_{k-1} into the form $\begin{bmatrix} F_k(\delta) \\ 0 \end{bmatrix}$, where $F_k(\delta)$ has full row rank equal to the rank of $S_{k-1}(\delta)$, that is, $\operatorname{rank}(F_k(\delta)) = \operatorname{rank}(S_{k-1}(\delta))$.

Thus the matrix $N_k(\delta)$ is build up by the concatenation of the matrices $\{\Delta_i(\delta)\}$, $i = 1, \ldots, k$, that is,

$$N_k(\delta) \triangleq \begin{bmatrix} \Delta_1(\delta) \\ \vdots \\ \Delta_k(\delta) \end{bmatrix}. \tag{6.6}$$

Let $\mathcal{N}_k(\delta)$ the free $\mathbb{R}[\delta]$-submodule of $\mathbb{R}^n[\delta]$ generated by the rows of the matrix $N_k(\delta)$ constructed before.

Lemma 6.1. *There exists a positive number k^* such that $\mathcal{N}_{k^*+i}(\delta) = \mathcal{N}_{k^*}(\delta)$ for all $i \geq 1$.*

Proof. Since $\mathbb{R}^n[\delta]$ is a Noetherian module, every ascending chain of submodules is stationary (see Fact 6.3). By the construction of $N_k(\delta)$ we precisely have an ascending chain of submodules of $\mathbb{R}^n[\delta]$, that is,

$$\mathcal{N}_1(\delta) \subseteq \mathcal{N}_2(\delta) \subseteq \cdots \subseteq \mathcal{N}_k(\delta) \subseteq \cdots.$$

Therefore there exists a positive number k^* satisfying the conclusion of the lemma. □

Remark 6.2. The lemma says that the integer k^* does exist, but it does not give a clue of how many iterations of the algorithm is needed to arrive at k^*. Despite that an exact estimate of such a number of iterations is difficult to calculate, Bejarano and Zheng [7] give a way of getting an upper estimate.

Remark 6.3. The matrix $N_{k^*}(\delta)$ is used by Silverman [58] and Molinari [46] to calculate the null-output weakly unobservable subspace, denoted by $\mathcal{V}^*$, which is the largest submodule $\mathcal{V}$ achieving the conditions

$A(\delta)\mathcal{V} \subseteq \mathcal{V} + \operatorname{Im} B(\delta)$ and $C(\delta)\mathcal{V} = 0$. In the case without delays, the identity kernel $N_{k^*} = \mathcal{V}$. However, this is not true in general for the class of systems considered in (6.2). For more details, we refer the reader to [5].

Moreover, a linear system (without delays) is strongly observable (= UIO according to our definition) if and only if $\mathcal{V}^* = \{0\}$ (= number of rows; see [60]). However, again, this is not true in general when considering systems with delays, as can be seen by the following simple example:

$$\dot{x}_1(t) = x_2(t) \text{ and } \dot{x}_2(t) = x_2(t) + w(t),$$
$$y_1(t) = x_1(t) - x_1(t-h) - x_2(t),$$
$$y_2(t) = x_2(t) - x_2(t-h),$$

with $x_1(0) = x_{1,0}$ and $x_2(0) = x_{2,0}$. Thus, if $w(t) = -x_2(t)$, then we get that $x_1(t) = x_{1,0} + x_{2,0}t$ and $x_2(t) = x_{2,0}$ for all $t \geq 0$. Hence, for $t \geq h$, the output can be rewritten as

$$y_1(t) = x_{2,0}h - x_{2,0},$$
$$y_2(t) = 0.$$

Therefore the identity $y_1(t) \equiv 0$ is obtained if we choose $x_{2,0} = 0$. Thus the dynamic equation takes the form $\dot{x}_1(t) = 0$, $x_2 \equiv 0$, that is, we get the *zero dynamics* equation $\dot{x}_1(t) = 0$. However, since $C(\delta) = \begin{bmatrix} 1-\delta & -1 \\ 0 & 1-\delta \end{bmatrix}$, the null space of $C(\delta)$ is $\{0\}$, which in turn implies that $\mathcal{V}^*(\delta) = \{0\}$. Therefore $\mathcal{V}^*(\delta) = \{0\}$ does not ensure the UI observability of the system.

Intuitively, the invariant factors of $\{N_k(\delta)\}$ should not depend on the matrices $T_0(\delta), \ldots, T_{k-1}(\delta)$, since they are unimodular over $\mathbb{R}[\delta]$. We prove such a claim in the following lemma.

Lemma 6.2. *The invariant factors of the matrix $N_{k^*}(\delta)$ are independent of the choice of $T_0(\delta), \ldots, T_{k^*-1}(\delta)$.*

Proof. We carry out the proof by induction. First, it is easy to verify that for two matrices $\Delta_1(\delta)$ and $\overline{\Delta}_1(\delta)$ calculated in (6.5) with two different matrices $T_0(\delta)$ and $\overline{T}_0(\delta)$, respectively, we obtain that

$$\overline{\Delta}_1(\delta) = J_1(\delta)\Delta_1(\delta) \tag{6.7}$$

for some unimodular matrix $J_1(\delta)$. Indeed, since $\overline{T}_0(\delta)$ and $T_0^{-1}(\delta)$ are invertible and $F_1(\delta)$ has full row rank, we obtain that

$$\overline{T}_0 T_0^{-1} = \begin{bmatrix} * & * \\ 0 & J_1 \end{bmatrix}, \tag{6.8}$$

which immediately implies that $J_1(\delta)$ is invertible over $\mathbb{R}[\delta]$. This applied to (6.5) yields (6.7).

Let $N_{k-1}(\delta)$ and $\overline{N}_{k-1}(\delta)$ be generated by the sets of matrices $\{T_0(\delta), \ldots, T_{k-2}(\delta)\}$ and $\left\{\overline{T}_0(\delta), \ldots, \overline{T}_{k-2}(\delta)\right\}$, respectively. Suppose that $\overline{N}_{k-1}(\delta) = J_{k-1}(\delta) N_{k-1}(\delta)$ for some unimodular matrix $J_{k-1}(\delta)$. We will prove that $\overline{N}_k(\delta) = J_k(\delta) N_k(\delta)$ for some unimodular matrix $J_k(\delta)$. By a straightforward calculation we can verify that N_k satisfies the following recursive algorithm:

$$\left[\begin{array}{c|c} W_k(\delta) & X_k(\delta) \\ \hline 0 & N_k(\delta) \end{array}\right] = Y_{k-1}(\delta) \left[\begin{array}{c|c} N_{k-1}(\delta) B(\delta) & N_{k-1}(\delta) A(\delta) \\ D(\delta) & C(\delta) \end{array}\right] \tag{6.9}$$

where $Y_k(\delta)$ is unimodular, and $W_k(\delta)$ has full row rank. Analogously, (6.9) is valid for $\overline{N}_k(\delta)$ with some matrices $\overline{Y}_{k-1}(\delta)$ and $\overline{W}_k(\delta)$. Thus we obtain that

$$\begin{aligned} \begin{bmatrix} W_k & X_k \\ 0 & N_k \end{bmatrix} &= Y_{k-1} \begin{bmatrix} J_{k-1}^{-1} & 0 \\ 0 & I \end{bmatrix} \overline{Y}_{k-1}^{-1} \overline{Y}_{k-1} \left[\begin{array}{c|c} J_{k-1} N_{k-1} B & J_{k-1} N_{k-1} A \\ D & C \end{array}\right] \\ &= Y_{k-1} \begin{bmatrix} D_{k-1}^{-1} & 0 \\ 0 & I \end{bmatrix} \overline{Y}_{k-1}^{-1} \begin{bmatrix} \overline{W}_k & \overline{X}_k \\ 0 & \overline{N}_k \end{bmatrix}. \end{aligned}$$

Therefore, using the same arguments to prove that J_1 is unimodular, we deduce the existence of a unimodular matrix J_k such that $\overline{N}_k = J_k N_k$. Therefore $\overline{N}_k$ and N_k have the same Smith form, that is, they have the same invariant factors. □

6.5. Sufficient conditions for UI observability

From now, it is possible to give sufficient observability conditions. In the way to arrive at such conditions, we will draw the connection of the recursive algorithm used to obtain $\Delta_k(\delta)$ and a way that may be used for the reconstruction of the state vector. Let us define $\hat{y}^1(t) = T_0(\delta) \begin{bmatrix} 0 \\ y(t) \end{bmatrix}$.

Then from (6.2) and (6.5) we obtain

$$\hat{y}^1(t) \triangleq \begin{bmatrix} \hat{y}_1^1(t) \\ \hat{y}_2^1(t) \end{bmatrix} = \begin{bmatrix} G_1(\delta)x(t) + F_1(\delta)w(t) \\ \Delta_1(\delta)x(t) \end{bmatrix}. \tag{6.10}$$

From here we will define a chain of vectors $\{\hat{y}^i(t)\}$. As for $\hat{y}_1^i(t)$ and $\hat{y}_2^i(t)$, they are the upper and lower subvectors of $\hat{y}^i(t)$, respectively, whose dimension will be implicitly defined.

We have already obtained a *virtual* output $\hat{y}_2^1(t)$ without the influence of the unknown input $w(t)$. Let us define $\alpha_1 = \deg \Delta_1(\delta)$. Then, for $t \geq \alpha_1 h$, $\hat{y}_2^1(t)$ is differentiable. Hence, according to (6.10) and (6.2), we obtain the following equation for $t \geq \alpha_1$:

$$\frac{d}{dt}\hat{y}_2^1(t) = \frac{d}{dt}\Delta_1(\delta)x(t) = \Delta_1(\delta)A(\delta)x(t) + \Delta_1(\delta)B(\delta)w(t). \tag{6.11}$$

Now let us generate an extended vector $\xi_1(t) \triangleq \begin{bmatrix} \frac{d}{dt}\hat{y}_2^1(t) \\ \hat{y}_1^1(t) \end{bmatrix}$. Thus, again, we define $\hat{y}^2(t) = T_1(\delta)\xi_1(t)$, $t \geq \alpha_1 h$. In view of (6.5) and (6.11), we obtain the identity

$$\hat{y}^2(t) \triangleq \begin{bmatrix} \hat{y}_1^2(t) \\ \hline \hat{y}_2^2(t) \end{bmatrix} = \begin{bmatrix} G_2(\delta)x(t) + F_2(\delta)w(t) \\ \hline \Delta_2(\delta)x(t) \end{bmatrix}.$$

Differentiation of $\hat{y}_2^2(t)$ gives, for $t \geq h\max(\alpha_1, \alpha_2)$ (where $\alpha_2 = \deg \Delta_2(\delta)$),

$$\frac{d}{dt}\hat{y}_2^2(t) = \Delta_2(\delta)A(\delta)x(t) + \Delta_2(\delta)B(\delta)w(t).$$

Likewise, we may define the second extended vector $\xi_2(t) \triangleq \begin{bmatrix} \frac{d}{dt}\hat{y}_2^2(t) \\ \hat{y}_1^2(t) \end{bmatrix}$. After defining $\hat{y}_3(t) \triangleq T_2(\delta)\xi_2(t)$, we obtain the identities $\hat{y}_1^3(t) = G_3(\delta)x(t) + F_3(\delta)w(t)$ and $\hat{y}_2^3(t) = \Delta_3(\delta)x(t)$.

The previous procedure allows us writing the following expressions, for $k \geq 1$ and $t \geq h\max_{1 \leq i \leq k-1} \deg \Delta_i(\delta)$:

$$\xi_0(t) \triangleq \begin{bmatrix} 0 \\ y(t) \end{bmatrix}, \quad \xi_k(t) \triangleq \begin{bmatrix} \frac{d}{dt}\hat{y}_2^k(t) \\ \hat{y}_1^k(t) \end{bmatrix},$$
$$\hat{y}^k(t) \triangleq \begin{bmatrix} \hat{y}_1^k(t) \\ \hat{y}_2^k(t) \end{bmatrix} \triangleq T_{k-1}(\delta)\xi_{k-1}(t). \tag{6.12}$$

Thus we obtain the identities

$$\Delta_k(\delta)\,x(t) = \hat{y}_2^k(t)\,,\ \ k \geq 1. \tag{6.13}$$

Hence we define $Y(t)$ as

$$Y(t) \triangleq \begin{bmatrix} \hat{y}_2^1(t) \\ \hat{y}_2^2(t) \\ \vdots \\ \hat{y}_2^{k^*}(t) \end{bmatrix}.$$

With the definition $t^* = h \max_{1\leq i \leq k^*-1} \deg \Delta_i(\delta)$, by (6.13) we have that

$$N_{k^*}(\delta)\,x(t) = Y(t) \ \text{ for all } t \geq t^*. \tag{6.14}$$

Theorem 6.1. *System (6.1) is BUIO on $[t^*, t_2]$ for every $t_2 > t^*$ if $N_{k^*}(\delta)$ has n constant invariant factors, that is, $\mathrm{Inv}_S[N_{k^*}(\delta)] = \{\psi_1 \cdots \psi_n\}$, where each $\psi_i \in \mathbb{R}_{\neq 0}$. Moreover, the state trajectories $x(t)$ can be reconstructed in finite time if the time derivatives of $y(t)$ can be calculated.*

Proof. If $y(t) = 0$ on the interval $[t^*, t_2]$, then also $Y(t)$ is equal to zero on that interval. Hence, by (6.14), $x(t) \equiv 0$, provided that $N_{k^*}(\delta)$ has n constant invariant factors.

Moreover, according to Fact 6.2, the matrix $N_{k^*}(\delta)$ is left invertible, that is, there exists a matrix $[N_{k^*}(\delta)]_L^{-1}$ such that $[N_{k^*}(\delta)]_L^{-1} N_{k^*}(\delta) = I$. Therefore we get that

$$x(t) = [N_{k^*}(\delta)]_L^{-1}\, Y(t)\,. \tag{6.15}$$

□

Remark 6.4. In the case where $w(t)$ is a smooth vector function, there exists a matrix $\Omega(\delta) \begin{bmatrix} y(t) \\ \dot{y}(t) \\ \vdots \\ y^{(k^*-1)}(t) \end{bmatrix} = Y(t)$. Thus in that way the state $x(t)$ can be reconstructed as

$$x(t) = [N_{k^*}(\delta)]_L^{-1}\, \Omega(\delta) \begin{bmatrix} y(t) \\ \dot{y}(t) \\ \vdots \\ y^{(k^*-1)}(t) \end{bmatrix}.$$

Remark 6.5. By defining the forward shift operator as $\delta^{-1}f(t) = f(t+h)$ we can extend the considered ring to $\mathbb{R}[\delta, \delta^{-1}]$, in which δ has an inverse. Then we can relaxed the condition of Theorem 6.1 by asking the matrix $N_{k^*}(\delta)$ to have n invariant factors of the form $a\delta$. If so, then the matrix $N_{k^*}(\delta)$ is left invertible over the ring $\mathbb{R}[\delta, \delta^{-1}]$. The physical meaning of that is that, according to (6.15), $x(t)$ may depend on the future values of the system output.

6.6. Observer design

The condition obtained for the observability analysis is also useful for the design of a Luenberger-like observer involving the past and actual values of the system output. One more condition will be required. We risk to say that the conditions required for the proposed observer are considerably relaxed since they coincide with the necessary and sufficient conditions for the unknown input observer design of linear systems without delays.

More precisely, we first try to decompose system (6.2) into a simpler form, provided that some conditions are satisfied, and then transform it into a higher-dimensional observer normal form with output (and the derivative of the output) injection and its delay. Finally, we can design an unknown input observer for the obtained observer normal form. The main idea comes from the observer design for linear time-delay systems without unknown inputs presented in [35].

To design the observer, the following assumptions will be done:

A1 $\text{Inv}_S[N_{k^*}(\delta)] = \{\psi_1 \cdots \psi_n\}$, where each $\psi_i \in \mathbb{R}_{\neq 0}$.

A2 $\text{Inv}_S[C_{BD}(\delta)] = \text{Inv}_S[C_D(\delta)]$, where

$$C_{BD}(\delta) = \begin{bmatrix} C(\delta)B(\delta) & D(\delta) \\ D(\delta) & 0 \\ B(\delta) & 0 \end{bmatrix} \text{ and } C_D(\delta) = \begin{bmatrix} C(\delta)B(\delta) & D(\delta) \\ D(\delta) & 0 \end{bmatrix}.$$

Remark 6.6. For the case without delays ($h = 0$), condition A1 is equivalent to $\text{rank}(N_{k^*}) = n$, and A2 can be written by changing Inv_S by rank. In that case, both conditions are necessary and sufficient for the design of a Luenberger-like observer with convergence of arbitrary rate.

Lemma 6.3. *If Assumption A2 is satisfied, then there exists a matrix* $W(\delta) \in \mathbb{R}^{(n+p)\times 2p}[\delta]$ *such that:*

1. $W(\delta)C_D(\delta) = \begin{bmatrix} B(\delta) & 0 \\ D(\delta) & 0 \end{bmatrix}$, *and*

2. *for any matrix* $J(\delta)$ *such that* $J(\delta)\begin{bmatrix} B(\delta) \\ D(\delta) \end{bmatrix} = 0$, *then* $J(\delta)W(\delta) = 0$.

Proof. We notice that $C_{BD}(\delta) = \begin{bmatrix} C_D(\delta) \\ \begin{bmatrix} B(\delta) & 0 \end{bmatrix} \end{bmatrix}$. Thus, according to Fact 6.1, assumption A2 implies that $\begin{bmatrix} B(\delta) & 0 \end{bmatrix}$ is spanned by $C_D(\delta)$. Moreover, it is clear that the $\begin{bmatrix} D(\delta) & 0 \end{bmatrix}$ is also spanned by $C_D(\delta)$. Therefore there exists a matrix $W(\delta)$ satisfying the claim of the lemma.. □

To prove the second condition of Lemma 6.3, let $U(\delta)$ be a unimodular matrix over $\mathbb{R}[\delta]$ that transforms the following matrix into its Hermite form:

$$U(\delta)C_D(\delta) = \begin{bmatrix} V(\delta) \\ 0 \end{bmatrix},$$

where $V(\delta)$ is full row rank over $\mathbb{R}[\delta]$. By the first claim of the lemma we can select a matrix $\bar{V}(\delta)$ such that $\bar{V}(\delta)V(\delta) = \begin{bmatrix} B(\delta) & 0 \\ D(\delta) & 0 \end{bmatrix}$. Thus we can select $W(\delta)$ as $W(\delta) = [\bar{V}(\delta), 0]U(\delta)$, with which we have $W(\delta)C_D(\delta) = \begin{bmatrix} B(\delta) & 0 \\ D(\delta) & 0 \end{bmatrix}$. Moreover, since $V(\delta)$ is full row rank over $\mathbb{R}[\delta]$, for any matrix $J(\delta)$ such that $J(\delta)\begin{bmatrix} B(\delta) \\ D(\delta) \end{bmatrix} = 0$, we always have $J(\delta)\bar{V}(\delta) = 0$, which implies $J(\delta)W(\delta) = 0$.

6.7. Observer normal form

6.7.1 System decomposition

Suppose that $W(\delta) \in \mathbb{R}^{(n+p)\times 2p}[\delta]$ is a matrix over $\mathbb{R}[\delta]$ such that Lemma 6.3 is satisfied. Decompose $W(\delta) = [W_1(\delta), W_2(\delta)]$ with $W_1(\delta), W_2(\delta) \in \mathbb{R}^{(n+p)\times m}[\delta]$. Then we must have $W_1(\delta)D(\delta) = 0$. Hence we obtain $W_1(\delta)y = W_1(\delta)C(\delta)x$, which in turn yields the equation

$$W_1(\delta)\dot{y} = W_1(\delta)C(\delta)A(\delta)x + W_1(\delta)C(\delta)B(\delta)u.$$

Thus, due to Lemma 6.3, we have that

$$W_1(\delta)\dot{y} + W_2(\delta)y = W(\delta)\begin{bmatrix} C(\delta)A(\delta) \\ C(\delta) \end{bmatrix} x + \begin{bmatrix} B(\delta) \\ D(\delta) \end{bmatrix} w.$$

Decompose again the matrix $W(\delta)$ as

$$W(\delta) = \left(\frac{K(\delta)}{\Gamma(\delta)}\right) = \begin{bmatrix} K_1(\delta) & K_2(\delta) \\ \hline \Gamma_1(\delta) & \Gamma_2(\delta) \end{bmatrix},$$

where $K_i(\delta) \in \mathbb{R}^{n\times p}[\delta]$ and $\Gamma_i(\delta) \in \mathbb{R}^{p\times p}[\delta]$ for $i = 1, 2$. Thereby we can substitute $\begin{bmatrix} B(\delta) \\ D(\delta) \end{bmatrix} w$ into system (6.2) to replace the unknown input w. That is, system (6.2) can be rewritten as follows:

$$\begin{aligned} \dot{x} &= \bar{A}(\delta)x + K_1(\delta)\dot{y} + K_2(\delta)y, \\ y &= \tilde{C}(\delta)x + \Gamma_1(\delta)\dot{y} + \Gamma_2(\delta)y, \end{aligned} \tag{6.16}$$

where $\bar{A}(\delta) = A(\delta) - K(\delta)\begin{bmatrix} C(\delta)A(\delta) \\ C(\delta) \end{bmatrix} \in \mathbb{R}^{n\times n}[\delta]$ and $\tilde{C}(\delta) = C(\delta) - \Gamma(\delta)\begin{bmatrix} C(\delta)A(\delta) \\ C(\delta) \end{bmatrix} \in \mathbb{R}^{p\times n}[\delta]$.

Suppose rank $\left(\tilde{C}(\delta)\right) p_1 \leq p$. Then there exists a unimodular matrix $\Lambda(\delta)$ over $\mathbb{R}[\delta]$ such that

$$\Lambda(\delta)\tilde{C}(\delta) = \begin{bmatrix} \bar{C}(\delta) \\ 0 \end{bmatrix} \tag{6.17}$$

with $\bar{C}(\delta) \in \mathbb{R}^{p_1\times n}[\delta]$ being full row rank over $\mathbb{R}[\delta]$. Finally, by noting $\bar{y} = \Lambda(\delta)y$, system (6.16) can be written in the following decomposed form:

$$\begin{aligned} \dot{x} &= \bar{A}(\delta)x + K_1(\delta)\Lambda^{-1}(\delta)\dot{\bar{y}} + K_2(\delta)\Lambda^{-1}(\delta)\bar{y}, \\ \bar{y} &= \begin{bmatrix} \bar{C}(\delta) \\ 0 \end{bmatrix} x + \bar{\Gamma}_1(\delta)\dot{\bar{y}} + \bar{\Gamma}_2(\delta)\bar{y}, \end{aligned} \tag{6.18}$$

where

$$\begin{aligned} \bar{\Gamma}_1(\delta) &= \Lambda(\delta)\Gamma_1(\delta)\Lambda^{-1}(\delta) \in \mathbb{R}^{p\times p}[\delta], \\ \bar{\Gamma}_2(\delta) &= \Lambda(\delta)\Gamma_2(\delta)\Lambda^{-1}(\delta) \in \mathbb{R}^{p\times p}[\delta]. \end{aligned} \tag{6.19}$$

6.7.2 Normal form

Hou et al. [35] proposed an observer normal form for time-delay systems without unknown inputs. Let us recall such a result.

Theorem 6.2 ([35]). *There exist a* bicausal generalized *change of coordinates* $z = T(\delta)x$ *($T(\delta)$ is left invertible) and a matrix* $\bar{F}(\delta)$ *such that the system*

$$\begin{aligned} \dot{\bar{x}} &= \bar{A}(\delta)\,\bar{x}, \\ \bar{y} &= \bar{C}(\delta)\bar{x} \end{aligned} \tag{6.20}$$

is transformed into the following observer normal form:

$$\begin{cases} \dot{\bar{z}} = A_0\bar{z} + \bar{F}(\delta)\bar{y}, \\ \bar{y} = C_0\bar{z}, \end{cases}$$

where

$$\begin{aligned} A_0 &= \begin{bmatrix} 0 & \mathrm{I}_{p_1} & \cdots & 0 \\ 0 & 0 & \ddots & \vdots \\ 0 & 0 & \cdots & I_{p_1} \\ 0 & 0 & \cdots & 0 \end{bmatrix} \in \mathbb{R}^{p_1 l^* \times p_1 l^*}, \\ C_0 &= \left[\mathrm{I}_{p_1}, 0, \cdots, 0\right] \in \mathbb{R}^{p_1 \times p_1 l^*}, \end{aligned} \tag{6.21}$$

if and only if there exists an integer $l^* \in \mathbb{N}_0$ *such that the matrix*

$$\bar{\mathcal{O}}_{l^*}(\delta) = \begin{bmatrix} \bar{C}(\delta) \\ \bar{C}(\delta)\bar{A}(\delta) \\ \vdots \\ \bar{C}(\delta)\bar{A}^{l^*-1}(\delta) \end{bmatrix} \in \mathbb{R}^{p_1 l^* \times n}[\delta] \tag{6.22}$$

is left unimodular over $\mathbb{R}[\delta]$.

Moreover, the matrix $\bar{T}(\delta) = \mathrm{col}\{\bar{T}_1(\delta), \cdots, \bar{T}_{l^*}(\delta)\}$ *is defined as follows:*

$$\begin{cases} \bar{T}_1(\delta) = \bar{C}(\delta), \\ \bar{T}_{i+1}(\delta) = \bar{T}_i(\delta)\bar{A}(\delta) - \bar{F}_i(\delta)\bar{C}(\delta) \text{ for } 1 \le i \le l^* - 1, \end{cases} \tag{6.23}$$

with $\bar{F}_i(\delta)$ *being determined by the following equations:*

$$\bar{F}(\delta) := \left[\bar{F}_{l^*}(\delta), \cdots, \bar{F}_1(\delta)\right] = \bar{C}(\delta)\bar{A}^{l^*}(\delta)\left[\bar{\mathcal{O}}_{l^*}(\delta)\right]_L^{-1}. \tag{6.24}$$

Thus, in this subsection, we describe how to obtain an analogous observer normal form for the class of systems considered in (6.2). We start with the following result.

Lemma 6.4. *If Assumption A1 is satisfied, then the matrix*

$$\mathcal{O}_{k^*}(\delta) = \begin{bmatrix} C(\delta) \\ C(\delta)A(\delta) \\ \vdots \\ C(\delta)A^{k^*-1}(\delta) \end{bmatrix} \in \mathbb{R}^{pk^* \times n}[\delta] \tag{6.25}$$

is left unimodular over $\mathbb{R}[\delta]$*, where the integer* k^* *is defined in Lemma 6.1.*

Proof. Let us assume that Assumption A1 is satisfied. Consider the first step of the algorithm to generate $N_{k^*}(\delta)$. Since $\Delta_0(\delta) = 0$, there exist matrices $Q_1(\delta)$ and $R_1(\delta)$ over $\mathbb{R}[\delta]$ such that $\Delta_1(\delta) = Q_1(\delta)C(\delta)$ and $G_1(\delta) = R_1(\delta)C(\delta)$. Therefore in the second iteration, we obtain

$$\begin{bmatrix} F_2(\delta) & G_2(\delta) \\ 0 & \Delta_2(\delta) \end{bmatrix} = T_1(\delta) \begin{bmatrix} Q_1(\delta)C(\delta)B(\delta) & Q_1(\delta)C(\delta)A(\delta) \\ F_1(\delta) & R_1(\delta)C(\delta) \end{bmatrix},$$

from which we can see that there exist the matrices $Q_2(\delta)$ and $R_2(\delta)$ over $\mathbb{R}[\delta]$ such that $\Delta_2(\delta) = Q_2(\delta)\begin{bmatrix} C(\delta) \\ C(\delta)A(\delta) \end{bmatrix}$ and $G_2(\delta) = R_2(\delta)\begin{bmatrix} C(\delta) \\ C(\delta)A(\delta) \end{bmatrix}$. Following in this way, we can see that in the kth iteration, there exists the matrix $Q_k(\delta)$ such that $\Delta_k(\delta) = Q_k(\delta)\begin{bmatrix} C(\delta) \\ \vdots \\ C(\delta)A^k(\delta) \end{bmatrix}$ with $k \in \mathbb{N}_0$. Hence the matrix $N_{k^*}(\delta)$ can be expressed as $N_{k^*}(\delta) = Q(\delta)\mathcal{O}_{k^*}(\delta)$ for some matrix $Q(\delta)$. Thus, under assumption A1, $N_{k^*}(\delta)$ admits a left inverse matrix $[N_{k^*}(\delta)]_L^{-1}$ such that

$$[N_{k^*}(\delta)]_L^{-1} N_{k^*}(\delta) = [N_{k^*}(\delta)]_L^{-1} Q(\delta)\mathcal{O}_{k^*}(\delta) = I,$$

which implies that $[N_{k^*}(\delta)]_L^{-1} Q(\delta)$ is a left inverse of $\mathcal{O}_{k^*}(\delta)$. □

Based on Lemma 6.4, we have the following result for the deduced system (6.16).

Lemma 6.5. *Let assumptions A1 and A2 be satisfied. Then the matrix*

$$\tilde{\mathcal{O}}_{k^*}(\delta) = \begin{bmatrix} \tilde{C}(\delta) \\ \tilde{C}(\delta)\bar{A}(\delta) \\ \vdots \\ \tilde{C}(\delta)\bar{A}^{k^*-1}(\delta) \end{bmatrix} \in \mathbb{R}^{pk^* \times n}[\delta] \tag{6.26}$$

is left unimodular over $\mathbb{R}[\delta]$.

Proof. According to Lemma 6.4, we need only to prove that if Assumption A2 is satisfied for the quadruple $(A(\delta), B(\delta), C(\delta), D(\delta))$, then this assumption is also satisfied for the quadruple $\left(\bar{A}(\delta), B(\delta), \tilde{C}(\delta), D(\delta)\right)$. □

Let T_0 be used in the first step to generate $N_{k^*}(\delta)$. Since $\Delta_0(\delta) = 0$, we have

$$\begin{aligned}
&T_0(\delta)\begin{bmatrix} \Delta_0(\delta)B(\delta) & \Delta_0(\delta)\bar{A}(\delta) \\ D(\delta) & \tilde{C}(\delta) \end{bmatrix} \\
&= T_0(\delta)\begin{bmatrix} \Delta_0(\delta)B(\delta) & \Delta_0(\delta)A(\delta) \\ D(\delta) & C(\delta) \end{bmatrix} - T_0(\delta)\begin{bmatrix} 0 & \Delta_0(\delta)K(\delta)\begin{bmatrix} C(\delta)A(\delta) \\ C(\delta) \end{bmatrix} \\ 0 & \Gamma(\delta)\begin{bmatrix} C(\delta)A(\delta) \\ C(\delta) \end{bmatrix} \end{bmatrix} \\
&= \begin{bmatrix} \tilde{Q}_1(\delta)D(\delta) & G_1(\delta) - \tilde{Q}_1(\delta)\Gamma(\delta)\begin{bmatrix} C(\delta)A(\delta) \\ C(\delta) \end{bmatrix} \\ 0 & \Delta_1(\delta) \end{bmatrix}
\end{aligned}$$

for some matrix $\tilde{Q}_1(\delta)$ over $\mathbb{R}[\delta]$. The last inequality is derived from the second claim of Lemma 6.3. Hence $\Delta_1(\delta)$ for $\left(\bar{A}(\delta), B(\delta), \tilde{C}(\delta), D(\delta)\right)$ is the same as for $(A(\delta), B(\delta), C(\delta), D(\delta))$. For the second step of the same algorithm, we obtain

$$\begin{aligned}
&T_1(\delta)\begin{bmatrix} \Delta_1(\delta)B(\delta) & \Delta_1(\delta)\bar{A}(\delta) \\ \tilde{Q}_1(\delta)D(\delta) & G_1(\delta) - \tilde{Q}_1(\delta)\Gamma(\delta)\begin{bmatrix} C(\delta)A(\delta) \\ C(\delta) \end{bmatrix} \end{bmatrix} \\
&= T_1(\delta)\begin{bmatrix} \Delta_1(\delta)B(\delta) & \Delta_1(\delta)A(\delta) \\ \tilde{Q}_1(\delta)D(\delta) & G_1(\delta) \end{bmatrix} - T_1(\delta)\begin{bmatrix} 0 & \Delta_1(\delta)K(\delta)\begin{bmatrix} C(\delta)A(\delta) \\ C(\delta) \end{bmatrix} \\ 0 & \tilde{Q}_1(\delta)\Gamma(\delta)\begin{bmatrix} C(\delta)A(\delta) \\ C(\delta) \end{bmatrix} \end{bmatrix}
\end{aligned}$$

$$= \begin{bmatrix} \tilde{Q}_2(\delta)\begin{bmatrix} B(\delta) \\ D(\delta) \end{bmatrix} & G_2(\delta) - \tilde{Q}_2(\delta)\begin{bmatrix} K(\delta) \\ \Gamma(\delta) \end{bmatrix}\begin{bmatrix} C(\delta)A(\delta) \\ C(\delta) \end{bmatrix} \\ 0 & \Delta_2(\delta) \end{bmatrix}.$$

Thus by induction it is easy to see that the matrix $N_{k^*}(\delta)$ for the quadruple $\left(\bar{A}(\delta), B(\delta), \tilde{C}(\delta), D(\delta)\right)$ is the same as the matrix $N_{k^*}(\delta)$ calculated for $(A(\delta), B(\delta), C(\delta), D(\delta))$.

Corollary 6.1. *Let assumptions A1 and A2 be satisfied. Then the matrix $\bar{\mathcal{O}}_{k^*}(\delta)$ defined in (6.22) (with $l = k^*$) is left unimodular.*

Proof. First, let $\Lambda(\delta)$ be a unimodular matrix such that $\Lambda(\delta)\tilde{C}(\delta) = \begin{bmatrix} \bar{C}(\delta) \\ 0 \end{bmatrix}$. Then after defining the unimodular matrix $\Lambda_d(\delta) = \mathrm{diag}\{\Lambda(\delta), \cdots, \Lambda(\delta)\} \in \mathbb{R}^{pk^* \times pk^*}[\delta]$, we get the equation

$$E\Lambda_d(\delta)\tilde{\mathcal{O}}_{k^*}(\delta) = \begin{bmatrix} \bar{\mathcal{O}}_{k^*}(\delta) \\ 0_{(p-r)k^* \times n} \end{bmatrix},$$

where $E \in \mathbb{R}^{pk^* \times pk^*}$ is a permutation matrix. Therefore the last equation yields the identity $\mathrm{Inv}_S\tilde{\mathcal{O}}_{k^*}(\delta) = \mathrm{Inv}_S\bar{\mathcal{O}}_{k^*}(\delta)$. Then the result follows by Lemma 6.5. □

All the results presented hitherto allow for transforming system (6.2) into a normal observer form similar to that described in (6.2).

Theorem 6.3. *Let assumptions A1 and A2 be satisfied. Then with the change of coordinates $z(t) = T(\delta)x(t) \in \mathbb{R}^{n_z}$ and with $T(\delta)$ defined in (6.23), system (6.2) can be transformed into the following observer normal form:*

$$\begin{cases} \dot{z}(t) = A_0 z(t) + \left(\begin{bmatrix} \bar{F}(\delta) & 0 \end{bmatrix} + \bar{K}_2(\delta)\right)\bar{y}(t) + \bar{K}_1(\delta)\dot{\bar{y}}(t), \\ \bar{y}(t) = \begin{bmatrix} C_0 \\ 0 \end{bmatrix} z(t) + \bar{\Gamma}_1(\delta)\dot{\bar{y}}(t) + \bar{\Gamma}_2(\delta)\bar{y}(t), \end{cases} \tag{6.27}$$

where $\bar{\Gamma}_1(\delta)$, $\bar{\Gamma}_2(\delta)$, A_0, C_0, and $\bar{F}(\delta)$ are defined in (6.19), (6.21), and (6.24), respectively, with

$$\begin{aligned} \bar{K}_1(\delta) &= \bar{T}(\delta)K_1(\delta)\Lambda^{-1}(\delta) \in \mathbb{R}^{n_z \times p}[\delta], \\ \bar{K}_2(\delta) &= \bar{T}(\delta)K_2(\delta)\Lambda^{-1}(\delta) \in \mathbb{R}^{n_z \times p}[\delta], \end{aligned} \tag{6.28}$$

where $n_z = rk^*$.

Proof. The proof is just the concatenation of the previously obtained results, so we summarize it in the following steps:

1. Compute a unimodular matrix $U(\delta)$ (over $\mathbb{R}[\delta]$) such that C_D is reduced to the form
$$\begin{bmatrix} V(\delta) \\ 0 \end{bmatrix} = U(\delta) C_D(\delta),$$
where $V(\delta)$ must be a full row rank matrix over $\mathbb{R}[\delta]$. Then calculate $\bar{V}(\delta)$ such that $\bar{V}(\delta) V(\delta) = \begin{bmatrix} B(\delta) & 0 \\ D(\delta) & 0 \end{bmatrix}$. Then by definition $W(\delta) = [\bar{V}(\delta) \quad 0] U(\delta)$.
2. Do a partition of $W(\delta)$ as $W(\delta) = \begin{pmatrix} K(\delta) \\ \Gamma(\delta) \end{pmatrix} = \begin{bmatrix} K_1(\delta) & K_2(\delta) \\ \Gamma_1(\delta) & \Gamma_2(\delta) \end{bmatrix}$, where $K_i(\delta) \in \mathbb{R}^{n \times p}[\delta]$ and $\Gamma_i(\delta) \in \mathbb{R}^{p \times p}[\delta]$ for $i = 1, 2$. Then rewrite system (6.2) as (6.18) with
$$\bar{A}(\delta) = A(\delta) - K(\delta) \begin{bmatrix} C(\delta) A(\delta) \\ C(\delta) \end{bmatrix}, \quad \tilde{C}(\delta) = C(\delta) - \Gamma \begin{bmatrix} C(\delta) A(\delta) \\ C(\delta) \end{bmatrix}.$$
Moreover, find a unimodular matrix $\Lambda(\delta)$ over $\mathbb{R}[\delta]$ such that $\Lambda(\delta)\tilde{C}(\delta) = \begin{bmatrix} \bar{C}(\delta) \\ 0 \end{bmatrix}$ such that $\bar{C}(\delta)$ is of full row rank.
3. Obtain $\bar{T}(\delta)$ defined in (6.23) and $\bar{F}(\delta)$ defined in (6.24).
4. Then the matrices of (6.27) are obtained: A_0 and C_0 are defined in (6.21), matrices $\bar{\Gamma}_1(\delta)$ and $\bar{\Gamma}_2(\delta)$ are defined in (6.19), and the matrices $\bar{K}_1(\delta)$ and $\bar{K}_2(\delta)$ are defined in (6.28).

□

6.8. Unknown input observer design

The design of the observer is straightforward once the observer normal form (6.27) has been obtained. Let us do it.

Theorem 6.4. *Let assumptions A1 and A2 be satisfied. Then the following system is an observer for system* (6.2) *and consequently an observer for system* (6.1)*:*

$$\begin{cases} \dot{\xi}(t) = L_0 \xi(t) + J(\delta)\Lambda(\delta) y(t), \\ \hat{z}(t) = \xi(t) + H(\delta)\Lambda(\delta) y(t), \\ \hat{x}(t) = T_L^{-1}(\delta)\hat{z}(t), \end{cases} \tag{6.29}$$

with $T_L^{-1}(\delta)$ *defined in* (6.23) *and*

$$\begin{aligned} L_0 &= A_0 - \bar{G}_0 C_0, \\ H(\delta) &= \bar{K}_1(\delta) - \begin{bmatrix} G_0 & 0 \end{bmatrix} \bar{\Gamma}_1(\delta), \\ J(\delta) &= L_0 H(\delta) + \begin{bmatrix} \bar{F}(\delta) & 0 \end{bmatrix} + \bar{K}_2(\delta) + \begin{bmatrix} \bar{G}_0 & 0 \end{bmatrix} \left(I - \bar{\Gamma}_2(\delta)\right), \end{aligned} \tag{6.30}$$

where $\bar{G}_0$ *is a constant matrix that makes* $\left(A_0 - \bar{G}_0 C_0\right)$ *a Hurwitz matrix.*

Proof. By Theorem 6.3 we can start our analysis by using in the system the observer normal form (6.27). Thus let us define $e_z(t) = z(t) - \hat{z}(t)$ and $e_x(t) = x(t) - \hat{x}(t)$. Then, due to the identity $\bar{y} = \Lambda(\delta)y$, we get the dynamic equation for $e_z(t)$:

$$\begin{aligned} \dot{e}_z(t) = A_0 z(t) - L_0 \hat{z}(t) &+ \begin{bmatrix} G_0 & 0 \end{bmatrix} \left(\bar{\Gamma}_2(\delta) - I\right) \bar{y}(t) \\ &+ \begin{bmatrix} G_0 & 0 \end{bmatrix} \bar{\Gamma}_1(\delta) \dot{\bar{y}}(t). \end{aligned}$$

Therefore, due to the equation for $\bar{y}(t)$ in (6.27), the above equation becomes

$$\dot{e}_z(t) = [A_0 - G_0 C_0] e_z(t).$$

Thus, by choosing G_0 so that $[A_0 - G_0 C_0]$ is Hurwitz, the error $e_z(t)$ will exponentially tend to zero. Consequently, the error $e_x(t) = T_L^{-1}(\delta) e_z(t)$ also exponentially tends to zero. □

Remark 6.7. The proposed method is based on the output injection (delayed) technique. We see that the observation error dynamics $\dot{e}_z(t) = [A_0 - G_0 C_0] e_z(t)$ is independent of the time-delay constant h. Thereby this method can be applied to any commensurate and constant delay. That is, conditions A1 and A2, depend completely on the structure of the system and not on the time-delay constant.

Remark 6.8. Finally, we would like to point out that the computations over polynomial matrices required for the design of the observer can be carried out using the existing packages already implemented in software tools like Matlab®, Mathematica, and Maple. A pseudocode is given in Appendix 6.A.

6.9. Illustrative example

Consider the following academic example:

$$A(\delta)=\begin{bmatrix}0 & -1 & 1 & 0\\ -1 & \delta & 0 & 0\\ 1 & 0 & 1 & 0\\ \delta & 1 & 1 & -1\end{bmatrix},\quad B(\delta)=\begin{bmatrix}0 & 0\\ 0 & 0\\ 1 & 1\\ 0 & \delta\end{bmatrix},$$

$$C(\delta)=\begin{bmatrix}\delta & 0 & 0 & 0\\ 0 & 1 & 0 & 0\\ 0 & 0 & 1 & 1\end{bmatrix},\quad D(\delta)=\begin{pmatrix}1 & \delta\\ 0 & 0\\ 1 & \delta\end{pmatrix}.$$

By applying iterative algorithm to calculate the matrix $N_{k^*}(\delta)$ we find that $k^*=3$ and $\text{Inv}_S N_{k^*}(\delta)=\{1,1,1,1\}$. Thus assumption A2 is satisfied. Furthermore we have the identity

$$\text{Inv}_S\begin{bmatrix}C(\delta)B(\delta) & D(\delta)\\ D(\delta) & 0\\ B(\delta) & 0\end{bmatrix}=\text{Inv}_S\begin{bmatrix}C(\delta)B(\delta) & D(\delta)\\ D(\delta) & 0\end{bmatrix}$$
$$=\{1,1,1\},$$

which implies that assumption A2 is also satisfied. According to Theorem 6.4, there exists a Luenberger-like observer to estimate the trajectories of the state variables.

Following the steps given in Theorem 6.3, we obtain the following matrices.

Step 1: The computed matrix $W(\delta)$ has the form

$$W(\delta)=\begin{bmatrix}K(\delta)\\ \Gamma(\delta)\end{bmatrix}=\left[\begin{array}{c|c}K_1(\delta) & K_2(\delta)\\ \hline \Gamma_1(\delta) & \Gamma_2(\delta)\end{array}\right]=\left[\begin{array}{ccc|ccc}0 & 0 & 0 & 0 & 0 & 0\\ 0 & 0 & 0 & 0 & 0 & 0\\ \delta-1 & 0 & 1-\delta & 1 & 0 & \delta-1\\ -\delta & 0 & \delta & 0 & 0 & -\delta\\ \hline 0 & 0 & 0 & 1 & 0 & 0\\ 0 & 0 & 0 & 0 & 0 & 0\\ 0 & 0 & 0 & 1 & 0 & 0\end{array}\right].$$

Step 2: With the above deduced $W(\delta)$, we obtain the matrices $\bar{A}(\delta)$ and $\tilde{C}(\delta)$:

$$\bar{A}(\delta)=\begin{bmatrix} 0 & -1 & 1 & 0 \\ -1 & \delta & 0 & 0 \\ \delta^2-\delta & \delta^2-1 & -\delta^2+2\delta & -2\delta+2 \\ -\delta^2 & 1-\delta-\delta^2 & 1-\delta+\delta^2 & 2\delta-1 \end{bmatrix},$$

$$\tilde{C}(\delta)=\begin{bmatrix} 0 & 0 & 0 & 0 \\ 0 & 1 & 0 & 0 \\ -\delta & 0 & 1 & 1 \end{bmatrix}.$$

Since the matrix $\tilde{C}(\delta)$ has rank equal to 2, we choose the unimodular matrix $\Lambda(\delta)=\begin{bmatrix} 0 & 1 & 0 \\ 0 & 0 & 1 \\ 1 & 0 & 0 \end{bmatrix}$ to obtain the matrix $\bar{C}(\delta)=\begin{bmatrix} 0 & 1 & 0 & 0 \\ -\delta & 0 & 1 & 1 \end{bmatrix}$.

Step 3: We can check that $\bar{\mathcal{O}}_3(\delta)$ is left unimodular over $\mathbb{R}[\delta]$. Then the matrices $\bar{T}(\delta)$ and $\bar{F}(\delta)$ take the following values:

$$\bar{T}(\delta)=\begin{bmatrix} 0 & 1 & 0 & 0 \\ -\delta & 0 & 1 & 1 \\ -1 & 2-4\delta+\delta^2 & 0 & 0 \\ -\delta & 0 & 1 & 1 \\ -2+4\delta-\delta^2 & -\delta+\delta^2 & -1 & 0 \\ -\delta & 0 & 1 & 1 \end{bmatrix},$$

$$\bar{F}(\delta)=\begin{bmatrix} \bar{F}_1(\delta) \\ \hline \bar{F}_2(\delta) \\ \hline \bar{F}_3(\delta) \end{bmatrix}=\begin{bmatrix} -2-\delta^2+5\delta & 0 \\ 0 & 0 \\ \hline 1+3\delta-5\delta^2+\delta^3 & 0 \\ 0 & 0 \\ \hline 3-4\delta-\delta^2+\delta^3 & 2\delta-2 \\ 0 & 1 \end{bmatrix}. \quad (6.31)$$

Step 4: Thus the system can be transformed into the simple observer form (6.27) with $F(\delta)$ given in (6.31) and

$$A_0=\begin{bmatrix} 0 & 0 & 1 & 0 & 0 & 0 \\ 0 & 0 & 0 & 1 & 0 & 0 \\ 0 & 0 & 0 & 0 & 1 & 0 \\ 0 & 0 & 0 & 0 & 0 & 1 \\ 0 & 0 & 0 & 0 & 0 & 0 \\ 0 & 0 & 0 & 0 & 0 & 0 \end{bmatrix},\ C_0=\begin{bmatrix} 1 & 0 & 0 & 0 & 0 & 0 \\ 0 & 1 & 0 & 0 & 0 & 0 \end{bmatrix},$$

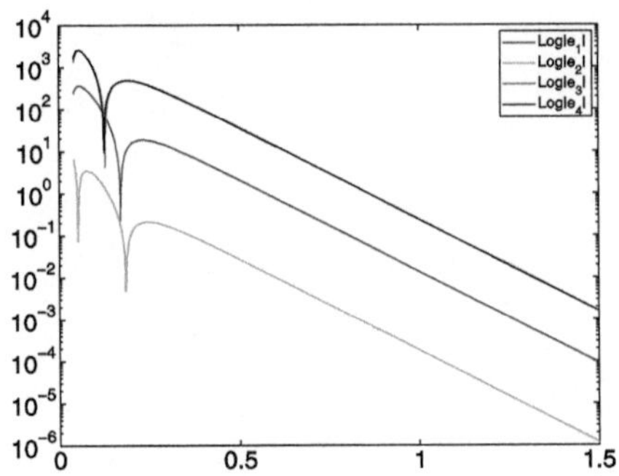

Figure 6.1 The observation error (in log scale) for $h = 0.01$ s.

$$\begin{bmatrix} \bar{K}_1(\delta) & \bar{K}_2(\delta) \end{bmatrix} = \begin{bmatrix} 0 & 0 & 0 & 0 & 0 & 0 \\ 0 & 1 & -1 & 0 & -1 & 1 \\ 0 & 0 & 0 & 0 & 0 & 0 \\ 0 & 1 & -1 & 0 & -1 & 1 \\ 0 & \delta - 1 & 1 - \delta & 0 & 1 - \delta & -1 \\ 0 & 1 & -1 & 0 & -1 & 1 \end{bmatrix},$$

$$\begin{bmatrix} \bar{\Gamma}_1(\delta) & \bar{\Gamma}_2(\delta) \end{bmatrix} = \begin{bmatrix} 0 & 0 & 0 & 0 & 0 & 0 \\ 0 & 0 & 0 & 0 & 0 & 1 \\ 0 & 0 & 0 & 0 & 0 & 1 \end{bmatrix}.$$

Then we can design the observer as that proposed in (6.29), for which the matrix $\bar{G}_0$ is selected as

$$\bar{G}_0 = \begin{bmatrix} 85 & 0 & 2000 & 0 & 12500 & 0 \\ 0 & 85 & 0 & 2000 & 0 & 12500 \end{bmatrix}^T.$$

Step 5: For simulations proposed, we have selected the unknown input as $w_1 = -10 \sin 100t$ and $w_2 = 20 \sin 20t$. The simulation step is 0.001 s, and the constant time-delay is $h = 0.01$ s. The observation errors are depicted in Fig. 6.1, from which we can notice not only the convergence of the proposed observer, but also the delay effect in the observer at the initial time, which is equal to 0.04 s and is due to the term δ^4 in $[T(\delta)]_L^{-1}$. The singularity in the figure is due to the fact that the observation error passes zero and changes the sign. To show that the proposed method is independent of the time-delay involved in the studied system, another simulation was made with the same gains and a greater delay $h = 0.1$ s, whose results are depicted in Fig. 6.2. We can see in Fig. 6.1 and Fig. 6.2 that the esti-

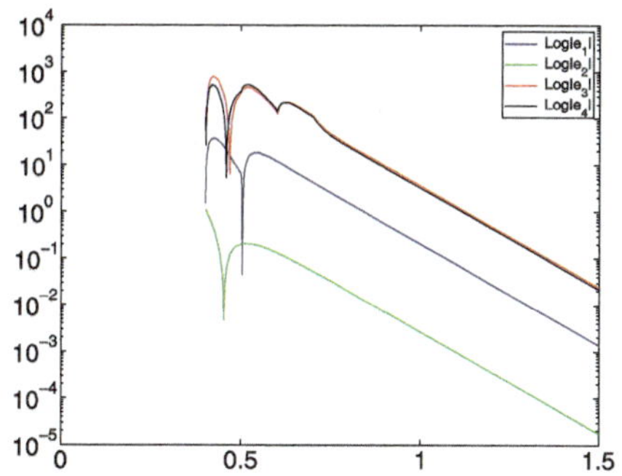

Figure 6.2 The observation error (in log scale) for $h = 0.1s$.

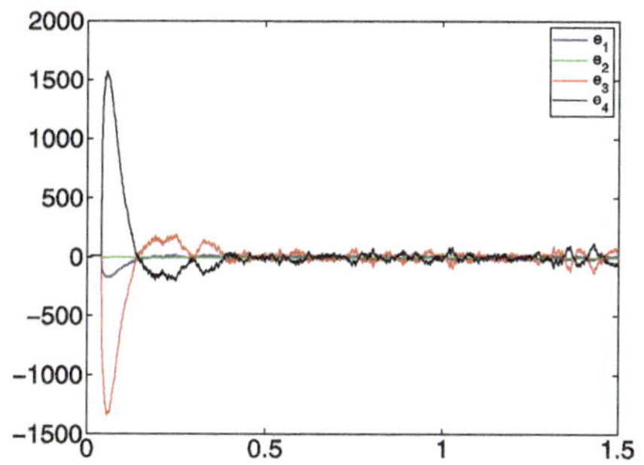

Figure 6.3 The observation error for $h = 0.01$ s with noisy measurement.

mation errors converge to zero with the same speed. The robustness of the observer respect to output noise can be seen in [67].

To show the robustness of the proposed observer, we have added to the system output a zero-mean random noise with an amplitude in the interval $[-2, 2]$. The estimation errors are depicted in Fig. 6.3, where we can notice that the estimation error is always bounded and around the origin.

6.10. Conclusions

1. In the presented chapter, the observability and observer design for a general class of linear time-delayed systems is affected by unknown inputs. We have shown the following results:
   - Sufficient conditions that guarantee the observability of the system, despite the unknown inputs, are given.

- A formula (6.15), which expresses the state as an algebraic function of the system output and some of its time derivatives. The formula allows estimation of the state vector; however, this method requires an algorithm to do the differentiation.
- A Luenberger-like observer proposed in (6.29). The robustness of the observer with respect to output noises is illustrated through a simulation.

2. Despite the conditions used for the observer design are only sufficient, we risk to say that they are not restrictive. Our claim arises from the fact that they become necessary and sufficient for the design of a linear observer, with a prescribed convergence rate, of linear systems *without* time delays.
3. Following the ideas described in this chapter, extensions can be developed for other types of systems, for instance, descriptor systems with commensurate time-delays; see, for example, [8,9,66].
4. For future work, we may consider that the system is not observable but detectable. This may be done by using the results presented in [5], where the detectability of a class of linear systems over a principal ideal domain is considered. Also, we can study the class of time-delay systems where the delays are not commensurate. In that case the system can be represented using the multivariable polynomial ring. Nonetheless, the multivariable polynomial ring is no more a principal ideal domain, and consequently, the Smith normal form cannot be used in the analysis.

Appendix 6.A. Pseudocode for the observability observer design

A computer code can be developed for, first, checking if the observability and observer design conditions are fulfilled, and then to design the observer gain required to implement the observer. The code may be written using Matlab, Mathematica, Maple, or any other software that works with symbolic computation. In Algorithm 6.1, we give a pseudocode for checking if the observability condition of Theorem 6.1 is fulfilled, that is, if condition **A1** is true, and also for checking if the system satisfies the additional condition **A2**, required for the observer design. If both conditions are fulfilled, then the pseudocode of Algorithm 6.2 describes how to calculate the matrices required to implement the observer given in (6.29).

Algorithm 6.1 Checking the observability and observer conditions.

procedure SILVERMAN-MOLINARI($A(\delta), B(\delta), C(\delta), D(\delta)$) ▷ Finding $N_{k^*}(\delta)$ and its invariant factors.

    $\Delta(\delta) \leftarrow 0_{1\times n}$

    $G(\delta) \leftarrow C(\delta)$

    $F(\delta) \leftarrow D(\delta)$ ▷ First, define the matrices $\Delta_0(\delta)$, $G_0(\delta)$, and $F_0(\delta)$

    $N(\delta) \leftarrow \Delta(\delta)$

    $inv1 \leftarrow 0$

    $inv2 \leftarrow 1/2$

    $k \leftarrow 0$

    while $inv1 \neq inv2$ **do** ▷ The cycle stops once the set of invariant factors is equal to the set of invariant factors of the previous iteration.

        $inv1 \leftarrow inv2$

        $S(\delta) \leftarrow \begin{bmatrix} \Delta(\delta)B(\delta) \\ F(\delta) \end{bmatrix}$

        $T(\delta) \leftarrow$ a unimodular matrix such that $T(\delta)S(\delta) = \begin{bmatrix} F_k(\delta) \\ 0 \end{bmatrix}$ ▷ $F_k(\delta)$ must be of full row rank. The Hermite form may be used.

        $F(\delta) \leftarrow F_k(\delta)$ ▷ Save the matrix obtained in the previous step

        $\begin{bmatrix} G(\delta) \\ \Delta(\delta) \end{bmatrix} \leftarrow T(\delta) \begin{bmatrix} \Delta(\delta)A(\delta) \\ G(\delta) \end{bmatrix}$

        $N(\delta) \leftarrow \begin{bmatrix} N(\delta) \\ \Delta(\delta) \end{bmatrix}$

        $inv2 \leftarrow$ the invariant factors of $N(\delta)$. Those can be obtained from the Smith normal form of $N(\delta)$.

        $k = k + 1$

    end while

    return $N(\delta)$, $inv2$, and k ▷ The matrix $N_{k^*}(\delta)$, a vector with the invariant factors of it, and k^*.

end procedure

$inv3 \leftarrow$ the invariant factors of $\begin{bmatrix} N(\delta) \\ I_n \end{bmatrix}$

if $inv2 = inv3$ **then** ▷ Check if $N_{k^*}(\delta)$ has n constant invariant factors.

    print $N_{k^*}(\delta)$ **has** n **constant invariant factors, and the system is BUIO**

    $C_{BD}(\delta) \leftarrow \begin{bmatrix} C(\delta)B(\delta) & D(\delta) \\ D(\delta) & 0 \\ B(\delta) & 0 \end{bmatrix}$ ▷ Define the matrix $C_{BD}(\delta)$.

    $inv1 \leftarrow$ the invariant factors of $C_{BD}(\delta)$ ▷ Use the Smith form.

    $C_D(\delta) \leftarrow \begin{bmatrix} C(\delta)B(\delta) & D(\delta) \\ D(\delta) & 0 \end{bmatrix}$ ▷ Define the matrix $C_D(\delta)$.

    $inv2 \leftarrow$ the invariant factors of $C_D(\delta)$ ▷ Use the Smith form.

    if $inv1 = inv2$ **then**

        Print A Luenberger-like observer can be designed

    end if

else

    The Luenberger-like observer design cannot be guaranteed

end if

Algorithm 6.2 Design of the observer matrices.

procedure OBSERVER GAINS($A(\delta), B(\delta), C(\delta), D(\delta), k^*, C_D$)

$U(\delta) \leftarrow$ a unimodular matrix such that $U(\delta)C_D(\delta) = \begin{bmatrix} V(\delta) \\ 0 \end{bmatrix}$ ▷ Use the Hermite form to have $V(\delta)$ of full row rank.

$V(\delta) \leftarrow$ Save the matrix $V(\delta)$ calculated in the previous step.

$\bar{V}(\delta) \leftarrow$ a matrix such that $\bar{V}(\delta)V(\delta) = \begin{bmatrix} B(\delta) & 0 \\ D(\delta) & 0 \end{bmatrix}$

$W(\delta) \leftarrow [\bar{V}(\delta), 0]U(\delta)$

$\begin{bmatrix} K_1(\delta) & K_2(\delta) \\ \Gamma_1(\delta) & \Gamma_2(\delta) \end{bmatrix} \leftarrow W(\delta)$ ▷ $K_i(\delta)$ is of $n \times p$ and $\Gamma_i(\delta)$ is of $p \times p$.

$K(\delta) \leftarrow \begin{bmatrix} K_1(\delta) & K_2(\delta) \end{bmatrix}$

$\Gamma \leftarrow \begin{bmatrix} \Gamma_1(\delta) & \Gamma_2(\delta) \end{bmatrix}$

$\bar{A}(\delta) \leftarrow A(\delta) - K(\delta)\begin{bmatrix} C(\delta)A(\delta) \\ C(\delta) \end{bmatrix}$

$\tilde{C}(\delta) \leftarrow C(\delta) - \Gamma\begin{bmatrix} C(\delta)A(\delta) \\ C(\delta) \end{bmatrix}$

$\Lambda(\delta) \leftarrow$ a matrix such that $\Lambda(\delta)\tilde{C}(\delta) = \begin{bmatrix} \bar{C}(\delta) \\ 0 \end{bmatrix}$ ▷ Use the Hermite form.

$\bar{C}(\delta) \leftarrow$ save the matrix $\bar{C}(\delta)$ obtained in the previous step

$\bar{\mathcal{O}}_{k^*}(\delta) \leftarrow \begin{bmatrix} \bar{C}(\delta)\bar{A}^0(\delta) \\ \vdots \\ \bar{C}(\delta)\bar{A}^{k^*-1}(\delta) \end{bmatrix}$

$\bar{F}(\delta) \leftarrow \bar{C}(\delta)\bar{A}^{k^*}(\delta)\left[\bar{\mathcal{O}}_{k^*}(\delta)\right]_L^{-1}$

$\left[\bar{F}_{k^*}(\delta), \cdots, \bar{F}_1(\delta)\right] \leftarrow \bar{F}(\delta)$ ▷ Each $\bar{F}_i$ is a square matrix.

$\bar{T}_1(\delta) \leftarrow \bar{C}(\delta)$ ▷ First step to generate $\bar{T}(\delta)$.

for $i = 2 \rightarrow k^*$ **do** ▷ Generate the matrices $\bar{T}_i(\delta)$

  $\bar{T}_i(\delta) \leftarrow \bar{T}_{i-1}(\delta))\bar{A}(\delta) - \bar{F}_{i-1}(\delta)\bar{C}(\delta)$

end for

$\bar{T}(\delta) \leftarrow \mathrm{col}(\bar{T}_1(\delta) \cdots \bar{T}_{k^*}(\delta))$ ▷ Construct the matrix $\bar{T}$.

$\bar{\Gamma}_1(\delta) \leftarrow \Lambda(\delta)\Gamma_1(\delta)\Lambda^{-1}(\delta)$

$\bar{\Gamma}_2(\delta) \leftarrow \Lambda(\delta)\Gamma_2(\delta)\Lambda^{-1}(\delta)$

$A_0 \leftarrow \begin{bmatrix} 0 & I_{p_1} & \cdots & 0 \\ 0 & 0 & \ddots & \vdots \\ 0 & 0 & \cdots & I_{p_1} \\ 0 & 0 & \cdots & 0 \end{bmatrix}$ ▷ A square matrix of $p_1 k^* \times p_1 k^*$.

$C_0 \leftarrow \left[I_{p_1}, 0, \cdots, 0\right]$ ▷ A matrix of $p_1 \times p_1 k^*$.

$\bar{G}_0 \leftarrow$ a matrix such that $\left(A_0 - \bar{G}_0 C_0\right)$ is Hurwitz

$L_0 \leftarrow A_0 - \bar{G}_0 C_0$

$H(\delta) \leftarrow \bar{K}_1(\delta) - \begin{bmatrix} G_0 & 0 \end{bmatrix}\bar{\Gamma}_1(\delta)$

$J(\delta) \leftarrow L_0 H(\delta) + \begin{bmatrix} \bar{F}(\delta) & 0 \end{bmatrix} + \bar{K}_2(\delta) + \begin{bmatrix} \bar{G}_0 & 0 \end{bmatrix}(I - \bar{\Gamma}_2(\delta))$

end procedure

return $L_0, J(\delta), \Lambda(\delta), H(\delta), \bar{T}(\delta)$ ▷ These are the matrices required to implement the observer.

References

[1] W.A. Adkins, Algebra: An Approach Via Module Theory, Springer-Verlag, 1999 (Chapter: Matrices over PIDs).
[2] W.A. Adkins, Algebra: An Approach Via Module Theory, Springer-Verlag, 1999 (Chapter: Modules and Vector Spaces).
[3] M. Anguelova, B. Wennberg, On analytic and algebraic observability of nonlinear delay systems, Automatica 46 (2010) 682–686.
[4] M. Anguelova, B. Wennberga, State elimination and identifiability of the delay parameter for nonlinear time-delay systems, Automatica 44 (2008) 1373–1378.
[5] F.J. Bejarano, Detectability of linear systems over a principal ideal domain with unknown inputs, Systems & Control Letters 78 (2015) 1–7.
[6] F.J. Bejarano, T. Floquet, W. Perruquetti, G. Zheng, Observability and detectability of singular linear systems with unknown inputs, Automatica 49 (2013) 793–800.
[7] F.J. Bejarano, G. Zheng, Observability of linear systems with commensurate delays and unknown inputs, Automatica 50 (2014) 2077–2083.
[8] F.J. Bejarano, G. Zheng, Observability of singular systems with commensurate time-delays and neutral terms, Automatica 85 (2017) 462–467.
[9] F.J. Bejarano, G. Zheng, Unknown input functional observability of descriptor systems with neutral and distributed delay effect, Automatica 85 (2017) 186–192.
[10] K. Bhat, H. Koivo, An observer theory for time-delay systems, IEEE Transactions on Automatic Control 21 (1976) 266–269.
[11] S. Bhattacharyya, Observer design for linear systems with unknown inputs, IEEE Transactions on Automatic Control 23 (1978) 483–484.
[12] J. Brewer, J. Bunce, F.V. Vleck, Linear Systems Over Commutative Rings, Marcel Dekker, New York, 1986.
[13] W. Brown, Matrices Over Commutative Rings, Pure and Applied Mathematics, Marcel Dekker, Inc., New York, Basel, Hong Kong, 1993.
[14] G. Conte, A.M. Perdon, G. Guidone-Peroli, Unknown input observers for linear delay systems: a geometric approach, in: Proceedings of the 42nd IEEE Conference on Decision and Control, Maui, Hawaii USA, 2003, pp. 6054–6059.
[15] M. Darouach, Linear functional observers for systems with delays in state variables, IEEE Transactions on Automatic Control 46 (2001) 491–496.
[16] M. Darouach, Full order unknown inputs observers design for delay systems, in: 14th Mediterranean Conference on Control and Automation, Ancona, Italy, 2006, pp. 1–5.
[17] M. Darouach, M. Zasadzinski, S. Xu, Full-order observers for linear systems with unknown inputs, IEEE Transactions on Automatic Control 39 (1994) 606–609.
[18] R. Eising, Realization and stabilization of 2-D systems, IEEE Transactions on Automatic Control AC-23 (1978) 793–799.
[19] R. Eising, M. Hautus, Realization algorithms for systems over a principal ideal domain, Mathematical Systems Theory 14 (1981) 353–366.
[20] H. Estrada-Garcia, C. Moog, L. Márquez-Martínez, Tracking problem for nonlinear time-delay multi-input multi-output systems, 2007, pp. 1812–1816.
[21] A. Fattouh, O. Sename, J.M. Dion, An unknown input observer design for linear time-delay systems, in: 38th IEEE Conference on Decision and Control, Phoenix, AZ, U.S., 1999, pp. 4222–4227.
[22] M. Fliess, H. Mounier, Controllability and observability of linear delay systems: an algebraic approach, ESAIM. Control, Optimisation and Calculus of Variations 3 (1998) 301–314.
[23] E. Fridman, Introduction to Time-Delay Systems: Analysis and Control, Systems & Control: Foundations & Applications, Birkhäuser, 2014.
[24] Y.M. Fu, G.R. Duan, S.M. Song, Design of unknown input observer for linear time-delay systems, International Journal of Control, Automation, and Systems 2 (2004) 530–535.

[25] A. Garate-Garcia, L. Marquez-Martinez, C. Moog, Equivalence of linear time-delay systems, IEEE Transactions on Automatic Control 56 (2011) 666–670.
[26] A. Germani, C. Manes, P. Pepe, An asymptotic state observer for a class of nonlinear delay systems, Kybernetika 37 (2001) 459–478.
[27] I. Gohberg, P. Lancaster, L. Rodman, Matrix Polynomials, Classics in Applied Mathematics, SIAM, USA, 2009 (Chapter S1, the Smith Form and Related Problems).
[28] I. Gohberg, P. Lancaster, L. Rodman, Matrix Polynomials, Classics in Applied Mathematics, SIAM, USA, 2009 (Chapter S2, the Matrix Equation AX-XB=C).
[29] M. Halás, An algebraic framework generalizing the concept of transfer functions to nonlinear systems, Automatica 44 (2008) 1181–1190.
[30] H. Hammouri, Z. Tmar, Unknown input observer for state affine systems: a necessary and sufficient condition, Automatica 46 (2010) 271–278.
[31] M. Hautus, Strong detectability and observers, Linear Algebra and Its Applications 50 (1983) 353–368.
[32] R. Hermann, A. Krener, Nonlinear controllability and observability, IEEE Transactions on Automatic Control 22 (1977) 728–740.
[33] G. Hostetter, J. Meditch, Observing systems with unmeasurable inputs, IEEE Transactions on Automatic Control 18 (1973) 307–308.
[34] M. Hou, P. Müller, Design of observers for linear systems with unknown inputs, IEEE Transactions on Automatic Control 37 (1992) 871–875.
[35] M. Hou, M.H. Zitek, R. Patton, An observer design for linear time-delay systems, IEEE Transactions on Automatic Control 47 (2002) 121–125.
[36] N. Ito, H. Inaba, Block triangular decoupling for linear systems over principal ideal domains, SIAM Journal on Control and Optimization 35 (1997) 744–765.
[37] R.E. Kalman, P.L. Falb, M.A. Arbib, Topics in Mathematical System Theory, McGraw-Hill, New York, 1969.
[38] E. Kamen, Mathematical System Theory, Springer Berlin Heidelberg, 1991, pp. 31–324 (Chapter: Linear systems over rings: from R.E. Kalman to the present).
[39] Y. Kawano, T. Ohtsuka, Observability at an initial state for polynomial systems, Automatica 49 (2013) 1126–1136.
[40] U. Kotta, M. Tõnso, Realization of discrete-time nonlinear input-output equations: polynomial approach, Automatica 48 (2012) 255–262.
[41] A. Krener, W. Respondek, Nonlinear observer with linearizable error dynamics, SIAM Journal on Control and Optimization 23 (1985) 197–216.
[42] P. Kudva, N. Viswanadham, A. Ramakrishna, Observers for linear systems with unknown inputs, IEEE Transactions on Automatic Control 25 (1980) 113–115.
[43] E. Lee, S. Neftci, A. Olbrot, Canonical forms for time delay system, IEEE Transactions on Automatic Control 27 (1982) 128–132.
[44] E. Lee, A. Olbrot, Observability and related structural results for linear hereditary systems, International Journal of Control 34 (1981) 1061–1078.
[45] D.G. Luenberger, Observing the state of a linear system, IEEE Transactions on Military Electronics (1964) 74–80.
[46] B. Molinari, A strong controllability and observability in linear multivariable control, IEEE Transactions on Automatic Control 21 (1976) 761–764.
[47] C. Moog, R. Castro-Linares, M. Velasco-Villa, L.A. Marque-Martinez, The disturbance decoupling problem for time-delay nonlinear systems, IEEE Transactions on Automatic Control 45 (2000).
[48] H. Mounier, P. Rouchon, R. Joachim, Some examples of linear systems with delays, Journal Européen Des Systèmes Automatisés 31 (1997) 911–925.
[49] C. Musili, Introduction to Rings and Modules, second ed., Narosa, New Delhi, 2009 (Chapter: Modules with Chain Conditions).
[50] T. Ohtsuka, Algebraic structures in nonlinear systems over rings obtained by immersion, SIAM Journal on Control and Optimization 47 (2008) 1961–1976.

[51] A.W. Olbrot, Observability and observers for a class of linear systems with delays, IEEE Transactions on Automatic Control 26 (1981) 513–517.
[52] A.M. Perdon, M. Anderlucci, M. Caboara, Computational algebra techniques for linear dynamical systems over rings: state estimations problems, in: IEEE International Symposium on Intelligent Control, IEEE, 2006, pp. 379–384.
[53] P. Picard, J. Lafay, V. Kucera, Model matching for linear systems with delays and 2d systems, Automatica 34 (1998) 183–191.
[54] J. Pommaret, Partial Differential Control Theory, Vol 2: Control Systems, Mathematics and Its Applications, vol. 530, Kluwer Academic Publishers, 2001 (Chapter 5).
[55] R. Rabah, On observability of linear delay systems with unknown inputs, in: 3rd IEEE Mediterranean Symposium on New Directions, Limassol, Cyprus, 1995.
[56] J.P. Richard, Time-delay systems: an overview of some recent advances and open problems, Automatica 39 (2003) 1667–1694.
[57] O. Sename, New trends in design of observers for time-delay systems, Kybernetika 37 (2001) 427–458.
[58] L. Silverman, Inversion of multivariable linear systems, IEEE Transactions on Automatic Control 14 (1969) 270–276.
[59] E. Sontag, Linear systems over commutative rings: a survey, Ricerche di Automatica 7 (1976) 1–34.
[60] H. Trentelman, A. Stoorvogel, M.L.J. Hautus, Control Theory for Linear Systems, Communications and Control Engineering, Springer, New York, London, 2001.
[61] S. Wang, E. Davison, P. Dorato, Observing the states of systems with unmeasurable disturbances, IEEE Transactions on Automatic Control 20 (1975) 716–717.
[62] X. Xia, L. Marquez, P. Zagalak, C. Moog, Analysis of nonlinear time-delay systems using modules over non-commutative rings, Automatica 38 (2002) 1549–1555.
[63] F. Yang, R.W. Wilde, Observers for linear systems with unknown inputs, IEEE Transactions on Automatic Control 33 (1988) 677–681.
[64] G. Zheng, J.P. Barbot, D. Boutat, Identification of the delay parameter for nonlinear time-delay systems with unknown inputs, Automatica 49 (2013) 1755–1760.
[65] G. Zheng, J.P. Barbot, D. Boutat, T. Floquet, J.P. Richard, On observation of time-delay systems with unknown inputs, IEEE Transactions on Automatic Control 56 (2011) 1973–1978.
[66] G. Zheng, F.J. Bejarano, Observer design for linear singular time delay systems, Automatica 80 (2017) 1–9.
[67] G. Zheng, F.J. Bejarano, W. Perruquetti, J.P. Richard, Unknown input observer for linear time-delay systems, Automatica 61 (2015) 35–43.

CHAPTER SEVEN

Adaptive-robust controller for a class of systems with time-varying input delay and state-dependent uncertainty

Spandan Roy and Harsh Shukla
Robotics Research Center, International Institute of Information Technology Hyderabad (IIIT-H), Hyderabad, India

Contents

7.1. Introduction

7.1.1 Background and motivation

Time delays are inherently encountered in many real-life scenarios such as power systems, robotics applications, chemical and biological processes, engine cycle delays in internal combustion engine, and so on [1–3]. Left uncompensated, time delays may potentially compromise the controller accuracy and system stability. Therefore, the study of stability and performance of dynamical systems under the influence of time delay have received

Control Strategy for Time-Delay Systems
https://doi.org/10.1016/B978-0-12-820599-0.00012-4

extensive attention for many years. The state of the art in this domain is well documented in [1,2,4–6] along with the existing challenges.

Amongst the earliest controllers developed for linear time-delay systems, a few notable works include the Smith predictor-based approach [7], Artstein model reduction [8], and finite spectrum assignment [9]. Thereafter, the predictor-based methodology was extensively investigated by many researchers [9–16]. A stabilizing controller for linear systems with state delay and neutral-type delay is reported in [17] with predefined uncertainty bound. Sampled-data control is utilized in [18,19] for input delay compensation considering known system uncertainty. The case of feedforward linear systems under both state and input delays is reported by [20]. Therein, the unknown time delay is predicted adaptively via a projection function, but without considering plant uncertainties. An adaptive control solution for linear systems with linearly parameterized uncertainty with known bounds is reported in [21]. Adaptive and robust control techniques are utilized in [22–24] for linear systems to address the effects of time delay and uncertainties. Adaptive backstepping method is used in [22] for robust stabilization problem of linear nonminimum phase systems subjected to unknown dynamics and input delay. Integral sliding-mode control is used by [23] for uncertain linear systems under time-varying delay at the input. However, this scheme relies on a priori known bounds of uncertainties and only suitable for slowly varying input delay. On the other hand, the situation of unknown upper bounds of uncertainty is tackled in [24] via adaptive sliding-mode control. Unfortunately, the adaptation mechanism yields a monotonically increasing switching gain, which may cause high gain instability [25,26].

In comparison, the literature for nonlinear systems with input delay is not vast. Furthermore, the control design for nonlinear systems is more challenging because the closed-loop stability analysis cannot rely on linear boundedness of a system model [6]. Some earlier works, based on the Razumikhin theorem, reported delay-dependent [27] and delay-independent [28] results considering the time delay to be known precisely. A backstepping-based predictor approach is proposed in [29] for delay compensation of forward complete and strict-feedforward nonlinear systems utilizing ODE–PDE cascade transformation. The controller reported in [30] aims for delay compensation in states and dead zone at the input. However, these control laws cannot handle parametric uncertainties. This issue was addressed in [31], where a predictor-based approach is proposed for constant input delay based on the Lyapunov–Krasvoskii method while considering uncertainty in the inertia matrix of Euler–Lagrange systems. The same approach was later extended for time-varying input delay in [32] and also for state delay [33]. However, the works in [31–33] require exact knowledge of input delay and can only negotiate slowly varying input delay. A robust controller is proposed in [34] to tackle time-varying input delay via Razumikhin-theorem-based analysis. However, a priori knowledge of uncertainty bounds is required to design any robust controller, which is not always available in practice due to unmodeled dynamics and unknown disturbances.

To handle the situation of unknown uncertainty bound, adaptive-robust control laws are developed (see [25,35–40] and references therein). To compensate time-varying network delays in uncertain complex network systems, adaptive sliding-mode-based control law is developed in [41]. In a different application, [42] used adaptive law to evaluate the coupling strength for synchronization of chaotic systems under network-induced delay. However, the adaptive laws proposed in [41,42] create overestimation of switching gains by not allowing these gains to decrease [43]. The issue of over- and under-estimation problems of switching gain for uncertain Euler–Lagrange systems under arbitrarily varying input delay was first addressed in [36]. On the other hand, this work considers the uncertainties to be bounded by an unknown constant, which is a conservative assumption [35] when system states are involved in the upper bound structure of uncertainty (cf. [37,43–46]).

7.1.2 Contributions

In view of the gaps in the state of the art as discussed before, in this work, we propose a new adaptive-robust control law named adaptive-robust time delay control (ARTDC) for tracking control of a class of systems subjected to arbitrarily varying input delay and state-dependent uncertainty. The highlights of the proposed scheme, especially compared to [36] are as follows:

- ARTDC avoids any a priori assumption of uncertainties being upper bounded by a constant. This enhances the applicability of the proposed scheme.
- Under similar selection of *nominal control parameters*,[1] ARTDC enhances the maximum allowable input delay for the system in comparison to

[1] Here nominal control parameters signify the gains that are not designated to tackle uncertainties.

[36] (Lemma 7.4) while still remaining insensitive toward variations in input delay.

The effectiveness of ARTDC is verified via simulation and further substantiated via comparative experiments using a wheeled mobile robot (WMR).

7.1.3 Notations

The following notations are used throughout this paper: $\lambda_{\min}(\cdot)$, $||(\cdot)||$, $(\cdot)^{\dagger}$ denote the minimum eigenvalue, Euclidean norm, and pseudoinverse of $(\cdot)$, respectively; $\Re^{+}$ defines the positive real numbers; $\vee$, $\wedge$ denote logical "or" and logical "and", respectively; 'sgn' is the standard signum function; a positive definite matrix Ξ is denoted as $\Xi > 0$, and I denotes the identity matrix of proper dimension.

7.1.4 Organization

The remainder of this chapter is organized as follows. The control problem is highlighted in Section 7.2. Section 7.3 details the proposed control scheme along with rigorous closed-loop stability analysis. This section also presents some key advantages of the proposed ARTDC law compared to the state of the art; Sections 7.4 and 7.5 present the simulation and experimental results. Section 7.6 presents concluding remarks with a future direction.

7.2. System dynamics and problem formulation

Let us consider the following class of systems with input delay:

$$\ddot{q} = f(q, \dot{q}) + g(q, \dot{q})\tau(t - h_{\text{in}}), \tag{7.1}$$

where $q, \dot{q} \in \Re^{n}$ are the system states, $\tau \in \Re^{m}$ is the control input, $f(q, \dot{q}) \in \Re^{n}$ and $g(q, \dot{q}) \in \Re^{n\times m}$ are uncertain system dynamics terms with $m \geq n$, and $h_{\text{in}}(t)$ is a time-varying input delay. The following assumptions define the uncertainties present in system (7.1).

Assumption 7.1. Let the dynamics term f be decomposed as $f = \hat{f} + \Delta f$, where $\hat{f}$ is the known (nominal) part, and Δf is the unknown part having the following upper bound structure [35]:

$$||\Delta f(q, \dot{q})|| \leq \gamma_0^{**} + \gamma_1^{**}||x||, \tag{7.2}$$

where $\gamma_0^{**}, \gamma_1^{**} \in \Re^{+}$ are unknown finite constants, and $x = [q,\ \dot{q}]$.

Assumption 7.2. $||g(q, \dot{q})|| \in \mathcal{L}_\infty$ for all q, $\dot{q}$, and the nominal part of g, that is, $\hat{g}$, can be designed so that

$$||(g\hat{g}^\dagger(t-h_{\text{in}})-I)|| \triangleq ||\bar{g}|| < 1 \quad \forall h_{\text{in}}. \tag{7.3}$$

Assumption 7.3. There exists a known scalar $\bar{h}_{\text{in}}$ such that $|h_{\text{in}}(t)| \leq \bar{h}_{\text{in}}$. However, instantaneous values of delay $h_{\text{in}}(t)$ and its time derivatives along with their bounds are unknown.

Remark 7.1. Assumptions 7.1 and 7.2 are quite common for practical electromechanical systems, stemming from the physical laws that govern the dynamics (see [44]), whereas Assumption 7.3 defines that the input delay can have unknown (arbitrary) variation, that is, no restriction is imposed on the rate of variation in h_{in}.

Let there exist constant matrices $P \in \Re^{2n\times 2n} > 0$, $Q \in \Re^{2n\times 2n} > 0$, and $K \in \Re^{n\times n} > 0$ which satisfy the following conditions:

(i) $P = \begin{bmatrix} P_1 & P_2 \\ P_2^T & P_3 \end{bmatrix}$, where $P_1 > 0$, $P_2 > 0$, $P_3 > 0$, $P_3^{-1}P_2 > 0$;

(ii) $A = \begin{bmatrix} 0 & I \\ -K & -2\Omega \end{bmatrix}$ is Hurwitz, where $\Omega = P_3^{-1}P_2 > 0$;

(iii) $A^TP + PA = -Q$.

In [36] the matrix P is defined as the solution of the following Lyapunov equation:

$$\bar{A}^TP + P\bar{A} = -Q, \tag{7.4}$$

where $\bar{A} = \begin{bmatrix} 0 & I \\ -K_1 & -K_2 \end{bmatrix}$, $K_1 > 0$, $K_2 > 0$. So, the values of P and Q in conditions (i) and (iii) will be similar to that of [36] if

$$K_1 = K \quad \text{and} \quad K_2 = 2\Omega. \tag{7.5}$$

Establishment of relation (7.5) is necessary for the comparison between the proposed controller and the work [36], as discussed later (Lemma 7.4).

Control Problem. Under Assumptions 7.1–7.3, to design an adaptive-robust control framework for system (7.1) to track a desired trajectory $q^d(t)$, where the selections of desired trajectories satisfy $q^d, \dot{q}^d, \ddot{q}^d \in \mathcal{L}_\infty$.

7.3. Adaptive-robust time delay controller (ARTDC): design and analysis

Let $e(t) = q(t) - q^d(t)$ be the tracking error, and let $\xi = [e^T \quad \dot{e}^T]^T$. The control input τ for ARTDC is defined as

$$\tau = \hat{g}^\dagger(\hat{u} + \Delta u - \hat{f}), \tag{7.6}$$

where $\hat{u}$ is the nominal control designed as

$$\hat{u} = \ddot{q}^d - \Omega\dot{e}, \tag{7.7}$$

and Δu is the switching control input responsible for negotiation of the uncertainties (defined later).

Further, $\tau(t)$ goes through an input delay of h_{in} as

$$\tau(t - h_{\text{in}}) = \hat{g}^\dagger(t - h_{\text{in}})\{\hat{u}(t - h_{\text{in}}) + \Delta u(t - h_{\text{in}}) - \hat{f}(t - h_{\text{in}})\}. \tag{7.8}$$

Taking $B = \begin{bmatrix} 0 & I \end{bmatrix}^T$, let us define an error variable as

$$\begin{aligned} s &= B^T P\xi = P_3\dot{e} + P_2 e \\ \Rightarrow P_3^{-1}s &= \dot{e} + \Omega e. \end{aligned} \tag{7.9}$$

The time derivative of (7.9) yields

$$\begin{aligned} P_3^{-1}\dot{s} &= \ddot{e} + \Omega\dot{e} \\ &= \ddot{q} - \ddot{q}^d + \Omega\dot{e} \\ &= f + g\tau(t - h_{\text{in}}) - \ddot{q}^d + \Omega\dot{e}. \end{aligned} \tag{7.10}$$

Substituting (7.8) into (7.10), we get

$$\begin{aligned} P_3^{-1}\dot{s} = &f - \hat{f}(t - h_{\text{in}}) + \Delta u(t - h_{\text{in}}) + \hat{u}(t - h_{\text{in}}) \\ &+ (g\hat{g}^\dagger(t - h_{\text{in}}) - I)\tau(t - h_{\text{in}}) - \ddot{q}^d + \Omega\dot{e}. \end{aligned} \tag{7.11}$$

Further, taking the time derivative of (7.9) and utilizing (7.11) yield the following:

$$\begin{aligned} \ddot{e} &= -Ke - \Omega\dot{e} + P_3^{-1}\dot{s} + Ke \\ &= -Ke - \Omega\dot{e} - \Omega\dot{e}(t - h_{\text{in}}) + \Delta u(t - h_{\text{in}}) + \sigma + Ke + \Omega\dot{e} + \bar{g}\Delta u(t - h_{\text{in}}), \end{aligned} \tag{7.12}$$

where $\bar{g} = (g\hat{g}^{\dagger}(t - h_{\text{in}}) - I)$, and

$$\sigma = f - \hat{f}(t - h_{\text{in}}) + \bar{g}(\hat{u}(t - h_{\text{in}}) - \hat{f}(t - h_{\text{in}})) - \ddot{q}^{d} + \ddot{q}^{d}(t - h_{\text{in}}) \quad (7.13)$$

is defined as the *overall uncertainty*.

Lemma 7.1. *If system* (7.1) *is locally Lipschitz continuous for all delays, then the upper bound of σ follows the structure*

$$||\sigma|| \leq \gamma_0^* + \gamma_1^* ||\xi||, \quad (7.14)$$

where $\gamma_0^*, \gamma_1^* \in \Re^+$ are unknown finite constants.

Proof. Based on Assumption 7.1 and following the derivations as in [37, 46,47] (e.g., Lemma 1 in [47]), condition (7.14) can be established and therefore left for the interested readers to verify. □

Remark 7.2. In [36], σ is assumed to be upper bounded by an unknown constant. Such an assumption is conservative in nature because due to the involvement of states in the upper bound of $||\sigma||$ as in (7.14), it puts an a priori restriction on the system states. A similar assumption is commonly found in the literature of (nondelayed) adaptive-robust designs (cf. discussions in [44–46]). In the proposed control framework we avoid such assumption.

After defining the structure of $||\sigma||$ as in (7.14), we now design Δu as

$$\Delta u(t) = -\varsigma(t)\text{sgn}(s(t)), \;\; \varsigma(t) = \frac{1}{(1 - ||\bar{g}||)}\{c(t) + \beta(t) + \rho(t)\}, \quad (7.15)$$

$$c(t) = \hat{\gamma}_0(t) + \hat{\gamma}_2(t) + \hat{\gamma}_1(t)||\xi||, \quad (7.16)$$

where the switching gains $\hat{\gamma}_0, \hat{\gamma}_1 \in \Re^+$ are responsible for tackling uncertainties, whereas the other switching gains $\beta, \rho, \hat{\gamma}_2 \in \Re^+$ play crucial role in closed-loop stability (see Remark 7.6). The switching gains are evaluated as

$$\dot{\hat{\gamma}}_j = \begin{cases} -\alpha_j||\xi||^j||s|| & \text{if } s^T\dot{s} \leq 0 \vee \beta \leq \underline{\beta} \vee \rho \leq \underline{\rho}, \\ \alpha_j||\xi||^j||s|| & \text{if } \hat{\gamma}_i \leq \underline{\gamma}_i \vee s^T\dot{s} > 0, \end{cases} \quad (7.17)$$

$$\dot{\hat{\gamma}}_2 = \begin{cases} -\varsigma\alpha_2||\xi||^3 & \text{if } s^T\dot{s} \leq 0 \vee \beta \leq \underline{\beta} \vee \rho \leq \underline{\rho}, \\ \alpha_2||\xi||||s|| & \text{if } \hat{\gamma}_i \leq \underline{\gamma}_i \vee s^T\dot{s} > 0, \end{cases} \quad (7.18)$$

$$\dot{\beta}=\begin{cases}-\frac{1}{\beta} & \text{if } \beta>\underline{\beta}\\ \delta & \text{if } \beta\leq\underline{\beta}\end{cases},\quad \dot{\rho}=\begin{cases}-\frac{||s||}{\rho} & \text{if } \rho>\underline{\rho},\\ \delta||s|| & \text{if } \rho\leq\underline{\rho},\end{cases} \tag{7.19}$$

$$j=0,1,\hat{\gamma}_{i0}>\underline{\gamma}_i, i=0,1,2, \rho_0>\underline{\rho}, \beta_0>\underline{\beta},$$

where $\hat{\gamma}_{i0}$, ρ_0, β_0 are the initial values of $\hat{\gamma}_i$, ρ, β, respectively, and $\underline{\gamma}_i, \underline{\beta}, \underline{\rho}, \alpha_i, \varsigma, \delta \in \Re^+$ are user-defined scalars.

From (7.17)–(7.19) we can infer that

$$\hat{\gamma}_i(t)\geq\underline{\gamma}_i,\ i=0,1,2,\quad \underline{\rho}\leq\rho(t)\leq\rho_0,\quad \text{and}\quad \underline{\beta}\leq\beta(t)\leq\beta_0\quad \forall t. \tag{7.20}$$

7.3.1 Stability analysis

Before presenting the closed-loop stability result, we state the following Razumikhin lemma, which is instrumental in deriving a tolerance of time delay for the system.

Lemma 7.2. *Using the Razumikhin-type theorem* [48], *let the following standard inequality hold for a Lyapunov function candidate* $V(\xi)$*:*

$$V(\xi(\nu))<rV(\xi(t)),\qquad t-2h_{\text{in}}\leq\nu\leq t, \tag{7.21}$$

where $r>1$ *is a constant.*

Remark 7.3. Condition (7.21) should not be considered as an assumption. On the contrary, it represents an analytical condition. We refer the interested readers to [49, Chap. 1] for a detailed discussion.

We analyze the stability using the following Lyapunov function candidate:

$$V=V_1+\sum_{i=0}^{1}\frac{1}{2\alpha_i}(\hat{\gamma}_i-\gamma_i^*)^2+\frac{1}{2\alpha_2}\hat{\gamma}_2^2+\frac{1}{2}\Gamma\beta^2+\frac{1}{2}\Gamma_1\rho^2, \tag{7.22}$$

where $V_1=\frac{1}{2}\xi^TP\xi$, and the terms $\Gamma, \Gamma_1\in\Re^+$ will be defined later.

Before inspecting the closed-loop system stability, we state two important lemmas. Lemma 7.3 simplifies $\dot{V}_1$, which consequently reveals the maximum allowable input delay $\bar{h}_{\text{in}}$, and Lemma 7.4 identifies the condition under which the proposed work can extend the allowable delay over the state of the art.

Lemma 7.3.

(i) *$\dot{V}_1(\xi)$ can be expressed in a simplified way as*

$$\dot{V}_1(\xi) \leq -\frac{1}{2}\xi^T (Q - hG)\xi + \Gamma + ||s||\Gamma_1 \\ + s^T(\Delta u + \sigma + (PB)^{\dagger} PC\xi + \bar{g}\Delta u). \tag{7.23}$$

(ii) *The upper bound of delay $\bar{h}_{\text{in}}$ for system (7.1) is found to be*

$$h_{\text{in}} < \frac{\lambda_{\min}(Q)}{||G||} := \bar{h}_{\text{in}}, \tag{7.24}$$

where

$$G = (\eta PB_1(A_1 P^{-1} A_1^T + B_1 P^{-1} B_1^T + P^{-1})B_1^T P + 2(r/\eta)P);$$

the matrices P, K, Q (cf. (7.4), (7.5)) and scalar design parameters $r > 1$, $\eta > 0$ are selected so that

$$\lambda_{\min}(Q) > h_{\text{in}}||G|| \quad \forall h_{\text{in}}.$$

Proof. The error dynamics (7.12) can be written in the state-space form as

$$\dot{\xi} = A_1\xi + B_1\xi(t - h_{\text{in}}) + B(\sigma + \Delta u(t - h_{\text{in}}) + \bar{g}\Delta u(t - h_{\text{in}})) + C\xi, \tag{7.25}$$

where $A_1 = \begin{bmatrix} 0 & I \\ -K & -\Omega \end{bmatrix}$, $B_1 = \begin{bmatrix} 0 & 0 \\ 0 & -\Omega \end{bmatrix}$, $C = \begin{bmatrix} 0 & 0 \\ K & \Omega \end{bmatrix}$.

Further, $\xi(t - h_{\text{in}}) = \xi(t) - \int_{-h_{\text{in}}}^{0} \dot{\xi}(t + \theta)\mathrm{d}\theta$, where the derivative inside the integral is with respect to θ. So, (7.25) can alternatively be expressed as

$$\dot{\xi}(t) = A\xi(t) - B_1 \int_{-h_{\text{in}}}^{0} \dot{\xi}(t + \theta)\mathrm{d}\theta + C\xi + B(\sigma + \Delta u(t - h_{\text{in}}) + \bar{g}\Delta u(t - h_{\text{in}})). \tag{7.26}$$

(i) Since $\xi^T PC\xi = s^T(PB)^{\dagger} PC\xi$, using (7.26), we have

$$\dot{V}_1 = -\frac{1}{2}\xi^T Q\xi - \int_{-h}^{0} \xi^T PB_1[A_c\xi(t + \theta) + B_1\xi(t - h_{\text{in}} + \theta)$$

$$+B\Phi(t+\theta)]d\theta+s^T\{(1+\bar{g}(t-h_{\text{in}}))\Delta u(t-h_{\text{in}})+\sigma+(PB)^{\dagger}PC\xi\}, \tag{7.27}$$

where $A_C=A_1+C$, $\Phi(t)=(1+\bar{g}(t-h_{\text{in}}))\Delta u(t-h_{\text{in}})+\sigma(t)$.

Applying (7.21) to (7.22), we achieve the following relation:

$$\xi^T(\nu)P\xi(\nu)<r\xi^T(t)P\xi(t)+\varphi(\nu), \tag{7.28}$$

where

$$\begin{aligned}\varphi(\nu)=&r\{\sum_{i=0}^{1}\frac{1}{2\alpha_i}(\hat{\gamma}_i(t)-\gamma_i^*)^2+\frac{1}{2\alpha_2}\hat{\gamma}(t)_2^2+\Gamma\beta^2(t)+\Gamma_1\rho^2(t)\}\\&-\{\sum_{i=0}^{1}\frac{1}{2\alpha_i}(\hat{\gamma}_i(\nu)-\gamma_i^*)^2+\frac{1}{2\alpha_2}\hat{\gamma}(\nu)_2^2+\Gamma\beta^2(\nu)+\Gamma_1\rho^2(\nu)\}.\end{aligned} \tag{7.29}$$

For any two nonzero vectors z and $\acute{z}$, there exist a constant $\eta>0$ and a matrix $D>0$ such that

$$-2z^T\acute{z}\le\eta z^TD^{-1}z+(1/\eta)\acute{z}^TD\acute{z}. \tag{7.30}$$

Then, applying (7.30) and taking $D=P$, we obtain the following inequalities:

$$\begin{aligned}-2\int_{-h_{\text{in}}}^{0}\xi^TPB_1A_c[\xi(t+\theta)]\,d\theta\le\bar{h}_{\text{in}}\xi^T[\eta PB_1A_cP^{-1}\\\times A_c^TB_1^TP+(r/\eta)P)]\xi+\int_{-h_{\text{in}}}^{0}(1/\eta)\varphi(t+\theta)d\theta,\end{aligned} \tag{7.31}$$

$$\begin{aligned}-2\int_{-h_{\text{in}}}^{0}\xi TPB_1B_1\xi(t-h+\theta)d\theta\le\bar{h}_{\text{in}}\xi^T[\eta PB_1B_1P^{-1}\\\times B_1^TB_1^TP+(r/\eta)P)]\xi+\int_{-h_{\text{in}}}^{0}(1/\eta)\varphi(t-h_{\text{in}}+\theta)d\theta,\end{aligned} \tag{7.32}$$

$$\begin{aligned}-2\int_{-h_{\text{in}}}^{0}\xi^TPB_1[B\Phi(t+\theta)]\,d\theta\le\bar{h}_{\text{in}}\xi^T\left[\eta PB_1P^{-1}B_1^TP\right]\xi\\+\int_{-h_{\text{in}}}^{0}(1/\eta)B^T\Phi(t+\theta)PB\Phi(t+\theta)d\theta.\end{aligned} \tag{7.33}$$

Since (7.17)–(7.19) are piecewise continuous, assuming that the system is locally Lipschitz for all $h_{\text{in}}\in[0\quad\bar{h}_{\text{in}}]$, we have that there exist $\Gamma,\Gamma_1\in\Re^+$

such that the following inequalities hold:

$$\frac{1}{2\eta}|| \int_{-h_{\text{in}}}^{0} [\varphi(t+\theta) + \varphi(t - h_{\text{in}} + \theta) \\ + (B\Phi(t+\theta)^T PB\Phi(t+\theta)]d\theta|| \leq \Gamma, \tag{7.34}$$

$$||(\Delta u - \Delta u(t - h_{\text{in}})) + (\bar{g}\Delta u - \bar{g}(t - h_{\text{in}}))\Delta u(t - h_{\text{in}}))|| \leq \Gamma_1. \tag{7.35}$$

Conditions (7.34)–(7.35) essentially imply that the rate of change in uncertainty is not faster than that of the input delay, which holds in practice for actuator delays [50].

Substituting (7.31)–(7.35) into (7.27), we have

$$\dot{V}_1(\xi) \leq -\frac{1}{2}\xi^T \left(Q - h_{\text{in}} G\right) \xi + \Gamma + ||s||\Gamma_1 \\ + s^T(\Delta u + \sigma + (PB)^{\dagger} PC\xi + \bar{g}\Delta u). \tag{7.36}$$

(ii) For the closed-loop stability of the system, the first term of (7.36) is required to be negative, that is,

$$Q - h_{\text{in}} G > 0 \Rightarrow \lambda_{\min}(Q) > h_{\text{in}}||G|| \quad \forall h_{\text{in}}.$$

Hence we obtain the maximum allowable delay as

$$h_{\text{in}} < \lambda_{\min}(Q)/||G|| := \bar{h}_{\text{in}}. \tag{7.37}$$

□

Remark 7.4. Condition (7.37) implies that the parameters Q, G are to be designed from an a priori given admissible/tolerable upper bound of input delay (cf. Assumption 7.3).

After obtaining an allowable range of delay, the following lemma highlights how ARTDC is able to extend this range over the existing adaptive-robust design [36].

Lemma 7.4. *For a similar choice of the matrices P and Q (when (7.5) is satisfied), ARTDC provides larger maximum allowable delay than AROLC as in [36] if the controller gains are designed to satisfy the following condition:*

$$\lambda_{\min}(\bar{P}_3) > ||\bar{P}_2 K \Omega^{-1}||, \tag{7.38}$$

where $\bar{P}_2$, $\bar{P}_3$ are defined during the proof.

Proof. The maximum allowable delay provided by AROLC [36] is given as

$$h_{\text{in}} < \lambda_{\min}(Q)/||G_1|| := \hat{h}_{\text{in}}, \tag{7.39}$$

where $G_1 = (\eta P\bar{B}_1(\bar{A}_1 P^{-1}\bar{A}_1^T + \bar{B}_1 P^{-1}\bar{B}_1^T + P^{-1})\bar{B}_1^T P + 2(r/\eta)P)$, $\bar{A}_1 = \begin{bmatrix} 0 & I \\ 0 & 0 \end{bmatrix}$, and $\bar{B}_1 = \begin{bmatrix} 0 & 0 \\ -K_1 & -K_2 \end{bmatrix}$.

Let us define

$$P^{-1} = \begin{bmatrix} \bar{P}_1 & \bar{P}_2 \\ \bar{P}_2^T & \bar{P}_3 \end{bmatrix}.$$

Then by (7.5) the matrices G (as in (7.24)) and G_1 (as in (7.39)) are simplified as

$$G = \eta P \begin{bmatrix} 0 & 0 \\ 0 & J \end{bmatrix} P + 2(r/\eta)P,\ G_1 = \eta P \begin{bmatrix} 0 & 0 \\ 0 & J_1 \end{bmatrix} P + 2(r/\eta)P, \tag{7.40}$$

$$J = (\Omega K\bar{P}_1 + 2\Omega^2\bar{P}_2^T)K\Omega + 2\Omega^2\bar{P}_3\Omega^2 + \Omega\bar{P}_3\Omega, \tag{7.41}$$

$$\begin{aligned} J_1 = {} & (4\Omega K\bar{P}_1 + 16\Omega^2\bar{P}_2^T)K\Omega + 16\Omega^2\bar{P}_3\Omega^2 + 4\Omega\bar{P}_3\Omega \\ & + K(\bar{P}_3 + \bar{P}_1)K + (K + 2\Omega)\bar{P}_2^T K. \end{aligned} \tag{7.42}$$

Moreover, $P > 0 \Rightarrow P^{-1} > 0$. Then applying Schur's complement to $P^{-1} > 0$, we have

$$\bar{P}_3 > 0, \tag{7.43}$$

$$\bar{P}_1 > \bar{P}_2\bar{P}_3^{-1}\bar{P}_2^T \Rightarrow \lambda_{\min}(\bar{P}_1) > \frac{||\bar{P}_2||^2}{\lambda_{\min}(\bar{P}_3)} \tag{7.44}$$

$$\Rightarrow 0 < ||\bar{P}_2|| < \sqrt{\lambda_{\min}(\bar{P}_1)\lambda_{\min}(\bar{P}_3)}. \tag{7.45}$$

Inspecting (7.37), (7.39), and (7.40), we can note that $\bar{h}_{\text{in}} > \hat{h}_{\text{in}}$ would be achieved if $J < J_1$. This implies that the following condition needs to be satisfied:

$$\begin{aligned} K(\bar{P}_3 + \bar{P}_1)K + 14\Omega^2\bar{P}_3\Omega^2 + 3\Omega\bar{P}_3\Omega > {} & -14\Omega^2\bar{P}_2^T K\Omega \\ & - K\bar{P}_2^T K - 2\Omega\bar{P}_2^T K. \end{aligned} \tag{7.46}$$

On the other hand, this condition will be satisfied if

$$\textbf{(i)} \quad \Omega^2\bar{P}_3\Omega^2 > -\Omega^2\bar{P}_2^T K\Omega \Rightarrow \bar{P}_3\Omega > -\bar{P}_2^T K, \tag{7.47}$$

$$\textbf{(ii)}\quad 3\Omega\bar{P}_3\Omega > -2\Omega\bar{P}_2^T K \Rightarrow \bar{P}_3\Omega > -(2/3)\bar{P}_2^T K, \tag{7.48}$$

$$\textbf{(iii)}\quad K(\bar{P}_3+\bar{P}_1)K > -K\bar{P}_2^T K \Rightarrow \bar{P}_3+\bar{P}_1 > -\bar{P}_2. \tag{7.49}$$

Using (7.45), condition (7.49) will hold if

$$\lambda_{\min}(\bar{P}_1)+\lambda_{\min}(\bar{P}_3) > ||\bar{P}_2|| > \sqrt{\lambda_{\min}(\bar{P}_1)\lambda_{\min}(\bar{P}_3)}. \tag{7.50}$$

Since $\lambda_{\min}(\bar{P}_1) > 0$, $\lambda_{\min}(\bar{P}_3) > 0$ from (7.43)–(7.44), condition (7.50) always holds. Hence taking into account conditions (7.47)–(7.48), we can achieve $\bar{h}_{\text{in}} > \hat{h}_{\text{in}}$ if

$$\lambda_{\min}(\bar{P}_3) > ||\bar{P}_2 K\Omega^{-1}||. \tag{7.51}$$

□

Theorem 7.1. *System (7.1) employing ARTDC with the control input (7.6)–(7.19) is asymptotically converging in tracking error* ξ *when* $\{\hat{\gamma}_i \le \underline{\gamma}_i \vee s^T\dot{s} > 0, i=0,1,2\}$ *and UUB (uniformly ultimately bounded) otherwise.*

Proof. Exploring various combinations of Δu and gains $\hat{\gamma}_i$, β, ρ in (7.15) and (7.17)–(7.19), respectively, the following two possible cases are identified:

Case (i): $\hat{\gamma}_i \le \underline{\gamma}_i \vee s^T\dot{s} > 0,\ i=0,1,2.$

Case (ii): $s^T\dot{s} \le 0 \vee \beta \le \underline{\beta} \vee \rho \le \underline{\rho}.$

Note that $||\sigma + (PB)^\dagger PC\xi||$ satisfies the upper bound structure (7.14) with the constant $||(PB)^\dagger PC||$ added to γ_1^*. However, for convenience, we will use $\gamma_1^* = \gamma_1^{*'} + ||(PB)^\dagger PC||$ with some abuse of notation, where $\gamma_1^{*'}$ is given in (7.14).

Using Lemma 7.3 and (7.15)–(7.19), we investigate the stability of the closed-loop system for various cases as follows:

Case (i): $\hat{\gamma}_i \le \underline{\gamma}_i \vee s^T\dot{s} > 0,\ i=0,1,2.$

Taking $\Upsilon = (Q - hG)$, the time derivative of (7.22) yields

$$\begin{aligned}\dot{V} &\le -\frac{1}{2}\xi^T\Upsilon\xi + \Gamma + s^T\{-(c+\rho+\beta)\frac{s}{||s||} + \sigma\} + ||s||\Gamma_1 \\ &\quad + \sum_{i=0}^{1}(\hat{\gamma}_i - \gamma_i^*)||\xi||^i||s|| + \frac{1}{\alpha_2}\hat{\gamma}_2\dot{\hat{\gamma}}_2 + \Gamma\beta\dot{\beta} + \Gamma_1\rho\dot{\rho} \\ &\le -\frac{1}{2}\lambda_{\min}(\Upsilon)||\xi||^2 \le 0. \end{aligned} \tag{7.52}$$

From (7.52) we can infer that $V(t) \in \mathcal{L}_\infty$. Thus by the definition of V we have $\xi(t), \hat{\gamma}_i(t), \rho(t), \beta(t) \in \mathcal{L}_\infty$ for $i=0,1,2$. This further implies that

$\tau(t), \sigma(t) \in \mathcal{L}_\infty$. Since the system follows the locally Lipschitz condition for all $h_{\text{in}} \in [0 \quad \bar{h}_{\text{in}}]$, from (7.26) we have $\dot{\xi}(t) \in \mathcal{L}_\infty$. The boundedness of $\xi(t)$ and $\dot{\xi}(t)$ ensures that $\ddot{V}(t) \in \mathcal{L}_\infty$ from (7.52). Thus using the Lyapunov like lemma [51] yields

$$\lim_{t\to\infty} \dot{V}(t) = 0 \Rightarrow \lim_{t\to\infty} \xi(t) = 0.$$

Case (ii): $s^T\dot{s} \leq 0 \vee \beta \leq \underline{\beta} \vee \rho \leq \underline{\rho}$.
Since $||s|| \leq ||B^T P|| ||\xi||$ and $\hat{\gamma}_2 \geq \underline{\gamma}_2$, by (7.15)–(7.19) the time derivative of (7.22) yields

$$\begin{aligned}\dot{V} &\leq -\frac{1}{2}\xi^T \Upsilon \xi + \Gamma + s^T\{-(c+\rho+\beta)\frac{s}{||s||} + \sigma\} + ||s||\Gamma_1 \\ &\quad - \sum_{i=0}^{1}(\hat{\gamma}_i - \gamma_i^*)||\xi||^i||s|| + \frac{1}{\alpha_2}\hat{\gamma}_2\dot{\hat{\gamma}}_2 + \Gamma\beta\dot{\beta} + \Gamma_1\rho\dot{\rho} \\ &\leq -\frac{1}{2}\lambda_{\min}(\Upsilon)||\xi||^2 + f(||\xi||),\end{aligned} \tag{7.53}$$

where

$$\begin{aligned}f(||\xi||) &= -\varsigma\underline{\gamma}_2||\xi||^3 + \{(1+\delta\rho_0)\Gamma_1 + 2\gamma_0^*\}||B^T P|| ||\xi|| + 2\gamma_1^*||B^T P|| ||\xi||^2 \\ &\quad + (1+\delta\beta_0)\Gamma\end{aligned}$$

is a 3rd-degree polynomial of $||\xi||$. Now, applying Descartes' rule of sign change, we can notice that the equation $f(||\xi||) = 0$ has exactly one positive real root. So, other two roots are either two negative real roots or a pair of complex conjugate roots. Let $\iota > 0$ be the positive real root of $f(||\xi||) = 0$. Moreover, since the coefficient of the cubic term (the highest degree term) of $f(||\xi||)$ is negative, $f(||\xi||) < 0$ for $||\xi|| > \iota$ (Bolzano's intermediate value theorem).

Following the lines of the proof in [40], UUB stability can be claimed when $||\xi|| > \iota$. Further, since ς is a design parameter (defined in (7.18)), we can select a high value of ς that helps to reduce ι and achieve better tracking accuracy. □

However, the control law (7.15) is discontinuous in nature. So, we modified it as follows with the help of a user-defined scalar $\epsilon \in \Re^+$:

$$\Delta u(t) = -\begin{cases}\zeta\frac{s}{||s||} & \text{if } ||s|| \geq \epsilon, \\ \zeta\frac{s}{\epsilon} & \text{if } ||s|| < \epsilon.\end{cases} \tag{7.54}$$

The modified closed-loop stability under (7.54) is stated as follows.

Lemma 7.5. *System* (7.1) *employing ARTDC with the switching control input* (7.54) *follows a similar stability notion mentioned in Theorem 7.1 when* $||s|| \geq \epsilon$, *whereas the system remains UUB when* $||s|| < \epsilon$.

Proof. For the condition $||s|| < \epsilon$, we identify the following two possible conditions:

Case (iii): $||s|| < \epsilon \wedge \{\hat{\gamma}_i \leq \underline{\gamma}_i \vee s^T\dot{s} > 0, i = 0, 1, 2\}$.

Case (iv): $||s|| < \epsilon \wedge \{s^T\dot{s} \leq 0 \vee \beta \leq \underline{\beta} \vee \rho \leq \underline{\rho}\}$.

The stability aspect of these two cases is analyzed using (7.15)–(7.19) and the results obtained from Lemma 7.3 and Theorem 7.1.

Case (iii): $||s|| < \epsilon \wedge \{\alpha \leq \gamma \vee s^T\dot{s} > 0, i = 0, 1, 2\}$.

Following the lines of the proof in [43], we can show that there exists γ_{iM} such that $\hat{\gamma}_i \leq \gamma_{iM}$ for $i = 0, 1, 2$ in this case. Using (7.15) and the fact that $||s|| < \epsilon$, for this case, we have

$$\begin{aligned}\dot{V} &\leq -\frac{1}{2}\xi^T\Upsilon\xi + \Gamma + s^T\{-(c+\rho+\beta)\frac{s}{\epsilon} + \sigma\} + ||s||\Gamma_1 \\ &\quad + \sum_{i=0}^{1}(\hat{\gamma}_i - \gamma_i^*)||\xi||^i||s|| + \frac{1}{\alpha_2}\hat{\gamma}_2\dot{\hat{\gamma}}_2 + \Gamma\beta\dot{\beta} + \Gamma_1\rho\dot{\rho} \\ &\leq -\frac{1}{2}\lambda_{\min}(\Upsilon)||\xi||^2 + \epsilon\gamma_{1M}||\xi|| + \epsilon(\gamma_{0M} + \gamma_{2M}).\end{aligned} \tag{7.55}$$

Case (iv): $||s|| < \epsilon \wedge \{s^T\dot{s} \leq 0 \vee \beta \leq \underline{\beta} \vee \rho \leq \underline{\rho}\}$.

Similarly, for this case,

$$\begin{aligned}\dot{V} &\leq -\frac{1}{2}\xi^T\Upsilon\xi + \Gamma + s^T\{-(c+\rho+\beta)\frac{s}{\epsilon} + \sigma\} + ||s||\Gamma_1 \\ &\quad - \sum_{i=0}^{1}(\hat{\gamma}_i - \gamma_i^*)||\xi||^i||s|| + \frac{1}{\alpha_2}\hat{\gamma}_2\dot{\hat{\gamma}}_2 + \Gamma\beta\dot{\beta} + \Gamma_1\rho\dot{\rho} \\ &\leq -\frac{1}{2}\lambda_{\min}(\Upsilon)||\xi||^2 - \varsigma\underline{\gamma}_2||\xi||^3 + \{(1+\delta\rho_0)\Gamma_1 + 2\gamma_1^*\}\epsilon||\xi|| \\ &\quad + (1+\delta\beta_0)\Gamma + 2\epsilon\gamma_0^*.\end{aligned} \tag{7.56}$$

Case (iv) can be analyzed exactly as Case (ii). With similar arguments as in Section II.A of [43], UUB can be claimed for the closed-loop system when $||s|| < \epsilon$. □

Remark 7.5 (Independence from the variation of input delay). The closed-loop stability analysis reveals that, utilizing ARTDC, no restriction is imposed on the rate of change of delay, that is, $\dot{h}_{in}$ and $\ddot{h}_{in}$. This allows ARTDC to tackle arbitrary variation of input delay.

Remark 7.6 (Importance of ρ and β). It can be noted from (7.52) that the negative semidefinite result was achieved by canceling the terms Γ and $||s||\Gamma_1$ via β and ρ, respectively. Compared to [36], such design helps ARTDC to provide the asymptotic convergence of tracking error compared to the UUB result obtained in the former when the tracking error moves away from switching surface (i.e., in Case (i)).

Remark 7.7. Note that various terms like γ_0^*, γ_1^*, Γ, Γ_1 and γ_{iM}, $i=0,1,2$, are only used for analytical purpose, and they are not utilized for controller design. Further, high values of α_i, $i=0,1,2$, help us to achieve better error convergence. On the other hand, (7.53) and (7.56) reveal that smaller values δ, β_0, and ρ_0 reduce the error bounds in the UUB conditions.

Remark 7.8 (On a generalization of ARTDC). The upper bound structure of uncertainty as in (7.2) holds for many practical systems such as wheeled mobile robot (reduced-order model), Monod kinetics, and so on [38]. Nevertheless, to encompass a larger class of uncertainty structures (cf. [44]), the proposed method can be extended by suitably modifying the control laws (7.15)–(7.18) following the guidelines outlined as in [37,38].

Comparison with the existing results: In comparison with the existing works in the literature, the advantages of the proposed ARTDC are summarized as follows:

- In contrast to the method AROLC in [36], the proposed ARTDC framework extends the maximum allowable input delay of the system for the same choice of *nominal control parameters* according to Lemma 7.4. A comparative maximum allowable delay for ARTDC and AROLC is provided in Table 7.1 for various η, keeping fixed P, Q, r values similar to those in [36]. Here, η, P, Q, and r are treated as the *nominal control parameters*.
- Unlike [31–33], ARTDC does not depend on $\dot{h}_{in}$ and $\ddot{h}_{in}$, which allows the system to be insensitive toward arbitrary variation of input delay. Specifically, fast variation of input delay is allowed by ARTDC compared to only slow variation allowed in those works. Moreover, ARTDC does not involve any threshold value in its adaptive laws (7.17)–(7.19) and allows the switching gains to decrease as error decreases (i.e., when $s^T\dot{s} \leq 0$). Hence ARTDC avoids the over- and underestimation problems of switching gain.

Table 7.1 Maximum allowable input delay (ms) comparison.

η	Maximum allowable delay (ms)	
	AROLC ([36])	ARTDC (proposed)
0.5	102.6	123.4
1.0	125	231.7
1.5	110.2	311.4
2.0	93.6	361.2
2.5	79.5	384

- ARTDC is specifically designed by considering a more generalized structure of the upper bound of uncertainty (as provided in (7.14)), which involves system states. Hence ARTDC removes the conservativeness regarding presumption of constant upper bound on uncertainties like in [36].
- Whereas [36] provided UUB results for all the conditions, an asymptotic convergence of tracking error ξ is achieved for Case (i) by the proposed ARTDC. This is due the introduction of the gains ρ and β. This case is important as it signifies the situation where the error is increasing and going away from the switching surface s. Hence an improved tracking accuracy is obtained in ARTDC. Moreover, with the modified adaptive law of ARTDC, Case (iii) of [36], which shows UUB result, is subsumed by ARTDC in Case (i). This further allows ARTDC to provide a better tracking accuracy through the asymptotic convergence of the error.

7.4. Simulation results

In this section, we verify the effectiveness of the proposed ARTDC scheme using the simple academic example

$$\ddot{q} = \underbrace{\frac{1}{2}(10|\dot{q}| + 5q)}_{f(q,\dot{q})} + \underbrace{\frac{1}{2}}_{g(q,\dot{q})} \tau(t - h_{\text{in}}(t)), \tag{7.57}$$

$$h_{\text{in}}(t) = 0.02|\sin(0.01t)|. \tag{7.58}$$

Note that system (7.57) satisfies the boundedness conditions in Assumptions 7.1–7.2. The nominal system parameters are selected as $\hat{f} = (|\dot{q}| + q)$, $\hat{g} = 0.4$. We apply the proposed ARTDC laws (7.6), (7.7), and (7.15) with

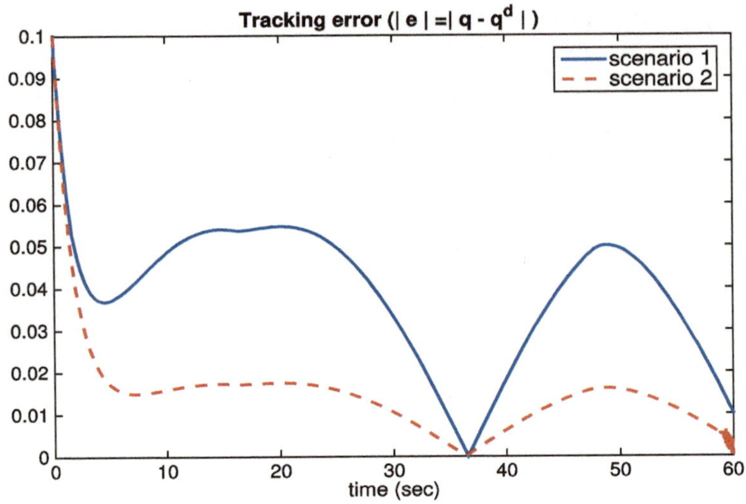

Figure 7.1 Trajectory tracking performance comparison of ARTDC under various scenarios.

adaptive laws (7.17)–(7.19) using the following control design parameters: $Q = I$, $K = 1$, $\Omega = 0.5$, $r = 1.01$, $\eta = 0.7$, $\bar{g} = 0.75$, $\delta = 10$, $\underline{\rho} = \underline{\beta} = 0.05$, $\varsigma = 0.1$, $\alpha_0 = \alpha_1 = 0.1$, $\alpha_2 = 1$, $\underline{\gamma}_i = 0.001$, $\hat{\gamma}_{i0} = \rho_0 = \beta_0 = 2$, $i = 0, 1, 2$, and $\epsilon = 0.1$. Such a choice of parameters satisfy conditions (7.3) and (7.24). The objective is to track the desired trajectory $q^d = \sin(0.1t)$.

To verify the effectiveness of various adaptive-robust gains, we choose the following two scenarios:

- **Scenario 1:** $\hat{\gamma}_1 = \hat{\gamma}_2 = \beta = \rho = 0$ in (7.15);
- **Scenario 2:** $\hat{\gamma}_j$, $j = 0, 1, 2$, and β, ρ are evaluated as in (7.15) and (7.17)–(7.19).

Scenario 1 can be considered to be equivalent to that of AROLC [36]. Fig. 7.1 demonstrates the benefit of additional adaptive-robust gains as in (7.16), stemming from the state-dependent uncertainty structure as in Lemma 7.1, where ARTDC in Scenario 2 clearly provides significant performance improvement. The state response under Scenario 2 against the desired trajectory is shown in Fig. 7.2. Further, evolutions of various adaptive-robust gains for the same scenario are depicted in Figs. 7.3 and 7.4.

To test the performance of ARTDC in noisy feedback scenario, we added a Gaussian noise of variance 0.001 (see Fig. 7.5) in the feedback of q, $\dot{q}$, and $\ddot{q}$ under Scenario 2. Tracking performance of ARTDC under such a scenario is demonstrated via Figs. 7.6–7.9.

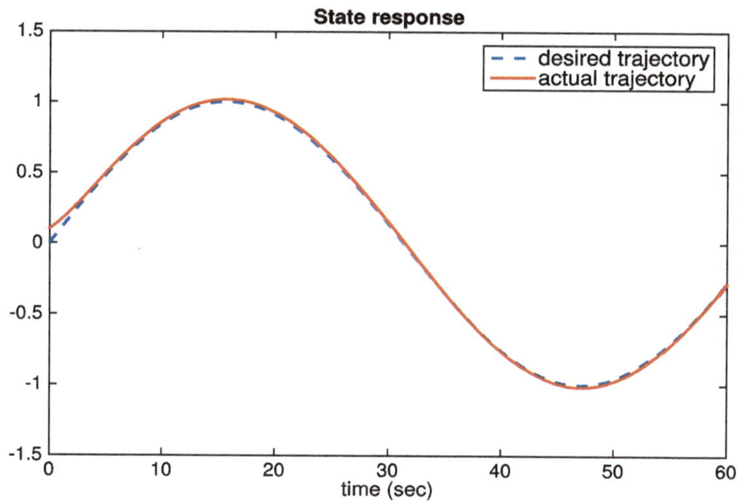

Figure 7.2 State response with ARTDC for Scenario 2.

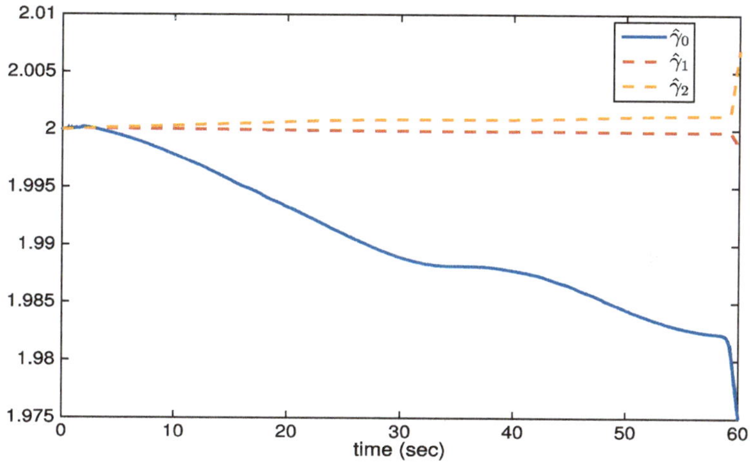

Figure 7.3 Evolution of various adaptive-robust gains for Scenario 2.

The Matlab® code for the simulation is provided in Appendix 7.A.

7.5. Application: nonholonomic WMR

In this section, we evaluate the performance of ARTDC compared to AROLC [36] via experimentation on a wheeled mobile robot. Non-

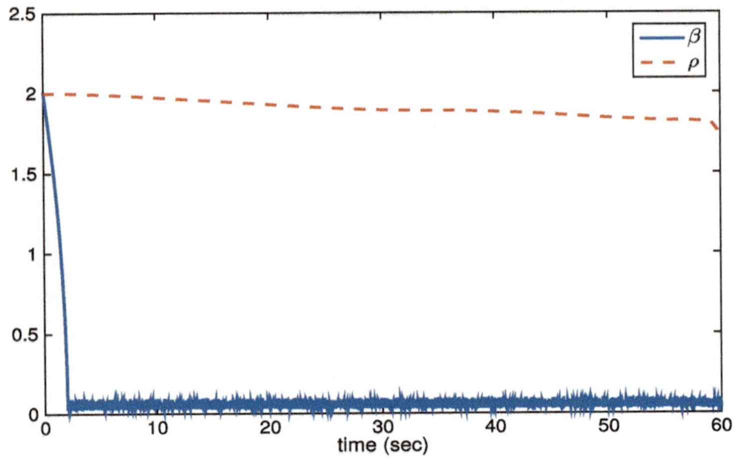

Figure 7.4 Evolution of gains β and ρ.

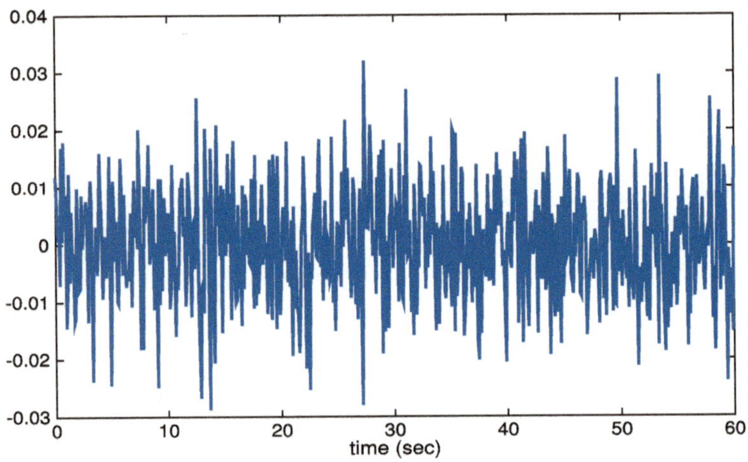

Figure 7.5 Noise profile.

holonomic WMR, which has vast applications in transportation, planetary exploration, surveillance, security, human–machine interfaces, and so on, provides a unique platform to test the proposed control law. Therefore, the performance of ARTDC is verified using a commercially available 'PIONEER 3' WMR under various sampling intervals.

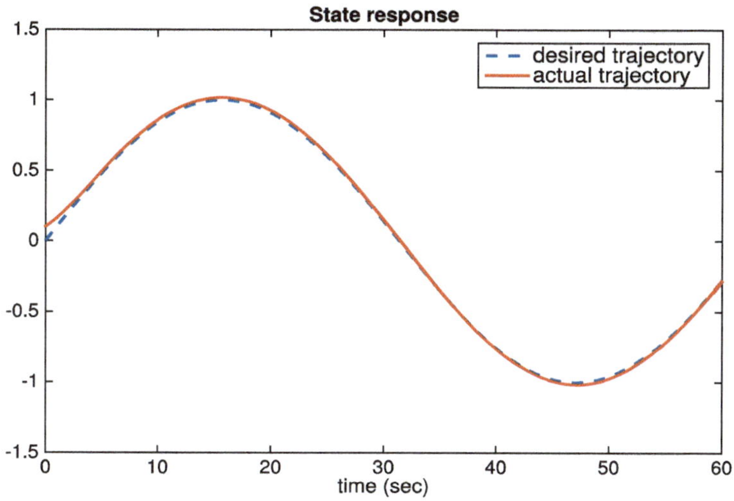

Figure 7.6 State response with ARTDC for Scenario 2 under noisy feedback.

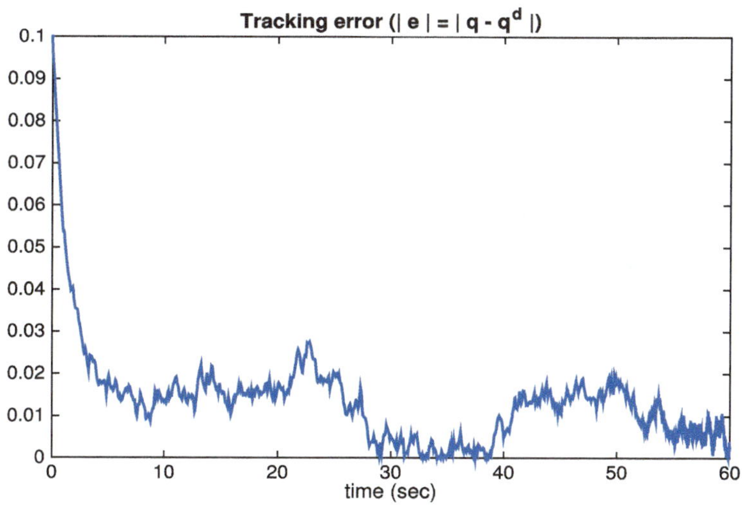

Figure 7.7 Tracking performance of ARTDC for Scenario 2 with noise.

The Euler–Lagrange formulation of a nonholonomic WMR (Fig. 7.10) is given as [52]

$$M(q)\ddot{q} + C(q, \dot{q})\dot{q} = L\tau - A^{*T}\lambda^{*}, \tag{7.59}$$

where

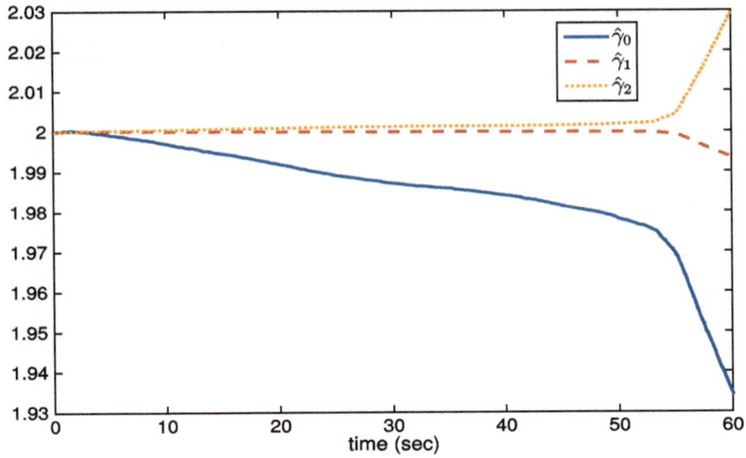

Figure 7.8 Evolution of various adaptive-robust gains for Scenario 2 with noise.

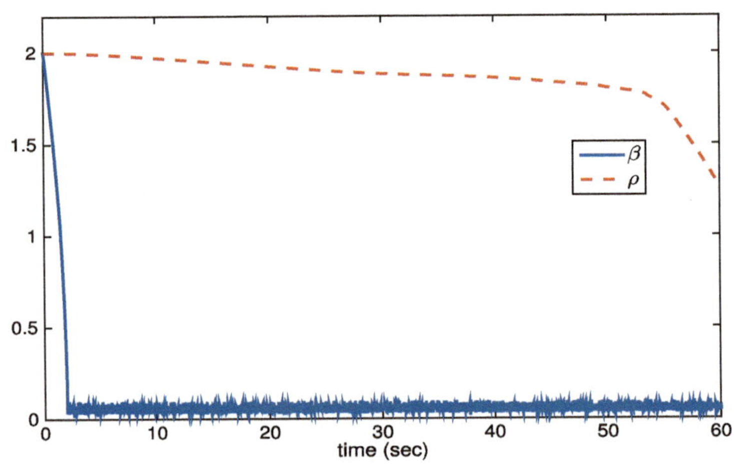

Figure 7.9 Evolution of gains β and ρ under noisy feedback case.

$$M = \begin{bmatrix} m & 0 & md\sin\varphi & 0 & 0 \\ 0 & m & -md\cos\varphi & 0 & 0 \\ md\sin\varphi & -md\cos\varphi & \bar{I} & 0 & 0 \\ 0 & 0 & 0 & I_w & 0 \\ 0 & 0 & 0 & 0 & I_w \end{bmatrix},$$

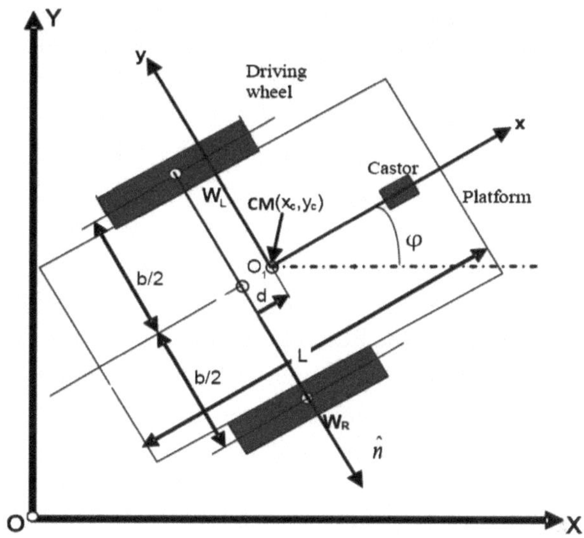

Figure 7.10 Schematic of a WMR.

$$L = \begin{bmatrix} 0 & 0 \\ 0 & 0 \\ 0 & 0 \\ 1 & 0 \\ 0 & 1 \end{bmatrix}, \; C(q,\dot{q})\dot{q} = \begin{bmatrix} md\dot{\varphi}^2 \cos\varphi \\ md\dot{\varphi}^2 \sin\varphi \\ 0 \\ 0 \\ 0 \end{bmatrix}, \; \tau = \begin{bmatrix} \tau_r \\ \tau_l \end{bmatrix}.$$

Here $q \in \Re^5 = \{x_c, y_c, \varphi, \theta_r, \theta_l\}$ is the generalized coordinate vector of the system, (x_c, y_c) are the coordinates of the center of mass (CM) of the system, φ is the heading angle, (θ_r, θ_l) and (τ_r, τ_l) are the rotation and torque inputs of the right and left wheels, respectively, m, $\bar{I}$, I_w, r_w, and b represent the system mass, system inertia, wheel inertia, wheel radius, and robot width, respectively, d is the distance to the CM from the center of the line joining the two wheel axis, and A^* and λ^* represent the constraint matrix and vector of constraint forces (Lagrange multipliers), respectively. Expressions of A^* and $\bar{I}$ can be found in [50,52].

It is worth noting that system (7.59) has only two control inputs although having five generalized coordinates. In fact, for WMR, we can only directly control wheel positions (θ_r, θ_l) rather than (x_c, y_c, φ). Further, for the available experimental setup, wheel encoders are the only source of odometric data. Therefore the system dynamics can be represented as a combination of a reduced-order dynamics and kinematic model for efficient controller design [38,52]:

$$M_R\ddot{q}_R + C_R\dot{q}_R = \tau, \tag{7.60}$$

$$\dot{q} = \underbrace{\begin{bmatrix} \frac{r_w}{b}\left(\frac{b}{2}\cos(\varphi) - d\sin(\varphi)\right) & \frac{r_w}{b}\left(\frac{b}{2}\cos(\varphi) + d\sin(\varphi)\right) \\ \frac{r_w}{b}\left(\frac{b}{2}\sin(\varphi) + d\cos(\varphi)\right) & \frac{r_w}{b}\left(\frac{b}{2}\sin(\varphi) - d\cos(\varphi)\right) \\ \frac{r_w}{b} & -\frac{r_w}{b} \\ 1 & 0 \\ 0 & 1 \end{bmatrix}}_{S(q)} \dot{q}_R, \tag{7.61}$$

where

$$M_R = S^T M S = \begin{bmatrix} k_1 & k_2 \\ k_2 & k_1 \end{bmatrix}, \tag{7.62}$$

$$k_1 = I_w + \{\bar{I} + m(b^2/4 - d^2)\}(r_w{}^2/b^2),$$

$$k_2 = \{m(b^2/4 + d^2) - \bar{I}\}(r_w{}^2/b^2),$$

$$C_R = S^T(M\dot{S} + CS) = \begin{bmatrix} 0 & 0 \\ 0 & 0 \end{bmatrix}, q_R = [\theta_r\ \theta_l]^T. \tag{7.63}$$

The WMR dynamics (7.59) is based on the rolling without slipping condition. However, in practical circumstances a WMR is always subjected to uncertainties like friction, external disturbance, and so on. Hence incorporating (7.63), the system dynamics (7.60) is modified as

$$M_R\ddot{q}_R + f(\dot{q}_R) + d_R = \tau, \tag{7.64}$$

where $f(\dot{q}_R)$ and d_R are considered to be the unmodeled dynamics and disturbance, respectively. Note that dynamics (7.64), when represented as in (7.1), satisfies the uncertainty structures defined by Assumptions 7.1–7.2 [38].

The objective is to apply the proposed ARTDC and AROLC [36] to the reduced-order WMR system (7.64) to track a desired $q_R^d(t)$, which in fact track a desired $q^d(t)$ via (7.61). To illustrate the fact, we can direct a WMR to move in a specified circular path by designing two suitable different and fixed wheel velocities or in a Lawn–Mower path by applying an approximated square-wave velocity profile to the wheels [38,52,53]. In the following subsection the performance of ARTDC and AROLC [36] is compared using the 'PIONEER-3' wheeled mobile robot.

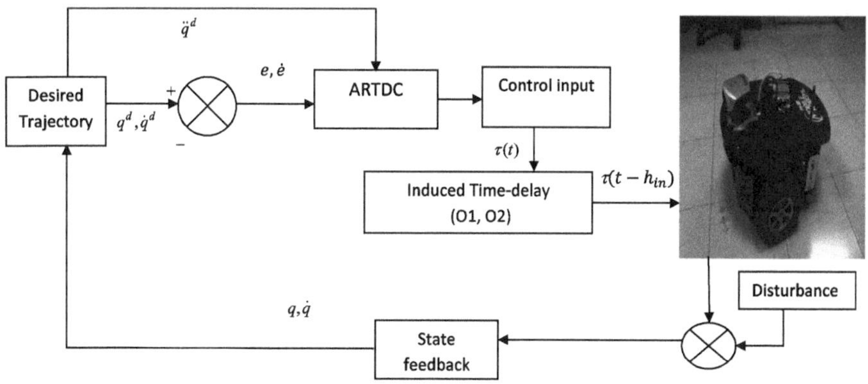

Figure 7.11 A schematic of the experimental arrangement.

7.5.1 Experimental scenarios

The value of *nominal control parameters* P, Q, r, K, and η are chosen the same as in [36], which provides $\hat{h}_{\text{in}} = 125$ ms for AROLC and $\bar{h}_{\text{in}} = 231.7$ ms for ARTDC. Further, $\delta = 0.1$, $\rho_0 = \beta_0 = 0.3$, $\underline{\rho} = \underline{\beta} = 0.05$, $\varsigma = 20$, $\alpha_i = 1$, $\underline{\gamma}_i = 0.001$, $i = 0, 1, 2$, are chosen for ARTDC, and $\epsilon = 0.1$ is selected for both ARTDC and AROLC. The following two time-varying input delays are chosen:

O1: $h_{\text{in}}(t) = 80 + 20\,\text{abs}(\sin(t))$ ms.

O2: $h_{\text{in}}(t) = 120 + 100\,\text{abs}(\sin(0.1t))$ ms.

A schematic of the experimental arrangement is shown in Fig. 7.11, depicting how input delay is introduced into the system. The robot is directed to move in a circular path as mentioned in [36].

7.5.2 Results and discussion

Fig. 7.12 depicts the desired circular path and the path tracked by ARTDC for **O1**. The tracking performance comparison for ARTDC and AROLC is illustrated through Figs. 7.13 and 7.14 in terms of x and y position errors (absolute value) for **O1**, respectively. Similarly, the comparative tracking performance of both controllers for **O2** are depicted through Figs. 7.15 and 7.16, respectively. For scenario **O2**, at some instances the input delay goes outside the permissible upper bound for AROLC, and hence its performance deteriorates against ARTDC. However, for the case **O1**, even when the delay is within its permissible $\bar{h}_{\text{in}}$ (i.e., 125 ms), ARTDC provides a better accuracy than AROLC due to its better stabilizing property.

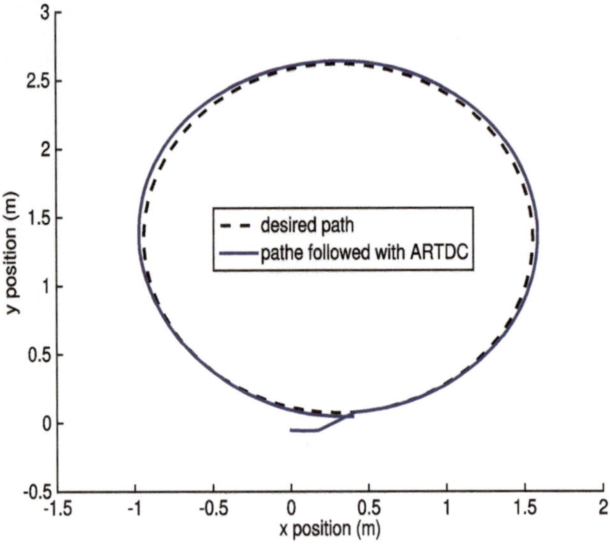

Figure 7.12 Trajectory tracking performance of ARTDC.

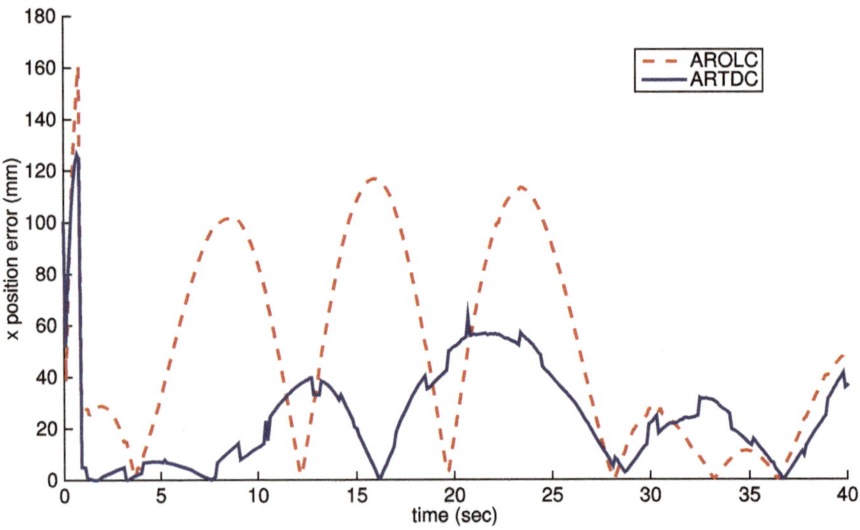

Figure 7.13 *x* position tracking error comparison of ARTDC and AROLC.

A summary of the performance of individual controller is provided in Table 7.2 along with their average x position error (AXE), average y position error (AYE), and total variation (TV) [36]. Besides indicating the amount of control input consumed, TV also captures the complexity of the controllers

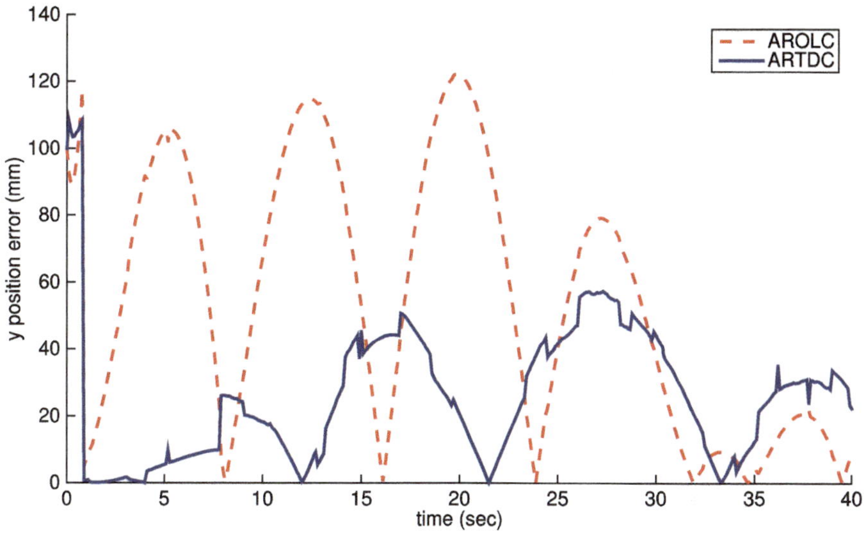

Figure 7.14 *y* position tracking error comparison of ARTDC and AROLC.

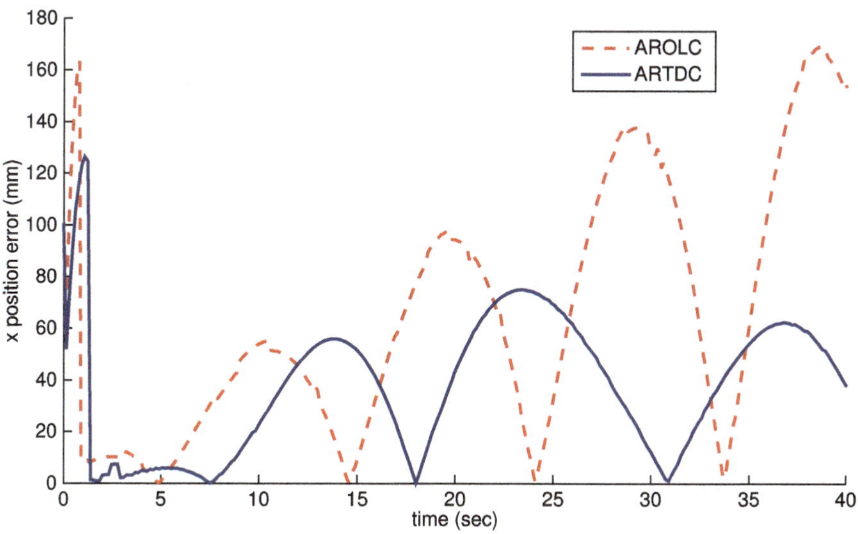

Figure 7.15 *x* position tracking error comparison of ARTDC and AROLC.

[54]. Hence, due to the added switching gains β, ρ, $\hat{\gamma}_1$, $\hat{\gamma}_2$, complexity in the control algorithm rises for ARTDC in comparison to AROLC. As a matter of fact, even though TV of ARTDC is less than that of AROLC, it is close to the numerical value of AROLC.

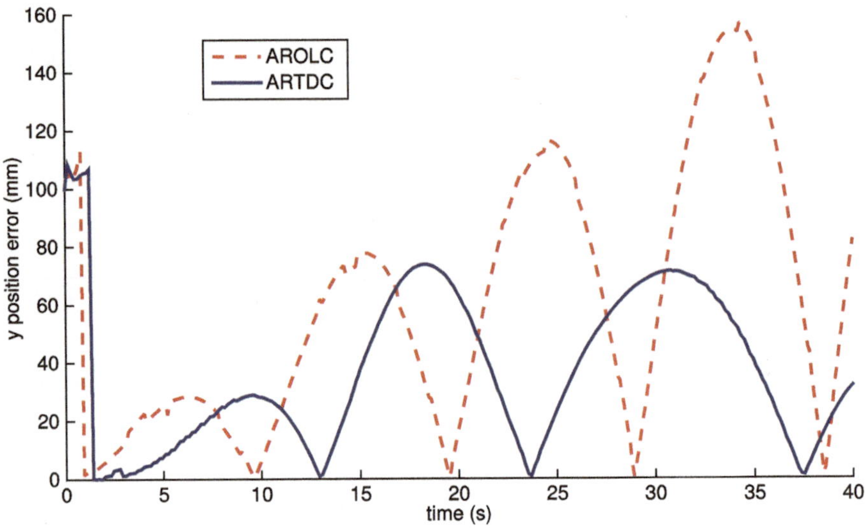

Figure 7.16 *y* position tracking error comparison of ARTDC and AROLC.

Table 7.2 Performance comparison of AROLC and ARTDC.

Delay	Controller	AXE (mm)	AYE (mm)	TV (Nm)
O1	AROLC ([36])	50.69	53.73	4.47
	ARTDC (proposed)	25.37	25.42	3.93
O2	AROLC ([36])	66.80	61.41	5.53
	ARTDC (proposed)	37.88	36.78	4.87

7.6. Conclusion and future work

In this work, we proposed a new adaptive-robust tracking controller for uncertain systems under the influences of time-varying input delay and state-dependent uncertainty. In contrast to the existing adaptive-robust strategies, the proposed framework avoided any restrictive assumption such as uncertainties being a priori upper bounded by a constant; instead, the control structure relied on a linear-in-parameters upper bound structure of uncertainty, stemming from physical properties of a system. In addition, compared to state of the art, we also achieved increased delay tolerability and (partially) improved stability result. Crucially, such features were achieved while being invariant with respect to the rate of change of input delay. The proposed scheme considered a simple uncertainty structure as a proof of concept; however, suggestive steps for a generalized solution were

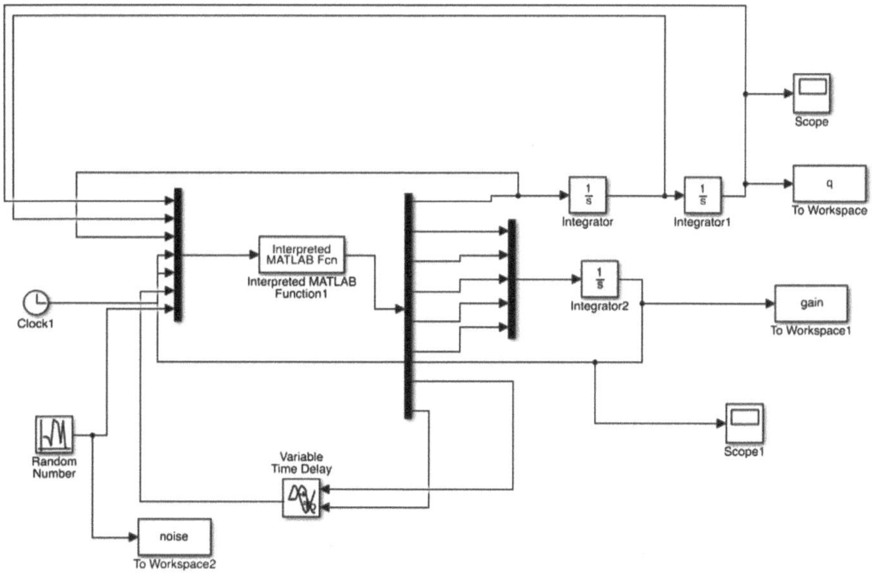

Figure 7.17 Simulink model.

also highlighted. The effectiveness of the proposed controller was substantiated via extensive simulation and comparative experimental studies.

Nonetheless, the present study primarily considered a small input delay at the actuator level. Therefore an important future work would be realizing the proposed concept for systems (e.g., wide-area power systems) that face comparatively large communication delays.

Appendix 7.A. Simulation code

Here we provide the Matlab–Simulink model used for simulating system (7.57)–(7.58). The provided code is a function to be used in conjunction with Simulink: the code is called in the Simulink "Matlab Interpreted Block". A diagram of the Simulink block is given in Fig. 7.17.

Interpreted MATLAB Function Code

```
function out = test_delay(U)
% U is a vector of inputs

%feedback
```

```
noise=U(11);
q = U(1)+noise;
q_dot = U(2)+noise;
q_ddot = U(3)+noise;

gamma_hat_0= U(4);
gamma_hat_1= U(5);
gamma_hat_2= U(6);
beta = U(7);
rho = U(8);

t = U(9); % this is time

tao_d = U(10); % this is delayed input

f_hat = (1*abs(q_dot) + 1*q);
g_hat = 0.4;

% generating reference trajectory
q_d = sin(0.1*t);
q_d_dot = 0.1*cos(0.1*t);
q_d_ddot = -0.1^2*sin(0.1*t);

% error calculation
e = q - q_d;
e_dot  = q_dot - q_d_dot;
e_ddot = q_ddot - q_d_ddot;

% Solving Lyapunov
Q = [1 0;0 1];
% Hurwitz matrix
A= [0 1 ; -1 -1];
%lypunov equation
P = lyap (A',Q);
omega = 0.5;

%calculating s
s = P(2,2)*e_dot + P(1,2)*e;
s_dot = P(2,2)*e_ddot + P(1,2)*e_dot;
```

```

%calculating zeeta
g_bar = 0.75;
norm_eita = norm([e, e_dot]);
c_t = gamma_hat_0 + gamma_hat_2 + (gamma_hat_1*norm_eita);
zeta = (1 /(1 - g_bar))*(c_t+beta+rho);

%calculating delta u using saturation function
epsilon = 0.2;
if norm(s) >= 0.2
    delta_u = - zeta*(s/norm(s));
else
    delta_u = - zeta*(s/epsilon);
end

%calculating tao
u = q_d_ddot - (omega * e_dot);
tao = 1/g_hat*( - f_hat +  delta_u + u);

%the induced delay
delay=0.02*abs(sin(0.01*t));

%system output
q_ddot = 1/2*(10*abs(q_dot) + 5*q) + 1/2*tao_d;

%calculation of gammas
rho_bar = 0.05;
beta_bar = 0.05;
alpha_0 = .1;
alpha_1 = .1;
alpha_2 = 1;

del = 10;
varsigma = .1;

flag = s*s_dot;

if flag <= 0 || beta <= beta_bar || rho <= rho_bar
    gamma_hat_0 = -alpha_0 * norm(s);
```

```
else
    gamma_hat_0 = alpha_0 * norm(s);
end

if flag <= 0 || beta <= beta_bar || rho <= rho_bar
    gamma_hat_1 = -alpha_1 * norm(s) * norm_eita;
else
    gamma_hat_1 = alpha_1* norm(s) * norm_eita;
end

if flag <= 0 || beta <= beta_bar || rho <= rho_bar
    gamma_hat_2 = - varsigma* alpha_2 * norm_eita^3;
else
    gamma_hat_2 = alpha_2 * norm_eita * norm(s);
end

%calculating beta
if beta > beta_bar
    beta_dot = -1/beta;
else
    beta_dot = del;
end

%calculating rho
if rho > rho_bar
   rho_dot = - norm(s)/rho;
else
   rho_dot =  del*norm(s);
end

% output
out = [ q_ddot, gamma_hat_0, gamma_hat_1 , gamma_hat_2 ,
     beta_dot ,rho_dot,tao,delay];
```

References

[1] C.T. Abdallah, R. Sipahi, S.-J. Niculescu, W. Michiels, K. Gu, Stability and stabilization of systems with time delay: limitations and opportunities, IEEE Transactions on Control Systems Technology 31 (1) (2011) 38.

[2] J.-P. Richard, Time-delay systems: an overview of some recent advances and open problems, Automatica 39 (10) (2003) 1667–1694.

[3] S. Roy, A. Patel, I.N. Kar, Analysis and design of a wide-area damping controller for inter-area oscillation with artificially induced time delay, IEEE Transactions on Smart Grid 10 (4) (2019) 3654–3663.
[4] K. Gu, S.-I. Niculescu, Survey on recent results in the stability and control of time-delay systems, Journal of Dynamic Systems, Measurement, and Control 125 (2) (2003) 158–165.
[5] V.L. Kharitonov, Robust stability analysis of time delay systems: a survey, Annual Reviews in Control 23 (1999) 185–196.
[6] M. Krstic, Delay Compensation for Nonlinear, Adaptive, and PDE Systems, Springer, 2009.
[7] K.J. Astrom, C.C. Hang, B. Lim, A new Smith predictor for controlling a process with an integrator and long dead-time, IEEE Transactions on Automatic Control 39 (2) (1994) 343–345.
[8] Z. Artstein, Linear systems with delayed controls: a reduction, IEEE Transactions on Automatic Control 27 (4) (1982) 869–879.
[9] A. Manitius, A. Olbrot, Finite spectrum assignment problem for systems with delays, IEEE Transactions on Automatic Control 24 (4) (1979) 541–552.
[10] R. Lozano, P. Castillo, P. Garcia, A. Dzul, Robust prediction-based control for unstable delay systems: application to the yaw control of a mini-helicopter, Automatica 40 (4) (2004) 603–612.
[11] D. Yue, Q.-L. Han, Delayed feedback control of uncertain systems with time-varying input delay, Automatica 41 (2) (2005) 233–240.
[12] Z. Wang, P. Goldsmith, D. Tan, Improvement on robust control of uncertain systems with time-varying input delays, IET Control Theory & Applications 1 (1) (2007) 189–194.
[13] A. Gonzalez, A. Sala, P. Garcia, P. Albertos, Robustness analysis of discrete predictor-based controllers for input-delay systems, International Journal of Systems Science 44 (2) (2013) 232–239.
[14] A. González, A. Sala, P. Albertos, Predictor-based stabilization of discrete time-varying input-delay systems, Automatica 48 (2) (2012) 454–457.
[15] Z.-Y. Li, B. Zhou, Z. Lin, On robustness of predictor feedback control of linear systems with input delays, Automatica 50 (5) (2014) 1497–1506.
[16] I. Karafyllis, M. Krstic, Delay-robustness of linear predictor feedback without restriction on delay rate, Automatica 49 (6) (2013) 1761–1767.
[17] R. Lu, H. Wu, J. Bai, New delay-dependent robust stability criteria for uncertain neutral systems with mixed delays, Journal of the Franklin Institute 351 (3) (2014) 1386–1399.
[18] T.H. Lee, J.H. Park, S.-M. Lee, O. Kwon, Robust sampled-data control with random missing data scenario, International Journal of Control 87 (9) (2014) 1957–1969.
[19] T.H. Lee, J.H. Park, S.-M. Lee, O.-M. Kwon, Robust synchronisation of chaotic systems with randomly occurring uncertainties via stochastic sampled-data control, International Journal of Control 86 (1) (2013) 107–119.
[20] N. Bekiaris-Liberis, M. Krstic, Delay-adaptive feedback for linear feedforward systems, Systems & Control Letters 59 (5) (2010) 277–283.
[21] D. Bresch-Pietri, M. Krstic, Adaptive trajectory tracking despite unknown input delay and plant parameters, Automatica 45 (9) (2009) 2074–2081.
[22] J. Zhou, C. Wen, W. Wang, Adaptive backstepping control of uncertain systems with unknown input time-delay, Automatica 45 (6) (2009) 1415–1422.
[23] X. Zhang, H. Su, R. Lu, Second-order integral sliding mode control for uncertain systems with control input time delay based on singular perturbation approach, IEEE Transactions on Automatic Control 60 (11) (2015) 3095–3100.
[24] Y.-H. Roh, J.-H. Oh, Sliding mode control with uncertainty adaptation for uncertain input-delay systems, International Journal of Control 73 (13) (2000) 1255–1260.

[25] S. Roy, I.N. Kar, J. Lee, M. Jin, Adaptive-robust time-delay control for a class of uncertain Euler–Lagrange systems, IEEE Transactions on Industrial Electronics 64 (9) (2017) 7109–7119.
[26] P.A. Ioannou, J. Sun, Robust Adaptive Control, Courier Corporation, 2012.
[27] M. Jankovic, Control Lyapunov–Razumikhin functions and robust stabilization of time delay systems, IEEE Transactions on Automatic Control 46 (7) (2001) 1048–1060.
[28] C. Hua, G. Feng, X. Guan, Robust controller design of a class of nonlinear time delay systems via backstepping method, Automatica 44 (2) (2008) 567–573.
[29] M. Krstic, Input delay compensation for forward complete and strict-feedforward nonlinear systems, IEEE Transactions on Automatic Control 55 (2) (2009) 287–303.
[30] Z. Zhang, S. Xu, B. Zhang, Exact tracking control of nonlinear systems with time delays and dead-zone input, Automatica 52 (2015) 272–276.
[31] N. Sharma, S. Bhasin, Q. Wang, W.E. Dixon, Predictor-based control for an uncertain Euler–Lagrange system with input delay, Automatica 47 (11) (2011) 2332–2342.
[32] N. Fischer, R. Kamalapurkar, N. Fitz-Coy, W.E. Dixon, Lyapunov-based control of an uncertain Euler–Lagrange system with time-varying input delay, in: 2012 American Control Conference (ACC), IEEE, 2012, pp. 3919–3924.
[33] R. Kamalapurkar, N. Fischer, S. Obuz, W.E. Dixon, Time-varying input and state delay compensation for uncertain nonlinear systems, IEEE Transactions on Automatic Control 61 (3) (2015) 834–839.
[34] S. Roy, I.N. Kar, Robust control of uncertain Euler–Lagrange systems with time-varying input delay, in: Proceedings of the Advances in Robotics, ACM, 2017, p. 16.
[35] V.I. Utkin, A.S. Poznyak, Adaptive sliding mode control with application to super-twist algorithm: equivalent control method, Automatica 49 (1) (2013) 39–47.
[36] S. Roy, I.N. Kar, Adaptive robust tracking control of a class of nonlinear systems with input delay, Nonlinear Dynamics 85 (2) (2016) 1127–1139.
[37] S. Roy, I.N. Kar, Adaptive–robust approach for time-delayed control of a class of uncertain EL systems, in: Adaptive-Robust Control With Limited Knowledge on Systems Dynamics, Springer, 2020, pp. 41–68.
[38] S. Roy, S.B. Roy, I.N. Kar, Adaptive–robust control of Euler–Lagrange systems with linearly parametrizable uncertainty bound, IEEE Transactions on Control Systems Technology 26 (5) (2018) 1842–1850.
[39] S. Roy, I.N. Kar, Adaptive sliding mode control of a class of nonlinear systems with artificial delay, Journal of the Franklin Institute 354 (18) (2017) 8156–8179.
[40] S. Roy, I.N. Kar, J. Lee, N.G. Tsagarakis, D.G. Caldwell, Adaptive-robust control of a class of el systems with parametric variations using artificially delayed input and position feedback, IEEE Transactions on Control Systems Technology 27 (2) (2019) 603–615.
[41] X.-Z. Jin, J.H. Park, Adaptive sliding-mode insensitive control of a class of non-ideal complex networked systems, Information Sciences 274 (2014) 273–285.
[42] X.-Z. Jin, W.-W. Che, D. Wang, Active coupling and its circuitry designs of chaotic systems against deteriorated and delayed networks, Nonlinear Dynamics 70 (3) (2012) 1867–1877.
[43] S. Roy, S.B. Roy, J. Lee, S. Baldi, Overcoming the underestimation and overestimation problems in adaptive sliding mode control, IEEE/ASME Transactions on Mechatronics 24 (5) (2019) 2031–2039.
[44] S. Roy, S. Baldi, A simultaneous adaptation law for a class of nonlinearly parametrized switched systems, IEEE Control Systems Letters 3 (3) (2019) 487–492.
[45] S. Roy, J. Lee, S. Baldi, A new continuous-time stability perspective of time-delay control: introducing a state-dependent upper bound structure, IEEE Control Systems Letters 3 (2) (2019) 475–480.
[46] S. Roy, S. Baldi, L.M. Fridman, On adaptive sliding mode control without a priori bounded uncertainty, Automatica (2019) 108650.

[47] S. Roy, S. Baldi, On reduced-complexity robust adaptive control of switched Euler–Lagrange systems, Nonlinear Analysis: Hybrid Systems 34 (2019) 226–237.
[48] G. Leitmann, On the efficacy of nonlinear control in uncertain linear systems, Journal of Dynamic Systems, Measurement, and Control 103 (2) (1981) 95–102.
[49] S. Roy, I.N. Kar, Adaptive-Robust Control With Limited Knowledge on Systems Dynamics: An Artificial Input Delay Approach and Beyond, vol. 257, Springer Nature, 2019.
[50] S. Roy, I.N. Kar, Time-delayed control (TDC): design issues and solutions, in: Adaptive-Robust Control With Limited Knowledge on Systems Dynamics, Springer, 2020, pp. 23–39.
[51] H.K. Khalil, Nonlinear Systems, Prentice-Hall, Upper Saddle River, 2002.
[52] S. Roy, S. Nandy, R. Ray, S.N. Shome, Robust path tracking control of nonholonomic wheeled mobile robot: experimental validation, International Journal of Control, Automation, and Systems 13 (4) (2015) 897–905.
[53] S. Roy, S. Nandy, R. Ray, S.N. Shome, Time delay sliding mode control of nonholonomic wheeled mobile robot: experimental validation, in: 2014 IEEE International Conference on Robotics and Automation (ICRA), IEEE, 2014, pp. 2886–2892.
[54] S. Skogestad, Simple analytic rules for model reduction and PID controller tuning, Journal of Process Control 13 (4) (2003) 291–309.

CHAPTER EIGHT

Adaptive control of large-scale systems with long input and state delays and time-varying delays in the uncertain nonlinear interconnections

S.H. Hashemipour[a], N. Vasegh[b], and A. Khaki Sedigh[c]
[a]Department of Electrical Engineering, Roudsar and Amlash Branch, Islamic Azad University, Roudsar, Iran
[b]Department of Electrical Engineering, Shahid Rajaee Teacher Training University, Tehran, Iran
[c]Department of Electrical Engineering, Khaje Nasir Toosi University of Technology, Tehran, Iran

Contents

8.1. Introduction

Large-scale interconnected systems are present in electrical power systems, chemical reactors, economic systems, and computer communication networks. In recent decades, researchers have focused on the decentralized control of large-scale systems, and numerous results are available [1,2]. Time delays are inevitably introduced in the information exchange among

Control Strategy for Time-Delay Systems
https://doi.org/10.1016/B978-0-12-820599-0.00013-6

subsystems of large-scale interconnected systems through communication networks. Thus investigation of decentralized control problems of nonlinear interconnected time-delayed systems is of practical importance. However, the analysis of time-delayed systems is inherently complex [3,4]. In many practical control problems the parameters and dynamics of the actual system are unknown. Also, the uncertainties bounds may be unknown or only partially known. Therefore adaptive schemes must be considered to update the partially known uncertainty bounds for large-scale systems that are dynamically interconnected with time delays.

It has been reported that adaptive control is an effective method for controlling systems with uncertainties and delays (see, e.g., [5–7]). Adaptive schemes are considered to update the partially known uncertainty bounds for large-scale systems that are dynamically interconnected with time delays. In [8] the plant model with linear interconnections was considered, where the nonlinear inputs are bounded by unknown bounds, and an adaptive controller was designed for the model. In [9] the plant includes uncertain delayed nonlinear interconnections bounded by a known function, and therefore there was no adaptation law. The nonlinear interconnection terms are generally assumed to be bounded by the linear functions of the states norm [6]. In a similar study [10], the nonlinear interconnection terms are assumed to be bounded by a higher-order polynomial of the states. In [11] the nonlinear interconnected terms for switched interconnected uncertain nonlinear systems are studied, and neural networks are employed to approximate the unknown interconnected terms. In nonlinear stochastic large-scale systems with actuator faults, fuzzy logic is used to identify the unknown lumped nonlinear functions in [12]. Due to physical restrictions of large-scale systems, designing a centralized controller is often difficult and impossible. Therefore in [13–16], decentralized controllers are applied to uncertain large-scale time-delay interconnected systems. In addition to the studies mentioned, many researchers have focused on the adaptive control of large-scale systems [6,7,13,17], but input delays, which are essential in practical applications, are neglected.

The Smith predictor was introduced for stable linear systems with pure input delay in [18]. The limitations of open-loop stability can be resolved using the predictor feedback approach [19]. For linear systems with input and state time delays, a stabilizing sliding mode control scheme is presented in [20]. Based on the low-gain feedback structure [21], a finite-dimensional feedback control from the truncated traditional predictor feedback was proposed [22]. To resolve the problems concerning the predictor feedback controllers for linear systems with both state and input delays, a truncated predictor feedback was introduced [23]. Also, the Smith predictor is applied to such systems in [5]. For delay compensation, two auxiliary dynamic adaptive filters with adjustable gains were included in the adaptive controller part. However, due to the presence of such filters, the tracking error cannot be minimized. The nested predictor is another method for stable ordinary systems with states and input delays [24]. In [14], interconnected systems without input and state delays are considered. The large-scale systems with delay in interconnection terms and no input delay were investigated in [25]. The proposed control law is very complex and hardly applied in practical systems. An adaptive observer should be designed for delay predictions in input delay compensations for large-scale systems with unknown interconnected. The adaptive observer in large-scale systems has received considerable attention [26–29]. In [26], it is claimed that nonlinear observers for a class of one-sided Lipschitz time-delay systems are designed in the presence of disturbances. In [27] a state observer for interconnected nonlinear systems with time-varying input delay is designed. In [28] a state observer is presented for large-scale systems, but input delays are not considered. However, the uncertainties upper bounds and the nonlinear interconnections are unknown. In [29], it is shown that infinite-dimensional observer-based output feedback controllers can be generalized to linear systems with both time-varying input and output delays. These papers deal with designing a suitable observer and do not consider exact prediction and input delay compensation in large-scale systems with both state and input delays.

In this chapter, we investigate the problem of adaptive control of the decentralized model reference for a class of uncertain large scale systems with constant state and input delays and varying delays in the interconnected terms. Three control methods are presented for these systems. In the first method, the interconnected terms are bounded by polynomials of states, and delayed states with undetermined coefficients and acceptable responses are derived. But the interconnected term is not considered, which deteriorates the desired performance. Because of the uncertainty of interconnected term, a state adaptive observer is designed, and the control law is applied to the observed system. In the observer design, it is assumed that the upper bound of the interconnected term is unknown. In both methods a nested predictor is used to compensate the input delay. This method is applied to the systems with simultaneous state and input delays, and the input delay can be arbitrarily larger than the state delay. The presented methods are only applicable to open-loop stable systems. The third presented method resolves this assumption with uncertain and linear interconnected terms. In this method, by adding integrators the delay in input is changed into a state delay, and then the delay in input is compensated by a dynamical compensator. The effectiveness of the proposed approaches is illustrated by simulation results.

8.2. Input delay compensation

In this section, we present methods for the input delay compensation in systems with both state and input delays, where the input delay can be arbitrarily large.

8.2.1 Input delay compensation via nested prediction

A nested predictor-based approach is established to predict the future states and compensate the input delays. Consider the following system with pure input delay:

$$\dot{x}(t) = Ax(t) + Bu(t - R). \tag{8.1}$$

This system can be stabilized by [19]

$$u(t) = Kx(t + R), \tag{8.2}$$

where

$$x(t + R) = e^{AR}x(t) + \int_{-R}^{0} e^{-A\theta} Bu(t + \theta)\,d\theta. \tag{8.3}$$

Controller (8.3) is known as the traditional predictor feedback. However, in the case of delayed state, we have:

$$\dot{x}(t) = Ax(t) + A_d x(t - d) + Bu(t - R), \tag{8.4}$$

where $u(t)$ can be chosen as

$$u(t) = K_d x(t - d + R) + Kx(t + R), \tag{8.5}$$

and the closed-loop system becomes

$$\dot{x}(t) = (A + BK)x(t) + (A_d + BK_d)x(t - d). \tag{8.6}$$

In these equations, $x \in R^n$ and $u \in R^m$ represent the state and control vectors, respectively, $A, A_d \in R^{n\times n}$ and $B \in R^{n\times m}$ are known constant matrices, and d and R are known constant delays. This system can be stabilized by proper selection of feedback gains K and K_d similar to [24]. If a delayed state is added to (8.1), then (8.3) can be written as

$$x(t+R) = \left(e^{AR}x(t) + \int_{-d}^{R-d} e^{A(R-d-\theta)} A_d x(t+\theta)d\theta + \int_{-R}^{0} e^{-A\theta} Bu(t+\theta)d\theta \right). \tag{8.7}$$

However, if $R > d$ in (8.5), then the second term on the right-hand side still involves the future state, which indicates that this predictor is not causal and not implementable. To solve this, a nested prediction approach is proposed in [24]. The predicted time R is achieved step by step. Initially, a small $\theta \in [0, d]$ is selected, and then (8.7) is recursively applied until the desired time is obtained. Letting $R = \theta_1$ in (8.7), we have:

$$\hat{x}(t+\theta_1) = \left(e^{A\theta_1}x(t) + \int_{-d}^{\theta_1-d} e^{A(\theta_1-d-\theta)} A_d x(t+\theta)d\theta + \int_{-R}^{\theta_1-R} e^{A(\theta_1-R-\theta)} Bu(t+\theta)d\theta \right). \tag{8.8}$$

Now if we substitute $\hat{x}(t+\theta_1)$ instead of $x(t)$ in (8.8), then we obtain $\hat{x}(t+2\theta_1)$, and this process continues up to when $n\theta_1$ is equal to R. For the nth step, we obtain the following approximation:

$$\hat{x}(t+n\theta_1) = e^{A\theta_1}x(t+(n-1)\theta_1) + \int_{-d}^{\theta_1-d} e^{A(\theta_1-d-\theta)} A_d x(t+(n-1)\theta_1+\theta)d\theta + \int_{-R}^{\theta_1-R} e^{A(\theta_1-R-\theta)} Bu(t+(n-1)\theta_1+\theta)d\theta. \tag{8.9}$$

Since $\theta \in [0, d]$ is small and we select θ_1 as small as possible, the integrals on the right-hand side of (8.9) can be approximated using the mean value theorem [30] as follows:

$$\int_a^b f(\theta)d\theta = (b-a)f(a). \tag{8.10}$$

Then, for $n = 1, 2, \ldots$, we have

$$\begin{cases} \hat{x}(t+\theta_1) = e^{A\theta_1}\left[x(t) + \theta_1\left(A_d x(t-d) + Bu(t-R)\right)\right], \\ \hat{x}(t+n\theta_1) = e^{A\theta_1}\left[\hat{x}(t+(n-1)\theta_1) \right. \\ \qquad \left. + \theta_1\left(A_d\hat{x}\left(t+(n-1)\theta_1 - d\right) + Bu(t+(n-1)\theta_1 - R)\right)\right]. \end{cases} \tag{8.11}$$

Letting $\theta_1 = \frac{d}{N'}$, by a proper choice of N', the accuracy can be increased. We can consider n such that $n\theta_1 = R$. Now, we will show that the sequence $\{\hat{x}(t+n\theta_1)\}$ approaches $\hat{x}(t+R)$. Define the prediction error as

$$E(t) = x(t+R) - \hat{x}(t+R). \tag{8.12}$$

Let $s = t + \theta$. Then (8.7) can be rewritten as

$$\hat{x}(t+R) = e^{AR}x(t) + \int_{t-d}^{t+R-d} e^{A(R-d+t-s)}A_d x(s)ds + \int_{t-R}^{t} e^{A(t-s)}Bu(s)ds. \tag{8.13}$$

The dynamic prediction of $\hat{x}(t+R)$ is derived as

$$\begin{aligned} \dot{\hat{x}}(t+R) &= e^{AR}\left(\dot{x}(t) - A_d x(t-d) - Bu(t-R)\right) + A_d x(t+R-d) + Bu(t) \\ &\quad + A\left(\int_{t-d}^{t+R-d} e^{A(R-d+t-s)}A_d x(s)ds + \int_{t-R}^{t} e^{A(t-s)}Bu(s)ds\right). \end{aligned} \tag{8.14}$$

Using (8.4) and $e^{AR}A = Ae^{AR}$, we have

$$\begin{aligned} \dot{\hat{x}}(t+R) &= A\left(e^{AR}x(t) + \int_{t-d}^{t+R-d} e^{A(R-d+t-s)}A_d x(s)ds + \int_{t-R}^{t} e^{A(t-s)}Bu(s)ds\right) \\ &\quad + A_d x(t+R-d) + Bu(t). \end{aligned} \tag{8.15}$$

By simple manipulations we obtain the following equation:

$$\dot{\hat{x}}(t+R) = A\hat{x}(t+R) + A_d x(t+R-d) + Bu(t). \tag{8.16}$$

Using (8.16), we derive the prediction error dynamics as

$$\dot{E}(t+R) = AE(t+R). \tag{8.17}$$

Since $E(R) = 0$, $E(t+R) = 0$, for all $t \geq 0$. On the other hand, if the interconnection $f_d(.)$ term is known, then the prediction accuracy can be increased by considering the interconnected term. Then, for prediction $x(t+R)$ instead of (8.7), we can use the following relation:

$$\hat{x}(t+R) = e^{AR}x(t) + \int_{-d}^{R-d} e^{A(R-d-\theta)}A_d x(t+\theta)d\theta + \int_{-R}^{0} e^{-A\theta}Bu(t+\theta)d\theta$$
$$+ \int_{-R}^{0} e^{-A\theta}Bf_d(t+R+\theta)d\theta. \tag{8.18}$$

Using (8.16), the prediction error is as in (8.17). But in many large-scale systems the interconnected term is uncertain, and for this reason, we did not consider it in the prediction.

8.2.2 Input delay compensation via inclusion of integrators

Consider the following system with both state and input delays:

$$\dot{x}(t) = Ax(t) + A_d x(t-d) + Bu(t-R) \tag{8.19}$$

with

$$R = (\alpha - 1)\,d + \Delta d, \tag{8.20}$$

where $\alpha \geq 1$ is an integer, and $\Delta d \in \left[0,\, d\right)$.

If we define

$$u(t) = \bar{u}(t-(d-\Delta d)), \tag{8.21}$$

then Eq. (8.19) can be written as

$$\dot{x}(t) = Ax(t) + A_d x(t-d) + B\bar{u}(t-\alpha d). \tag{8.22}$$

Choosing an auxiliary state vector

$$G_1(t) = \bar{u}(t-\alpha d + d), \tag{8.23}$$

(8.22) can be rewritten as

$$\dot{x}(t) = Ax(t) + A_d x(t-d) + BG_1(t-d). \tag{8.24}$$

Defining the states

$$\begin{cases} \dot{G}_i(t) = G_{i+1}\left(t-d\right), & i \in [1, \alpha-1], \\ \dot{G}_\alpha(t) = G_{\alpha+1}(t), \end{cases} \tag{8.25}$$

and introducing the augmented state vector $\omega(t) = \left[x^T(t), G_1^T(t), \ldots, G_\alpha^T(t)\right]^T \in \Re^{n+m\alpha}$ give

$$\dot{\omega}(t) = A'\omega(t) + A'_d\omega(t-d) + B'G_{\alpha+1}(t), \tag{8.26}$$

where $G_{\alpha+1}(t)$ is the input vector, and

$$A' = \begin{bmatrix} A & 0_{n\times\alpha m} \\ 0_{\alpha m\times n} & 0_{\alpha m\times\alpha m} \end{bmatrix}, \quad A'_d = \begin{bmatrix} [A_d, B] & 0_{n\times m(\alpha-1)} \\ 0_{(\alpha-1)m\times(n+m)} & I_{m(\alpha-1)} \\ 0_{m\times(n+m)} & 0_{m\times m(\alpha-1)} \end{bmatrix},$$
$$B' = \begin{bmatrix} 0_{(n+(\alpha-1)m)\times m} \\ I_m \end{bmatrix}. \tag{8.27}$$

In (8.26), there is no input delay, and only a delay in the state vector is present. By inclusion of integrators, the delay in input is compensated, and the problem reduces to the augmented time-delay system with only delay in the state vector. In the following, we design a controller for the augmented time-delay system (8.26) that can stabilize system (8.19) according to [31].

For the augmented time-delay system, we apply the state feedback

$$G_{\alpha+1}(t) = K_0\omega(t) + K_d\omega(t-d), \tag{8.28}$$

where K_0 and K_d are constant matrices, and by (8.26) and (8.28) the closed-loop system is

$$\dot{\omega}(t) = (A' + B'K_0)\omega(t) + (A'_d + B'K_d)\omega(t-d). \tag{8.29}$$

The control law (8.28) can be obtained by integrating $G_i(t)$. Also, $\bar{u}(t)$ is related to $G_1(t)$ by

$$\bar{u}(t) = G_1(t + (\alpha - 1)d). \tag{8.30}$$

Since this relation is noncausal, the new states are defined as

$$\begin{cases} g_i(t) = G_i\left(t + (\alpha - i)\,d\right), & i \in [1, \alpha], \\ g_{\alpha+1}(t) = G_{\alpha+1}(t). \end{cases} \tag{8.31}$$

The augmented state vector $\omega(t)$ can be defined as

$$\omega(t) = \begin{bmatrix} x(t) \\ G_1(t) \\ G_2(t) \\ \vdots \\ G_\alpha(t) \end{bmatrix} = \begin{bmatrix} x(t) \\ g_1(t-(\alpha-1)d) \\ g_2(t-(\alpha-2)d) \\ \vdots \\ g_\alpha(t) \end{bmatrix}. \tag{8.32}$$

Therefore compensator (8.25) can be written as

$$\begin{cases} \dot{g}_i(t) = g_{i+1}(t), & i \in [1, \alpha-1], \\ \dot{g}_\alpha(t) = K_0\omega(t) + K_d\omega\left(t-d\right), \\ \overline{u}(t) = g_1(t), & t \geq 0, \end{cases} \tag{8.33}$$

where these equations are causal. The initial condition of this compensator is

$$g(\theta), \quad \forall\theta \in \left[-\alpha d, 0\right], \quad \text{where } g = [g_1, \ldots, g_\alpha]^T.$$

To obtain the gains K_0 and K_d in (8.28) and to achieve stability, we use the LMI method. Therefore the gains K_0 and K_d are obtained as

$$\begin{bmatrix} K_0 \\ K_d \end{bmatrix} = \begin{bmatrix} K_{0,x} & K_{0,1} & K_{0,2} & \ldots & K_{0,\alpha} \\ K_{d,x} & K_{d,1} & K_{d,2} & \ldots & K_{d,\alpha} \end{bmatrix}, \tag{8.34}$$

where $K_{0,x}, K_{d,x} \in \Re^{m\times n}$ and $K_{0,i}, K_{\tau,i} \in \Re^{m\times m}$, $i \in [1, \alpha]$. Controller (8.33) is written as [31]

$$\begin{cases} \dot{g}_i(t) = g_{i+1}(t), & i \in [1, \alpha-1], \\ \dot{g}_\alpha(t) = K_{0,x}x(t) + \sum\limits_{i=1}^{\alpha} K_{0,i}g_i(t-(\alpha-i)d) + K_{d,x}x(t-\tau) \\ \qquad + \sum\limits_{i=1}^{\alpha} K_{d,i}g_i\left(t-(\alpha-i+1)d\right), \\ \overline{u}(t) = g_1(t), & t \geq 0. \end{cases} \tag{8.35}$$

8.2.3 Input delay compensation in the stable systems with periodic input

Define a periodic signal as

$$x(t) = x(t+T) = x(t+\beta T), \tag{8.36}$$

where β is a positive integer. Let

$$\beta T = R + b, \tag{8.37}$$

where $b \geq 0$ and $\beta = 1$; if $T \geq R$, then $\beta \geq 1$.

So by a time delay and without any predictions the signal $x(t+R)$ can be produced. So using (8.36) and (8.37), we have

$$x(t) = x(t+\beta T) = x(t+R+b), \tag{8.38}$$

and then

$$x(t-b) = x(t+R). \tag{8.39}$$

This relation shows that using the periodicity property, time delay can be used instead of prediction. This simple and practical feature can be applied for input delay compensation in the stable systems with periodic input.

8.3. Problem formulation and assumptions

Consider a class of large-scale systems composed of N interconnected subsystems with delays in states and inputs, described by the following equations:

$$\begin{aligned} S_i: \quad \dot{x}_i(t) &= A_i x_i(t) + A_{di} x_i(t-d_i) + B_i u_i(t-R_i) \\ &+ B_i \sum_{j=1}^{N} \xi_{ij}\left(x_j(t-h_{ij}(t)), t-h_{ij}(t)\right), \end{aligned} \tag{8.40}$$

where $x_i \in \Re^{n_i}$ and $u_i \in \Re^{m_i}$ represent the state and control vectors of the ith subsystem, respectively, $A_i, A_{di} \in \Re^{n_i \times n_i}$, and $B_i \in \Re^{n_i \times m_i}$ are known scalar matrices, d_i and R_i are known constant delays, $\xi_{ij}\left(x_j(t-h_{ij}(t)), t-h_{ij}(t)\right)$ are uncertain interconnections, which indicate the interconnections among the current states and the delayed states of systems S_i and S_j, and $h_{ij}(t)$ are bounded time-varying delays, which are differentiable and satisfy

$$0 \leq \bar{f}_{ij} \leq h_{ij}(t) \leq \bar{h}_{ij} < \infty, \tag{8.41}$$

where $\bar{f}_{ij}$ and $\bar{h}_{ij}$ are positive constants. In the literature [6,10], it is generally assumed that $h_{ij}(t)$ are positive and their derivatives are less than one. In this paper, we remove this assumption and assume that time-varying delays $h_{ij}(t)$ are bounded nonnegative continuous functions. The initial conditions are

$$x_i(t) = \Xi_i(t), \quad t \in [t_0 - \tau_i, t_0], \; i = 1, 2, ..., N,$$

where $\tau_i = \max\left\{\bar{h}_{ij}, d_i\right\}$ and $\Xi_i(t)$ are continuous functions.

The stable nondelayed reference model is defined by the equation

$$\dot{x}_{mi}(t) = A_{mi}x_{mi}(t) + B_{mi}D_i(t), \tag{8.42}$$

where $x_{mi} \in \Re^{n_i}$ is the state vector, $D_i(t) \in \Re^{m_i}$ is the piecewise continuous reference input to the ith reference model, and $A_{mi} \in \Re^{n_i \times n_i}$ and $B_{mi} \in \Re^{n_i \times m_i}$ are known matrices.

For system (8.41) and model reference system (8.42), we introduce the following assumptions.

Assumption 8.1. There are positive definite matrices P_i satisfying the equation

$$A_{mi}^T P_i + P_i A_{mi} = -Q_i, \quad i = 1, 2, ..., N, \tag{8.43}$$

where Q_i are positive definite matrices.

Assumption 8.2. There are constant matrices $z_i \in \Re^{n_i}$, $m_i \in \Re^{n_i}$ and a nonzero scalar θ_{ri} such that the following equations are satisfied:

$$A_i - A_{mi} = B_i z_i, \quad A_{di} + B_i m_i = 0, \quad B_{mi} = B_i \theta_{ri}.$$

Remark 8.1. The so-called matching condition widely employed in the strenuous filtering and controlling problems satisfies Assumption 8.2 (see [5,10]).

These assumptions are used for all methods; however, for the controller design in each section, we add new assumptions.

8.4. Design controller

8.4.1 The interconnections are nonlinear and bounded by polynomials with unknown gains

In this section, we design a decentralized model reference adaptive control for a class of stable large-scale systems with long delays in the inputs and delays in the states and time-varying delays in the nonlinear interconnection terms. Moreover, the ultimate uniformly bounded stability of the closed-loop system error dynamics is proven using the Lyapunov stability theory. The controller is designed in two steps. First, a nested predictor is established to predict the future states for compensation of input delays. Then, assuming that the predicted state $\hat{x}(t+R)$ is available, we will design an adaptive controller. Also, we assume the nonlinear interconnection terms to be bounded by polynomials with unknown gains.

Assumption 8.3. The interconnections term satisfy the following inequalities:

$$\begin{aligned}\sum_{j=1}^{N}\left\|\xi_{ij}\left(x_j(t),x_j(t-h_{ij}(t)),t\right)\right\| &\le \sum_{j=1}^{N}\sum_{s=1}^{p_{ij}}\alpha_{ijs}\left\|x_j(t)\right\|^{s}\\ &\quad+\sum_{j=1}^{N}\sum_{l=1}^{q_{ij}}\beta_{ijl}\left\|x_j(t-h_{ij}(t))\right\|^{l},\end{aligned} \tag{8.44}$$

where $\|\,\|$ is the Euclidean norm, p_{ij} and q_{ij} are known positive scalars, and α_{ijs} and β_{ijl} are unknown scalars. The bound of the interconnected term is limited.

Assumption 8.4. All subsystems are stable.

In this section, we design a decentralized adaptive feedback controller for input delay system (8.40) that satisfies the above assumptions. The goal is to ensure that all closed-loop signals remain bounded and the tracking error becomes small enough. Define the tracking error

$$e_i(t)=x_i(t)-x_{mi}(t). \tag{8.45}$$

The error dynamics is obtained as

$$\begin{aligned}\dot{e}_i(t)&=A_ix_i(t)+A_{di}x_i(t-d_i)+B_iu_i(t-R_i)+B_i\sum_{j=1}^{N}\xi_{ij}\left(x_j(t),x_j(t-h_{ij}(t)),t\right)\\ &\quad-A_{mi}x_{mi}(t)-B_{mi}D_i(t).\end{aligned} \tag{8.46}$$

Since the reference model (8.42) is stable, by Assumption 8.3 we obtain the following inequalities:

$$
\begin{aligned}
&\sum_{j=1}^{N} \left\| \xi_{ij}\left(x_j(t), x_j(t-h_{ij}(t)), t\right) \right\| \\
&\quad \le \sum_{j=1}^{N}\sum_{s=1}^{p_{ij}} \alpha_{ijs} \left\| x_{mj}(t) + e_j(t) \right\|^{s} + \sum_{j=1}^{N}\sum_{l=1}^{q_{ij}} \beta_{ijl} \left\| x_{mj}(t-h_{ij}(t)) + e_j(t-h_{ij}(t)) \right\|^{l} \\
&\quad \le \sum_{j=1}^{N}\sum_{s=1}^{p_{ij}} \bar{\alpha}_{ijs} \left\| e_j(t) \right\|^{s} + \sum_{j=1}^{N}\sum_{l=1}^{q_{ij}} \bar{\beta}_{ijl} \left\| e_j(t-h_{ij}(t)) \right\|^{l} + \delta_i \\
&\quad = \sum_{j=1}^{N} \bar{\alpha}_{ij}^{T} U_{ij}\left(\left\| e_j(t) \right\|\right) + \sum_{j=1}^{N} \bar{\beta}_{ij}^{T} W_{ij}\left(\left\| e_j(t-h_{ij}(t)) \right\|\right) + \delta_i,
\end{aligned}
\tag{8.47}
$$

where $\bar{\alpha}_{ijs}$, $\bar{\beta}_{ijl}$, and δ_i are unknown positive scalars, and

$$
\begin{aligned}
&\overline{\alpha}_{ij} = \left(\overline{\alpha}_{ij1}, \overline{\alpha}_{ij2}, \ldots, \overline{\alpha}_{ij\,p_{ij}}\right)^{T}, \quad \overline{\beta}_{ij} = \left(\overline{\beta}_{ij1}, \overline{\beta}_{ij2}, \ldots, \overline{\beta}_{ij\,q_{ij}}\right)^{T}, \\
&U_{ij}(.) = \left(\|e_j(t)\|, \|e_j(t)\|^2, \ldots, \|e_j(t)\|^{p_{ij}}\right)^{T}, \\
&W_{ij}(.) = \left(\left\| e_j\left(t-h_{ij}(t)\right)\right\|, \left\| e_j\left(t-h_{ij}(t)\right)\right\|^2, \ldots, \left\| e_j\left(t-h_{ij}(t)\right)\right\|^{q_{ij}}\right)^{T}.
\end{aligned}
$$

The main result is presented as follows.

Theorem 8.1. *Consider system* (8.40)*. Let the decentralized adaptive feedback controller be as follows:*

$$
u_i(t) = u_{i1}(t) + u_{i2}(t) + u_{i3}(t), \tag{8.48}
$$

where

$$
u_{i1}(t) = -z_i \hat{x}_i(t+R_i) + \theta_{ri} D_i(t+R_i) + m_i \hat{x}_i(t+R_i-d_i), \tag{8.49}
$$

$$
u_{i2}(t) = -\theta_i(t+R_i) B_i^{T} \frac{\partial V_i^{T}(e_i(t+R_i))}{\partial e_i(t+R_i)}, \tag{8.50}
$$

$$
u_{i3}(t) = \frac{-\eta_i(t+R_i)\, B_i^{T} \frac{\partial V_i^{T}(e_i(t+R_i))}{\partial e_i(t+R_i)}}{\left\| B_i^{T} \frac{\partial V_i^{T}(e_i(t+R_i))}{\partial e_i(t+R_i)} \right\|}, \tag{8.51}
$$

and $e_i(t+R_i) = \hat{x}_i(t+R_i) - x_{mi}(t+R_i)$ *and* $\theta_i(t+R_i)$, $\eta_i(t+R_i)$ *are adaptive parameters with the following adaption laws:*

$$
\dot{\theta}_i(t+R_i) = \frac{1}{2}\Gamma_i \left\| B_i^{T} \frac{\partial V_i^{T}(e_i(t+R_i))}{\partial e_i(t+R_i)} \right\|^2 - \frac{1}{2}\Gamma_i \sigma_{i1} \theta_i(t+R_i), \tag{8.52}
$$

$$\dot{\eta}_i(t+R_i) = \frac{1}{2}\phi_i \left\| B_i^T \frac{\partial V_i^T(e_i(t+R_i))}{\partial e_i(t+R_i)} \right\| - \frac{1}{2}\phi_i \sigma_{i2} \eta_i(t+R_i). \tag{8.53}$$

Then the tracking error converges uniformly and is ultimately boundedly stable, and the large-scale system is stable.

Proof. See [32]. □

The predicted error equation with interconnected term is

$$\begin{aligned}\dot{E}_i(t+R_i) &= A_i E_i(t+R_i) + B_i \sum_{j=1}^{N} \xi_{ij}\left(x_{ij}(t+R_i), x_{ij}(t+R_i-h_{ij}(t)), t+R_i\right) \\ &\quad - e^{A_i R_i} B_i \sum_{j=1}^{N} \xi_{ij}\left(x_{ij}(t), x_{ij}(t-h_{ij}(t)), t\right).\end{aligned} \tag{8.54}$$

By considering Assumption 8.3 and the proof of uniformly bounded stability the error remains limited and bounded. Note that m_i, z_i, and Ξ_i are chosen to keep the tracking error within a specified bound.

8.4.2 Decentralized adaptive state observer design

For exact prediction, the interconnection term is required to be known. However, we assume that this term is unknown. Therefore in this section, we propose an adaptive observer with a new and simple structure for an unknown large scale system to estimate the states of the system and the unknown interconnected terms. Hence the output vector $y_i(t)$ is defined as

$$y_i(t) = C_i x_i(t). \tag{8.55}$$

For the observer design, we use the following assumptions.

Assumption 8.5. The Euclidean norm of the nonlinear interconnection term $\xi_{ij}(.)$ is bounded by a nonlinear function $\rho_{ij}(.)$ multiplied by an unknown constant vector θ_{ij}^* as follows:

$$\left\| \xi_{ij}\left(x_j(t-h_{ij}(t)), t-h_{ij}(t)\right) \right\| \le \left(\theta_{ij}^*\right)^T \rho_{ij}\left(x_j(t-h_{ij}(t)), t-h_{ij}(t)\right). \tag{8.56}$$

We assume that $\rho_{ji}(.) > 0$ and these functions are unknown, continuous, and uniformly bounded with respect to the state x_j and time t.

Assumption 8.6. The pair (A_i, C_i) is completely observable.

Assumption 8.7. The matching condition for designing state observer is described as $A_i - A_{mi} = K_i C_i$, $A_{di} - K_i' C_i = 0$, where the matrices K_i and K_i' are the observer gains.

Theorem 8.2. *Consider system* (8.45). *Let the adaptive state observer be given as follows:*

$$\begin{aligned}\dot{\hat{x}}_i(t) &= A_i\hat{x}_i(t) + A_{di}\hat{x}_i(t-d_i) + B_i u_i(t-R_i) \\ &\quad + B_i\sum_{j=1}^{N} F_{ij}\left(\hat{\varphi}_i(t), e_i'(t), t\right) + K_i\left(y_i(t) - C_i\hat{x}_i(t)\right) \\ &\quad + K_i'\left(y_i(t-d_i) - C_i\hat{x}_i(t-d_i)\right),\end{aligned} \tag{8.57}$$

where

$$F_{ij}\left(\hat{\varphi}_i(t), e_i'(t), t\right) = \frac{1}{2}\eta_i\hat{\varphi}_i(t)B_i^T P_i e_i'(t) \tag{8.58}$$

and

$$\frac{d\hat{\varphi}_i(t)}{dt} = -\gamma_i\sigma_i\hat{\varphi}_i(t) + \gamma_i\eta_i\left\|B_i^T P_i e_i'(t)\right\|^2. \tag{8.59}$$

By Assumptions 8.1, 8.5, 8.6, and 8.7 the observation error converges uniformly exponentially to a ball.

Proof. See [33]. □

Using Theorem 8.2 and convergence states in the proposed observer, (8.57) can be used instead of (8.40) in designing the model reference adaptive controller.

8.4.2.1 Model reference adaptive controller design

In this section, we design an MRAC for the estimated system.

Theorem 8.3. *For system* (8.57), *we have the following decentralized adaptive feedback controller*

$$u_i(t) = u_{i1}(t) + u_{i2}(t), \tag{8.60}$$

where

$$u_{i1}(t) = -z_i\hat{x}_{pi}(t+R_i) + \theta_{ri}D_i(t+R_i) + m_i\hat{x}_{pi}(t+R_i-d_i), \tag{8.61}$$

$$u_{i2}(t) = -\frac{1}{2}\eta_i\hat{\varphi}_i(t+R_i)B_{mi}^T P_i e_i'(t+R_i), \tag{8.62}$$

in which η_i *are positive constants, and* $\hat{\varphi}_i(t+R_i)$ *is obtained from* (8.59) *with input* $e_i'(t+R_i)$*. So the adaptive law is*

$$\frac{d\hat{\varphi}_i(t+R_i)}{dt} = -\gamma_i\sigma_i\hat{\varphi}_i(t+R_i) + \eta_i\gamma_i \left\| B_{mi}^T P_i e_i'(t+R_i) \right\|^2. \tag{8.63}$$

Then we will prove that the estimated signal converges to the original signal. Prior to the proof, we note several points. The tracking error is defined as

$$e_i'(t+R_i) = x_{pi}(t+R_i) - \hat{x}_i(t+R_i). \tag{8.64}$$

Also, the predicted state $x_{pi}(t+R_i)$ *is*

$$\begin{aligned} x_{pi}(t+R_i) = e^{A_i R_i}\hat{x}_i(t) &+ \int_{-d_i}^{R_i-d_i} e^{A_i(R_i-d_i-\lambda)} A_{di}\hat{x}_i(t+\lambda)d\lambda \\ &+ \int_{-R_i}^{0} e^{-A_i\lambda} B_i\left(u_i(t+\lambda) + F_{ij}\left(t+R_i+\lambda\right)\right) d\lambda. \end{aligned} \tag{8.65}$$

As mentioned before, by designing the proposed observer the interconnected term was estimated, and it can be used for prediction and approximated by recursive relationships (8.65)*. Using* (8.12) *and* (8.18), *the dynamics of the prediction error is*

$$\dot{E}(t) = AE(t+R) + B_i \sum_{j=1}^{N}\left[\xi_{ij}\left(x_{ij}(t), x_{ij}(t-h_{ij}(t))\right) - F_{ij}\left(\hat{\varphi}_i(t), e_i'(t), t\right)\right]. \tag{8.66}$$

By Assumption 8.4, if (8.67) *is bounded, then the prediction is suitable:*

$$B_i \sum_{j=1}^{N}\left[\xi_{ij}\left(x_{ij}(t), x_{ij}(t-h_{ij}(t))\right) - F_{ij}\left(\hat{\varphi}_i(t), e_i'(t), t\right)\right] = \dot{e}'_i(t) - A_{mi}e_i'(t). \tag{8.67}$$

Since it was proved in the previous section that the observation error $e_i'(t)$ *is bounded and that by Assumption 8.5 the interconnected term is continuous and bounded, without loss of generality, we can assume that* $x_p(t) = \hat{x}(t)$.

Proof. See [33]. □

8.4.3 Input delay compensation by adding integrators in a large-scale system

In this section, for input delay compensation, we design a dynamic input delay compensator by inclusion of integrators. Using this compensator, by removing the input delay the system becomes an augmented time-delay system. In this system, the sum of delays in the states is equal to the delays in the inputs. To control the obtained system, we use the LMI approach. Also, we assume that the open-loop system is unstable. To reduce the effect of the interconnected terms, the state $x(t+R)$ is required in the adaptive controller. This can be simply obtained if the reference signals are periodic. By this assumption, which is true in many practical cases, we design the controllers by employing LMI. Furthermore, we assume that the upper bounds of the external disturbances and interconnection terms are completely unknown.

In this section, we assume that $h_{ij}(t)$ in (8.40) are bounded time-varying delays, which are differentiable and satisfy

$$0 \le h_{ij}(t) \le \overline{h}_{ij} < \infty, \quad 0 \le \dot{h}_{ij}(t) \le \overset{*}{h}_{ij} < 1. \tag{8.68}$$

Assumption 8.8. Subsystems can be unstable.

Remark 8.2. The following definitions can be used for upper bounds of the interconnection terms and external disturbances:

$$\rho_{ij}(t) = \max_{\theta_i} \lambda_{\max}\left(\xi_{ij}(\theta_i, t)\,\xi_{ij}^T(\theta_i, t)\right) \tag{8.69}$$

and

$$\Omega_i(t) = \max_{\delta_i} \lambda_{\max}\left(\bar{v}_i(\delta_i, t)\bar{v}_i^T(\delta_i, t)\right), \tag{8.70}$$

where $\rho_{ij}(t)$ and $\Omega_i(t)$ are assumed to be known bounded continuous functions. Also, we assume that there exist unknown positive constants ρ_{ij}^* and Ω_i^* such that

$$\rho_{ij}^* = \max\left(\rho_{ij}(t)\right), \quad t \ge 0, \tag{8.71}$$

and

$$\Omega_i^* = \max\left(\Omega_i(t)\right), \quad t \ge 0. \tag{8.72}$$

Furthermore, we introduce the following lemma from [23] to use for designing decentralized state feedback controllers.

In this section, by inclusion of integrators and the LMI method, we first compensate the input delay. Then by a simple and practical method, we obtain the states $x_i(t+R_i)$ and design a decentralized feedback controller. Now we are ready to design decentralized feedback controllers for system (8.40), which satisfy Assumptions 8.2 and 8.8. The goal is to ensure that all the closed-loop signals remain bounded and the tracking error becomes small enough. Define the tracking error as

$$e_i(t) = x_i(t) - x_{mi}(t). \tag{8.73}$$

The error dynamics is obtained as

$$\begin{aligned}\dot{e}_i(t) &= A_i x_i(t) + A_{di} x_i(t-d_i) + B_i u_i(t-R_i) \\ &\quad + B_i \left(\sum_{j=1}^{N} \xi_{ij}(\theta_i, t)\, x_j(t-h_{ij}(t)) + \bar{v}_i(\delta_i, t) \right) - A_{mi} x_{mi}(t) - B_{mi} D_i(t).\end{aligned} \tag{8.74}$$

Since the reference model (8.42) is stable, we obtain the following inequalities:

$$\begin{aligned}\sum_{j=1}^{N} \xi_{ij}(\theta_i, t)\, x_j(t-h_{ij}(t)) &= \sum_{j=1}^{N} \xi_{ij}(\theta_i, t) \left\| x_{mj}(t-h_{ij}(t)) + e_j(t-h_{ij}(t)) \right\| \\ &\le \sum_{j=1}^{N} \xi_{ij}(\theta_i, t) \left\| e_j(t-h_{ij}(t)) \right\| + o_i,\end{aligned} \tag{8.75}$$

where o_i is an unknown positive scalar. The main result is as follows.

Theorem 8.4. *For system* (8.40), *let the feedback controller be given by*

$$u_i(t) = u_{i1}(t) + u_{i2}(t) + u_{i3}(t), \tag{8.76}$$

where

$$u_{i1}(t) = -z_i x_{mi}(t+R_i) + \theta_{ri} D_i(t+R_i) + m_i x_{mi}(t+R_i-d_i), \tag{8.77}$$

$$u_{i2}(t) = \bar{u}_i(t-(d_i-\Delta d_i)), \tag{8.78}$$

$$u_{i3}(t) = -\frac{1}{2}\pi_i(\omega_i(t+R_i), t+R_i)\, B_i''^{T} P_i \omega_i(t+R_i), \tag{8.79}$$

$\pi_i(\omega_i(t+R_i), t+R_i)$ is a known nonlinear function defined as

$$\pi_i(\omega_i(t+R_i), t+R_i) = c_{i1}\eta_i + c_{i2}\left\| B_i''^T P_i \omega_i(t+R_i) \right\|^2, \tag{8.80}$$

and

$$\omega_i(t+R_i) = \left[e_i^T(t+R_i) \quad g_{i1}^T(t+R_i-(\alpha_i-1)d_i) \quad g_{i2}^T(t+R_i-(\alpha_i-2)d_i) \quad \dots \quad g_{i\alpha}^T(t+R_i) \right]^T \tag{8.81}$$

with $e_i(t+R_i) = x_i(t+R_i) - x_{mi}(t+R_i)$, and $x_i(t+R_i)$ can be obtained by applying time delay in the periodic state $x_i(t)$.

Then the tracking error converges uniformly ultimately and is boundedly stable, and the large-scale system is stable.

Proof. See [34]. □

8.5. Simulation results

To show the effectiveness of the proposed approaches, we present a numerical example. The interconnected terms are assumed unknown and bounded. In the first method the interconnected term is not considered in the prediction, and in the second method, the interconnected term is estimated by an adaptive observer, and then this term is used for exact prediction. A large-scale time-delay system with interconnected terms and delays in both state and input composed of two subsystems is described by the following example.

Example 8.1.

$$\begin{aligned}
\dot{x}_1(t) &= \begin{bmatrix} -1 & 1 \\ -2 & -3 \end{bmatrix} x_1(t) + \begin{bmatrix} 0 & 0 \\ -2 & -3 \end{bmatrix} x_1(t-d_1) + \begin{bmatrix} 0 \\ 1 \end{bmatrix} u_1(t-R_1) \\
&\quad + \begin{bmatrix} 0 \\ 1 \end{bmatrix} \begin{pmatrix} \delta_{11}x_{11}(t) + \delta_{12}x_{12}(t)x_{21}(t-h_{13}) \\ +\delta_{13}x_{22}^2(t) + \delta_{14}x_{11}^2(t-h_{14}(t)) \end{pmatrix} \\
\dot{x}_2(t) &= \begin{bmatrix} -1 & 1 \\ -3 & -3 \end{bmatrix} x_2(t) + \begin{bmatrix} 0 & 0 \\ -2 & -3 \end{bmatrix} x_2(t-d_2) + \begin{bmatrix} 0 \\ 1 \end{bmatrix} u_2(t-R_2) \\
&\quad + \begin{bmatrix} 0 \\ 1 \end{bmatrix} \begin{pmatrix} \delta_{21}x_{21}(t) + \delta_{22}x_{21}^2(t-h_{23}) \\ +\delta_{23}x_{12}^2(t) \\ +\delta_{24}x_{12}(t)x_{22}(t-h_{24}(t)) \end{pmatrix}.
\end{aligned} \tag{8.82}$$

The reference model is selected as

$$\dot{x}_{m1}(t) = \begin{bmatrix} -1 & 1 \\ -6 & -5 \end{bmatrix} x_{m1}(t) + \begin{bmatrix} 0 \\ 1 \end{bmatrix} D_1(t),$$

$$\dot{x}_{m2}(t) = \begin{bmatrix} -1 & 1 \\ -7 & -5 \end{bmatrix} x_{m2}(t) + \begin{bmatrix} 0 \\ 1 \end{bmatrix} D_2(t), \tag{8.83}$$

where $D_1(t) = 10 \sin \frac{\pi}{4} t$, $D_2(t) = 10 \cos \frac{\pi}{4} t$.

Therefore by Theorem 8.1 the controller and the adaptive laws are as follows:

$$\begin{aligned} u_1(t) = {} & [-4 \quad -2] x_1(t+R_1) + [2 \quad 3] x_1(t-d_1+R_1) \\ & - 2\theta_1(t+R_1) B_1^T \frac{\partial V_1^T(e_1(t+R_1))}{\partial e_1(t+R_1)} \\ & + D_1(t+R_1) - 2\eta_1(t+R_1) \operatorname{sign}\left(B_1^T \frac{\partial V_1^T(e_1(t+R_1))}{\partial e_1(t+R_1)} \right), \end{aligned} \tag{8.84}$$

$$\begin{aligned} u_2(t) = {} & [-4 \quad -2] x_2(t+R_2) + [2 \quad 3] x_2(t-d_2+R_2) \\ & - 2\theta_2(t+R_2) B_2^T \frac{\partial V_2^T(e_2(t+R_2))}{\partial e_2(t+R_2)} \\ & + D_2(t+R_2) - 2\eta_2(t+R_2) \operatorname{sign}\left(B_2^T \frac{\partial V_2^T(e_2(t+R_2))}{\partial e_2(t+R_2)} \right), \end{aligned} \tag{8.85}$$

with $e_i(t+R_i) = \hat{x}_i(t+R_i) - x_{mi}(t+R_i)$.

The future $x_{mi}(t+R_i)$ can be obtained by applying the input $D_i(t+R_i)$ to (8.83). Also, the state $x_i(t+R_i)$ can be approximated by recursive relationships (8.7). The controller parameters are chosen as

$$\begin{aligned} & d_1 = d_2 = 1, \; R_1 = R_2 = 3.4, \; h_{13} = 0.5, \; h_{23} = 1, \\ & h_{14}(t) = 0.25(1 + \sin t), \; h_{24}(t) = 0.5(1 + \cos t), \\ & \delta_{11} = 0.2 + 0.3 \sin \pi t, \; \delta_{21} = 0.2 + 0.3 \cos \pi t, \; \delta_{ij} = 0.5, \; Q_i = 100I. \end{aligned}$$

Also, the initial conditions are

$$x_i(0) = \begin{bmatrix} -1 \\ 1 \end{bmatrix}, \quad x_{mi}(0) = \begin{bmatrix} 0 \\ 0 \end{bmatrix}.$$

Simulation results reveal the following points. Figs. 8.1–8.2 show that the predicted signal converges to the original signal. In Figs. 8.3–8.6 the plant and reference model states $x_i(t)$ and $x_{mi}(t)$ are plotted. Also, in

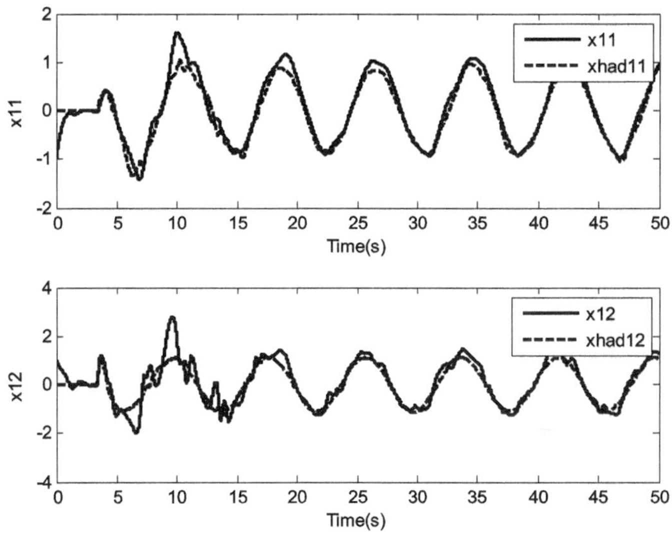

Figure 8.1 Predicted and original signal x_1.

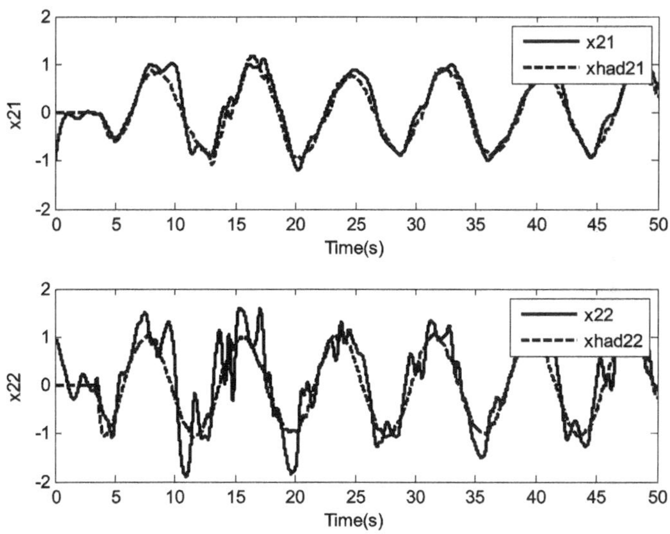

Figure 8.2 Predicted and original signal x_2.

Figs. 8.7–8.8 the errors $e_i(t)$ and control signals are demonstrated. Clearly, despite the state and input delay, the tracking error converges uniformly ultimately and boundedly stable in the presence of nonlinear interconnected terms with time-varying delays.

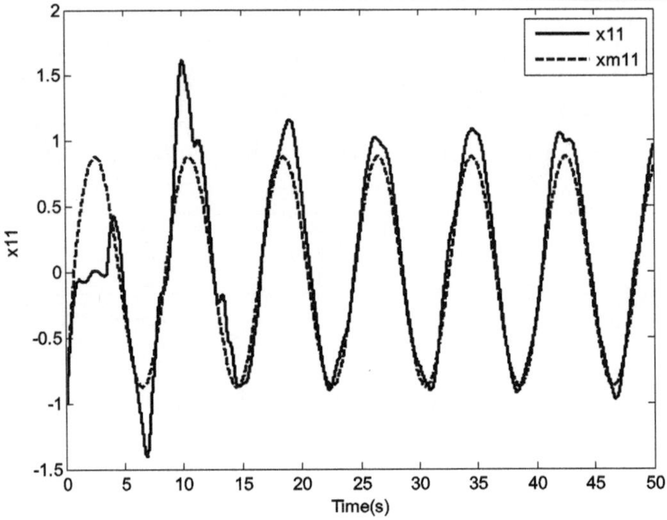

Figure 8.3 The control result of states x_{11} and x_{m11}.

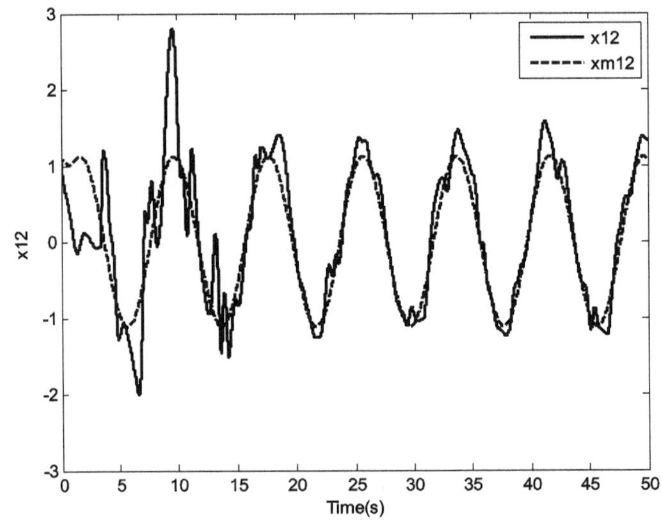

Figure 8.4 The control result of states x_{12} and x_{m12}.

Example 8.2.

$$\dot{x}_1(t) = \begin{bmatrix} -1 & 1 \\ -2 & -3 \end{bmatrix} x_1(t) + \begin{bmatrix} 0 & 0 \\ -2 & -3 \end{bmatrix} x_1(t - d_1) + \begin{bmatrix} 0 \\ 1 \end{bmatrix} u_1(t - R_1)$$
$$+ \begin{bmatrix} 0 \\ 1 \end{bmatrix} \left(\sum_{j=1}^{N} \xi_{1j}(x_j(t - h_{1j}(t)), t - h_{1j}(t)) \right)$$

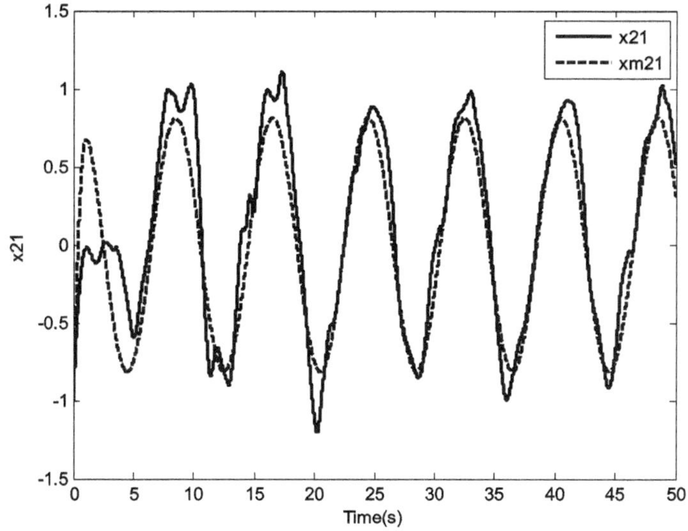

Figure 8.5 The control result of states x_{21} and x_{m21}.

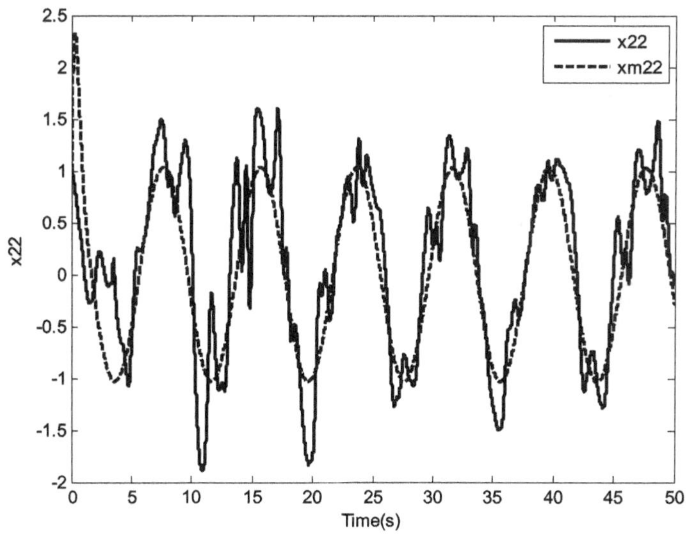

Figure 8.6 The control result of states x_{22} and x_{m22}.

$$\dot{x}_2(t) = \begin{bmatrix} -1 & 1 \\ -3 & -3 \end{bmatrix} x_2(t) + \begin{bmatrix} 0 & 0 \\ -2 & -3 \end{bmatrix} x_2(t-d_2) + \begin{bmatrix} 0 \\ 1 \end{bmatrix} u_2(t-R_2) + \begin{bmatrix} 0 \\ 1 \end{bmatrix} \left(\sum_{j=1}^{N} \xi_{2j}(x_j(t-h_{2j}(t)), t-h_{2j}(t)) \right) \tag{8.86}$$

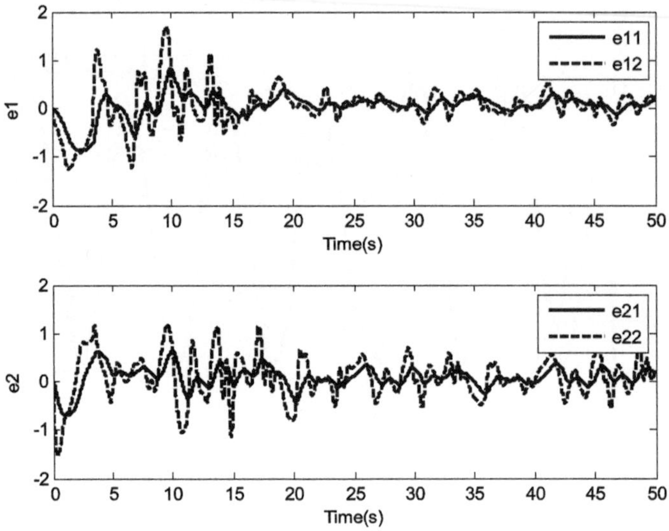

Figure 8.7 Errors e_1 and e_2.

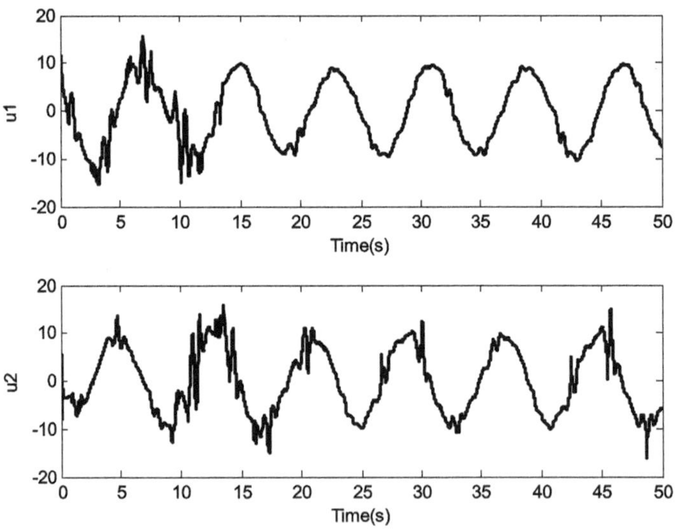

Figure 8.8 Signal controls u_1 and u_2.

with $\xi_{ij}(x_j(t-h_{ij}(t)), t-h_{ij}(t)) = \zeta_{ij}(x_i(t))e^{0.5x_{j1}(t-h_{ij}(t))+0.6x_{j2}(t-h_{ij}(t))}$, where, according to Assumption 8.5, $\zeta_{ij}(t)$ are unknown positive bounded functions. As described before, for exact prediction, by Theorem 8.2 we design the adaptive observer

$$\dot{\hat{x}}_1(t) = A_1\hat{x}_1(t) + A_{d1}\hat{x}_1(t-d_1) + B_1\left(u_1(t-R_1) + \sum_{j=1}^{2} F_{1j}\left(\hat{\varphi}_1(t), e_1'(t), t\right)\right)$$
$$+\begin{bmatrix} 0 & 0 \\ 4 & 2 \end{bmatrix}\left(x_1(t) - \hat{x}_1(t)\right) + \begin{bmatrix} 0 & 0 \\ -2 & -3 \end{bmatrix}\left(x_1(t-d_1) - \hat{x}_1(t-d_1)\right)$$
$$\dot{\hat{x}}_2(t) = A_2\hat{x}_2(t) + A_{d2}\hat{x}_2(t-d_2) + B_2\left(u_2(t-R_2) + \sum_{j=1}^{2} F_{2j}\left(\hat{\varphi}_2(t), e_2'(t), t\right)\right)$$
$$+\begin{bmatrix} 0 & 0 \\ 4 & 2 \end{bmatrix}\left(x_2(t) - \hat{x}_2(t)\right) + \begin{bmatrix} 0 & 0 \\ -2 & -3 \end{bmatrix}\left(x_2(t-d_2) - \hat{x}_2(t-d_2)\right) \tag{8.87}$$

where $F_{ij}(.)$ of (8.58) and $\hat{\varphi}_i(t)$ of (8.59) are obtained, and $e_i'(t)$ is the observer error. By (8.83) and Theorem 8.3 the controllers are

$$u_1(t) = \begin{bmatrix} -4 & -2 \end{bmatrix}\hat{x}_1(t+R_1) + \begin{bmatrix} 2 & 3 \end{bmatrix}\hat{x}_1(t+R_1-d_1) + D_1(t+R_1)$$
$$-\frac{1}{2}\eta_1\hat{\varphi}_1(t+R_1)B_{m1}^T P_1 e_1'(t+R_1), \tag{8.88}$$
$$u_2(t) = \begin{bmatrix} -4 & -2 \end{bmatrix}\hat{x}_2(t+R_2) + \begin{bmatrix} 2 & 3 \end{bmatrix}\hat{x}_2(t+R_2-d_2) + D_2(t+R_2)$$
$$-\frac{1}{2}\eta_2\hat{\varphi}_2(t+R_2)B_{m2}^T P_2 e_2'(t+R_2), \tag{8.89}$$

where $e_i'(.)$ is obtained by (8.64). The parameter values of the controller and systems are chosen as

$$d_1 = d_2 = 1,\ R_1 = R_2 = 3.4,\ \eta_i = 2,\ \gamma_i = 5,\ \sigma_i = 0.1,\ Q_i = 10I,$$
$$\zeta_{1j}(x_1(t)) = 0.2\sin(x_{11}(t)),\ \zeta_{2j}(x_2(t)) = 0.3\sin(x_{21}(t)),$$
$$h_{i1}(t) = 1 + 0.5\sin(\pi t),\ h_{i2}(t) = 1 + 0.4\sin(\pi t).$$

The functions $h_{i1}(t)$ and $h_{i2}(t)$ are bounded and continuous, and their derivatives are not necessarily less than one. Delays in state d_i and input R_i are constant, η_i, γ_i, σ_i, Q_i are controller parameters, and $\zeta_{ij}(x_i(t))$ are the coefficients of the interconnected terms. The initial conditions are

$$x_1(0) = \begin{bmatrix} -1 \\ 1 \end{bmatrix},\quad x_2(0) = \begin{bmatrix} -2 \\ 1.5 \end{bmatrix},\quad x_{mi}(0) = \begin{bmatrix} 0 \\ 0 \end{bmatrix}.$$

The simulation results showed that proper estimation was achieved with long input delay and unknown nonlinear interconnected terms in the large-scale systems. The input delay is appropriately compensated by nested

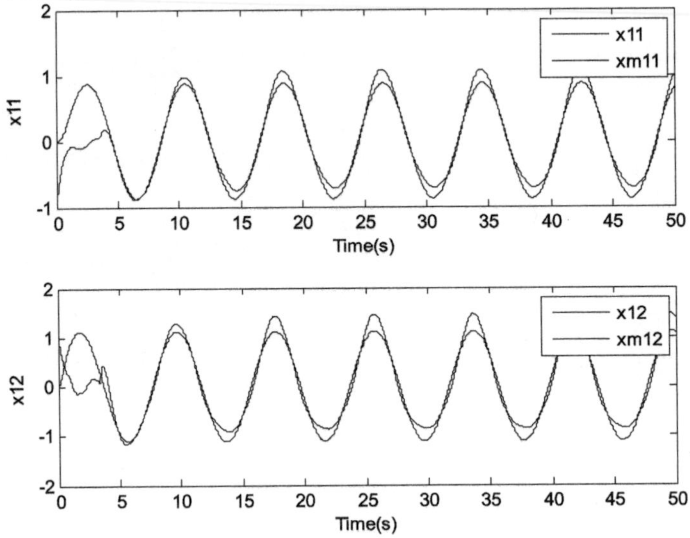

Figure 8.9 The control result of states x_1 and x_{m1}.

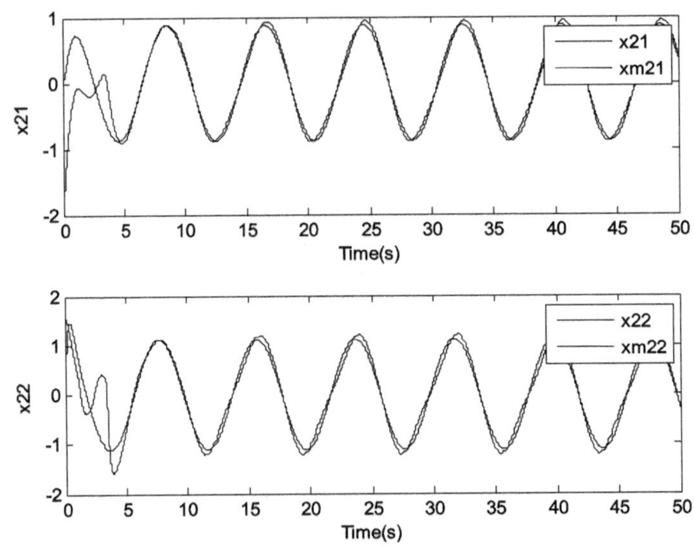

Figure 8.10 The control result of states x_2 and x_{m2}.

predictors, so that a model reference adaptive control can be designed for the system. Figs. 8.9–8.10 illustrate the plant states and reference model states $x_i(t)$ and $x_{mi}(t)$, and in Figs. 8.11–8.14 the tracking and observation errors and the control signals are shown.

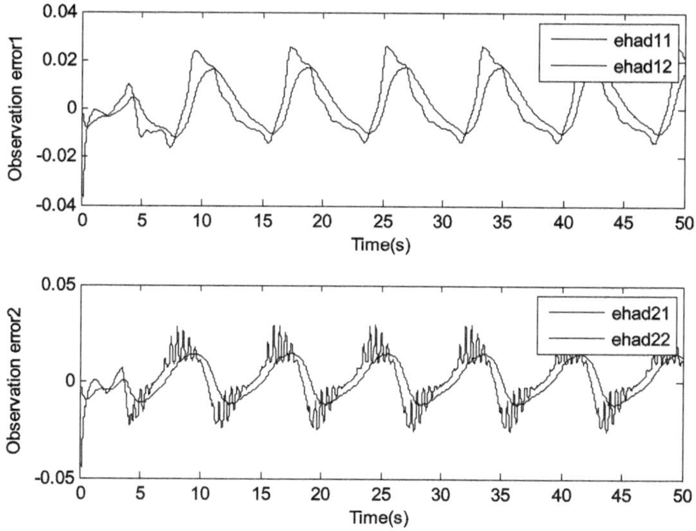

Figure 8.11 Observation error.

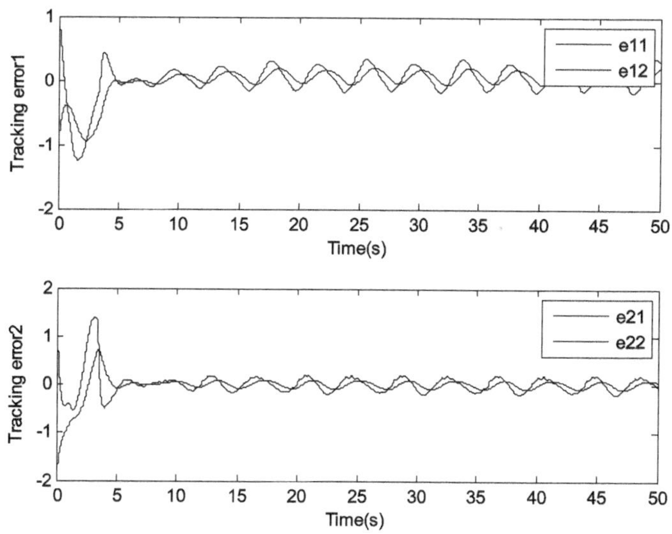

Figure 8.12 Tracking error.

Example 8.3. Unstable open loop system

$$\dot{x}_1(t) = \begin{bmatrix} 0 & 1 \\ 0 & 0.3 \end{bmatrix} x_1(t) + \begin{bmatrix} 0 & 0 \\ -0.1 & -0.2 \end{bmatrix} x_1(t - d_1) + \begin{bmatrix} 0 \\ 1 \end{bmatrix} u_1(t - R_1)$$
$$+ \begin{bmatrix} 0 \\ 1 \end{bmatrix} \begin{pmatrix} \xi_{11}(\theta_1, t)\, x_{11}(t - h_{11}(t)) \\ +\xi_{12}(\theta_1, t)\, x_{21}(t - h_{12}(t)) + \bar{v}_1(\delta_1, t) \end{pmatrix}$$

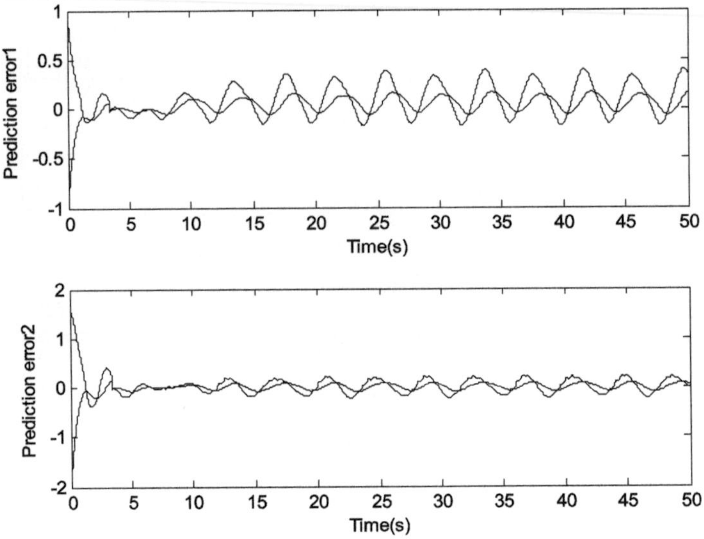

Figure 8.13 Prediction error.

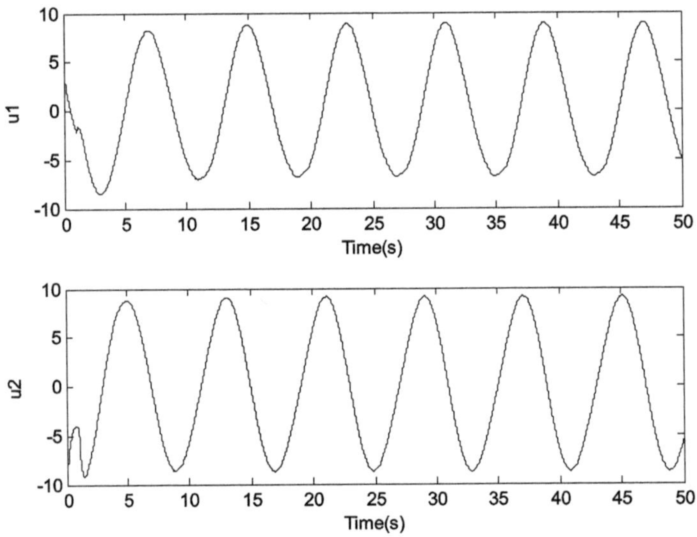

Figure 8.14 Signal controls u_1 and u_2.

$$\dot{x}_2(t) = \begin{bmatrix} 0 & 1 \\ 0.1 & 0.2 \end{bmatrix} x_2(t) + \begin{bmatrix} 0 & 0 \\ -0.2 & -0.4 \end{bmatrix} x_2(t-d_2) + \begin{bmatrix} 0 \\ 1 \end{bmatrix} u_2(t-R_2) + \begin{bmatrix} 0 \\ 1 \end{bmatrix} \begin{pmatrix} \xi_{21}(\theta_2, t)\, x_{12}(t-h_{21}(t)) \\ +\xi_{22}(\theta_2, t)\, x_{22}(t-h_{22}(t)) + \bar{v}_2(\delta_2, t) \end{pmatrix} \tag{8.90}$$

To design an adaptive controller, the reference model is chosen as

$$\dot{x}_{mi}(t) = \begin{bmatrix} 0 & 1 \\ -2 & -3 \end{bmatrix} x_{mi}(t) + \begin{bmatrix} 0 \\ 1 \end{bmatrix} D_i(t), \quad i = 1, 2, \tag{8.91}$$

where $D_1(t) = 10 \sin \frac{2\pi}{3} t$, $D_2(t) = 10 \cos \frac{2\pi}{3} t$.

Therefore by Theorem 8.4 the controller and the adaptive law are

$$\begin{aligned} u_1(t) &= \begin{bmatrix} -2 & -3.3 \end{bmatrix} x_{m1}(t + R_1) + \begin{bmatrix} 0.1 & 0.2 \end{bmatrix} x_{m1}(t - d_1 + R_1) \\ &\quad + D_1(t + R_1), \\ g_{11}&\left(t - d_1 + \Delta d_1\right) - \frac{1}{2} \pi_1 \left(\omega_1(t + R_1), t + R_1\right) B_1''^{T} P_1 \omega_1(t + R_1), \end{aligned} \tag{8.92}$$

$$\begin{aligned} u_2(t) &= \begin{bmatrix} -2.1 & -3.2 \end{bmatrix} x_{m2}(t + R_2) + \begin{bmatrix} 0.2 & 0.4 \end{bmatrix} x_{m2}(t - d_2 + R_2) \\ &\quad + D_2(t + R_2), \\ g_{21}&\left(t - d_2 + \Delta d_2\right) - \frac{1}{2} \pi_2 \left(\omega_2(t + R_2), t + R_2\right) B_2''^{T} P_2 \omega_2(t + R_2), \end{aligned} \tag{8.93}$$

where $g_{11}(t) = \bar{u}_1(t)$ and $g_{21}(t) = \bar{u}_2(t)$. Also, using controller (8.35), the gains K_0 and K_d are

$$\begin{aligned} K_{10} &= [0.0123 \quad 0.0558 \quad -0.1314 \quad -0.3915 \quad -0.3173], \\ K_{1d} &= [-0.0016 \quad -0.0068 \quad 0.0145 \quad -0.4376 \quad -1.2494], \\ K_{20} &= [0.0073 \quad 0.0354 \quad -0.0389 \quad -0.2042 \quad -0.1993], \\ K_{2d} &= \begin{bmatrix} -0.0078 & -0.0159 & 0.0262 & -0.3083 & -1.1365 \end{bmatrix}, \end{aligned} \tag{8.94}$$

and $\pi_i\left(\omega_i(t + R_i), t + R_i\right)$ is obtained by (8.79). Since P_i is selected as diagonal, we have

$$B_i''^{T} P_i \omega_i(t + R_i) = B_i^{T} P_i e_i(t + R_i), \tag{8.95}$$

where

$$e_i(t + R_i) = x_i(t + R_i) - x_{mi}(t + R_i). \tag{8.96}$$

The future states $x_{mi}(t+R_i)$ can be obtained by applying the input $D_i(t+R_i)$ to (8.91). Also, the state $x_i(t + R_i)$ can be obtained by applying time delay in the periodic signal $x_i(t)$:

$$x_i(t + R_i) = x_i(t - b_i). \tag{8.97}$$

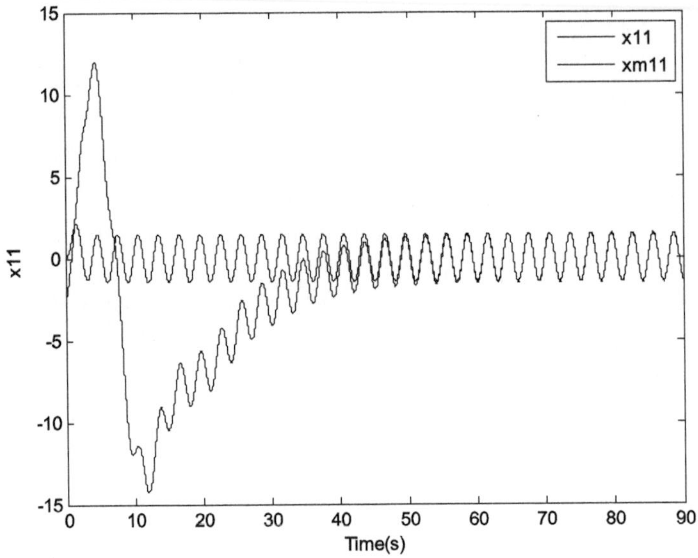

Figure 8.15 The time response of the state variable x_{11} and reference model x_{m11}.

The parameter values of the controller are chosen as

$$d=1,\ R=2.3,\ \alpha=3,\ \Delta d=0.3,\ T=3,\ b=0.7,$$
$$\xi_{ij}(\theta_i, t)=0.5\sin(5t),\ \zeta_{2j}(t)=0.3\sin(3t),$$
$$h_{i1}(t)=0.5+0.5\sin(t),\ h_{i2}(t)=0.2+0.8\sin(t),\ \bar{v}_i(\delta_i, t)=0.$$

It is worth noting that $h_{i1}(t)$ and $h_{i2}(t)$ are continuous and bounded functions. The initial conditions are set to be

$$x_i(0)=\begin{bmatrix}-3\\3\end{bmatrix},\quad x_{mi}(0)=\begin{bmatrix}0\\0\end{bmatrix}.$$

Simulation results are shown below. Figs. 8.15–8.18 illustrate that the plant and reference model states are $x_i(t)$ and $x_{mi}(t)$. In Figs. 8.19–8.20 the errors $e_i(t)$ and control signals are demonstrated. The designed controller signals and the closed-loop state responses are shown in the figures. We see that despite the state and input delays, the closed-loop system error can be guaranteed to be uniformly ultimately bounded in the presence of interconnected terms with time-varying delays.

The features and limitations of each method can be determined by examining the response of numerical examples. A nested predictor can present

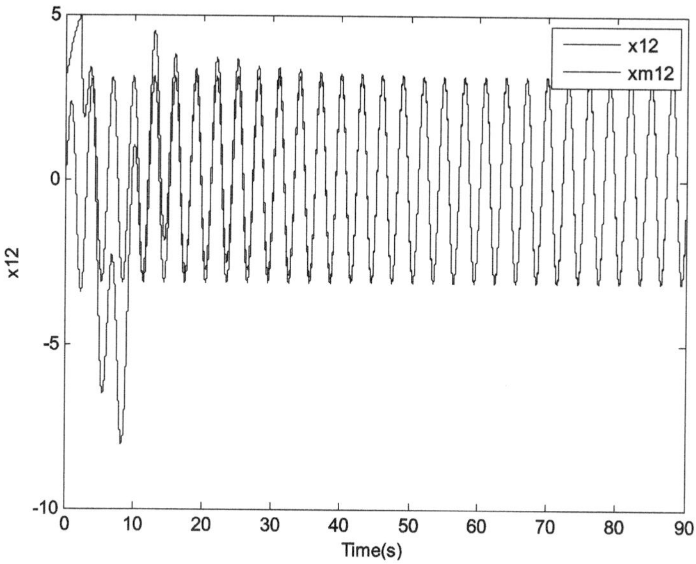

Figure 8.16 The time response of the state variable x_{12} and reference model x_{m12}.

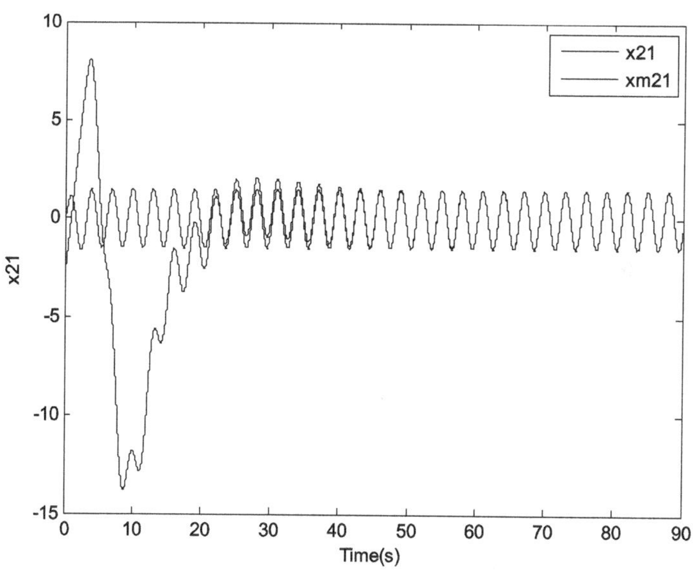

Figure 8.17 The time response of the state variable x_{21} and reference model x_{m21}.

a suitable response to compensate input delay, but it may cause a delay in practical applications when numerically solving problems. The method of adding integrator is a more appropriate solution than a nested predictor

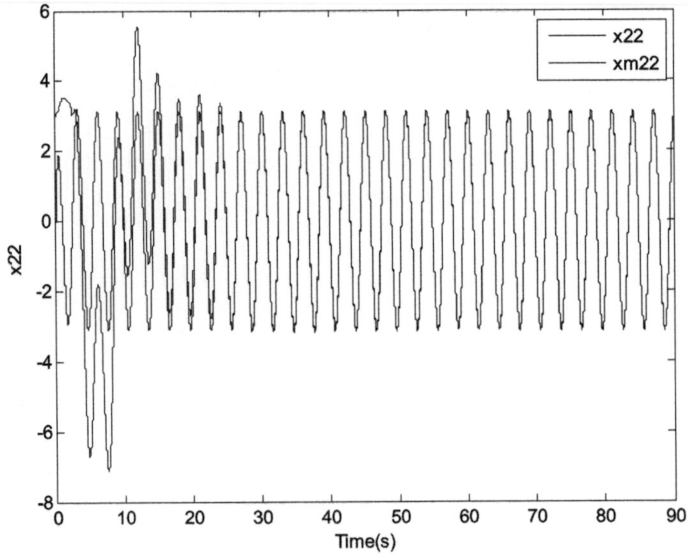

Figure 8.18 The time response of the state variable x_{22} and reference model x_{m22}.

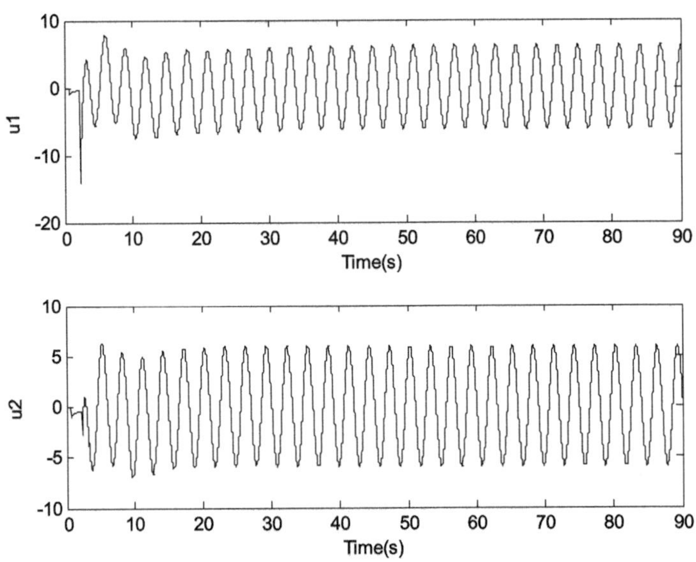

Figure 8.19 Signal controls u_1 and u_2.

because it can be used in open-loop unstable systems and avoids solving system problems numerically, but it adds new poles to the system and diminishes the delay compensation speed. Nonetheless, each method has its

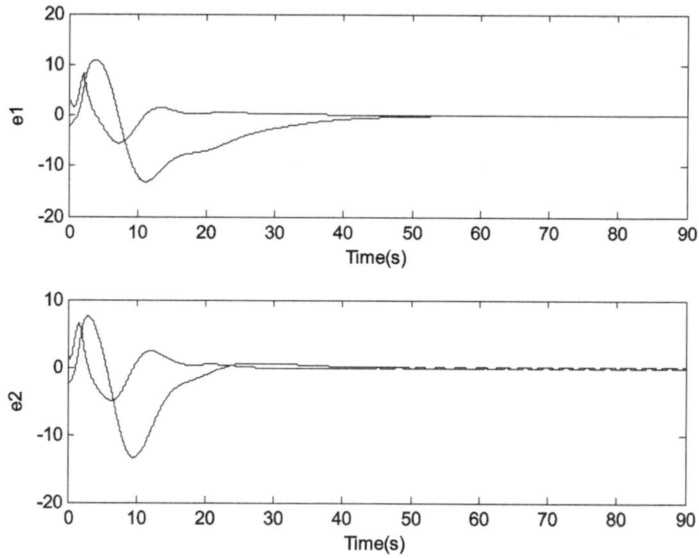

Figure 8.20 Errors e_1 and e_2.

own advantages, which depend on the system characteristics in practical applications. Interconnected term bounding is a characteristic index that separates such approaches and solutions, where the second method has superior performance because it can bound this term using a function with an unknown upper bound. Time-varying delays in nonlinear interconnected terms are bounded nonnegative continuous functions, and their derivatives are not necessarily less than one. Accordingly, it can be argued that the second method is more widely applicable.

The importance of the proposed methods is that large-scale systems having delay and nonlinearity in the interconnected term ultimately tend to follow a linear stable system with low tracking error.

8.6. Conclusions

In this chapter, we discussed the decentralized control of the large-scale system using a model reference adaptive control strategy considering constant state and input delays and time-varying delay associated with the interconnected term. We studied three methods for input delay compensation in a large-scale system. In the first method the interconnected term is bounded using a polynomial function with unknown coefficients. In the second method the interconnected term is bounded using a function with an unknown upper bound, and the third method examines an unstable open loop with a linear and unknown interconnected term.

In this study, we proposed two new solutions to compensate the input delay; first, a nested predictor is established to predict the future states for input delays compensation. Then by assuming that the predicted state is available the adaptive controller is designed. Also, the nonlinear interconnection terms are assumed to be bounded by polynomials with unknown gains. This is a step-by-step approach adopted to find a proper prediction state. The approach can be extended for using in the large-scale systems because, in such systems, the interconnected term is practically unknown, although it needs to be known for exact prediction. To solve this problem, we proposed an adaptive observer assuming that the interconnected term is bounded with a function that can be unknown. The purpose is to reach more accurate predictions and, finally, suitable tracing by knowing the interconnected term in the observed system. The next method uses dynamic compensation in which the input delay problem is converted into the state delay problem. Also, for systems that use model reference adaptive control, it proposes a simple and practical method. A delay can be used instead of prediction.

The features and limitations of each method with respect to simulation results can be summarized as follows. A nested predictor can present a suitable response to compensate input delay, but it may cause a delay in practical applications when solving problems numerically. The "adding integrator" method is a more appropriate solution than a nested predictor because it can be used in open-loop unstable systems and avoid solving system problems numerically, but it adds new poles to the system and diminishes the delay compensation speed. Nonetheless, each method has its own advantages, which depend on the system characteristics in practical applications. Interconnected term bounding is a characteristic index that separates such approaches and solutions, where the second method has superior performance because it can bound this term using a function with an unknown upper bound. Time-varying delays in nonlinear interconnected terms are bounded nonnegative continuous functions, and their derivatives are not necessarily less than one. Accordingly, we can say that the second method can cover more systems in practical applications. Adopting such choices gives the designer more authority in a practical system so that the operator can first determine for which items proposed in this research the system shows more adaptation and then gain more acceptable results by adopting an optimal strategy.

Some points of distinction between this and the previous studies are that the input delay and bounding interconnected term with the time-varying delay were considered simultaneously. Using the adaptive control strategy is another important aspect of the present work. Due to self-learning capability, this controller is a proper solution for many systems, which makes them robust. In cases where there is uncertainty in the system robust control is another choice depending on the system type. The reasons for choosing an adaptive controller for large-scale systems are the feasibility of tracking the reference model and estimating unknown upper bound in the interconnected term with less conservative conditions in comparison with other robust methods. On the other hand, robust control consistently deals with the worst state, and therefore it has larger dimensions and thereby more complexity for practical applications. The validity of the main results was verified through several numerical examples. Therefore the proposed methodologies apply to a class of interconnected large-scale systems with time delays.

References

[1] M. Elloumi, S. Kamoun, Adaptive control scheme for large-scale interconnected systems described by Hammerstein models, Asian Journal of Control 19 (2017) 1075–1088.

[2] K.S. Haider, A. Ghafoor, M. Imran, F.M. Malik, Model reduction of large scale descriptor systems using time limited gramians, Asian Journal of Control 19 (2017) 1217–1227.

[3] D. Wang, Z. Wang, B. Shen, Y. Li, F.E. Alsaadi, An event-triggered approach to robust recursive filtering for stochastic discrete time-varying spatial-temporal systems, Signal Processing 145 (2018) 91–98.

[4] H. Liu, Z. Wang, B. Shen, T. Huang, F.E. Alsaadi, Stability analysis for discrete-time stochastic memristive neural networks with both leakage and probabilistic delays, Neural Networks 102 (2018) 1–9.

[5] B. Mirkin, P.-O. Gutman, Adaptive following of perturbed plants with input and state delays, in: 9th IEEE International Conference on Control and Automation (ICCA), Santiago, Chile, 2011.

[6] Chang-Chun Hua, Jing Leng, Xin-Ping Guan, Decentralized MRAC for large-scale interconnected systems with time-varying delays and applications to chemical reactor systems, Journal of Process Control 22 (10) (2012) 1985–1996.

[7] B. Mirkin, P.-O. Gutman, Y. Shtessel, Decentralized continuous MRAC with local asymptotic sliding modes of nonlinear delayed interconnected systems, Journal of the Franklin Institute 351 (4) (2014) 2076–2088.

[8] H.-T. Yau, J.-J. Yan, Robust decentralized adaptive control for uncertain large-scale delayed systems with input nonlinearity, Chaos, Solitons and Fractals 39 (4) (2009) 1515–1521.

[9] X.-G. Yan, S.K. Spurgeon, C. Edwards, Global time-delay dependent decentralized sliding mode control using only output information, in: 48th IEEE Conference on

Decision and Control and 28th Chinese Control Conference, Shanghai, China, 2009, pp. 6709–6714.
[10] Changchun Hua, Xinping Guan, Peng Shib, Decentralized robust model reference adaptive control for interconnected time-delay systems, Journal of Computational and Applied Mathematics 193 (2) (2006) 383–396.
[11] Y. Li, S. Tong, Adaptive neural networks prescribed performance control design for switched interconnected uncertain nonlinear systems, IEEE Transactions on Neural Networks and Learning Systems 29 (2018) 3059–3068.
[12] Y. Li, Z. Ma, S. Tong, Adaptive fuzzy output-constrained fault-tolerant control of nonlinear stochastic large-scale systems with actuator faults, IEEE Transactions on Cybernetics 47 (2017) 2362–2376.
[13] H. Wu, Decentralized adaptive robust tracking and model following for large-scale systems including delayed state perturbations in the interconnections, Journal of Optimization Theory and Applications 137 (2008) 231–253.
[14] H. Wu, Decentralised adaptive robust control of uncertain large-scale non-linear dynamical systems with time-varying delays, IET Control Theory & Applications 6 (2012) 629–640.
[15] M.S. Mahmoud, Decentralized reliable control of interconnected systems with time-varying delays, Journal of Optimization Theory and Applications 143 (2009) 497.
[16] M.S. Mahmoud, N.B. Almutairi, Decentralized stabilization of interconnected systems with time-varying delays, European Journal of Control 15 (2009) 624–633.
[17] H. Wu, M. Deng, Robust adaptive control scheme for uncertain non-linear model reference adaptive control systems with time-varying delays, IET Control Theory & Applications 9 (2015) 1181–1189.
[18] O.J. Smith, A controller to overcome dead time, ISA Journal 6 (1959) 28–33.
[19] A.Z. Manitius, A.W. Olbrot, Finite spectrum assignment problem for systems with delays, IEEE Transactions on Automatic Control 24 (1979) 541–552.
[20] S.A. Al-Shamali, O.D. Crisalle, H.A. Latchman, An approach to stabilize linear systems with state and input delay, in: American Control Conference, 2003. Proceedings of the 2003, 2003, pp. 875–880.
[21] Z. Lin, Low Gain Feedback, Springer, London, UK, 1988.
[22] Z. Lin, H. Fang, On asymptotic stability of linear systems with delayed input, IEEE Transactions on Automatic Control 52 (2007) 998–1013.
[23] B. Zhou, Z. Lin, G. Duan, Truncated predictor feedback for linear systems with long time-varying input delays, Automatica (2012) 2387–2399.
[24] B. Zhou, Input delay compensation of linear systems with both state and input delays by nested prediction, Automatica 50 (2014) 1434–1443.
[25] X. Zhang, Y. Lin, Adaptive output feedback control for a class of large-scale nonlinear time-delay systems, Automatica 52 (2015) 87–94.
[26] M.C. Nguyen, H. Trinh, Reduced-order observer design for one-sided Lipschitz time-delay systems subject to unknown inputs, IET Control Theory & Applications 10 (2016) 1097–1105.
[27] V. Léchappé, E. Moulay, F. Plestan, A. Glumineau, Interconnected delay and state observer for nonlinear systems with time-varying input delay, in: American Control Conference (ACC 2016), 2016, pp. 72–77.
[28] H. Wu, Decentralized local adaptive robust state observers of large scale interconnected systems with uncertainties, in: Control and Automation (MED), 2014 22nd Mediterranean Conference of, 2014, pp. 1117–1122.
[29] B. Zhou, Z.-Y. Li, Z. Lin, Observer based output feedback control of linear systems with input and output delays, Automatica 49 (2013) 2039–2052.
[30] R.L. Finney, G.B. Thomas, M.D. Weir, Calculus and Analytic Geometry, Addison-Wesley, 1992.

[31] S. Oucheriah, Decentralized stabilization of large scale systems with multiple delays in the interconnections, International Journal of Control 73 (2000) 1213–1223.
[32] S.H. Hashemipour, N. Vasegh, A.K. Sedigh, Decentralized model reference adaptive control for interconnected time delay systems with delay in state and compensation of long delay in input by nested prediction, International Journal of Control, Automation, and Systems 15 (2017) 1198–1206.
[33] S.H. Hashemipour, N. Vasegh, A. Khaki Sedigh, Control of large scale interconnected systems with input and state delays using decentralized adaptive state observers, Asian Journal of Control 22 (4) (2020) 1458–1470.
[34] S.H. Hashemipour, N. Vasegh, A.K. Sedigh, Decentralized MRAC for large-scale interconnected systems with state and input delays by integrators inclusion, Journal of Dynamic Systems, Measurement, and Control 139 (2017) 091009.

CHAPTER NINE

Lambert W function approach to analysis and stabilization of time-delay systems

Niraj Choudhary, S. Janardhanan, and Indra Narayan Kar
Department of Electrical Engineering, Indian Institute of Technology Delhi, New Delhi, India

Contents

9.1. Introduction

Most of practical responses of dynamic processes to exogenous actions and signals never show up spontaneously. A similar concept holds for interconnected models or processes where transmission (information or energy) may occur because of some chemical or physical laws. In some cases the mathematical model obtained by assuming that the future evolution of

https://doi.org/10.1016/B978-0-12-820599-0.00014-8

the system does not rely on the previous states and depends on the present state only would be inadequate for complete characterization of the system model. The most possible way to sweep over such erroneous models is to consider some knowledge of the previous states while modeling a system. This type of systems is generically known as time-delay systems, also best known as hereditary, aftereffects, deviating arguments, dead-time, post actions, or systems with time lag [1]. A TDS represents a group of infinite-dimensional systems that can be used to characterize delays in the states, control input, or in the measurements of a system. A TDS can be described by delay differential equations or difference-differential equations (DDEs), which come under the category of functional differential equations (FDEs) [2,3].

The achievable performance of a controlled process can be degraded or bounded by the existence of time lags, or it can even bring forth a system toward instability. The infinite-dimensional nature of a delay system introduces a transcendental characteristic equation (TCE) having an infinite number of roots. Thus transcendental behavior makes a system challenging to examine with traditional techniques, primarily, in examining stability and synthesizing controllers [4]. The simplest method to solve such problems is using an approximation. The Pade approximation is widely used for this purpose. It is a rational approximation and turns into a shortened fraction as a replacement for the dead time in the TCE. Yet, such an approximation comprises a restriction on accuracy, can lead the actual system toward instability, and induce a nonminimum phase [3].

Moreover, prediction-based methods, for example, finite spectrum assignment (FSA) [5] or the Smith predictor [6], are exploited to stabilize a TDS by converting the delayed system into a delay-free system. These approaches require calculations based on a system model, which may account into unforeseen errors when applied to an actual system. Therefore, due to computational concerns, a safe usage of such techniques is as yet an open issue. Synthesis has also been carried out using the Lyapunov-based methods by solving algebraic Riccati equations (AREs) or linear matrix inequalities (LMIs) [7]. The main issue in utilizing these methods is that they involve intricate formulations and can prompt possibly redundant control and may produce conservative results.

Time delays appear in diverse natural and engineered system models, like traffic flow problems, processes involving transport delays, teleoperation, biological models, and many more. The literature on TDSs is exceptionally comprehensive and comprises various good textbooks and survey papers. A decent survey on the recent issues and challenges in this domain is covered in [8]. Numerous pathways have been ensued by researchers to obtain explications to the stability and design issues. Among these, some efforts are made for exploring the absolute stability regions in terms of time delay [9,10] and direct to design techniques that employ time delay as a stabilizing agent [11]. Some studies are focused on the numerical computation of the eigenvalues of the system. These researches involve techniques based upon root-finding of the characteristic polynomial [12] and methods based on a discretization of the solution operator [13,14] or its infinitesimal generator [15,16]. Here onward approaches were introduced to optimize the location of the dominant poles to ensure the desired performance [17,18]. In [19], Krylov's method is employed to obtain the roots of large-scale problems. In the previous decade a framework based on the Lambert W function was established for analyzing delay differential equations (DDEs) [20]. This chapter focuses on this Lambert W function (LWF) based approach for design and analysis of linear time-invariant (LTI) multidimensional TDSs with a single time delay.

9.1.1 Motivation and background

Consider the nondelayed LTI system

$$\begin{aligned} \dot{x}(t) &= Ax(t) + Bu(t), \\ y(t) &= Cx(t) + Du(t). \end{aligned} \tag{9.1}$$

The closed-form solution of (9.1) can be expressed using a convolution integral and notions of the state transition matrix (STM) [21]. For systems having finite eigenspectrum as in (9.1), stability can be analyzed by inspecting the positions of the fixed number of characteristics roots in the complex plane.

The obtained solution is further exploited to deduce the controllability and observability Gramians, which aids in determining the controllability and observability of the system. If the controllability condition is satisfied, then it is possible to stabilize the system using state feedback control. Likewise, if the observability condition holds, then an observer-based design can be carried out.

For a system represented in the form of ordinary differential equations (ODEs) (9.1), the procedure discussed above is standard, because the closed-form solution can be obtained analytically. However, unlike for ODEs, these methods are challenging to implement straightway for delay differential equations (DDEs) because of the lack of a priori time-domain solution. Asl et al. presented a novel method based on the LWF concept, which is used to analyze TDSs analogously to a system of ODEs [20,22]. The distinctive feature of the method is that the solution of DDEs is expressed as a series sum of infinitely many exponential functions. Moreover, the solution obtained explicitly in the form of a series enables examining structural properties such as observability and controllability for TDSs [23], the exploitation of eigenvalue assignment methods for control synthesis [22,24–26], and other applications like spectrum design [27] and the estimation of decay rates [28].

For scalar cases, that is, first-order TDSs, the LWF framework can be efficiently utilized. In [29–31] an algorithm is presented to extend and generalize the idea to the multidimensional cases with the help of the matrix LWF. However, some degenerate cases were investigated in the spectrum analysis of multidimensional TDSs using the matrix LWF. In the LWF framework the analysis is carried out by considering that a one-to-one correspondence holds between the roots of the system and the branches of the LWF. Cepeda-Gomez et al. [32] reported that, in general, for a second-order TDS, there is no such correspondence as mentioned before between the branches of the LWF and the eigenvalues of the system.

This chapter explores the LWF method for the analysis and design of multidimensional TDSs. We show that the formulation carried out using the matrix LWF suffers from some limitations and does not produce satisfactory results for all the systems under consideration. We also provide some counterexamples to show that we need to be very careful in drawing conclusions about the spectrum of the higher-order system and the consecutive analysis using this approach. In particular, the method presented by Yi et al. [33] does not produce satisfactory results when the system is not of full rank or modes have multiplicity (repeated roots). In some cases the algorithm produces spurious characteristic roots (poles), which are not the actual modes of the system under consideration.

Thus the primary contribution of this chapter involves the investigation of the spectrum of higher-order linear time-delay systems in the framework of LWF and identification of a particular class (a common canonical form of TDSs), from which some counterexamples are provided to show that the already existing well-established literature suffers from some discrepancies and limitations. By introducing a more generalized class of TDSs we show that there is no one-to-one correspondence between the eigenvalues of TDS and the branches of the LWF. Moreover, an algorithm is developed to synthesize a state feedback controller for a class of systems that are in common canonical form.

9.1.2 Purpose and scope

The aim of this chapter is providing a new insight toward the LWF-based framework for the investigation and stabilization of LTI multidimensional systems comprising single time delay. Though the approach is well established for scalar TDSs, the intuitiveness presented in this chapter provides a better perspective toward exploitation of the approach for multidimensional systems so that appropriate analysis can be carried out.

9.2. Preliminaries

In this section, we provide a brief tutorial on the theoretical background of the LWF and its utilization for studying TDSs.

9.2.1 Background

We denote the LWF by W; it is also called the Omega function or the product log. But what is in fact the LWF? To explain this, let us think of the equation

$$xe^x = z. \tag{9.2}$$

For this equation, given x, can z be found? Or, given z, can x be found? So, Eq. (9.2) has no any solution for x in terms of elementary functions. Therefore the LWF is used to solve such type of equations, which is basically a transcendental equation. Transcendental equations are those involving exponentials or logarithmic functions including the variable to be solved for and having the specialty that cannot be described with the help of elementary functions. The general strategy to solve such equations is shifting all occurrences of the unknown variable to one side of the algebraic equation and arrange it in a similar fashion of $z = xe^x$ at that instance the LWF returns the value of the variable in x. Further, the process of finding x in (9.2) can be mapped as $z \xrightarrow{f^{-1}} x$. So, the LWF is the relation f^{-1} inputting z and outputting x:

$$x = W(z).$$

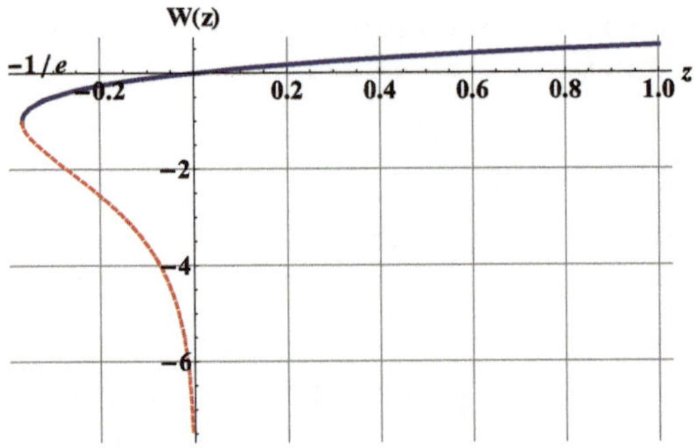

Figure 9.1 Real branches of the LWF. The upper branch $W_0(z)$ for $z \in [-1/e, 1]$ is the thick solid line; $W_{-1}(z)$ for $z \in [-1/e, 0]$ is the dashed lower curve.

For more clarification, let us consider the equation

$$c + dx^x = 0.$$

For this equation, a solution for x cannot be obtained in the form of elementary functions. But it has a solution by means of the LWF given as

$$\begin{aligned} \ln(-c/d) &= x\ln(x), \\ \ln(-c/d) &= \ln(x)e^{\ln(x)}, \\ W(\ln(-c/d)) &= \ln(x) \\ &\Rightarrow x = \exp(W(\ln(-c/d))). \end{aligned}$$

Formally, the LWF $W(z)$ is described as the inverse function of we^w. It satisfies the equation

$$W(z)e^{W(z)} = z \tag{9.3}$$

for all $z \in \mathbb{C}$. The LWF is a multivalued function with infinitely many branches [34]. The concept of the branches is like the way the branches of the logarithm are indexed by a multiple of $2k\pi$. The branches are indexed by a subscript k as W_k for $k \in \mathbb{Z}$ and give a distinct solution of the expression $z = we^w$. By convention the branch W_0 is addressed as the principal branch [35]. The following Mathematica code generates the real branches of the LWF as shown in Fig. 9.1.

```
graph2 = Plot[ProductLog[-1, x], {x, -1/E, 0},
PlotStyle -> {RGBColor[1, 0, 0], Dashed}];
graph1 = Plot[ProductLog[0, x], {x, -1/E, 1}];
```

```
Show[graph1, graph2, PlotRange -> All, GridLines -> ...
    Automatic,
AxesLabel -> {z, "W(z)"}, LabelStyle -> Directive[Bold, ...
    Large]]
```

For $z \in \mathbb{R}$, the function $W(z)$ has two real branches in the interval $-1/e \leq z < 0$, which join at $z = -1/e$ as shown in Fig. 9.1. Concretely, this means that the equation $z = we^w$ has two real solutions for $-1/e \leq z < 0$. The branch satisfying $W(z) \geq -1$ is denoted as $W_0(z)$, whereas the branch satisfying $W(z) \leq -1$ is referred as $W_{-1}(z)$. The principal branch is defined for all argument values of $z \geq -1/e$, whereas $W_{-1}(z)$ is only defined for the interval $-1/e \leq z < 0$ [36]. A branch is referred to as the real branch if it has a real-valued solution of (9.3) for real z. The MATLAB code is given below.

```
syms x
fplot(lambertw(x))
hold on
fplot(lambertw(-1,x))
hold off
axis([-0.5 4 -4 2])
title('Lambert W function, two main branches')
legend('k=0','k=1','Location','best')
```

Expressing $w = u + iv$ and $z = \xi + i\eta$, we can deduce the following:

$$\xi = e^u(u\cos v - v\sin v), \tag{9.4}$$

$$\eta = e^u(v\cos v + u\sin v). \tag{9.5}$$

For $z \in \mathbb{R}$, we have $\eta = 0$ in (9.5), and then

$$v = 0 \quad \text{or} \quad u = -v\cot v.$$

Subsequently,

$$W(z) = u \quad \text{or} \quad W(z) = -v\cot v + iv$$

A plot of the function $u = -v\cot v$ is shown in Fig. 9.2(a). Based on Fig. 9.2(a) and by convention, (9.6) characterizes the principal branch $W_0(z)$, (9.7) describes the $W_{-1}(z)$, and (9.8) characterizes the *k*th branch

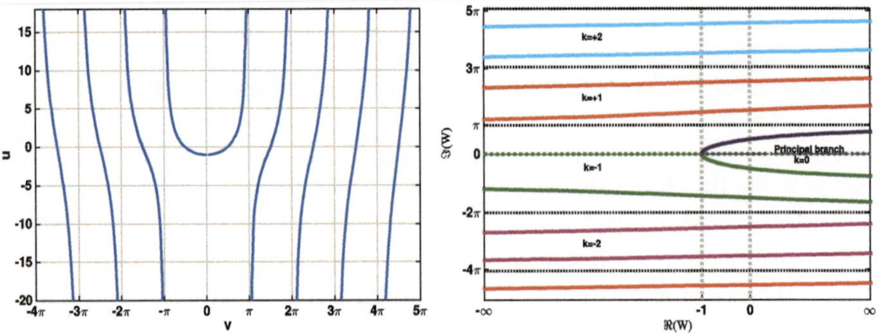

(a) (b)

Figure 9.2 (a) Plot of $u = -v \cot v$. (b) A few branches of the LWF.

[34]. Fig. 9.2(b) shows the ranges of the branches of $W_k(z)$,

$$W_0(z) = \begin{cases} -v\cot v + iv & \text{for } v \in (0, \pi), \\ u & \text{for } u \geq -1, \end{cases} \tag{9.6}$$

$$W_{-1}(z) = \begin{cases} u & \text{for } u < -1, \\ -v\cot v + iv & \text{for } v \in (-2\pi, -\pi) \cup (-\pi, 0), \end{cases} \tag{9.7}$$

$$W_k(z) = \begin{cases} -v\cot v + iv & \text{for } v \in ((2k-1)\pi, 2k\pi) \cup (2k\pi, (2k+1)\pi), \\ & k = 1, 2, \ldots, \\ -v\cot v + iv, & \text{for } v \in (2k\pi, (2k+1)\pi) \cup ((2k+1)\pi, 2(k+1)\pi), \\ & k = -2, -3, \ldots. \end{cases} \tag{9.8}$$

The following MATLAB® code generates Fig. 9.2(b).

```
clc
clear all
y = 0.9*linspace(0,pi,100);
z = -y.*cot(y)+1j*y;
plot(real(z),imag(z),'b')
xlabel('real part')
ylabel('imaginary part')
hold all

x = linspace(-1,3,100);
z = x;
plot(real(z),imag(z),'b')
```

```

x = linspace(-1,-10,100);
z = x;
plot(real(z),imag(z),'g')

y1 = linspace(-2*pi,-pi,100);
y2 = linspace(-pi,0,100);
y = union(y1,y2);
z = -y.*cot(y)+1j*y;
z(find(abs(z)>100))=NaN;
plot(real(z),imag(z),'g')

for k=1:2
    hold on
    y1 = linspace((2*k-1)*pi,2*k*pi,100);
    y2 = linspace(2*k*pi,(2*k+1)*pi,100);
    y = union(y1,y2);
    z = -y.*cot(y)+1j*y;
    z(find(abs(z)>50))=NaN;
    plot(real(z),imag(z),'r')
end

for k=-3:-2
    hold on
    y1 = linspace(2*k*pi,(2*k+1)*pi,100);
    y2 = linspace((2*k+1)*pi,(2*(k+1))*pi,100);
    y = union(y1,y2);
    z = y.*cot(y)+1j*y;
    z(find(abs(z)>50))=NaN;
    plot(real(z),imag(z),'m')
end

axis([-5.5 2.5 -5*pi 5*pi])
```

9.2.2 Matrix Lambert W function

To define the matrix LWF of an arbitrary matrix A, we first need to transform it into the Jordan canonical form. A matrix can be represented in its Jordan canonical form as [37,38]

$$J = \begin{bmatrix} J_{n_1}(\lambda_1) & & & \\ & J_{n_2}(\lambda_2) & & \\ & & \ddots & \\ & & & J_{n_p}(\lambda_p) \end{bmatrix},$$

where $J_n(\lambda)$ denotes the Jordan block of order n-by-n corresponding to the eigenvalue λ of multiplicity n. Then the matrix LWF of the matrix J can be defined as follows:

$$W_k(J) = \begin{bmatrix} W_{k_1}(J_{n_1}(\lambda_1)) & & & \\ & W_{k_2}(J_{n_2}(\lambda_2)) & & \\ & & \ddots & \\ & & & W_{k_p}(J_{n_p}(\lambda_p)) \end{bmatrix}.$$

It is worth noting that a different branch can be considered for every Jordan block. If there are p Jordan blocks in J and the index set for the branches of the scalar LWF is $\mathbb{Z}$, then $\mathbb{Z}^p$ would be the index set for the branches of $W_k(J)$. For distinct eigenvalues, that is, Jordan blocks of dimension 1, we can use the scalar LWF. For the remaining cases, the LWF (for a fixed branch) of a Jordan block can be defined using matrix functions (e.g., see [2,38]), that is,

$$W_k(J_n(\lambda)) = \begin{bmatrix} W_k(\lambda) & W_k^1(\lambda) & \dots & \frac{1}{(n-1)!}W_k^{(n-1)}(\lambda) \\ 0 & W_k(\lambda) & \dots & \frac{1}{(n-2)!}W_k^{(n-2)}(\lambda) \\ \vdots & \vdots & \dots & \vdots \\ 0 & 0 & \dots & W_k(\lambda) \end{bmatrix},$$

where $W_k^{(n-1)}(\lambda) = \frac{d^{(n-1)}}{d\lambda^{(n-1)}} W_k(\lambda)$. If $k=0$, then we need to assume that $\lambda \neq -e^{-1}$ (since $W_0^1(\lambda)|_{\lambda=-e^{-1}}$ is undefined). Therefore, for an arbitrary matrix A that can be transformed to the Jordan canonical form using similarity transformation $A = HJH^{-1}$, the matrix LWF can be expressed as

$$W_k(A) = HW_k(J)H^{-1}.$$

9.2.3 Scalar case with a single delay

Consider a scalar DDE with single time delay

$$\dot{x}(t) = ax(t) + a_d x(t-h) + bu(t), \tag{9.9}$$

where a, a_d, and b are scalars. $x(t) \in \mathbb{R}$, and $h > 0$ is the amount of delay. The boundary condition is specified as $x(t=0) = x_0$, and the history or preshape function is $x(t) = g(t)$ for $-h \le t < 0$. We employ the LWF to solve the TCE of (9.9) given as follows:

$$\triangle(s) = s - a - a_d e^{-sh} = 0. \tag{9.10}$$

After rearranging and multiplying both sides of (9.10) by he^{-ah}, we get

$$h(s-a)e^{h(s-a)} = a_d he^{-ah}.$$

Accordingly, the solution of the TCE (9.10) can be expressed using the LWF as

$$s = \frac{1}{h} W(a_d he^{-ah}) + a. \tag{9.11}$$

From (9.11) it is evident that the solution can be described explicitly as a function of the system parameters a, a_d, and h. Thus the infinite spectrum of a system of scalar DDEs as in (9.9) can be found by solving (9.11) for different branches of the LWF. Therefore, for $k \in \mathbb{Z}$, the solution can be given as

$$s_k = \frac{1}{h} W_k(a_d he^{-ah}) + a. \tag{9.12}$$

It is worth noting that, unlike the existing methodologies reported in [39] and the references therein, the LWF approach gives an analytical solution of DDEs in terms of its parameters.

9.2.3.1 Stability analysis

Stability of a scalar DDE as in (9.9) can be examined by inspecting the rightmost root in the complex plane. Mori et al. [40] reported that only the root associated with the principal branch is sufficient for stability determination. Consequently, the investigation of the roots associated with the rest of the branches is not necessary to ensure the stability. The following lemma presents the stability result reported in [40].

Lemma 9.1. *For arbitrary $z \in \mathbb{C}$, we have*

$$\max[Re\{\lambda(S_0)\}] \geq Re\{\lambda(S_k)|_{k\neq 0}\}.$$

9.2.3.2 Response of a scalar DDE

The complete (free response and forced response) response of (9.9) can be described by an infinite series as a function of the roots of the TCE (9.10) as in [33]:

$$x(t) = \underbrace{\sum_{k=-\infty}^{\infty} e^{s_k t} \beta_k^I}_{\text{free response}} + \underbrace{\int_0^t \sum_{k=-\infty}^{\infty} e^{s_k(t-\xi)} \beta_k^N bu(\xi) d\xi}_{\text{forced response}}, \tag{9.13}$$

where the coefficient β_k^I is a function of DDE parameters, the initial condition x_0, and the history function:

$$\beta_k^I = \frac{(x_0 + a_d e^{-s_k h} G(s))}{(1 + a_d h e^{-s_k h})},$$

where $G(s)$ is the Laplace transform of $g(t)$,

$$G(s) = \int_0^h e^{-s_k t} g(t-h)dt.$$

Similarly, β_k^N is a function of DDE parameters only and depends neither on $g(t)$ nor on x_0 and is given by

$$\beta_k^N = \frac{1}{(1 + a_d h e^{-s_k h})}.$$

Thus the solution form of DDEs using LWF is similar to that of ODEs. Bellman et al. [3] described the convergence conditions for such series solution forms. A practically significant and useful feature of this series description for $x(t)$ is that the series truncation, for example, $k = 0, \pm 1, \pm 2, \ldots, \pm n$, gives an approximated result in terms of the $(2n+1)$ most significant characteristic roots [20]. Therefore the overall system dynamics can be represented explicitly by using its finite-dimensional approximation.

Example 9.1. Consider (9.9) with parameters $a = -3$, $a_d = -1$, and $h = 0.5$ s. The corresponding eigenspectrum is shown in Fig. 9.3.

The following MATLAB code generates Fig. 9.3.

```
a = -3; ad = -1; h=0.5;
for k = -10:10
s = (1/h)*lambertw(k,ad*h*exp(-a*h))+a;
hold on
plot(real(s),imag(s),'b*')
end
xlabel('Re$(S_k)$','interpreter','latex')
ylabel('Imag$(S_k)$','interpreter','latex')
title('Eigenspectrum of S_k for k = -10:10')
```

9.2.3.3 Placement of dominant roots

In TDSs, one of the major concerns while synthesizing a feedback controller is the allocation of the closed-loop eigenvalues. The reason behind

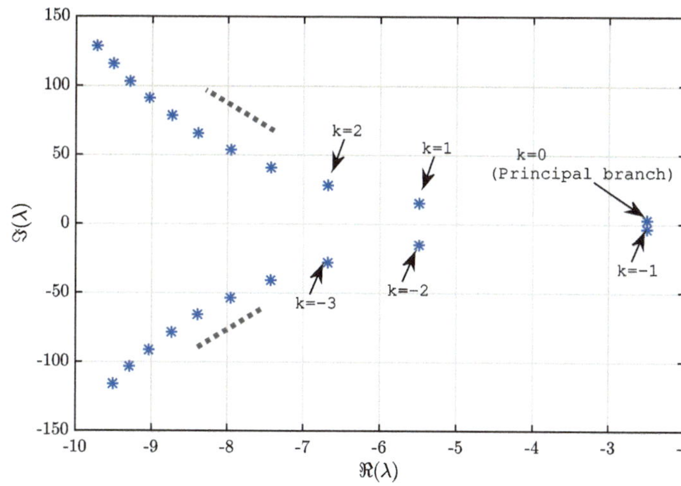

Figure 9.3 Eigenvalues of the system in Example 9.1 for $k = -10 : 10$.

this is an infinite number of eigenvalues associated with the TCE and a limited number of control parameters. Application of the conventional tools for ODEs [21] to assign a chosen subset of characteristic roots at desired locations may prompt other uncontrolled eigenvalues shifting toward the right of the origin in the complex plane [41]. However, the procedure of control design using the LWF concept, based on Lemma 9.1, equips with the acceptable control gains without such a loss of stability. The rightmost root of the closed-loop dynamics depends on the system parameters and can be assigned to the desired location in the complex plane by adjusting them. Yi et al. [22] presented such a method for eigenvalue assignment based on the LWF, which we revisit here.

Consider a scalar TDS as in (9.9) with the controller

$$u(t) = -Kx(t) - K_d x(t-h). \tag{9.14}$$

The control law (9.14) brings up the closed-loop dynamics as

$$\dot{x}(t) = (a - bK)x(t) + (a_d - bK_d)x(t-h). \tag{9.15}$$

Algorithm 9.1 describes the method for obtaining the control gains to assign the rightmost root.

Every branch of the LWF has a range limitation, as shown in Fig. 9.2(b), and that is why the rightmost root cannot be placed to any arbitrary location. For a scalar TDS, as in (9.15), this is evident by inspecting that for the

Algorithm 9.1

1: Select desired characteristic root $\lambda_{i,\text{des}}$
2: Set initial conditions $K = K_0$ and $K_d = K_{d0}$
3: While $\lambda_i(s_0) - \lambda_{i,\text{des}} \geq$ tolerance
4: Select $K = K_{\text{new}}$ and $K_d = K_{d\_\text{new}}$.
5: Let $a_{\text{new}} = (a - bK_{\text{new}})$ and $a_{d\_\text{new}} = (a_d - bK_{d\_\text{new}})$.
6: Compute $s_{0_{\text{new}}} = \frac{1}{h} W_0(a_{d\_\text{new}} h e^{-a_{\text{new}} h}) + a_{\text{new}}$.
7: End.

principal branch,

$$\frac{1}{h} Re\{W_0((a_d - bK_d)he^{-(a-bK)h}) + a - bK\}$$
$$\geq Re\{-\frac{1}{h} + a - bK\} \geq -\frac{1}{h} + a - bK \tag{9.16}$$

since $Re\{W_0(z)\} \geq -1$. Therefore (9.16) poses a restriction on assigning the rightmost pole of system (9.15). From (9.16) it is apparent that it cannot be less than $-\frac{1}{h} + a - bK$.

Example 9.2. Consider system (9.9) with control law (9.14), where $a = -1$, $a_d = 2.5$, $b = 1$, and delay $h = 1$ s. For these parameters, the rightmost root can be attributed to any value > -2 (as per condition (9.16)). Let $\lambda_{\text{des}} = -1.5$. Then the corresponding rightmost root can be obtained using $k = 0$, and the eigenspectrum is shown in Fig. 9.4.

9.2.4 A general system of DDEs: single-delay case

Consider a general system of DDEs in matrix-vector form

$$\begin{aligned} \dot{x}(t) &= Ax(t) + A_d x(t-h) + Bu(t), && t > 0, \\ x(t) &= x_0, && t = 0, \\ x(t) &= g(t), && t \in [-h, 0), \end{aligned} \tag{9.17}$$

where A, $A_d \in \mathbb{R}^{n\times n}$, $x(t) \in \mathbb{R}^{n\times 1}$ is the state vector, and x_0 and $g(t)$ denote the initial state and preshape function, respectively. The characteristic polynomial of system (9.17) is

$$\Delta(S) = S - A - A_d e^{-Sh} = 0, \tag{9.18}$$

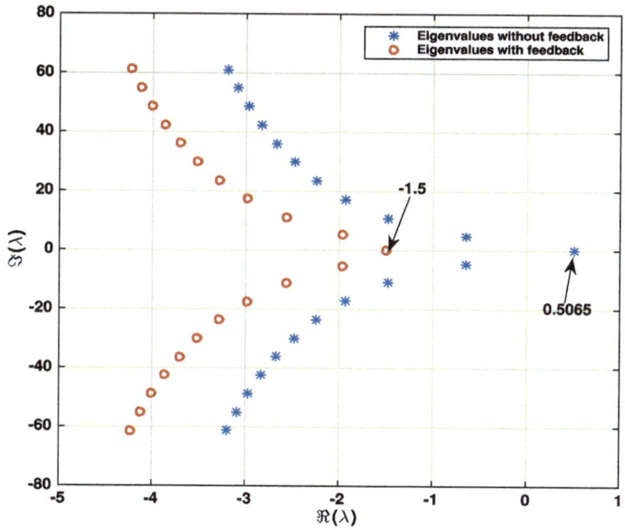

Figure 9.4 Eigenvalues of the system in Example 9.2.

where $S \in \mathbb{C}^{n\times n}$. After some manipulation and rearrangements, (9.18) can be rewritten as

$$h(S-A)e^{(S-A)h} = hA_d e^{-Ah}.$$

Since the matrices S and A do not commute in general, we introduce an auxiliary matrix Θ:

$$h(S-A)e^{(S-A)h} = A_d h\Theta. \tag{9.19}$$

Denote $M = hA_d\Theta$. Then we can obtain the solution matrix S_k for the kth branch of the LWF:

$$S_k = \frac{1}{h}W_k(M_k) + A. \tag{9.20}$$

Substitution of (9.20) into (9.18) results in the following nonlinear equation, from which we obtain the matrix M_k:

$$W_k(M_k)e^{W_k(M_k)+Ah} = A_d h. \tag{9.21}$$

The characteristic roots of (9.17) can be computed by following the steps in Algorithm 9.2 for branches $k \in \mathbb{Z}$ [33].

Algorithm 9.2

1: Solve Eq. (9.21) for $k \in \mathbb{Z}$.
2: Compute S_k by substituting M_k into (9.20).
3: Compute the eigenvalues of S_k.

The literature shows that the use of the LWF-based approach for solving DDEs is not new, and the methodology is already utilized in several derivative works [29,33,38,42–44]. The authors utilized the matrix LWF-based approach to find the characteristic roots and spectrum. The stability is then determined by earmarking the dominant roots of the system corresponding to the branches $k = 0, \pm 1, \ldots, \pm m$, where m is the nullity of the matrix A_d in (9.17), and this is reported in the form of conjecture in [33]. Thus the stability can be examined by considering only a few branches, and consecutively controllability and observability can be analyzed. This methodology is general and is not restricted to a commuting pair of A and A_d matrices.

In [32], some particular cases of this class of systems were considered, and it was found that the one-to-one correspondence does not generally exist between the branches of the LWF and the roots of the system. It is demonstrated that the formulation carried out in the book [33] and the works [22,24,29–31], and so on, which is about the analysis of systems of linear DDEs using the matrix LWF, suffers from some severe limitations.

Cepeda-Gomez et al. [32,45] disproved this correspondence by exploring a counterexample of a second-order system (9.17) defined by the system parameters as

$$A = \begin{bmatrix} 0 & 1 \\ -5 & -1 \end{bmatrix}, \quad A_d = \begin{bmatrix} 0 & 0 \\ -3 & -0.6 \end{bmatrix}, \quad h = 5.$$

This counterexample shows that we need to be very careful in drawing conclusions about the spectrum of multidimensional TDSs using the LWF framework, in particular, when the matrix A_d does not have full rank. Furthermore, in some cases the approach produces spurious roots, which are not the actual modes of the system under study and hence do not constitute the actual roots of the system.

In this chapter, we show that the assumption of one-to-one correspondence does not hold in general. By considering an nth-order TDS of the form usually common in applications it is shown that the complete eigenspectrum of a linear time-invariant TDS of this class can be obtained by using only two real branches of the LWF. This reason leads to the statement that one-to-one correspondence does not exist between the roots of the system and the branches of the LWF. The objective of this chapter is investigation and extension of the formulation carried out in [32,45] for more general classes of time-delay systems of arbitrary order n. Since the eigenvalues of a TDS are either real or complex conjugate pairs, it is possible to associate all the eigenvalues with only the real branches of the LWF. The fact behind this is that the nonlinear equation (9.21) associated with this method does not have a unique solution. Moreover, it is shown that besides the dominant roots, the principal branch can be also used to find some other roots.

9.3. Utilization of the Lambert W function for multidimensional time delay systems

Here we show that the stability conjecture presented in [33] does not hold in general. By considering a particular, yet ubiquitous, system of nth order, we can show that (9.21) may have multiple solutions and there is no one-to-one correspondence between the branches of the LWF and the roots of the system. We also show that for such classes of TDSs, all the characteristic roots can be found by using only the real branches corresponding to $k = 0$ and $k = -1$. These findings confront the conjecture presented in [33] regarding stability.

9.3.1 Extension to nth-order TDS

Let us first define the most common form of TDSs, the so-called common canonical (CC) form.

Definition 9.1. A system given by (9.17) is said to be in common canonical form if $A \in \mathbb{R}^{n\times n}$ is in companion form and $A_d \in \mathbb{R}^{n\times n}$ has all its rows zero except for the nth row and is represented as [46]

$$A = \begin{bmatrix} 0 & 1 & \dots & 0 \\ \vdots & \vdots & \ddots & \vdots \\ 0 & 0 & \dots & 1 \\ a_1 & a_2 & \dots & a_n \end{bmatrix}, \quad A_d = \begin{bmatrix} 0 & 0 & \dots & 0 \\ \vdots & \vdots & \ddots & \vdots \\ 0 & 0 & \dots & 0 \\ a_{d1} & a_{d2} & \dots & a_{dn} \end{bmatrix}. \tag{9.22}$$

Cepeda-Gomez et al. [32] proclaimed that the eigenspectrum of a second-order TDS in the CC form could be associated with only the real branches of the LWF. On similar lines, these results can be extended for an

nth-order system and stated in the form of Theorem 9.1. To achieve this, the underlying assumption is that S_k is real and in companion form.

Theorem 9.1. *The whole eigenspectrum of a time-delay system represented by (9.17) can be associated with only the two real branches of the LWG, that is, $k=0$ and $k=-1$, provided that it is in CC form.*

Proof. Considering the structure of A_d in (9.22), we can easily show that $M_k = hA_d\Theta_k$ for any given Θ_k has the form

$$M_k = \begin{bmatrix} 0 & 0 & \dots & 0 \\ \vdots & \vdots & \ddots & \vdots \\ 0 & 0 & \dots & 0 \\ m_1 & m_2 & \dots & m_n \end{bmatrix}, \tag{9.23}$$

where m_i, $i=1,2,\dots,n$, are scalars. Based on the value of the element m_n, there are two possible cases.

Case 9.1.1. $m_n \neq 0$. Then we obtain the matrix LWF of M_k as

$$W_k(M_k) = \begin{bmatrix} 0 & \dots & 0 & 0 \\ \vdots & \ddots & \vdots & \vdots \\ 0 & \dots & 0 & 0 \\ \frac{m_1}{m_n} W_k(m_n) & \dots & \frac{m_{n-1}}{m_n} W_k(m_n) & W_k(m_n) \end{bmatrix} = \frac{W_k(m_n)}{(m_n)} M_k. \tag{9.24}$$

Therefore the matrix LWF of M_k is the matrix M_k itself multiplied by a scalar constant $\frac{W_k(m_n)}{hm_n}$. Using (9.24), we can rewrite S_k as

$$S_k = \frac{W_k(m_n)}{hm_n} M_k + A. \tag{9.25}$$

Further,

$$S_k = \begin{bmatrix} 0 & 1 & \dots & 0 & 0 \\ \vdots & \vdots & \ddots & \vdots & \vdots \\ 0 & 0 & \dots & 0 & 1 \\ \frac{m_1}{hm_n} W_k(m_n) + a_1 & \frac{m_2}{hm_n} W_k(m_n) + a_2 & \dots & \frac{m_{n-1}}{hm_n} W_k(m_n) + a_{n-1} & \frac{1}{h} W_k(m_n) + a_n \end{bmatrix}. \tag{9.26}$$

Case 9.1.2. $m_n = 0$. Then

$$\lim_{m_n \to 0} \frac{W_k(m_n)}{m_n} \Rightarrow \lim_{x \to 0} \frac{x}{xe^x} = 1,$$

which gives $W_k(M_k) = M_k$. Hence, in this case, S_k can be given as

$$S_k = \begin{bmatrix} 0 & 1 & \dots & 0 & 0 \\ \vdots & \vdots & \ddots & \vdots & \vdots \\ 0 & 0 & \dots & 0 & 1 \\ \frac{m_1}{h} + a_1 & \frac{m_2}{h} + a_2 & \dots & \frac{m_{n-1}}{h} + a_{n-1} & \frac{1}{h} + a_n \end{bmatrix}.$$

Furthermore, the idea here is to execute the steps given in Algorithm 9.2 in a reverse manner to attribute a branch index k with a set of eigenvalues. To implement this, the first step is forming a real matrix S_k such that it gives the chosen eigenvalues. Let $\lambda_1, \lambda_2, \dots, \lambda_n$ be the eigenvalues of S_k. Then S_k can be represented in terms of its eigenvalues as follows:

$$S_k = \begin{bmatrix} 0 & 1 & 0 & \dots & 0 \\ \vdots & \vdots & \vdots & \ddots & \vdots \\ 0 & 0 & 0 & \dots & 1 \\ s_n & s_{n-1} & s_{n-2} & \dots & s_1 \end{bmatrix}, \tag{9.27}$$

where

$$\begin{aligned} s_n &= (-1)^n \text{ product of all } n \text{ roots}, \\ s_{n-1} &= +\text{sum of the product of the roots taken } n-1 \text{ at a time}, \\ s_{n-2} &= -\text{sum of the product of the roots taken } n-2 \text{ at a time}, \\ &\vdots \\ s_1 &= -\text{sum of all the roots}. \end{aligned}$$

Comparison of (9.26) and (9.27) results in the following:

$$W_k(m_n) = h(s_1 - a_n), \tag{9.28}$$

$$\begin{aligned} &\vdots \quad \vdots \\ m_2 &= \frac{(s_{n-1} - a_2)m_n}{(s_1 - a_n)}, \\ m_1 &= \frac{(s_n - a_1)m_n}{(s_1 - a_n)}. \end{aligned} \tag{9.29}$$

For a real matrix S_k, M_k is also real. Therefore both sides of Eq. (9.28) are real, which correspond to one of the real branches, depending on the value of the scalar LWF element $W_k(m_n)$. If $W_k(m_n) \in [-1, \infty)$, then it corresponds to $k=0$, and when $W_k(m_n) \in (-\infty, -1]$, it corresponds to $k=-1$. □

9.3.2 Proposed generalized class

The applicability of Theorem 9.1 can be extended through Corollary 9.1 for the cases where the TDS given by (9.17) is not in the CC form.

Remark 9.1. Theorem 9.1 is a generalization of the method presented in [32] from a second-order system to an nth-order one. However, the applicability of Theorem 9.1 is limited to a particular class of systems that are in the CC form. To remove this restriction, we present Corollary 9.1, which involves a broader class of TDS that are not explicitly in the CC form but can be transformed into it.

Corollary 9.1. *Given a TDS of the form* (9.17) *with A of arbitrary structure and $A_d = bc^T$, where $b, c \in \mathbb{R}^{n\times 1}$, there exists a nonsingular state transformation matrix Ξ such that $\Xi^{-1}b = [0\ldots 1]^T$. Then there exists a real solution of S_k satisfying $\Delta(S_k) = (S_k - \bar{A} - \bar{A}_d e^{-S_k h}) = 0$ such that*

$$(S_k - \bar{A}) = \begin{bmatrix} 0 & 0 & \ldots & 0 \\ \vdots & \vdots & \ddots & \vdots \\ 0 & 0 & \ldots & 0 \\ * & * & \ldots & * \end{bmatrix},$$

where $\bar{A} = \Xi^{-1}A\Xi$ *and* $\bar{A}_d = \Xi^{-1}bc^T\Xi = \begin{bmatrix} 0 & 0 & \ldots & 0 \\ \vdots & \vdots & \ddots & \vdots \\ 0 & 0 & \ldots & 0 \\ * & * & \ldots & * \end{bmatrix}$, *where $*$ represents any numeric value.*

Proof. Let A and $A_d = bc^T$ be a priori given as

$$A = \begin{bmatrix} A_1 \\ \vdots \\ A_{n-1} \\ A_n \end{bmatrix}, \quad b = \begin{bmatrix} * \\ \vdots \\ * \end{bmatrix}, \quad c = \begin{bmatrix} * \\ \vdots \\ * \end{bmatrix},$$

where $*$ represents any numerical value. Subsequently, in the absence of the control input, (9.17) can be rewritten as

$$\dot{x}(t) = Ax(t) + bc^T x(t-h). \tag{9.30}$$

Let there exist a nonsingular state transformation matrix Ξ such that $\Xi^{-1}b = [0\ldots1]^T$. The change of coordinates is given by a linear transformation $x = \Xi z$, where z represents states in the transformed domain. The then transformed system can be presented as

$$\dot{z}(t) = \bar{A}z(t) + \bar{A}_d z(t-h).$$

Further, our objective is to choose the structure of S_k such that $(S_k - \bar{A})$ has a structure similar to that of M_k. Therefore S_k must have the form

$$S_k = \begin{bmatrix} \bar{A}_1 \\ \vdots \\ \bar{A}_{n-1} \\ \hline * \ldots * \end{bmatrix}, \tag{9.31}$$

where numerical values representing $*$ can be obtained by using the Caylay–Hamilton theorem [21] and comparing the coefficients with its characteristic function. Here it is important to mention that $\bar{A}$ has a regular structure and $\bar{A}_d$ has its structure in the CC form; thus it can be considered as the system partially transformed into the CC form of TDS. □

Remark 9.2. The above results are also valid when A is in companion form. Then system (9.17) with $A_d = bc^T$ can be transformed into the CC form if (A, b) is controllable.

9.3.3 Generalized case

The results presented can also be extended for system (9.17) when $rank(A_d) = m$.

Theorem 9.2. *Let A, $A_d \in \mathbb{R}^{n\times n}$, and let $rank(A_d) = m$. Let A_d be in the common canonical form. Then the matrix $M = hA_d\Theta$ has the structure $M = \begin{bmatrix} 0_{(n-m)\times n} \\ X \end{bmatrix}$, where the submatrix $0_{(n-m)\times n}$ represents the zero matrix of*

$\mathbb{R}^{(n-m)\times n}$, *and* $X = [M_1 \ \ M_2]$. *The submatrix* $M_2 \in \mathbb{R}^{m\times n}$ *must be invertible. Then the Lambert W matrix* $W_k(M)$ *is equal to*

$$W_k(M) = \begin{bmatrix} 0_{(n-m)\times n} \\ W_k(M_2)M_2^{-1}X \end{bmatrix}.$$

Proof. Consider an $\mathbb{R}^{n\times n}$ matrix M of the structure

$$M = \begin{bmatrix} 0 & 0 \\ M_1 & M_2 \end{bmatrix},$$

where $M_1, M_2 \in \mathbb{R}^{m\times n}$ with M_2 being invertible. Now consider the similarity transformation matrix

$$V = \begin{bmatrix} I & 0 \\ -M_2^{-1}M_1 & I \end{bmatrix},$$

where I is the identity matrix of appropriate dimension. Further, we can clearly see that $V^{-1}MV$ is a block diagonal matrix of the structure

$$V^{-1}MV = \Lambda = \begin{bmatrix} 0 & 0 \\ 0 & M_2 \end{bmatrix}.$$

Given that $\Lambda = V^{-1}MV$, we have

$$W_k(M) = \begin{bmatrix} 0 & 0 \\ W_k(M_2)M_2^{-1}M_1 & W_k(M_2) \end{bmatrix}.$$

Now consider the case where $M \in \mathbb{R}^{mn\times mn}$ is of the structure

$$M = \begin{bmatrix} 0_{(n-m)\times n} \\ 0_{(n-m)\times n} \\ \vdots \\ 0_{(n-m)\times n} \\ X \end{bmatrix},$$

where, $0_{(n-m)\times n}$ is the zero $(n-m)\times n$ matrix, and

$$X = \begin{bmatrix} M_1 & M_2 & \dots & M_n \end{bmatrix}$$

with each M_i being an $m \times n$ matrix, and M_n is invertible.

In such a case,

$$W_k(M) = \begin{bmatrix} 0_{(n-m)\times n} \\ 0_{(n-m)\times n} \\ \vdots \\ 0_{(n-m)\times n} \\ W_k(M_n)M_n^{-1}X \end{bmatrix}.$$

□

Based on Theorem 9.2, the findings similar to Corollary 9.1 can be easily derived for the generalized class of systems where $rank(A_d) = m$, with the above structure of $W_k(M_k)$.

Once A_d is transformed into the CC form, all the characteristics roots of system (9.30) can be associated exclusively with the real branches of the LWF. Since system (9.17) is linear with real coefficients, its characteristic polynomial has all its roots either real or in complex conjugate pairs, and hence it is always possible to attribute all the roots using real S_k or, in other words, branches $k=0$ and $k=-1$ only.

Example 9.3. Eigenspectrum association with only the real branches.

Consider system (9.17) with delay $h=2$ s and matrices

$$A = \begin{bmatrix} -1 & 2 & -1 \\ -4 & -1 & -3 \\ -2 & -3 & -2 \end{bmatrix}, \quad A_d = \begin{bmatrix} 1 & -1 & 2 \\ 0 & 0 & 0 \\ -1 & 1 & -2 \end{bmatrix}, \quad b = \begin{bmatrix} -1 \\ 0 \\ 1 \end{bmatrix}, \quad c = \begin{bmatrix} -1 \\ 1 \\ -2 \end{bmatrix}.$$

The transformation matrix Ξ is given by

$$\Xi = \begin{bmatrix} 1 & -2 & -1 \\ 2 & -1 & 0 \\ -2 & 2 & 1 \end{bmatrix},$$

and the corresponding transformed system is obtained as

$$\bar{A} = \begin{bmatrix} -1 & -1 & 0 \\ -2 & -5 & -1 \\ -2 & 11 & 2 \end{bmatrix}, \quad \bar{A}_d = \begin{bmatrix} 0 & 0 & 0 \\ 0 & 0 & 0 \\ 5 & -3 & -1 \end{bmatrix}.$$

To achieve an analytical estimate of Θ_k, we use a "reverse-engineering" approach, as demonstrated in the proof of Theorem 9.1. For this, the first characteristic roots of the system are computed using QPmR algorithm [12].

Consider the roots $\lambda_1 = -0.12$ and $\lambda_{2,3} = 0.27 \pm 1.56i$. The matrices corresponding to these roots are

$$S_k = \begin{bmatrix} -1 & -1 & 0 \\ -2 & -5 & -1 \\ 18.42 & 38.01 & 6.43 \end{bmatrix}, \quad W_k(M_k) = \begin{bmatrix} 0 & 0 & 0 \\ 0 & 0 & 0 \\ 40.84 & 54.01 & 8.86 \end{bmatrix}. \tag{9.32}$$

From (9.32), $W(m_3) = 8.86 \in [-1, \infty)$, which is the range of the principal branch of the LWF. Therefore there exists a matrix M_k for which (9.32) is satisfied for $k = 0$. Since M_0 is singular, there are an infinite number of matrices Θ_0 that satisfy $M_0 = A_d h \Theta_0$. One of such matrices is

$$\Theta_0 = \begin{bmatrix} 1 & 1 & 1 \\ 1 & 1 & 1 \\ -143160 & -189340 & -31040 \end{bmatrix}.$$

If this value of Θ_0 is passed as an initial guess for solving M_k for $k = 0$, then the corresponding S_k matrix and its eigenvalues are obtained after few iterations, and the computation time is minimal, less than 2–3 s.

Again, let another set of characteristic roots be $\lambda_1 = -0.12$, $\lambda_{2,3} = -2.26 \pm 91.88i$. Correspondingly,

$$S_k = \begin{bmatrix} -1 & -1 & 0 \\ -2 & -5 & -1 \\ 7426.4 & 8453.6 & 1.4 \end{bmatrix}, \quad W_k(M_k) = \begin{bmatrix} 0 & 0 & 0 \\ 0 & 0 & 0 \\ 14857 & 16885 & -1.2 \end{bmatrix}.$$

In this case, $W(m_3) = -1.2 \in (-\infty, -1]$, which falls into the range of the $k = -1$ branch. Using this branch, Θ_{-1} is given as

$$\Theta_{-1} = \begin{bmatrix} 1 & 1 & 1 \\ 1 & 1 & 1 \\ -2058.1 & -2339.4 & 2.2 \end{bmatrix}.$$

The selection of initial conditions close to this matrix ensures the convergence to the solution. Fig. 9.5 shows some of the characteristic roots of the system under consideration, computed using $k = 0$ and $k = -1$ branches.

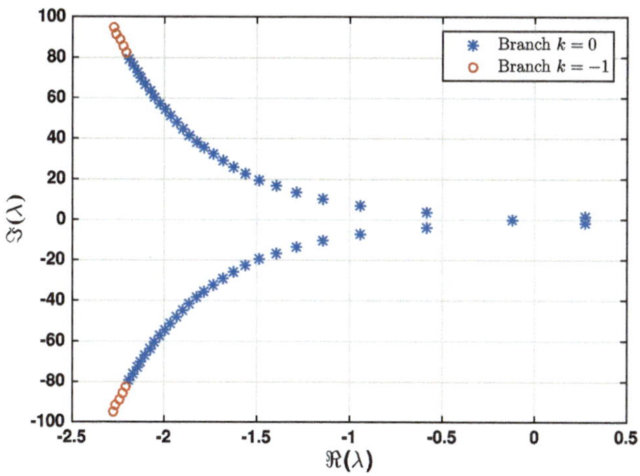

Figure 9.5 Eigenspectrum of the system in Example 9.3.

9.3.4 Discussion

The aim of this section is to show that there is no one-to-one correspondence between the branches of the LWF and eigenvalues of the system. As it is evident from Fig. 9.5, the set of eigenvalues marked with blue color are associated with the principal branch, and the set of eigenvalues marked with red color are associated with the lower branch of the LWF. Thus the illustration in this example shows that there is no one-to-one correspondence between the branches of the LWF and the eigenvalues of the system; moreover, the principal branch can be used to compute several roots rather than just the rightmost roots. The system considered in Example 9.3 has only one real eigenvalue $\lambda = -0.12$, and it has to be used in every combination with either $k = 0$ or $k = -1$ to form a real S_k. It is marked with asterisk ($*$) in Fig. 9.5; however, it can be computed by using both branches $k = 0$ and $k = -1$.

Remark 9.3. There are different methods to study time-delay systems, namely, the finite spectrum analysis, matrix measure method, Lyapunov–Krasovskii (L-K) functional approach, Razumikhin approach, and so forth. Due to the transcendental behavior of TDSs, each method has its own limitations. For example, the L-K functional approach is useful for stability analysis of both linear and nonlinear time-varying delay systems. However, choosing an appropriate L-K functional is not straightforward. Moreover, it is not trivial to use an L-K functional for closed-loop pole placement.

Duan et al. [28] showed that the LWF-based method is less conservative compared to the other methods. The approach is practical as it gives an analytical solution of a scalar TDS in terms of system parameters, and the principle branch alone is sufficient to analyze the stability as it gives the rightmost root. However, for higher-order systems, the approach is somewhat complicated.

The above results can be exploited to design a stabilizing controller by assigning a subset of desired eigenvalues of the closed-loop system as given in the next section.

9.4. Stabilization of time-delay systems

One of the major issues while synthesizing a controller for time-delay systems is the assignment of eigenvalues of the closed-loop system. The difficulty arises because of an infinite number of roots and a finite number of control parameters. The assignment of a subset of desired roots by conventional pole placement techniques for ODEs [21] may cause other uncontrolled eigenvalues to move to the right half of the complex plane [41]. However, the approach presented by Yi et al. [22] for control design using the LWF, based on Lemma 9.1, provides efficient control laws without loss of stability and is based on the hit and trial method. Moreover, getting a solution for the auxiliary matrix Θ is also challenging as Θ is not unique. In this chapter, we discuss the pertinent issues mentioned above and presents a reverse engineering algorithm that bypasses these issues.

Consider a TDS of the form

$$\dot{x}(t) = Ax(t) + Bu(t-h) \tag{9.33}$$

with feedback law

$$u(t) = Kx(t). \tag{9.34}$$

The closed-loop system is given as

$$\dot{x}(t) = Ax(t) + BKx(t-h). \tag{9.35}$$

The solution of the closed-loop system (9.35) can be expressed by (9.20) with $M_k = hBK\Theta_k$ and is given as

$$S_k = \frac{1}{h} W_k(hBK\Theta_k) + A.$$

The modified approach that assigns a subset of the closed-loop eigenvalues at desired locations in the complex plane is described in Algorithm 9.3.

Algorithm 9.3

1: Select a subset of desired characteristic roots $\lambda_{i,\text{des}}$ for $i = 1, ..., n$.
2: Formulate a matrix S_k using (9.27) such that it has selected eigenvalues.
3: Obtain $W_k(M_k) = h(S_k - A)$.
4: Using the definition of LWF, compute M_k from step 3.
5: Substitute $\Theta_k = e^{-S_k h} e^{h(S_k - A)}$ into $M_k = hBK\Theta_k$ and then compare both sides of it to obtain K.

9.4.1 Discussion and comparison

The approach in Algorithm 9.3 fixes Θ_k for a subset of eigenvalues and hence resolves the trouble of selecting an appropriate auxiliary matrix. Moreover, here the controller gain is the only parameter to be found instead of K and Θ_k, as it was in Yi et al. approach [22,33]. Thus it is a straightforward method, instead of the hit and trial approach depending on a set of the initial conditions.

Using Algorithm 9.3, we can place a subset of closed-loop characteristic roots at desired locations. However, it does not guarantee the overall stability of the system. Mori et al. [40] reported that the roots corresponding to the principal branch of LWF are the rightmost root, which determines stability. However, the findings in Section 9.3 show that for TDS as in (9.35) with A_d (equivalent to BK) being rank deficient, several roots are associated with the principal branch and practically it is hard enough to identify which one is the rightmost. Hence, in the present study, the stability is ensured using the Nyquist stability criterion.

9.4.2 Stability test using Nyquist criterion

The characteristic function of system (9.35) is a quasipolynomial of the form

$$p(s) = \det(sI - A - BKe^{-sh}). \tag{9.36}$$

Further, a TDS of the form (9.35) is stable if all the roots of its characteristic function (9.36) have negative real parts. A graphical method for stability check using Nyquist plots is presented in [47,48] and is recalled here.

A quasipolynomial (9.36) is Hurwitz if and only if the Nyquist plot of

$$\frac{p(j\omega)}{(1+j\omega)^n} \tag{9.37}$$

does not encircle the origin of the complex plane, where n is the order of the system.

Considering this statement, let $s=\sigma_s+j\omega+\eta$. Then $Re(s)\le\eta$ if and only if $\sigma_s\le 0$. Furthermore, $s=s_0$ is the rightmost root if it satisfies the following conditions.

1. The Nyquist plot of

$$\frac{p(j\omega+Re(s_0))}{(j\omega+1)^n} \tag{9.38}$$

passes through the origin.

2. The Nyquist plot of

$$\frac{p(j\omega+Re(s_0)+\epsilon)}{(j\omega+1)^n} \tag{9.39}$$

does not encounter the origin of the complex plane for every small $\epsilon>0$.

The first condition implies that $p(\sigma_s+j\omega+Re(s_0))$ has a root with zero real part, that is, $p(s)$ has a root with real part $Re(s_0)$. The second condition indicates that all the roots of $p(\sigma_s+j\omega+Re(s_0)+\epsilon)$ have negative real parts, that is, $p(s)$ has roots with real parts less than $Re(s_0)+\epsilon$ only, and, consequently, $p(s)$ has roots with $Re(s)\le Re(s_0)$ only, because ϵ is an arbitrary small number.

The designed controller is implemented on a two degree-of-freedom (DOF) mechanical vibration control problem to show the efficacy of the control algorithm.

Example 9.4. Consider the two-DOF mechanical vibration control problem shown in Fig. 9.6.

This mechanical system can be described by the following state equations with an input delay of $h=0.5$ s:

$$\dot{x}(t)=\begin{bmatrix}0 & 0 & 1 & 0\\ 0 & 0 & 0 & 1\\ -\frac{(k_1+k_2)}{m_1} & \frac{k_2}{m_1} & -\frac{(b_1+b_2)}{m_1} & \frac{b_2}{m_1}\\ \frac{k_2}{m_2} & -\frac{(k_2+k_3)}{m_2} & \frac{b_2}{m_2} & -\frac{b_2}{m_2}\end{bmatrix}x(t)+\begin{bmatrix}0 & 0\\ 0 & 0\\ \frac{1}{m_1} & 0\\ 0 & \frac{1}{m_2}\end{bmatrix}u(t-h).$$

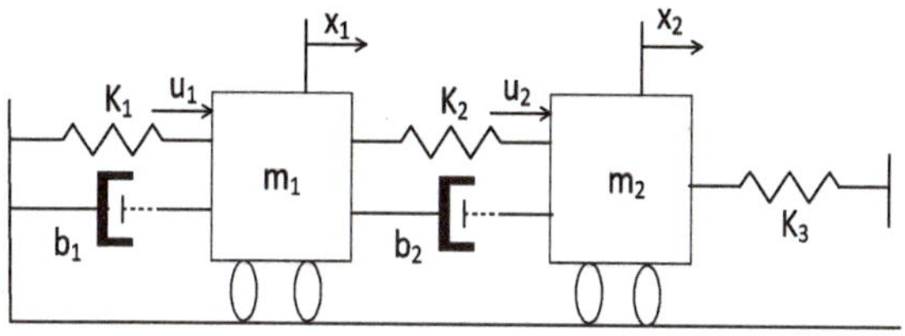

Figure 9.6 Schematic of a two-DOF mass-spring-damper system.

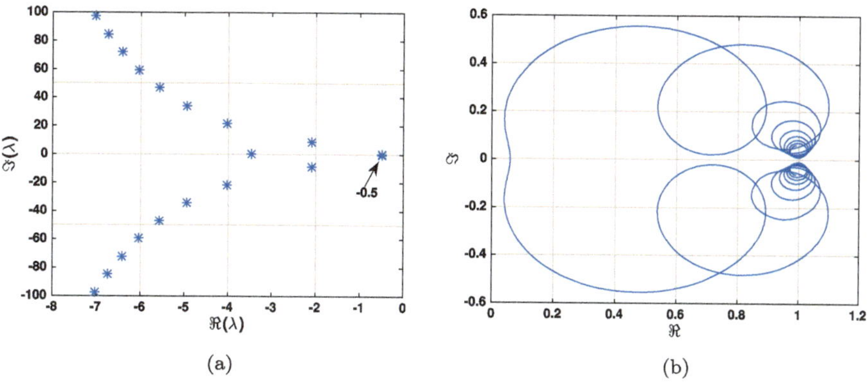

Figure 9.7 Simulation results for the mass-spring-damper system shown in Fig. 9.6. (a) Eigenspectrum. (b) Plot of $\frac{p(j\omega)}{(1+j\omega)^4}$.

We consider the system parameters $m_1 = m_2 = 0.5$, $k_1 = k_2 = k_3 = 1$, $b_1 = 0$, and $b_2 = 0.1$. If a subset of desired eigenvalues are placed at -0.5, then the required gains can be found using Algorithm 9.3 and come out to be

$$K = \begin{bmatrix} 1.45 & -1.52 & 0.53 & -0.99 \\ -0.75 & 2.21 & -0.63 & 0.78 \end{bmatrix}.$$

Fig. 9.7(a) shows that the designed controller places the selected eigenvalues of the closed-loop system exactly at desired locations. To ensure the stability of the closed-loop system, the Nyquist criterion is used, and no encirclement around zero is obtained, as shown in Fig. 9.7(b). Hence it is guaranteed that the closed-loop system is stable.

9.5. Conclusions

The method for the stability analysis and stabilization of TDSs presented by Yi et al. [30,33] and based on the matrix LWF has some limitations, and we need to be very careful in drawing conclusions about the spectrum of multidimensional systems using this approach. In particular, the approach does not produce satisfactory results when $rank(A_d) = 1$. It also assumes that there exists a one-to-one correspondence such that using the principal branch always leads to the rightmost roots of the system. In this chapter, we investigated a particular class of time-delay systems where A_d is rank deficient and found out that:

- No one-to-one correspondence exists in general between the branches of the Lambert W function and the roots of the system.
- The principal branch can be used for computing not only the dominant roots but also some other roots.
- Only the real branches of the Lambert W function, that is, $k = 0$ and $k = -1$, are sufficient to produce all the characteristic roots of this class of time-delay systems.
- The said correspondence is also established for the more general class of TDSs where the system is not explicitly in the CC form and $rank(A_d) = m$.
- The control algorithm discussed in this chapter is less computationally exhaustive than the other existing techniques as it fixes Θ_k for a subset of eigenvalues and hence resolves the trouble of selecting an appropriate auxiliary matrix.
- The controller gain is the only parameter to be found instead of K and Θ_k, as it was in Yi et al. approach [22,33]. Thus it is a straightforward method, instead of the hit and trial approach depending on a set of initial conditions.

Stability of the closed-loop system is ensured with the help of the Nyquist plot. The simulation examples verify the theoretical developments made in this chapter.

References

[1] S.I. Niculescu, E.I. Verriest, L. Dugard, J.M. Dion, Stability and Robust Stability of Time-Delay Systems: A Guided Tour, Lecture Notes in Control and Information Sciences, Springer Berlin Heidelberg, 1998.
[2] J.K. Hale, S.M. Verduyn Lunel, Introduction to Functional Differential Equations, Springer, New York, 1993.
[3] R.E. Bellman, L.C. Kenneth, Differential-Difference Equations, 1963.
[4] G. Stepan, Retarded Dynamical Systems: Stability and Characteristic Functions, Research Notes in Mathematics Series, Longman Sc. & Tech, 1990.
[5] A. Olbrot, Stabilizability, detectability, and spectrum assignment for linear autonomous systems with general time delays, IEEE Transactions on Automatic Control 23 (5) (1978) 887–890.

[6] O.J.M. Smith, Closed control of loop with dead time, Chemical Engineering Progress 53 (1957) 217–219.
[7] J. Loiseau, W. Michiels, S.I. Niculescu, R. Sipahi, Topics in Time Delay Systems: Analysis, Algorithms and Control, Springer-Verlag, Berlin, Heidelberg, 2009.
[8] R. Sipahi, S.I. Niculescu, C.T. Abdallah, W. Michiels, G. Keqin, Stability and stabilization of systems with time delay, IEEE Control Systems 31 (1) (2011) 38–65.
[9] N. Olgac, R. Sipahi, An exact method for the stability analysis of time-delayed linear time-invariant (LTI) systems, IEEE Transactions on Automatic Control 47 (5) (2002) 793–797.
[10] R. Sipahi, N. Olgac, A unique methodology for the stability robustness of multiple time delay systems, Systems & Control Letters 55 (10) (2006) 819–825.
[11] N. Olgac, A.F. Ergenc, R. Sipahi, "Delay scheduling": a new concept for stabilization in multiple delay systems, Journal of Vibration and Control 11 (9) (2005) 1159–1172.
[12] T. Vyhlidal, P. Zitek, Mapping based algorithm for large scale computation of quasi-polynomial zeros, IEEE Transactions on Automatic Control 54 (2009) 171–177.
[13] E.A. Butcher, O.A. Bobrenkov, On the Chebyshev spectral continuous time approximation for constant and periodic delay differential equations, Communications in Nonlinear Science and Numerical Simulation 16 (3) (2011) 1541–1554.
[14] D. Breda, Solution operator approximations for characteristic roots of delay differential equations, Applied Numerical Mathematics 56 (3) (2006) 305–317.
[15] D. Breda, S. Maset, R. Vermiglio, TRACE-DDE: A Tool for Robust Analysis and Characteristic Equations for Delay Differential Equations, Springer Berlin Heidelberg, Berlin, Heidelberg, 2009, pp. 145–155.
[16] Z. Wu, W. Michiels, Reliably computing all characteristic roots of delay differential equations in a given right half plane using a spectral method, Journal of Computational and Applied Mathematics 236 (9) (2012) 2499–2514.
[17] W. Michiels, K. Engelborghs, P. Vansevenant, D. Roose, Continuous pole placement for delay equations, Automatica 38 (5) (2002) 747–761.
[18] S. Mondie, W. Michiels, Finite spectrum assignment of unstable time-delay systems with a safe implementation, IEEE Transactions on Automatic Control 48 (12) (2003) 2207–2212.
[19] E. Jarlebring, K. Meerbergen, W. Michiels, A Krylov method for the delay eigenvalue problem, SIAM Journal on Scientific Computing 32 (6) (2010) 3278–3300.
[20] F.M. Asl, A.G. Ulsoy, Analysis of a system of linear delay differential equations, Journal of Dynamic Systems, Measurement, and Control 125 (2) (2003) 215–223.
[21] C.T. Chen, Linear System Theory and Design, Oxford University Press, Inc., New York, NY, USA, 1995.
[22] S. Yi, P.W. Nelson, A.G. Ulsoy, Eigenvalue assignment via the Lambert W function for control of time-delay systems, Journal of Vibration and Control 16 (78) (2010) 961–982.
[23] S. Yi, P.W. Nelson, A.G. Ulsoy, Controllability and observability of systems of linear delay differential equations via the matrix Lambert W function, IEEE Transactions on Automatic Control 53 (3) (2008) 854–860.
[24] S. Yi, A.G. Ulsoy, P.W. Nelson, Design of observer-based feedback control for time-delay systems with application to automotive powertrain control, Journal of the Franklin Institute 347 (2010) 358–376.
[25] S. Yi, P.W. Nelson, A.G. Ulsoy, Robust control and time domain specifications for systems of delay differential equations via eigenvalue assignment, in: Transactions of the ASME, Journal of Dynamic Systems, Measurement, and Control 132 (3) (2010) 1–7.
[26] S. Yi, P.W. Nelson, A.G. Ulsoy, DC motor control using the Lambert W function approach, in: Proceedings of the 10th IFAC Workshop on Time Delay Systems, vol. 10, 2012, pp. 49–54.

[27] F. Wei, D. Bachrathy, G. Orosz, A.G. Ulsoy, Spectrum design using distributed delay, International Journal of Dynamics and Control 2 (2) (2014) 234–246.
[28] S. Duan, J. Ni, A.G. Ulsoy, Decay function estimation for linear time delay systems via the Lambert W function, Journal of Vibration and Control 18 (10) (2012) 1462–1473.
[29] F.M. Asl, A.G. Ulsoy, Survey on analysis of time delayed systems via the Lambert W function, in: DCDIS A Supplement, Advances in Dynamical Systems 14 (S2) (2007) 296–301.
[30] S. Yi, A.G. Ulsoy, Solution of a system of linear delay differential equations using the matrix Lambert function, in: Proceedings of the American Control Conference, Minneapolis, MN, 2006, pp. 2433–2438.
[31] S. Yi, S. Duan, P.W. Nelson, A.G. Ulsoy, Analysis and control of time delay systems using the Lambert W DDE toolbox, in: Delay Systems, Advances in Delays and Dynamics, Springer International Publishing, 2014, pp. 271–284.
[32] R. Cepeda-Gomez, W. Michiels, Some special cases in the stability analysis of multi-dimensional time-delay systems using the matrix Lambert W function, Automatica 53 (3) (2015) 339–345.
[33] S. Yi, P.W. Nelson, A.G. Ulsoy, Time-Delay Systems: Analysis and Control Using the Lambert W Function, World Scientific, Singapore, 2010.
[34] R.M. Corless, G.H. Gonnet, D.E.G. Hare, D.J. Jeffrey, D.E. Knuth, On the Lambert W function, Advances in Computational Mathematics 5 (4) (1996) 329–359.
[35] R.M. Corless, G.H. Gonnet, D.E.G. Hare, D.J. Jeffrey, D.E. Knuth, Lambert's W function in Maple, The Maple Technical Newsletter 9 (1993) 12–22.
[36] D.J. Jeffrey, D.E.G. Hare, R.M. Corless, Unwinding the branches of the Lambert's W function in Maple, Mathematical Sciences 21 (1996) 1–7.
[37] N. Higham, Functions of Matrices, Society for Industrial and Applied Mathematics, 2008.
[38] J. Elias, D. Tobias, The Lambert W function and the spectrum of some multidimensional time-delay systems, Automatica 43 (12) (2007) 2124–2128.
[39] J.P. Richard, Time-delay systems: an overview of some recent advances and open problems, Automatica 39 (10) (2003) 1667–1694.
[40] H. Shinozaki, T. Mori, Robust stability analysis of linear time-delay systems by Lambert W function: some extreme point results, Automatica 42 (2006) 1791–1799.
[41] W. Michiels, S.I. Niculescu, Stability, Control, and Computation for Time-Delay Systems: An Eigenvalue-Based Approach, Advances in Design and Control, SIAM, 2014.
[42] C. Hwang, Y.C. Cheng, A note on the use of the Lambert W function in the stability analysis of time-delay systems, Automatica 41 (11) (2005) 1979–1985.
[43] A.G. Ulsoy, Time-delayed control of SISO systems for improved stability margins, Journal of Dynamic Systems, Measurement, and Control 137 (4) (2015) 041014.
[44] S. Yi, W. Choi, T. Abu-Lebdeh, Time-delay estimation using the characteristic roots of delay differential equations, American Journal of Applied Sciences 9 (6) (2012) 955–960.
[45] R. Cepeda-Gomez, W. Michiels, Special cases in using the matrix Lambert W function for the stability analysis of high-order linear systems with time delay, in: Proceedings of 12th IFAC Workshop on Time Delay Systems, Ann Arbor, MI, USA, 2015, pp. 7–12.
[46] N. Choudhary, J. Sivaramakrishnan, I.N. Kar, Analysis of multidimensional time delay systems using Lambert W function, Journal of Dynamic Systems, Measurement, and Control 139 (11) (2017) 114506.
[47] M. Fu, A.W. Olbrot, M.P. Polis, Robust stability for time-delay systems: the edge theorem and graphical tests, IEEE Transactions on Automatic Control 34 (8) (1989) 813–820.
[48] Z.H. Wang, H.Y. Hu, Calculation of the rightmost characteristic root of retarded type time delay systems via Lambert W function, Journal of Sound and Vibration 318 (2008) 757–767.

CHAPTER TEN

Control of constrained discrete-time systems with time-varying state delay[☆]

Valter J.S. Leite[a], Eugenio B. Castelan[b], Luis F.P. Silva[a], and Carla de Souza[b]

[a]Department of Mechatronics Engineering, *Campus* Divinópolis — CEFET–MG, Divinópolis, MG, Brazil
[b]Department of Automation and Systems Engineering, PPGEAS — DAS/CTC/UFSC, Florianópolis, SC, Brazil

Contents

☆ This work was partially supported by Brazilian National Council for Scientific and Technological Development (CNPq), Grant/Award numbers: 311208/2019-3 and 306927/2017-9, Coordination for the Improvement of Higher Education Personnel (CAPES), project Print CAPES-UFSC "Automation 4.0", 88887.364794/2019-00, and CEFET–MG.

Control Strategy for Time-Delay Systems
https://doi.org/10.1016/B978-0-12-820599-0.00015-X

10.1. Introduction

In the engineering literature, we may trace problems caused by delay, for instance, in [1] an early investigation of delay effects on toll telephone traffic. The main goal was to determine a relationship between telephone circuit loads and the corresponding delay to traffic. This investigation shows the delay varied during the day and, as expected, that there was a maximum delay from which calls were lost, impacting on the revenue per circuit mile. Also, some early papers like [2] concern the delay effects in the control-loop stability. A standard tool in this context is the celebrated Smith predictor [3], which has found to be an appealing tool, especially for the control of industrial processes. A remarkable characteristic in this case is that the engineer can design controllers almost forgetting the delay present in the process. Of course, both the delay and the process must attain some constraints like time-invariance and open-loop stability characteristics. Otherwise, some modified feedback structure also applies to deal with varying parameters, unstable processes, or the presence of disturbances by the inclusion of low-pass filters [3].

A main aspect concerning time delay is that a constant one introduces a phase-lag proportional to the frequency into the closed-loop control system. As a consequence, even for small delays, system with large bandwidth may have its performance or even its stability affected because of the decreasing on the stability margin. Therefore the delay can render unstable a control design that does not take the delay into account. Moreover, some classical frequency-based controller design techniques use some delay approximation instead of handling it directly, which is the case of the direct synthesis method (DSM) [4]. In such a method, the design of fixed-order controller, such as proportional-integral-derivative (PID) controller, bases on the system model and the desired closed-loop transfer function. Whenever the model has a delay, the DSM requires Taylor's approximation of an exponential term (the Laplace transform of the dead time), which, indeed, corresponds to a delayed state structure. Due to such an approximation, the improvement in transitory characteristics may be quite limited in the DSM approach. However, even for single-input-single-output systems, the analysis and design of control loops involving delays may become more involved in the presence of unknown and time-varying delays and process uncertainties and nonlinearities.

Delayed state systems received lots of attention in the last three decades [5,6] due to their incidence over relevant systems. See, for instance, its applications in robotics and epidemiological processes [5,7]. Also, the presence of delays may induce some undesirable behaviors, such as oscillations, degenerated performance, and instability [6], thus motivating the study of state-delayed systems. However, such a study can be a difficult task in both theory and practice. Additionally, the delay can be time varying, as in the case of the tubular reactor investigated in [8].

Furthermore, the presence of delay leads to more involved conditions in the case of saturating actuators, which appear in the majority of control systems, such a saturation coming from power limitation or due to operational and security reasons. A consequence of the saturating actuators is that even if a linear control law stabilizes a system, the origin of the closed-loop system can become unstable for some initial conditions [9,10]. Additionally, as well recognized in the literature, obtaining an estimate of the set of initial conditions ensuring closed-loop trajectories to converge to the origin can be a challenging task [9].

Besides the local stability issue, the closed-loop performance is a fundamental issue. Notably, the presence of saturating actuators requires a characterization of the admissible set of exogenous signals and the respective set of allowed initial conditions yielding to bounded trajectories. In the approach developed in this chapter, we use the input-to-state stability (ISS) concept [9,11] to handle these issues with the aid of the ℓ_2-gain from the disturbance signal and regulated output.

Discrete-time models are a natural way to handle delay in digital-based controllers [12,13]. See, for instance, the works [11,14–20], where the proposed conditions concern the robust stability and stabilization of state-delayed discrete-time systems. However, none of these papers can deal with the delay variation between two consecutive instants. Such an improvement was first achieved in [21,22] in the context of robust stabilization and in [23] for LPV state-delayed systems. It is worth noting that handling the variation rate of the delay may impact the stability and performance of the closed-loop system, as well as the size of the concerned estimate region of attraction, because this extra information may lead to less conservative design conditions.

Although recognized as a relevant issue, the investigations concerning the stability and stabilization of discrete-time state-delayed systems were put aside during a period, mainly by the argument that we can always rewrite them as an augmented delay-free system [24]. However, the presence of uncertain delays brought some criticisms to such an idea, because different (and growing) orders of delay-free systems are necessary to thoroughly investigate the stability (or stabilization) of such a system. Then, inspired by the fruitful works in the continuous-time counterpart, the discrete-time state-delayed systems become to receive more attention by the end of the 1990s [25]. Although the Lyapunov–Krasovskii-based approach is the most frequent choice to investigate delayed-state systems [11,26], we can find others based, for instance, on the Lyapunov–Razumikhin [27,28], set invariance properties [29], or matrix-norm approaches [30]. More recently, [19] revisited the idea of the augmented system discussed in [24] to handle both the stability analysis and the stabilization of time-varying discrete-time state-delayed systems. The authors in [19] show that their approach encompasses several specialized Lyapunov–Krasovskii functions because the augmented space leads to a more general Lyapunov–Krasovskii equivalent function. Also, the augmented model corresponds to a delay-free system switched by the delay. Based on such an approach, [20] presents a predictive control scheme dealing with input time-varying delay. In the context of networked control systems the work [31] also employs the augmented approach to handle delayed-control signals.

Although we find several works investigating the effects of saturating actuators in the literature, such as robust stabilization [32], dynamic output feedback stabilization with amplitude and rate saturation [33], and nested saturations [34], only a few studies concern state-delayed systems subject to saturating actuators. Most of them concern with the continuous-time systems [35–39]. In [40], both continuous- and discrete-time cases are investigated. Delayed discrete-time systems are investigated in [41,42], the fuzzy Takagi–Sugeno approach is taken in [43] by using the Lyapunov–Krasovskii approach to provide delay-dependent conditions for the synthesis of static state feedback controllers, and in [44] the authors investigate the $\mathcal{H}_\infty$ control. In all these cases, it is worth noting that the variation of the delay has not been considered. Other investigations include nonlinear discrete-time state-delayed systems as, for instance, in [45] and [46], where the system is rewritten in terms of a Takagi–Sugeno fuzzy representation.

A main drawback in the papers mentioned is that the geometrical choice to characterize the region of attraction is based on the norm of the sequence of initial conditions leading to conservative estimates. Such a limitation can be understood with a simple example. Consider a delayed-state system of order 2 and a maximum delay of $\bar{\tau}$ samples. In the approach proposed in the early works the initial conditions are characterized by a sequence of vectors $x_k = \begin{bmatrix} x_1 & x_2 \end{bmatrix}^T$ taken from a disc region with radius r of the state space at the instants $k = 0$, $k = -1, \ldots, k = -(\bar{\tau} - 1)$, and $k = -\bar{\tau}$. Fig. 10.1

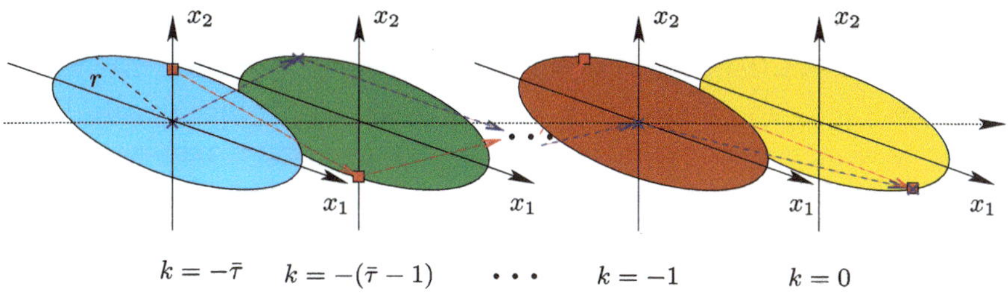

Figure 10.1 Initial conditions of a state-delayed system of second order, where $\|x_k\| \leq r$ at each instant $-\bar{\tau} \leq k \leq 0$.

illustrates such a region over current and past instants $\bar{\tau} \leq k \leq 0$. Note that in Fig. 10.1 two possible sequences are highlighted, both taken in the allowed set of initial conditions. The first one is marked with red (□), and the second one is marked with blue (×). Therefore from the instant $k = -\bar{\tau}$ to $k = -(\bar{\tau} - 1)$, the square (□) sequence varies from $(0, 1)$ to $(0, -1)$, whereas the cross (×) sequence varies from $(0, 0)$ to $(-0.5, 0.8)$. Thus the blue (×) sequence has a variation smaller than that in the red (□) sequence. Such a behavior can be noted in all transitions of both sequences. Thus, although both sequences of initial conditions satisfy the allowed set, the blue (×) sequence leads to a more smooth time variation.

To increase the value of r in Fig. 10.1, the authors in [40,45,47,48] take into account the state variation norm between two consecutive instants. With such an idea, smoother initial conditions, as the blue (×) sequence, allows us to compute larger discs, that is, to calculate greater values of r, thus yielding to less conservative estimates. However, the norm-based paradigm still brings conservatism to the estimates.

An approach using Lyapunov–Razumikhin functions can be found in [28], where the authors investigate polytopic delay difference inclusions and constraints on the control signal. However, such an approach does not take into account the delay variation. Indeed, the first works considering such a delay variation on a discrete-time state-delayed system are [21,45].

In the approach developed in this chapter, we propose to characterize the initial conditions in an augmented space, allowing a tailored representation of initial conditions of state-delayed systems. Moreover, we present convex conditions in terms of linear matrix inequalities (LMIs) to design state-feedback gains for time-varying discrete-time state-delayed systems subject to both saturating actuators and energy (ℓ_2) bounded disturbances. By using an approach based on ISS tools we provide delay-dependent conditions to design state feedback control laws. This approach ensures the local stability of a time-varying closed-loop system for an estimate set of initial conditions and admissible energy bounded disturbance signals. With the aid of some convex optimization procedures, we can maximize the initial set of admissible conditions, maximize the energy bound of admissible perturbations, or minimize the ℓ_2 gain between the disturbance signal and measurable output. Furthermore, we present a set of extensions matching with linear parameter-varying systems (LPV), quasi-LPV systems, switching systems, and robust control. In all cases, we suppose an unknown time-varying delay, which leads to state-feedback control law using only the current state. Some examples illustrate the approach and, also, how the estimates of the region of attraction relate to the considered kind of system (LPV, quasi-LPV, switching, or robust).

Notation: The symbol $\star$ represents the transpose blocks in relation to the diagonal of real square and symmetric matrices. The set of matrices with real entries and dimensions $m \times n$ is $\mathbb{R}^{m\times n}$. The matrices $\mathbf{I}$ and $\mathbf{0}$ denote, respectively, the identity and the null matrices of appropriate dimensions. The sets of real numbers, nonnegative real numbers, integer numbers, and integer numbers excluding the zero are denoted by $\mathbb{R}$, $\mathbb{R}^+$, $\mathbb{N}$, and $\mathbb{N}^*$, respectively. The ℓth line of a vector or matrix G is denoted by $G_{(\ell)}$. The set of integer numbers belonging to the interval from $a \in \mathbb{N}$ to $b \in \mathbb{N}$, $b \geq a$, is denoted by $\mathcal{I}[a, b]$.

10.2. State-delayed systems and stabilization problems

We consider discrete-time systems under time-varying delay whose dynamics is captured by the state-space equation

$$x_{k+1} = A(\alpha_k)x_k + A_d(\alpha_k)x_{k-\tau_k} + B(\alpha_k)\mathtt{sat}(u_k) + B_w(\alpha_k)w_k, \tag{10.1}$$

where the current state vector is denoted by $x_k \in \mathbb{R}^n$, $x_{k-\tau_k}$ is the state delayed by τ_k samples, and $w_k \in \mathbb{R}^{n_w}$ is the exogenous signal belonging to

$$\mathcal{W}(\delta) = \left\{w_k \in \mathbb{R}^{n_w} \;:\; \|w_k\|_2^2 \leq \delta^{-1}\right\}, \tag{10.2}$$

where $\delta^{-1} \geq 0$ gives the energy bound on $\|w_k\|_2 = \lim_{t\to\infty} \sqrt{\sum_{i=0}^{t} w_i^T w_i}$. The computed linear control signal is denoted by $u_k \in \mathbb{R}^{n_u}$, which, because of the available power from the actuators or because of operational security bounds, we consider amplitude saturated and denote by $\mathtt{sat}(u_k)$:

$$\mathtt{sat}(u_k) = \mathtt{sign}(u_k)\min\{|u_k|, \bar{u}\} \tag{10.3}$$

with $\bar{u} > 0$ denoting the maximum absolute deviation allowed to the control signal. Note that the saturation function is assumed symmetric around zero, that is, $-\bar{u} \leq \mathtt{sat}(u_k) \leq \bar{u}$, and decentralized, that is, each input control signal $u_{k(\ell)}$, $\ell \in \mathcal{I}[1,\ n_u]$, is handled independently. Therefore we must read Eq. (10.3) line by line. Furthermore, the time-varying delay τ_k belongs to the interval $\mathcal{I}[0,\ \bar{\tau}]$, where $\bar{\tau}$ corresponds to the maximum delay value. Because in physical systems the delay may vary under bounded rates, we assume that

$$|\tau_{k+1} - \tau_k| \leq \Delta_\tau. \tag{10.4}$$

Therefore Δ_τ is the absolute maximum amount of variation of the delay between two consecutive instants. As a matter of fact, it must satisfy $0 \leq \Delta_\tau \leq \bar{\tau}$. In particular, the delay is time-invariant if $\Delta_\tau = 0$.

We also suppose a performance or regulated output signal $z_k \in \mathbb{R}^{n_y}$ for system (10.1), which linearly depends of the state vector given by

$$z_k = C(\alpha_k)x_k. \tag{10.5}$$

The dynamic matrices $A(\alpha_k) \in \mathbb{R}^{n\times n}$ and $A_d(\alpha_k) \in \mathbb{R}^{n\times n}$, the input matrices $B(\alpha_k) \in \mathbb{R}^{n\times n_u}$ and $B_w(\alpha_k) \in \mathbb{R}^{n\times n_w}$, and the output matrix $C(\alpha_k) \in \mathbb{R}^{n_y\times n}$ depend on the parameter vector $\alpha_k \in \mathbb{R}^N$ that captures physical or operational changes in the system. We assume that they are represented by a polytopic set such that

$$\begin{aligned}&\begin{bmatrix} A(\alpha_k) & A_d(\alpha_k) & B(\alpha_k) & B_w(\alpha_k) & C^T(\alpha_k)\end{bmatrix}\\ &\quad = \sum_{i=1}^{N} \alpha_{k(i)} \begin{bmatrix} A_i & A_{di} & B_i & B_{wi} & C_i^T \end{bmatrix}\end{aligned} \tag{10.6}$$

with $\alpha_{k(i)}$ meaning the ith row of $\alpha_k \in \Lambda$:

$$\Lambda = \left\{ \alpha_k \in \mathbb{R}^N \ : \ \alpha_{k(i)} \geq 0 \text{ and } \sum_{i=1}^{N} \alpha_{k(i)} = 1 \right\}. \tag{10.7}$$

The polytopic representation (10.6)–(10.7) is quite general, allowing us to incorporate several dynamical phenomena such as time-varying parameter dependencies, uncertainties, and nonlinear dynamics. Some cases are of particular interest:

1. If the parameter α_k is measured, then we have the dynamics of the system described by an LPV model. In such a case, a parameter-dependent control law
$$u_k = K(\alpha_k)x_k \tag{10.8}$$
can be used. It is worth noting that the parameter dependency of controller gain $K(\alpha_k)$ may be nonlinear, even though the system has an LPV model.
2. In case we obtain α_k from some endogenous signals, such as a (sub)set of the states or inputs, we say that (10.1) is a quasiparameter-dependent system model. Therefore the control law (10.8) can still be used to locally stabilize the quasi-LPV system (10.1). The approaches based on Takagi–Sugeno modeling are examples of such a case.
3. Another case of interest is where the parameter vector α_k does not vary continuously but takes values where only the ith element of α_k is equal to one and all others are null. In this case, system (10.1) operates on modes corresponding to the vertices of the polytope. If the switching rule between modes is an a priori unknown induced signal, but its current value can be available, then the control law (10.8) can control the switching system (10.1). Furthermore, (10.8) can be written as
$$u_k = K_i x_k, \tag{10.9}$$
where i is related to the active mode. Note that in some cases the switching function is a control input to be designed, which usually leads to dwell-time-based approaches. Such a case yields a switched closed-loop system, which we do not consider in this chapter.
4. If α_k is unknown, then a robust controller approach can be adopted to stabilize (10.1) with a robust control law given by
$$u_k = Kx_k. \tag{10.10}$$
Such a control law must locally stabilize the parameter-dependent closed-loop system under the saturating control signal in (10.1). Note that the robust control law (10.10) may be designed by assuming either a time-varying or time-invariant α_k.

In all cases, we assume that the actual value of the delay d_k is not known, but rather its bounds $\underline{\tau}$ and $\bar{\tau}$ and its variation rate Δ_τ.

A fundamental issue concerning system (10.1) is its initial condition φ_0. Because we deal with a delayed-state system, a sequence of state-vectors displaced on time characterizes the initial condition. Such a characterization is required to ensure the uniqueness of the solutions of (10.1) for $k > 0$. So, we define the sequence φ_0 as follows:

$$\varphi_0 = \left\{x_{-\bar{\tau}},\ x_{-(\bar{\tau}-1)},\ \ldots,\ x_{-1},\ x_0\right\}. \tag{10.11}$$

In what follows, we present a first-order example to illustrate how saturation may affect the stability of system (10.1). We keep the system simple, so we can concentrate on the saturating actuator consequences.

Example 10.1. Consider a scalar and unperturbed state-delayed system as in (10.1), with $N = 2$, where $A_1 = 1.8$, $A_2 = 2.2$, $A_{d1} = -0.09$, $A_{d2} = -0.11$, $B_1 = 0.9$, $B_2 = 1.1$, and $B_{w1} = B_{w2} = 0$, meaning that there is no exogenous input ($w_k = 0$, $k \geq 0$). The delay satisfies $0 \leq \tau_k \leq 1 = \bar{\tau}$.

Firstly, we assume that there is no constraint on control signal, and then we calculate a (linear) state feedback control law as in (10.8) to stabilize it. By using the approach in [23, Th. 1] and removing the rows and columns of LMI related to the saturating actuator, we get the following control gains:

$$K_1 = \begin{bmatrix}-1.9488\end{bmatrix} \text{ and } K_2 = \begin{bmatrix}-1.9490\end{bmatrix}, \tag{10.12}$$

which yield to the LPV the control gain $K(\alpha_k) = \alpha_{k1}K_1 + \alpha_{k2}K_2$, with $\alpha_k \in \Lambda$. We simulate the closed-loop system under three initial conditions (i) $x_0 = -x_{-1} = 0.4$, (ii) $x_0 = -x_{-1} = 2$, and (iii) $x_0 = -x_{-1} = 5$. We used $\alpha_{1k} = \{0,\ 0,\ 0.4845,\ 0.2888,\ 0.8011,\ 0.1927,\ 0.7382\}$ and $\tau_k = \{1,\ 1,\ 1,\ 0,\ 1,\ 0,\ 1,\ 1\}$. The closed-loop trajectories of x_k (top) and the control signal u_k (bottom) are shown in Fig. 10.2. In both plots, we mark the obtained signals from the initial conditions (i), (ii), and (iii) with $*$, $\circ$, and $\square$, respectively. It is clear that all three trajectories converge to the origin in the absence of saturation limits.

Next, we assume the saturating actuator with $\bar{u} = 0.5$ with the same initial conditions mentioned before. The resulting closed-loop state trajectories and the (saturating) control signals are presented repectively in top and bottom plots of Fig. 10.3 with the same conventions already used. Note that only the trajectory emanating from the initial condition (i) converges

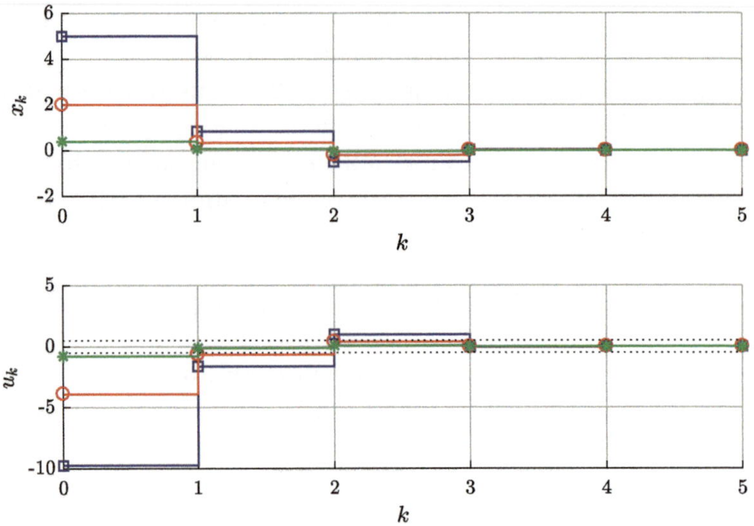

Figure 10.2 State and unbounded control signal of system (10.1) for initial conditions $x_0 = -x_{-1} = 0.4$ (∗), 2 (○), and 5 (□).

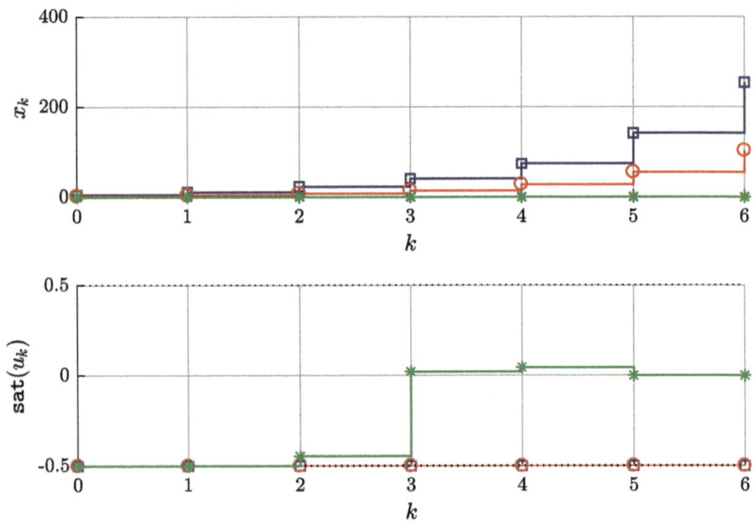

Figure 10.3 State (top) and control signal (bottom) of system (10.1) for initial conditions $x_0 = -x_{-1} = 0.4$ ((i), ∗), 2 ((ii), ○), and 5 ((iii), □).

asymptotically to the origin. Also, note that the control signals concerning the initial conditions (ii) and (iii) remain in the lower bound of the saturation value.

As illustrated by Example 10.1, the saturation may lead to an unstable trajectory of a closed-loop system for some initial conditions. Note that once the open-loop is unstable, whenever the actuator reaches its limit, the energy necessary to bring some trajectories back to the origin may be insufficient, allowing divergent behavior. However, such an unstable behavior may also arise for open-loop stable systems.

Thus let us initially assume that the disturbance signal is null, that is, $w_k = 0$ for $k \geq 0$ (meaning that $\delta \to \infty$). In such a case the convergence of the closed-loop trajectories depends only on the initial state of the system [9]. As a consequence, we need to consider that there exists a region of attraction associated with the saturating system, which consists of all initial conditions such that the corresponding trajectories converge to the origin. We denote such a region by $\mathcal{R}_{\mathcal{A}}$. Its geometrical characterization is not an easy task even for nondelayed systems. Among the reasons, we can mention the nonconvexity of this region, which can also be open and unlimited. Additionally, in the case of delayed systems, such a region is even more complex because of the required sequence of states φ_0 given by (10.11). Instead of $\mathcal{R}_{\mathcal{A}}$, we compute its estimate, given by $\mathcal{R}_{\mathcal{E}} \subseteq \mathcal{R}_{\mathcal{A}}$, such that all trajectories starting in $\mathcal{R}_{\mathcal{E}}$, with $w_k = 0$ for $k \geq 0$, are ensured to converge to the origin.

Moreover, we need to take into account the energy provided by the disturbance or exogenous signal w_k to ensure that all the trajectories remain inside $\mathcal{R}_{\mathcal{A}}$. So, we denote by $\mathcal{R}_0 \subseteq \mathcal{R}_{\mathcal{A}}$ the set of all initial states yielding a convergent closed-loop trajectory that does not leave $\mathcal{R}_{\mathcal{A}}$ for given $\delta > 0$ and $w_k \in \mathcal{W}(\delta)$. Similarly to $\mathcal{R}_{\mathcal{A}}$, the set $\mathcal{R}_0$ may be hard to characterize. Therefore we look for an estimate $\mathcal{R}_{\mathcal{E}0} \subseteq \mathcal{R}_0$ as large as possible. Note that, as $\delta \to \infty$, the estimate $\mathcal{R}_{\mathcal{E}0}$ converges to the estimate $\mathcal{R}_{\mathcal{E}}$.

The following definition handles the local stability of system (10.1) under the saturating control law u_k with the aid of the input-to-state stability.

Definition 10.1. The saturating closed-loop system (10.1)–(10.7) with the control law u_k is said to be input-to-state stable (ISS) if for all $w_k \in \mathcal{W}(\delta)$, any sequence φ_0 taken from $\mathcal{R}_0$ and all $\alpha_k \in \Lambda$, the corresponding state trajectories are confined in $\mathcal{R}_{\mathcal{A}}$, and if the disturbance is vanishing, then $\lim_{k\to\infty} \varphi_k = \underbrace{\{\mathbf{0}, \ldots, \mathbf{0}\}}_{(\bar{\tau}+1)\text{ times}}$.

Thus the main problem investigated in this paper concerning the LPV system (10.1) is as follows.

Problem 10.1. *Determine a parameter-dependent control gain $K(\alpha_k)$ that for all $w_k \in \mathcal{W}(\delta)$, $\alpha_k \in \Lambda$, and $\varphi_0 \in \mathcal{R}_{\mathcal{E}0}$, ensures the input-to-state stability of the closed-loop system* (10.1)–(10.8) *and, in addition, ensures a certain upper limit of the ℓ_2 gain, denoted by γ, between the perturbation w_k and the regulated output z_k such that*

$$\|z_k\|_2^2 \leq \gamma(\|w_k\|_2^2 + \mathrm{c}_\varphi), \tag{10.13}$$

where $\mathrm{c}_\varphi \geq 0$ is the bias term due to the initial conditions given by φ_0.

It is worth noting that Problem 10.1 matches the case of polytopic representation discussed in item 1. The other cases 2–4 can be equally treated by adequate changes in this problem formulation. For instance, in case 4, we should consider the control law (10.10) instead of (10.8) and the *robust* input-to-state stability, meaning that the conditions in Definition 10.1 must hold for all $\alpha_k \in \Lambda$ and the same (constant) state feedback control gain K.

10.3. Mathematical preliminaries

In this section, we introduce an augmented model representation used to handle system (10.1) and some auxiliary results used to establish our main design conditions.

10.3.1 Augmented model description

Inspired by the approach exploited in [19,21,22], we rewrite system (10.1) in an augmented space by using a stacked state-vector

$$\bar{x}_k = \begin{bmatrix} x_k \\ x_{k-1} \\ \vdots \\ x_{k-\bar{\tau}} \end{bmatrix} \in \mathbb{R}^{(\bar{\tau}+1)n}, \tag{10.14}$$

which is related to the sequence φ_k at time k. We also define the following vector and matrices, which help to represent the augmented system with compact notation:

$$\nu_i = \begin{bmatrix} \mathbf{0}_{1\times i-1} & 1 & \mathbf{0}_{1\times \bar{\tau}+1-i} \end{bmatrix}, \quad \theta_{ij} = \nu_i^T \nu_j, \tag{10.15}$$

where both i and j can take values from 1 to $\bar{\tau}$. With these definitions, we can write the dynamic equation (10.1) for $\bar{x}_{k+1}$ given in (10.14) and the regulated output (10.5) as follows:

$$\bar{x}_{k+1} = \mathbb{A}(\alpha_k, \tau_k)\bar{x}_k + \mathbb{B}(\alpha_k)\mathtt{sat}(\mathbb{K}(\alpha_k)\bar{x}_k) + \mathbb{B}_w(\alpha_k)w_k, \tag{10.16}$$

$$y_k = v_1 \otimes C(\alpha_k)\bar{x}_k, \tag{10.17}$$

where

$$\mathbb{A}(\alpha_k, \tau_k) = \theta_{11} \otimes A(\alpha_k) + \left(\sum_{j=1}^{\bar{\tau}-1} \theta_{(j+1)j}\right) \otimes \mathbf{I} + \theta_{1(\tau_k+1)} \otimes A_d(\alpha_k), \tag{10.18}$$

$$\mathbb{B}(\alpha_k) = v_1^T \otimes B(\alpha_k), \tag{10.19}$$

$$\mathbb{B}_w(\alpha_k) = v_1^T \otimes B_w(\alpha_k), \tag{10.20}$$

$$\mathbb{K}(\alpha_k) = v_1 \otimes K(\alpha_k). \tag{10.21}$$

To handle the saturation function $\mathtt{sat}(\cdot)$, we rewrite the closed-loop function (10.16) with the aid of the (symmetric) dead-zone nonlinearity:

$$\Psi(u_k) = u_k - \mathtt{sat}(u_k). \tag{10.22}$$

By summing and subtracting $\mathbb{B}(\alpha_k)\mathbb{K}(\alpha_k)\bar{x}_k = \left(v_1^T \otimes B(\alpha_k)\right)\mathbb{K}(\alpha_k)\bar{x}_k$ in the right-hand side of (10.16) and using the dead-zone function (10.22), we get

$$\bar{x}_{k+1} = \mathcal{A}(\alpha_k, \tau_k)\bar{x}_k - \mathbb{B}(\alpha_k)\Psi(\mathbb{K}(\alpha_k)\bar{x}_k) + \mathbb{B}_w(\alpha_k)w_k, \tag{10.23}$$

where the closed-loop matrix is given by

$$\begin{aligned}\mathcal{A}(\alpha_k, \tau_k) = \mathbb{A}(\alpha_k, \tau_k) + \mathbb{B}(\alpha_k)\mathbb{K}(\alpha_k) &= \theta_{11} \otimes \Big(A(\alpha_k) + B(\alpha_k)K(\alpha_k)\Big) \\ &+ \left(\sum_{j=1}^{\bar{\tau}-1} \theta_{(j+1)j}\right) \otimes \mathbf{I} + \theta_{1(\tau_k+1)} \otimes A_d(\alpha_k).\end{aligned} \tag{10.24}$$

It is possible to verify from (10.24) that the closed-loop equation (10.23) has a structure based on three main parts: one depending on α_k, the second one constant (terms on $\mathbf{I}$), and the third one depending on α_k and τ_k. Due to the varying but bounded nature of τ_k, the closed-loop system (10.23) is LPV for the vector parameter α_k, and it switches due to τ_k. Therefore we can view it as a switching system where the modes activated by τ_k are LPV. Moreover, as discussed in item 3 of Section 10.2, if α_k is a known switching function, then the closed-loop function (10.23) can be viewed as a switching system with partial knowledge of a switching function $\sigma_k = \begin{bmatrix}\alpha_k^T & \tau_k\end{bmatrix}^T$.

The augmented matrices $\hat{A}$, $\mathbb{B}$, $\mathbb{B}_w$, and $\mathbb{C}$ in (10.18)–(10.20) can be easily constructed with the aid of the following code:

```
function [A_aug,B_aug,Bw_aug,Cz_aug] ...
          = augmented_matrices(A,Ad,B,Bw,C,maxtau,N,n,m,q,p)
% input: matrices A, Ad, B, Bw, and C, with cell structure
%        N -> number of vertex
%        maxtau -> max value of tau(k)
% output: augmented matrices
Aaux   = cell(N,1);
A_aug  = cell(N,maxtau);
B_aug  = cell(N,1);
Bw_aug = cell(N,1);
Cz_aug = cell(N,1);

for vertice = 1:N
    B_aug{vertice}  = [B{vertice}' zeros(m,n*maxtau)]';
    Bw_aug{vertice} = [Bw{vertice}' zeros(q,n*maxtau)]';
    Cz_aug{vertice} = [Cz{vertice} zeros(p,n*maxtau)];
    for tau = 1:maxtau+1
        Aaux{vertice} = [A{vertice} zeros(n,n*maxtau);...
                  eye(n*maxtau,n*maxtau) zeros(n*maxtau,n)];
        for i = 1:maxtau+1
            if i==tau
                for row = 1:n
                    for col = 1:n
                        Aaux{vertice}(row,(col+(i-1)*n))...
                = Aaux{vertice}(row,(col+(i-1)*n))...
                              +Ad{vertice}(row,col);
                    end
                end
            end
        end
        A_aug{vertice,tau} = Aaux{vertice};
    end
end
end
```

10.3.2 Auxiliary results

To handle the saturation effect in the closed-loop system (10.23), we follow [9] and define the polyhedral set depending on an auxiliary signal $\omega_k \in \mathbb{R}^{n_u}$ and on the amplitude bound on the control signal $\bar{u}$ given by

$$\mathbb{S}(u-\omega,\bar{u}) = \{u \in \mathbb{R}^{n_u},\ \omega \in \mathbb{R}^{n_u}\ :\ |u-\omega| \le \bar{u}\}. \tag{10.25}$$

As we can see, the set $\mathbb{S}$ does not directly consider the bounds $\pm\bar{u}$ over u; instead, the difference between u and ω is bounded. As a consequence, the auxiliary signal ω becomes an extra degree of freedom for the design of u, allowing us to enlarge the control signal u beyond the bounds $\pm\bar{u}$. Therefore further values of $\bar{x}_k$ can be allowed to compute the state feedback control law u belonging to $\mathbb{S}$, reducing the conservatism of the estimates $\mathcal{R}_{\mathcal{E}}$ and $\mathcal{R}_{\mathcal{E}0}$. Furthermore, an augmented state $\bar{x}_k$ used to compute $u_k = \mathbb{K}(\alpha_k)\bar{x}_k$ yields an admissible control signal belonging to the set $\mathbb{S}$ if such a control signal satisfies $|u-\omega| \leq \bar{u}$ for each actuator of the system and all possible values of $\alpha_k \in \Lambda$ (in case of LPV systems) or for $\alpha_k = v_i^T$, $i = 1, \ldots, N$, in case of switching systems. We can guess that the latter case leads to more relaxed formulations if compared with the first one because no intermediate value of α_k is concerned with.

Lemma 10.1 (Generalized sector condition [9]). *If u and ω are signals belonging to $\mathbb{S}$, then the function $\Psi(u)$ defined in (10.22) satisfies the inequality*

$$\Psi(u)^T\mathcal{T}(\alpha_k)[\Psi(u)-\omega] \leq 0 \tag{10.26}$$

for any positive defined diagonal matrix $\mathcal{T}(\alpha_k) \in \mathbb{R}^{n_u \times n_u}$.

Thus from the saturating closed-loop system (10.23) we can investigate its input-to-state stability by using the Lyapunov theory with the aid of the above generalized sector condition and assuming the following Lyapunov candidate function [19,21,22]:

$$V_{\bar{x}_k} = \bar{x}_k^T W_{\alpha_k,\tau_k}^{-1} \bar{x}_k : \mathbb{R}^{(\bar{\tau}+1)n} \longrightarrow \mathbb{R}^+, \tag{10.27}$$

where $\alpha_k \in \Xi$, τ_k satisfies (10.4), and

$$W_{\alpha_k,\tau_k} = \sum_{i=1}^{N} \alpha_{k(i)} W_{i,\tau_k} \tag{10.28}$$

with $\mathbf{0} < W_{i,\tau_k} = W_{i,\tau_k}^T \in \mathbb{R}^{(\bar{\tau}+1)n\times(\bar{\tau}+1)n}$. Moreover, if (10.27) is a Lyapunov function ensuring the local stability of the closed-loop system, then we associate with it the following level set $\mathcal{L}_{\mathcal{V}}(\mu)$ for some real scalar $0 < \mu < \infty$:

$$\mathcal{L}_{\mathcal{V}}(\mu) = \left\{\bar{x}_k \in \mathbb{R}^{(\bar{\tau}+1)n} : V_{\bar{x}_k} \leq \mu^{-1}\right\}. \tag{10.29}$$

Lemma 10.2. *Assume that $V_{\bar{x}_k}$ in (10.27) is a Lyapunov function, $\alpha_k \in \Lambda$, $\tau_k \in \mathcal{I}[\underline{\tau}, \bar{\tau}]$, and $w_k \in \mathcal{W}$ for a given δ, $\beta > 0$ such that $\mathcal{L}_\mathcal{V}(\beta) \subset \mathcal{R}_0 \subset \mathcal{R}_\mathcal{A}$. If*

$$\Delta V_{\bar{x}_k} - \omega_k^T \omega_k + \gamma^{-1} z_k^T z_k < 0 \tag{10.30}$$

along the trajectories of system (10.23) emanating from $\mathcal{L}_\mathcal{V}(\beta)$, then for all $k^* > 0$,

$$V_{x_{k^*}} - V_{\bar{x}_0} - \sum_{i=0}^{k^*} \omega_i^T \omega_i + \frac{1}{\gamma} \sum_{i=0}^{k^*} z_i^T z_i < 0. \tag{10.31}$$

Hence for all $\bar{x}_0 \in \mathcal{L}_\mathcal{V}(\beta)$ and $w_k \in \mathcal{W}$, it follows that

1. *$V_{\bar{x}_k} < V_{\bar{x}_0} + \|w_k\|_2^2 \le \beta^{-1} + \delta^{-1} = \mu^{-1}$ for all $k \ge 0$, and thus the trajectories of the system remain bounded in $\mathcal{L}_\mathcal{V}(\mu) \subset \mathcal{R}_\mathcal{A}$.*
2. *As $k \to \infty$, $\|z_k\|_2^2 \le \gamma(\|w_k\|_2^2 + V_{\bar{x}_0})$.*
3. *If $w_k = 0$ for all $k \ge \bar{k} \ge 0$ as $k \to \infty$, then $\bar{x}_k \to 0$ without leaving $\mathcal{L}_\mathcal{V}(\mu) \subset \mathcal{R}_\mathcal{A}$.*

Proof. By summing up (10.30) from 0 to $k^* > 0$, we can verify that (10.30) implies (10.31), from which the three above items are now deduced.

Firstly, from (10.2) we have that $\sum_{k=0}^{\infty} \omega_k^T \omega_k \le \delta^{-1}$ for all $k > 0$, and therefore $V_{\bar{x}_k} \le V_{\bar{x}_0} + \delta^{-1}$ for all $k > 0$. Moreover, by considering that $\bar{x}_0 \in \mathcal{L}_\mathcal{V}(\beta)$, $V_{\bar{x}_0} \le \beta^{-1}$, and then $V_{\bar{x}_k} \le \beta^{-1} + \delta^{-1} = \mu^{-1}$ for some $\mu > 0$ we ensure that the trajectories starting in $\mathcal{L}_\mathcal{V}(\beta)$ remain in the set $\mathcal{L}_\mathcal{V}(\mu) \subseteq \mathcal{R}_\mathcal{A}$ according to statement *1*.

Statement *2* is obtained by assuming that $k^* \to \infty$ and therefore $V_{x_{k^*}} \to 0$ in (10.31). If $\bar{x}_0 = 0$, then it follows that $\|z_k\|_2^2 < \gamma \|\omega_k\|_2^2$, that is, γ is a limiting factor for ℓ_2-gain. At last, if the disturbance vanishes at an instant $k = \bar{k}$, that is, $\omega_k = 0$ for all $k > \bar{k}$, ten statement *2* implies that $V_{\bar{x}_k} < \mu^{-1}$ and, consequently, $\bar{x}_{\bar{k}} \in \mathcal{L}_\mathcal{V}(\mu)$.

From (10.30), for any $k > \bar{k}$, we have $\Delta V_{\bar{x}_k} < 0$, meaning that, by Lyapunov theory arguments, $\bar{x}_k$ asymptotically goes to the origin without leaving $\mathcal{L}_\mathcal{V}(\mu)$, ensuring statement *3*. □

Note that the candidate Lyapunov function presented in (10.27)–(10.28) slightly differs from that in [19], allowing us to handle the associated level set (10.29), as it will be clear later. Moreover, the augmented Lyapunov candidate function (10.27) shares the similar general Lyapunov–Krasovskii structure discussed in [19], that is, $V_{\bar{x}_k} = \sum_{i=1}^{\bar{\tau}+1} \sum_{j=1}^{\bar{\tau}+1} x_{k-i}^T \left[W_{\alpha_k,\tau_k}^{-1}\right]_{ij} x_{k-j}$, where $\left[W_{\alpha_k,\tau_k}^{-1}\right]_{ij}$ stands for the (i,j)-block of W_{α_k,τ_k}^{-1} with dimensions $n \times n$.

Such a choice of a Lyapunov candidate function, which is by definition delay-dependent, has the nice property of taking into account all possible combinations among the state vectors in the sequence φ_k, thus allowing us to look for less-conservative results than those based on other Lyapunov–Krasovskii functionals.

Aiming at dealing with the delay variation $\tau_k \in \mathcal{I}[\underline{\tau}, \bar{\tau}]$, we define the auxiliary set $\mathcal{C}(\tau_k)$ giving all possible values that may be assumed by the delay in the instant $k+1$:

$$\mathcal{C}(\tau_k) = \left\{\tau_{k+1} \in \mathbb{I} : \mathbb{I} \triangleq \mathcal{I}[\max(0, \tau_k - \Delta_\tau),\ \min(\bar{\tau}, \tau_k + \Delta_\tau)]\right\}, \qquad (10.32)$$

where τ^+ is a shorthand for τ_{k+1}. Then, for $\bar{\tau} = 7$, $\tau_k = 6$, and $\Delta_\tau = 2$, we have $\tau^+ \in \{4, 5, 6, 7\} = \mathcal{C}(2)$, characterizing the set of all possible values for τ_{k+1}. This set can be handled by the following code:

```
function [tau_inf,tau_sup] = C_tau(tau,maxtau,maxvar)
% output: maximum (tau_sup) and minimal (tau_min) values of the delay
% input
% tau     -> current value of tau(k).
% maxtau -> maximum allowed value for tau(k).
% maxvar -> maximum variation of the delay between two consecutive instants.

tau_inf = max(0,tau-maxvar)+1;
tau_sup = min(maxtau,tau+maxvar)+1;

end
```

It is worth mentioning that the definition of the set $\mathcal{C}(\tau_k)$ can be adjusted to capture specific requirements of the application. Furthermore, if the delay behavior is unknown and it can jump to any value between 0 and $\bar{\tau}$, then the set $\mathcal{C}(\tau_k)$ recovers the usual approach found in the literature where the delay may arbitrarily vary in the interval $[\underline{\tau}, \bar{\tau}]$, leading to more conservative design conditions.

10.4. Stabilization conditions

First, we consider the general case where the parameter α_k is time-varying and measured, leading to the parameter-dependent control law (10.8). The following theorem states a convex procedure allowing us to synthesize the gains ensuring local asymptotic stability for the closed-loop system (10.1)–(10.8). We further use the inferior block triangular matrix U

(or U_i) belonging to $\mathbb{R}^{(\bar{\tau}+1)n\times(\bar{\tau}+1)n}$ given as follows:

$$U = \begin{bmatrix} U_{11} & \mathbf{0} & \mathbf{0} & \cdots & \mathbf{0} \\ U_{21} & U_{22} & U_{23} & \cdots & U_{2(\bar{\tau}+1)} \\ \vdots & \vdots & \vdots & \ddots & \vdots \\ U_{(\bar{\tau}+1)1} & U_{(\bar{\tau}+1)2} & U_{(\bar{\tau}+1)3} & \cdots & U_{(\bar{\tau}+1)(\bar{\tau}+1)} \end{bmatrix} \in \mathbb{R}^{(\bar{\tau}+1)n\times(\bar{\tau}+1)n},$$

where each block U_{ij} has dimensions $n \times n$.

10.4.1 Control gain synthesis conditions

Theorem 10.3. *Suppose that there exist symmetric positive definite matrices* $W_{i,\tau_k} \in \mathbb{R}^{(\bar{\tau}+1)n\times(\bar{\tau}+1)n}$, *diagonal positive definite matrices* $S \in \mathbb{R}^{n_u\times n_u}$, *matrices* $Y_i \in \mathbb{R}^{n_u\times n_u}$, $Z_i \in \mathbb{R}^{n_u\times(\bar{\tau}+1)n}$, $i \in \mathcal{I}[1,N]$, $\tau_k \in \mathcal{I}[0,\bar{\tau}]$, *block triangular inferior matrices* $U \in \mathbb{R}^{(\bar{\tau}+1)n\times(\bar{\tau}+1)n}$, *and positive scalars* μ, γ, *and* δ *such that the following LMIs are feasible:*

$$\begin{bmatrix} -W_{q,\tau^+} & 0.5[(\mathbb{A}_{i,\tau_k}+\mathbb{A}_{j,\tau_k})U + \mathbb{B}_i(v_1\otimes Y_j) + \mathbb{B}_j(v_1\otimes Y_i)] \\ \star & 0.5(W_{i,\tau_k}+W_{j,\tau_k}) - U^T - U \\ \star & \star \\ \star & \star \\ \star & \star \end{bmatrix}$$

$$\left.\begin{matrix} -0.5(\mathbb{B}_i+\mathbb{B}_j)S & 0.5(\mathbb{B}_{wi}+\mathbb{B}_{wj}) & \mathbf{0} \\ 0.5(Z_i+Z_j)^T) & \mathbf{0} & 0.5U^T(\mathbb{C}_i+\mathbb{C}_j)^T \\ -2S & \mathbf{0} & \mathbf{0} \\ \star & -\mathbf{I} & \mathbf{0} \\ \star & \star & -\gamma\mathbf{I} \end{matrix}\right] < \mathbf{0},$$

$$i,q \in \mathcal{I}[1,N];\quad j \in \mathcal{I}[i,N];\quad \tau_k \in \mathcal{I}[\underline{\tau},\bar{\tau}];\quad \tau^+ \in \mathcal{C}(\tau_k), \tag{10.33}$$

$$\begin{bmatrix} W_{i,\tau_k} - U^T - U & v_1\otimes Y_{i(t)}^T - Z_{i(t)}^T \\ \star & -\mu\bar{u}_{(t)}^2 \end{bmatrix} < \mathbf{0}, \tag{10.34}$$

$$t \in \mathcal{I}[1,m];\quad i \in \mathcal{I}[1,N];\quad \tau_k \in \mathcal{I}[\underline{\tau},\bar{\tau}],$$

and

$$\mu - \delta < 0. \tag{10.35}$$

Then the linear parameter-dependent control gain

$$K(\alpha_k) = \sum_{i=1}^{N} \alpha_{k(i)} Y_i U_{11}^{-1}, \tag{10.36}$$

employed in the control law (10.8), *ensures that:*

1. *the trajectories of system* (10.23) *do not leave the set* $\mathcal{R}_\mathcal{E} = \mathcal{L}_\mathcal{V}(\mu) \subseteq \mathcal{R}_A$ *for all* $\bar{x}_0$ *belonging to the set* $\mathcal{R}_{\mathcal{E}0} = \mathcal{L}_\mathcal{V}(\beta) \subseteq \mathcal{R}_0$ *with* $\beta = (\mu^{-1} - \delta^{-1})^{-1}$ *for all* $k > 0$, $w_k \in \mathcal{W}$, *and* $w_k \neq 0$;
2. $\|z_k\|_2^2 < \gamma(\|w_k\|_2^2 + c_\varphi)$ *as* $k \to \infty$;
3. *the set* $\mathcal{R}_\mathcal{E}$ *is a region of asymptotic stability of system* (10.16)–(10.17) *for all* $k \geq 0$ *and* $w_k = 0$.

Moreover, an estimate of the region of attraction, $\mathcal{L}_\mathcal{V}(\mu)$ *is calculated as*

$$\mathcal{L}_\mathcal{V}(\mu) = \bigcap_{\substack{\tau_k \in \mathcal{I}[\underline{\tau}, \bar{\tau}] \\ i \in \mathcal{I}[1,N]}} \mathcal{E}(W_{i,\tau_k}^{-1}, \mu) \tag{10.37}$$

with $\mathcal{E}(W_{i,\tau_k}^{-1}, \mu) = \{\bar{x}_k \in \mathbb{R}^{(\bar{\tau}+1)n};\ \bar{x}_k^T W_{i,\tau_k}^{-1} \bar{x}_k \leq \mu^{-1}\}$.

Proof. The proof consists of two parts. In the first one, we show that if the LMIs (10.33) are satisfied under the hypothesis that Lemma 10.1 is fulfilled, then Lemma 10.2 is ensured, and $V_{\bar{x}_k}$ in (10.27) is a Lyapunov function. In the second part, we show that the proposed set of initial conditions $\mathcal{R}_\mathcal{E} = \mathcal{L}_\mathcal{V}(\beta)$ given by (10.37) ensures that the generalized sector condition is attained, and thus Lemma 10.1 is verified.

Part I. Supposing the feasibility of (10.33), multiply its left-hand side by $\alpha_{k(q)}$, $\alpha_{k(i)}$, $\alpha_{k(j)}$ and summing it up for i, $q \in \mathcal{I}[1, N]$ and $j \in \mathcal{I}[i, N]$, replace $v_1 \otimes Y(\alpha_k)$ and $Z(\alpha_k)$ by $\mathbb{K}(\alpha_k)U$ and $G(\alpha_k)U$, respectively. Using the fact that $[W(\alpha_k, \tau_k) - U]^T W^{-1}(\alpha_k, \tau_k)[W(\alpha_k, \tau_k) - U] \geq \mathbf{0} \Rightarrow -U^T W^{-1}(\alpha_k, \tau_k)U \leq W(\alpha_k, \tau_k) - U^T - U$ to replace the block (2, 2) of the resulting inequality, we obtain

$$\begin{bmatrix} -W(\alpha^+, \tau^+) & \mathbb{A}(\alpha_k, \tau_k)U + \mathbb{B}(\alpha_k)\mathbb{K}(\alpha_k)U & -\mathbb{B}(\alpha_k)S \\ \star & -U^T W^{-1}(\alpha_k, \tau_k)U & Z(\alpha_k)^T \\ \star & \star & -2S \\ \star & \star & \star \\ \star & \star & \star \end{bmatrix}$$

$$\begin{array}{cc} \mathbb{B}_w(\alpha_k) & \mathbf{0} \\ \mathbf{0} & U^T\mathbb{C}(\alpha_k)^T \\ \mathbf{0} & \mathbf{0} \\ -\mathbf{I} & \mathbf{0} \\ \star & -\gamma\mathbf{I} \end{array}\Bigg] < \mathbf{0}, \tag{10.38}$$

where $\alpha^+ = \alpha_{k+1}$. With the regularity of U from block $(2, 2)$ of (10.38), pre- and postmultiplying the resulting inequality by $\mathcal{T}_1 = \mathtt{diag}\{\mathbf{I}, U^{-T}, S^{-T}, \mathbf{I}, \mathbf{I}\}$ and $\mathcal{T}_1^T$, respectively, we have

$$\begin{bmatrix} -W(\alpha^+, \tau^+) & \mathbb{A}(\alpha_k, \tau_k) + \mathbb{B}(\alpha_k)\mathbb{K}(\alpha_k) & -\mathbb{B}(\alpha_k) & \mathbb{B}_w(\alpha_k) & \mathbf{0} \\ \star & -W^{-1}(\alpha_k, \tau_k) & G(\alpha_k)^T S^{-1} & \mathbf{0} & \mathbb{C}(\alpha_k)^T \\ \star & \star & -2S^{-1} & \mathbf{0} & \mathbf{0} \\ \star & \star & \star & -\mathbf{I} & \mathbf{0} \\ \star & \star & \star & \star & -\gamma\mathbf{I} \end{bmatrix} < \mathbf{0}. \tag{10.39}$$

Replacing $\mathbb{A}(\alpha_k, \tau_k) + \mathbb{B}(\alpha_k)\mathbb{K}(\alpha_k)$ by $\bar{\mathbb{A}}(\alpha_k, \tau_k)$ and applying Schur's complement twice in (10.39), we obtain

$$\begin{bmatrix} -W^{-1}(\alpha_k, \tau_k) + \gamma^{-1}\mathbb{C}(\alpha_k)^T\mathbb{C}(\alpha_k) & G(\alpha_k)^T S^{-1} & \mathbf{0} \\ \star & -2S^{-1} & \mathbf{0} \\ \star & \star & -\mathbf{I} \end{bmatrix} + \begin{bmatrix} \bar{\mathbb{A}}(\alpha_k, \tau_k) \\ -\mathbb{B}(\alpha_k) \\ \mathbb{B}_w(\alpha_k) \end{bmatrix} W(\alpha^+, \tau^+) \begin{bmatrix} \bar{\mathbb{A}}(\alpha_k, \tau_k) & -\mathbb{B}(\alpha_k) & \mathbb{B}_w(\alpha_k) \end{bmatrix} < \mathbf{0}. \tag{10.40}$$

By pre- and postmultiplying (10.40) by $X_k^T = \begin{bmatrix} \bar{x}_k^T & \Psi(\mathbb{K}\bar{x}_k)^T & \omega_k^T \end{bmatrix}^T$ and X_k, respectively, and replacing $\bar{\mathbb{A}}(\alpha_k, \tau_k)\bar{x}_k - \mathbb{B}(\alpha_k)\Psi(\mathbb{K}(\alpha_k)\bar{x}_k) + \mathbb{B}_w(\alpha_k)\omega_k$ and $\mathbb{C}(\alpha_k)\bar{x}_k$ by $\bar{x}_{k+1}$ and z_k, we get

$$\begin{aligned} &\bar{x}_{k+1}^T W^{-1}(\alpha^+, \tau^+)\bar{x}_{k+1} - \bar{x}_k^T W^{-1}(\alpha_k, \tau_k)\bar{x}_k - 2\Psi(\mathbb{K}(\alpha_k)\bar{x}_k)^T S^{-1} \\ &\quad \times (\Psi(\mathbb{K}(\alpha_k)\bar{x}_k) - G(\alpha_k)\bar{x}_k) - \omega_k^T\omega_k + \gamma^{-1} z_k^T z_k < 0, \end{aligned} \tag{10.41}$$

where α^+ represents α_{k+1}.

Denoting $T = S^{-1}$, assuming that $\Delta V_{\bar{x}_k} = \bar{x}_{k+1}^T W^{-1}(\alpha^+, \tau^+)\bar{x}_{k+1} - \bar{x}_k^T W^{-1}(\alpha_k, \tau_k)\bar{x}_k$, and supposing $\bar{x}_k \in \mathbb{S}$ defined in (10.25), we obtain

$$\begin{aligned}&\Delta V_{\bar{x}_k} - 2\Psi(\mathbb{K}(\alpha_k)\bar{x}_k)^T T(\Psi(\mathbb{K}(\alpha_k)\bar{x}_k) - G(\alpha_k)\bar{x}_k)\\&- \omega_k^T\omega_k + \gamma^{-1}z_k^T z_k \leq 0.\end{aligned} \tag{10.42}$$

If $\bar{x}_k \in \mathbb{S}$, then the generalized sector condition presented in Lemma 10.1 ensures the nonpositivity of $\Psi(\mathbb{K}(\alpha_k)\bar{x}_k)^T T(\Psi(\mathbb{K}(\alpha_k)\bar{x}_k) - G(\alpha_k)\bar{x}_k)$, which implies $\Delta V_{\bar{x}_k} - \omega_k^T\omega_k + \gamma^{-1}z_k^T z_k < 0$. Because of (10.13) and the positivity of $W(\alpha_k, \tau_k)$, we can conclude that there exist a sufficiently small $\epsilon_0 > 0$ such that

$$\epsilon_0\|\bar{x}_k\|^2 \leq V_{\bar{x}_k} \leq \epsilon_1\|\bar{x}_k\|^2, \text{ with } \epsilon_1^{-1} = \min_{i=1,\ldots,N}\ \lambda_{\min_{\tau_k\in\mathcal{I}[0,\bar{\tau}]}} W_{i,\tau_k} > 0.$$

Furthermore, by (10.42) with $\omega_k = 0$ we get

$$\begin{aligned}\Delta V_{\bar{x}_k} &< 2\Psi(\mathbb{K}(\alpha_k)\bar{x}_k)^T T(\Psi(\mathbb{K}(\alpha_k)\bar{x}_k) - G(\alpha_k)\bar{x}_k) - \gamma^{-1}z_k^T z_k\\&\leq -\epsilon_3\|\bar{x}_k\|^2 < 0\end{aligned}$$

for some $\epsilon_3 > 0$. Thus $V_{\bar{x}_k}$ given by (10.27)–(10.28) is a Lyapunov function for system (10.16)–(10.17). Then by Lemma 10.2 we get the local stability of the uncertain discrete-time linear system with time-varying state delay subject to actuator saturations whenever $\bar{x}_k$ belongs to the set $\mathbb{S}$.

Part II. In this part, we show that by the verification of LMIs (10.34) with the initial condition given by $\mathcal{R}_{\mathcal{E}} = \mathcal{L}_{\mathcal{V}}(\beta)$ (see (10.37)) we ensure that the signals u and ω belong to set $\mathbb{S}$; therefore the generalized sector condition, Lemma 10.1, is verified, certifying the results in the first part of this proof.

If (10.34) is also satisfied, then multiplying it by $\alpha_{k(i)}$, summing it up for $i \in \mathcal{I}[1, N]$, replacing $v_1 \otimes Y(\alpha_k)$ and $Z(\alpha_k)$ by $\mathbb{K}(\alpha_k)U$ and $G(\alpha_k)U$, respectively, using again the fact $-U^T W^{-1}(\alpha_k, \tau_k)U \leq W(\alpha_k, \tau_k) - U^T - U$, and pre- and postmultiplying the resulting matrix by $\mathcal{T}_2 = \mathtt{diag}\{U^{-T}, 1\}$ and $\mathcal{T}_2^T$, respectively, we get

$$\begin{bmatrix} -W^{-1}(\alpha_k, \tau_k) & (\mathbb{K}(\alpha_k)_{(\iota)} - G(\alpha_k)_{(\iota)})^T \\ \star & -\mu\bar{u}_{(\iota)}^2 \end{bmatrix} \leq \mathbf{0}. \tag{10.43}$$

Applying Schur's complement and pre- and postmultiplying (10.43) by $\bar{x}_k^T$ and $\bar{x}_k$, we obtain

$$-\bar{x}_k^T W^{-1}(\alpha_k, \tau_k)\bar{x}_k + \bar{x}_k^T(\mathbb{K}(\alpha_k)_{(\iota)} - G(\alpha_k)_{(\iota)})^T(\mu\bar{u}_{(\iota)}^2)^{-1}$$

$$\times\,(\mathbb{K}(\alpha_k)_{(t)} - G(\alpha_k)_{(t)})\bar{x}_k \le 0, \tag{10.44}$$

thus ensuring that $\mathcal{L}_\mathcal{V} \subseteq \mathbb{S}$. As a consequence, the generalized sector condition considered before is valid for any trajectory of the closed-loop system starting in $\mathcal{L}_\mathcal{V}$ and thus remaining in $\mathbb{S}$. □

Now we consider the case where the parameter vector α_k takes values where only the ith element is equal to one and all others are null. In this case, we have a switching system. Thus the following theorem computes the gains for the switching control law (10.9).

Theorem 10.4. *Suppose that there exist symmetric positive definite matrices $W_{i,\tau_k} \in \mathbb{R}^{(\bar{\tau}+1)n\times(\bar{\tau}+1)n}$, diagonal positive definite matrices $S \in \mathbb{R}^{n_u\times n_u}$, matrices $Y_i \in \mathbb{R}^{n_u\times(\bar{\tau}+1)n}$, $Z_i \in \mathbb{R}^{n_u\times(\bar{\tau}+1)n}$, $i \in \mathcal{I}[1,N]$, $\tau_k \in \mathcal{I}[0,\bar{\tau}]$, block triangular inferior matrices $U_i \in \mathbb{R}^{(\bar{\tau}+1)n\times(\bar{\tau}+1)n}$, and positive scalars μ, γ, and δ such that the LMIs* (10.35),

$$\begin{bmatrix} -W_{q,\tau^+} & \mathbb{A}_{i,\tau_k}U_i + \mathbb{B}_i(\nu_1\otimes Y_i) & -\mathbb{B}_iS & \mathbb{B}_{wi} & \mathbf{0} \\ \star & W_{i,\tau_k} - U_i^T - U_i & Z_i^T & \mathbf{0} & U_i^T\mathbb{C}_i^T \\ \star & \star & -2S & \mathbf{0} & \mathbf{0} \\ \star & \star & \star & -\mathbf{I} & \mathbf{0} \\ \star & \star & \star & \star & -\gamma\mathbf{I} \end{bmatrix} < \mathbf{0}, \tag{10.45}$$

$$i, q \in \mathcal{I}[1,N];\quad \tau_k \in \mathcal{I}[\underline{\tau},\bar{\tau}];\quad \tau^+ \in \mathcal{C}(\tau_k),$$

and

$$\begin{bmatrix} W_{i,\tau_k} - U_i^T - U_i & \nu_1\otimes Y_{i(t)}^T - Z_{i(t)}^T \\ \star & -\mu\bar{u}_{(t)}^2 \end{bmatrix} < \mathbf{0}, \tag{10.46}$$

$$t \in \mathcal{I}[1,m];\quad i \in \mathcal{I}[1,N];\quad \tau_k \in \mathcal{I}[\underline{\tau},\bar{\tau}],$$

are feasible. Then, the switching control gains in (10.9) *computed as*

$$K_i = Y_iU_{i,11}^{-1} \tag{10.47}$$

ensure that:

1. *the trajectories of system* (10.23) *do not leave the set $\mathcal{R}_\mathcal{E} = \mathcal{L}_\mathcal{V}(\mu) \subseteq \mathcal{R}_\mathcal{A}$ for all $\bar{x}_0$ belonging to the set $\mathcal{R}_{\mathcal{E}0} = \mathcal{L}_\mathcal{V}(\beta) \subseteq \mathcal{R}_0$ with $\beta = (\mu^{-1} - \delta^{-1})^{-1}$ for all $k > 0$, $w_k \in \mathcal{W}$, and $w_k \neq 0$;*
2. *$\|z_k\|_2^2 < \gamma(\|w_k\|_2^2 + \mathrm{c}_\varphi)$ as $k \to \infty$;*
3. *the set $\mathcal{R}_\mathcal{E}$ is a region of asymptotic stability of system* (10.16)–(10.17) *for all $k \ge 0$ and $w_k = 0$.*

Moreover, the estimate of the region of attraction $\mathcal{L}_{\mathcal{V}}(\mu)$ *is calculated as in* (10.37).

Proof. The proof follows similar lines as in Theorem 10.3, noting that due to the switching nature of the system, only the vertices are concerned. As a consequence, the matrices U and Y can be replaced by U_i and Y_i and thus be dependent on the mode i. Such a procedure exploits the switching yielding a more relaxed design condition. □

For the last case, we consider that α_k is not available. This results in the robust control law given by (10.10). Therefore the next theorem computes robust control gains that stabilize locally system (10.16)–(10.17) with $\mathbb{K}(\alpha_k) = \mathbb{K}$. It is interesting to mention that similar results have already been presented in [22].

Theorem 10.5. *Suppose that there exist symmetric positive definite matrices* $W_{i,\tau_k} \in \mathbb{R}^{(\bar{\tau}+1)n\times(\bar{\tau}+1)n}$*, diagonal positive definite matrices* $S \in \mathbb{R}^{n_u\times n_u}$*, matrices* $Y \in \mathbb{R}^{n_u\times(\bar{\tau}+1)n}$*,* $Z \in \mathbb{R}^{n_u\times(\bar{\tau}+1)n}$*,* $i \in \mathcal{I}[1,N]$*,* $\tau_k \in \mathcal{I}[0,\bar{\tau}]$*, block triangular inferior matrices* $U \in \mathbb{R}^{(\bar{\tau}+1)n\times(\bar{\tau}+1)n}$*, and positive scalars* μ*,* γ*, and* δ *such that the LMIs* (10.35),

$$\begin{bmatrix} -W_{q,\tau^+} & \mathbb{A}_{i,\tau_k}U + \mathbb{B}_i(\nu_1 \otimes Y) & -\mathbb{B}_i S & \mathbb{B}_{wi} & \mathbf{0} \\ \star & W_{i,\tau_k} - U^T - U & Z^T & \mathbf{0} & U^T\mathbb{C}_i^T \\ \star & \star & -2S & \mathbf{0} & \mathbf{0} \\ \star & \star & \star & -\mathbf{I} & \mathbf{0} \\ \star & \star & \star & \star & -\gamma\mathbf{I} \end{bmatrix} < \mathbf{0}, \tag{10.48}$$

$$i, q \in \mathcal{I}[1,N]; \quad \tau_k \in \mathcal{I}[\underline{\tau},\bar{\tau}]; \quad \tau^+ \in \mathcal{C}(\tau_k),$$

and

$$\begin{bmatrix} W_{i,\tau_k} - U^T - U & \nu_1 \otimes Y_{(\iota)}^T - Z_{(\iota)}^T \\ \star & -\mu\bar{u}_{(\iota)}^2 \end{bmatrix} < \mathbf{0}, \tag{10.49}$$

$$\iota \in \mathcal{I}[1,m]; \quad i \in \mathcal{I}[1,N]; \quad \tau_k \in \mathcal{I}[\underline{\tau},\bar{\tau}],$$

are feasible. Then the robust control gains in (10.16)–(10.17)*, with* $\mathbb{K}(\alpha_k) = \mathbb{K}$*, computed as*

$$K = YU_{11}^{-1} \tag{10.50}$$

are such that:

1. *for* $w_k \in \mathcal{W}$ *and* $w_k \neq 0$, *the trajectories of system* (10.23) *do not leave the set* $\mathcal{R}_\mathcal{E} = \mathcal{L}_\mathcal{V}(\mu) \subseteq \mathcal{R}_\mathcal{A}$ *for all* $\bar{x}_0$ *belonging to the set* $\mathcal{R}_{\mathcal{E}0} = \mathcal{L}_\mathcal{V}(\beta) \subseteq \mathcal{R}_0$ *with* $\beta = (\mu^{-1} - \delta^{-1})^{-1}$ *for all* $k > 0$;
2. *as* $k \to \infty$, $\|z_k\|_2^2 < \gamma(\|w_k\|_2^2 + V_{\bar{x}_0})$;
3. *for* $w_k = 0$, *the set* $\mathcal{R}_\mathcal{E}$ *is a region of asymptotic stability of system* (10.16)–(10.17) *for all* $k \geq 0$.

Moreover, the estimate of the region of attraction $\mathcal{L}_\mathcal{V}(\mu)$ *is calculated as in* (10.37).

Proof. If in conditions (10.45) and (10.46) of Theorem 10.4, we have $U_i = U$, $Y_i = Y$, and $Z_i = Z$, then we obtain conditions (10.48) and (10.49). Therefore, taking into account these modifications, the proof of Theorem 10.5 follows the same steps of the proof of Theorem 10.4. □

Remark 10.1. As mentioned in [23], the estimate of the region of attraction proposed here, $\mathcal{L}_\mathcal{V}(\mu) = \mathcal{R}_\mathcal{E} \subset \mathcal{R}_\mathcal{A}$, differs from those usually found in the literature, as those in [39], [38], [37], [36], and [35] for the continuous-time cases and in [40], [47], and [48] for discrete-time ones. These papers characterize the set of initial conditions by using the norm approach. Thus in the discrete-time case the sequence $\phi_{\bar{\tau},0}$ is given by

$$\mathcal{W}_\rho \triangleq \{\phi_0 : \|\varphi_0\|_2 \leq \rho\} \tag{10.51}$$

or by

$$\mathcal{X}(\rho, \rho_1) \triangleq \left\{\phi_0 : \|\varphi_0\|_2^2 \leq \rho^2,\ \|\Delta\varphi_0\|_2^2 \leq \rho_1^2\right\}, \tag{10.52}$$

where $\Delta\phi_0 = \phi_0 - \phi_{-1}$, and the positive scalars ρ and ρ_1 are usually to be found. Note that in the characterization given by $\mathcal{W}_\rho$, equation (10.51), the variation between two consecutive delayed states can be any variation such that each state vector belongs to the set $\mathcal{W}_\rho$. In the continuous-time case, it is necessary to replace $\Delta\phi_0$ by its time-derivative counterpart. On the other hand, the characterization of initial conditions through the set $\mathcal{X}(\rho, \rho_1)$ requires additionally that the variation $\Delta\varphi_0$ remains norm bounded by ρ_1. Thus if $\rho_1 = 2\rho$, then the set $\mathcal{W}_\rho$ is recovered, and the smaller the ρ_1, the larger the estimate of the region of attraction.

Remark 10.2. Note that the inferior block triangular form of the matrix U was motivated by the control law (10.8),which feeds back only the actual state x_k. Other control structures can be thought, as, for instance, by using also the samples $x_{k-\bar{\tau}}$ state delayed by $\bar{\tau}$, or the most general case of feeding back the information of the whole augmented state $\bar{x}_k$. In such cases the matrices U and Y should be accordingly structured. See, for instance, [22].

Remark 10.3. In some quasi-LPV systems, such as the Takagi–Sugeno ones, the estimate $\mathcal{R}_\mathcal{E}$ can be enlarged because we know the dependency of α_k on, for instance, the state x_k. In such a case the conditions verifying $\mathcal{L}_\mathcal{V}(1)$ can be relaxed: instead of verifying (10.37) for $\alpha_k \in \Lambda$, the known expression of $\alpha_k(x_k)$ can be replaced, and $\mathcal{R}_\mathcal{E}$ is computed directly over x_k. An example of such a case is given in Section 10.5.

10.4.2 Optimization design procedures

The developed convex conditions for synthesis of robust control gains and estimation of the region of attraction can be exploited to optimize some characteristics of interest for the closed-loop system as we present further.

10.4.2.1 Maximization of the estimation of the region of attraction ($\mathcal{R}_\mathcal{E}$)

Suppose that $\omega_k = 0$ for $k > 0$, that is, no perturbation is present in (10.1). In this case the objective is to design a control gain maximizing the estimate $\mathcal{R}_\mathcal{E}$. Note that Theorems 10.3, 10.4, and 10.5 deal with the four cases discussed in Section 10.2.

A first possibility is maximizing the volume of an ellipsoidal set, say $\mathcal{E}(R^{-1},1) = \{\bar{x}_k \in \mathbb{R}^{n(\bar{\tau}+1)}; \bar{x}_k R^{-1} \bar{x}_k \leq 1\}$, such that $\mathcal{E}(R^{-1},1) \subseteq \mathcal{L}_\mathcal{V}(\mu)$. Such an inclusion can be ensured by

$$\mu - 1 \leq 0, \begin{bmatrix} R & R \\ \star & W_{i,\tau_k} \end{bmatrix} \geq \mathbf{0}, \; i \in \mathcal{I}[1,N], \; \tau_k \in \mathcal{I}[\underline{\tau},\bar{\tau}]. \tag{10.53}$$

Because the volume of $\mathcal{E}(R^{-1},1)$ is proportional to $\sqrt{\det(R)}$ (see [49]), such a maximization of $\mathcal{R}_\mathcal{E}$ can be done through the convex optimization procedures:

$$\mathcal{J}_1 : \begin{cases} \min & -\ln(\det(R)) \\ \text{subject to} & \text{LMIs } (i),\ (j), \text{ and } (10.53), \end{cases} \tag{10.54}$$

where the parameters i and j can be chosen to deal with the three conditions mentioned before, which are associated with the theorems presented. Therefore the pairs (i,j) can be chosen from the list $\{((10.33), (10.34)), ((10.45), (10.46)), ((10.48), (10.49))\}$.

A similar possibility consists of maximizing the volume of $\mathcal{E}(R, 1) \subset \mathcal{L}_{\mathcal{V}}(\mu)$, which can ensured by

$$\mu - 1 \le 0, \begin{bmatrix} R & \mathbf{I} \\ \star & W_{i,\tau_k} \end{bmatrix} \ge \mathbf{0}, \ i \in \mathcal{I}[1, N], \tau_k \in \mathcal{I}[\underline{\tau}, \bar{\tau}]. \tag{10.55}$$

Therefore the convex optimization procedures can be proposed as in (10.54) but replacing (10.53) by (10.55). Moreover, in both cases the objective function presented in $\mathcal{J}_1$ can be replaced by a simpler one given by trace(R).

10.4.2.2 Maximization of disturbance tolerance (δ^{-1}) or minimization of the ℓ_2-gain

The maximization of the disturbance energy δ^{-1} consists of designing a state feedback gain that tolerates such a disturbance and ensures given admissible initial states $\mathcal{R}_0$. In particular, if $\varphi_0 = \mathbf{0}$ (i.e., the system is in equilibrium), then it follows that $\delta^{-1} = \mu^{-1}$ and the maximization of the δ^{-1} of the set $\mathcal{W}$ is a convex optimization procedure (see also [9, p. 73]):

$$\mathcal{J}_2 : \begin{cases} \min & \mu \\ \text{subject to} & \text{LMIs } (i) \text{ and } (j), \end{cases} \tag{10.56}$$

with the pair (i, j) chosen from the list $\{((10.33), (10.34)), ((10.45), (10.46)), ((10.48), (10.49))\}$.

Similarly, for a given energy limit of the disturbances δ^{-1}, if the objective is designing the control gain minimizing the ℓ_2 gain between ω_k and z_k, the optimization procedure $\mathcal{J}_2$ still can be used by just replacing the objective function μ by γ. Note that, in particular, if $\varphi_0 = \mathbf{0}$ (i.e., the system is in equilibrium), then it follows that $\delta^{-1} = \mu^{-1}$.

10.5. Numerical examples

We present two numerical examples to illustrate the use of the proposed methods and to clarify some differences between the kinds of systems considered in this chapter. In the first example, we use the same scalar system shown in Example 10.1, Section 10.2. The second example concerns with a second-order state delayed system.

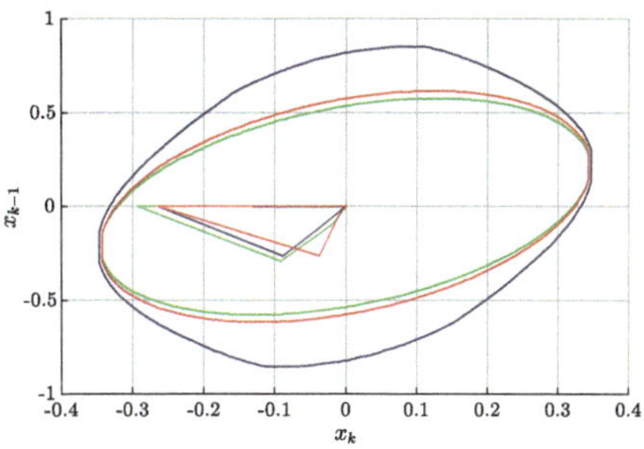

(a) Responses for disturbance and regions of attraction estimated for LPV system (green), Switching system (red), and Fuzzy system (blue).

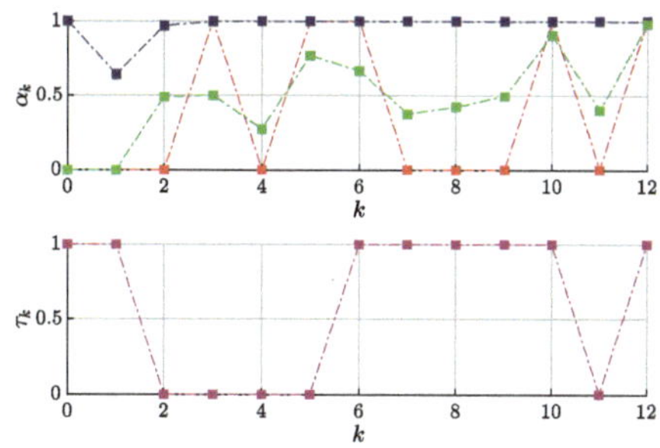

(b) τ_k (bottom) and α_k (top) where the colors are related to the respective regions at left.

Figure 10.4 Estimate regions of attractions and state trajectories for the considered systems assuming LPV, quasi-LPV, and switching cases.

10.5.1 Example 1

For the scalar system used in Example 10.1 in Section 10.2, we design a controller maximizing the tolerable disturbance energy δ^{-1}. With the optimization procedure $\mathcal{J}_2$ in (10.56), we computed the controller gains for the cases quasi-LPV (or fuzzy), LPV, and switching. We obtain the same value for δ in the three cases, that is, $\delta = 0.1279$, $\delta^{-1} = 7.8189$, which result in a disturbance signal with energy $\|\omega_k\|^2 = 2.7962$.

The three regions of attraction obtained are presented in Fig. 10.4a. The regions associated with LPV and switching systems were computed as (10.37), that is, by using the intersections among the level sets of the Lyapunov function. For the fuzzy case, we supposed that $\alpha_{1,k} = \exp(-\frac{x_1}{0.4}^2)$ and $\alpha_{2,k} = 1 - \alpha_{1,k}$ are the membership functions (e.g., see [43]). Therefore the region of attraction can be computed directly through the level set $\mathcal{L}_{\mathcal{V}}(\mu)$ in (10.29) by replacing the mentioned expressions for the vector α_k. The resulting inequality depends on the states x_k, and we search for the values of x_k and x_{k-1} such that $\mathcal{L}_{\mathcal{V}}(\mu)$ is satisfied. We can see in Fig. 10.4a that the fuzzy region of attraction for the fuzzy system is larger than the switched and LPV ones. This last one, the LPV system, leads to the most conservative estimate, because the parameter vector α_k is a priori unknown.

By using a fine grid over the regions we can quantify that the estimate of $\mathcal{R}_{\mathcal{E}}$ for the fuzzy system is about 24% larger than that for the LPV system. Compared with the $\mathcal{R}_{\mathcal{E}}$ for the switching system, the region of the fuzzy system increases by approximately 15%. The improvement in the fuzzy region is due to the possibility of embedding the knowledge on α_k in the $\mathcal{R}_{\mathcal{E}}$ computation.

In addition, for each investigated case, we simulated the closed-loop system response for the disturbance signal $\omega_k = \begin{bmatrix} -2.7962 & 0_{1,11} \end{bmatrix}$, which has the maximum allowed energy computed by our proposed methods. The state trajectories are presented in Fig. 10.4a, whereas Fig. 10.4b presents the considered values of α_k and τ_k (the same for all cases). We can observe that the trajectories remain within the $\mathcal{R}_{\mathcal{E}}$ estimates, and they converge to the origin when the disturbance vanishes, as expected.

The following code allows simulating the closed-loop time response for the quasi-LPV case (fuzzy T-S). For the LPV or switching cases, just the generation of the sequence α_k must be adjusted.

```
function  [x,v,u,z] = Trajectory(A,Ad,B,Bw,Cz,K,...
                                tmax,uo,N,n,m,q,p)
%
% Inputs:
% matrices A, Ad, B, Bw, Cz, and K with cell structure
% System matrices: A, Ad, B, Bw, Cz
% n, m, p, and q: columns of A, B, Bw, and lines of Cz.
% N    -> number of vertex
% tmax -> max value of tau(k)
% uo   -> positive valye of the symmetric saturation limit
%
% Outputs
% x -> state vector
% v -> linear control signal
% u -> saturated control signal
% z -> output

% Set the sequence of delay values
tau = [1 1 0 0 0 0 1 1 1 1 1 0 1 1];

% Set the sequence of disturbance signal
wk = [-2.7962 zeros(1,tmax)];

% Initial conditions
x = zeros((maxtau+1)*n,tmax+1);
z = zeros(p,tmax+1);

% memory pre-allocation
v      = zeros(m,tmax+1);
u      = zeros(m,tmax+1);
A_cx   = cell(tmax+1,1);
B_cx   = cell(tmax+1,1);
```

```
Bw_cx = cell(tmax+1,1);
Cx_cx = cell(tmax+1,1);
K_cx  = cell(tmax+1,1);

% Augmented system
[A_aug,B_aug,Bw_aug,Cz_aug] = augmented_matrices(A,...
                              Ad,B,Bw,Cz,tmax,N,n,m,q,p)

for k = 1:tmax+1
    A_cx{k}  = zeros(n*(maxtau+1),n*(maxtau+1));
    B_cx{k}  = zeros(n*(maxtau+1),m);
    Bw_cx{k} = zeros(n*(maxtau+1),q);
    Cz_cx{k} = zeros(p,n*(maxtau+1));
    K_cx{k}  = zeros(m,n*(maxtau+1));

    % Set the membership fucntions
    alpha_1 =  exp(-(x(1,k)/0.4)^2);
    alpha_2 = 1-alpha_1;

    for vertice  = 1:N
        A_cx{k}  = A_cx{k} + alpha(vertice,k)*...
                         A_aug{vertice,tau(k)+1};
        B_cx{k}  = B_cx{k} + alpha(vertice,k)*B_aug{vertice};
        Bw_cx{k} = Bws{k} + alpha(vertice,k)*Bw_aug{vertice};
        Cz_cx{k} = Cz_cx{k} + alpha(vertice,k)*...
                               Cz_aug{vertice};
        K_cx{k} = K_cx{k} + alpha(vertice,k)*K{vertice};
    end
    z(:,k) = Cz_cx{k}*x(:,k);
    v(:,k) = K_cx{k}*x(:,k);

    % Control signal saturation
    u(:,k) = sign(v(:,k))*min(abs(v(:,k)),uo);
    % State evolution
    x(:,k+1) = A_cx{k}*x(:,k) +...
             B_cx{k}*u(:,k) + Bw_cx{k}*wk(:,k);
end
end
```

For comparison with our results, we used the conditions proposed in [48] to estimate the maximum disturbance tolerance for the switched system case. By solving the optimization procedure proposed in [48, Eq. (44)], with $\beta_1 = \beta_2 = 0.01$ and $\delta_1 = \delta_2 = 0.01$, we could find 25.5680 for the maximum tolerance disturbance. On the other hand, if we use the optimization procedure $\mathcal{J}_2$ proposed here, then we can find 31.2754, showing that our proposals are less conservative.

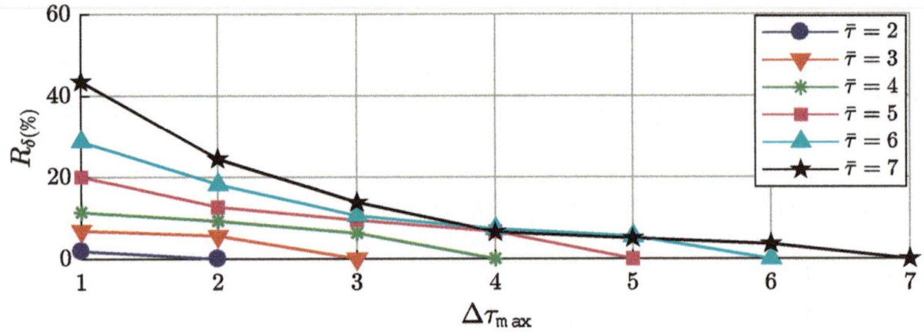

Figure 10.5 The augmentation percentage of disturbance energy ($R_{\delta(\%)}$, see the text) as a function of the maximum delay variation rate for $\bar{\tau}=2$ (•), $\bar{\tau}=3$ (▼), $\bar{\tau}=4$ (∗), $\bar{\tau}=5$ (■), $\bar{\tau}=6$ (▲), and $\bar{\tau}=7$ (★).

10.5.2 Example 2

Consider the system

$$A=(1+\rho)\begin{bmatrix}0.9 & -0.4\\ -0.3 & -0.7\end{bmatrix},\ A_d=(1+\rho)\begin{bmatrix}0.2 & -0.1\\ -0.2 & 0.1\end{bmatrix}, \tag{10.57}$$
$$B=(1+\rho)\begin{bmatrix}5\\ -1\end{bmatrix},\ B_w=(1+\rho)\begin{bmatrix}0.5\\ -0.1\end{bmatrix},\ C=\begin{bmatrix}0.5 & 1\end{bmatrix},$$

with the parameter $|\rho|\leq 0.1$, the time-varying delay $\tau_k\in\mathcal{I}[0,7]$, and the symmetric saturation limit $\bar{u}=1$.

Firstly, we assume ρ as an uncertain parameter and use optimization procedure $\mathcal{J}_2$ in (10.56) with $(i,j)=((10.48),(10.49))$. We consider different values of Δ_τ and computed the percentage ratio of increase of the energy with respect to the case of delay variation equal to the value of the maximum delay (which is the usual case in the literature). Such a computation is given by $R_{\delta(\%)}=(\delta^{-1}_{\Delta\tau\mathrm{max}}/\delta^{-1}_{\Delta\tau\mathrm{max}=\bar{\tau}}-1)\times 100\%$. Fig. 10.5 shows the obtained results, where it is clear that the better ones (larger percentage ratios) are achieved with the slower delay variations Δ_τ. With $\bar{\tau}=7$ and $\Delta_\tau=1$, we get the maximum tolerance improvement with a increase about 41%.

Similarly to the experiment performed for the tolerable disturbance, we apply the optimization procedure $\mathcal{J}_1$ in (10.54) with $(i,j)=((10.45),(10.46))$, thus assuming a switching case. Thus we assume that ρ is a measurable value that jumps on the extreme values ± 0.1, yielding a switching system. We also consider $\delta=0.020$, $\delta^{-1}=50$, and $\|\omega_k\|^2=7.0711$. Fig. 10.6 presents the ratio $R_{S(\%)}=(S_{\Delta_\tau}/S_{\Delta_\tau=\bar{\tau}}-1)\times 100\%$, where $S_{\Delta_\tau=\bar{\tau}}$

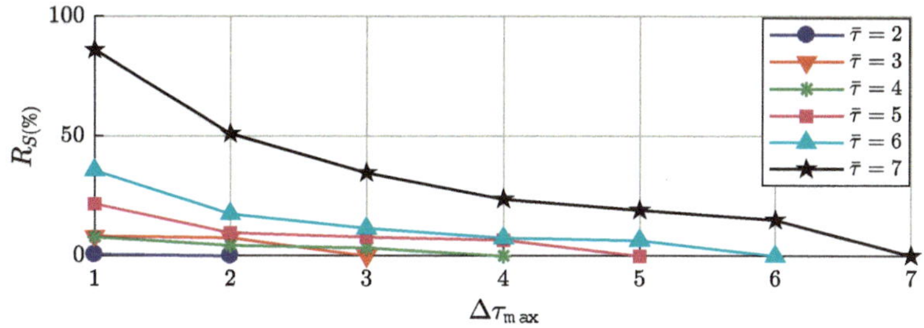

Figure 10.6 The augmentation percentage of the projection area on the x_k axis ($R_{S(\%)}$, see the text) as a function of the maximum delay variation rate for $\bar{\tau} = 2$ (•), $\bar{\tau} = 3$ (▼), $\bar{\tau} = 4$ (∗), $\bar{\tau} = 5$ (■), $\bar{\tau} = 6$ (▲) and $\bar{\tau} = 7$ (★).

is the area of the projection of $\mathcal{R}_{\mathcal{E}}$ over the plan x_k with $\Delta_\tau = \bar{\tau}$, and $S_{\Delta\tau_{max}}$ is the area of the projection of $\mathcal{R}_{\mathcal{E}}$ over x_k with Δ_τ picked in the interval $[1, \ \bar{\tau}]$. We note that the better ratio (highest value) is achieved with the slowest delay variation $\Delta_\tau = 1$, and the bigger delay $\bar{\tau} = 7$. Therefore the tests presented illustrate the relevance of handling the variation rate of the delay to achieve less conservative results.

Furthermore, we compare the region of attraction obtained through our proposed results for the T-S fuzzy systems with that obtained from the results proposed in [43]. Hence we used the optimization procedure $\mathcal{J}_1$ proposed here and the optimization procedure $\mathcal{P}_{\Upsilon_1}$ proposed in [43]. It is important to notice that $\Delta_\tau = 1$, and the result proposed in [43] concerns a state-feedback controller that uses information about both current and delayed states. By using a fine grid over the regions, we can quantify that the estimated region of attraction obtained from our proposal is about 16.86% higher than the estimated region of attraction from [43].

10.6. Future directions

A key issue handling delay systems under saturating actuators consists in developing stability analysis and synthesis conditions yielding less conservative results as possible in terms of allowable initial conditions or tolerable energy of exogenous disturbances. The use of polynomial matrix functions may be an interesting direction. In particular, the use of homogeneous polynomial functions [50] can improve the presented results, as they can enlarge the domain of stabilizable systems and also delivery control gains depending on a ratio of homogeneous polynomial functions. As a consequence, the size of $\mathcal{R}_{\mathcal{E}}$ may augment, allowing more rich families of initial conditions.

Moreover, the design of observers and filters for saturating state-delayed systems is a promising field of investigations, which can take advantage of the presented approach [51]. Equally attractive is the design of output-feedback controllers for the considered class of systems, especially, for dynamic-output-feedback controlled systems, which are usually welcome in industry and process applications and can profit from this approach to achieve less conservative controller design conditions.

The stability investigation of delayed systems under saturating actuators requires careful attention to characterization of the initial conditions. The proposed method, based on an augmented space, finds a set that is quite more significant than the other ones usually employed in the literature at the expense of more computational demanding conditions. Therefore a key issue is investigating other forms of characterizing the initial conditions in delayed systems, for instance, by considering families of specific functions over time.

It is worth mentioning that our proposal can be adapted to taking into account the region of validity of the used models or the safe operating region of the system. Particular applications such as gas/oil furnaces may require specific models (with their respective saturation limits on the control signal). Moreover, the rate of variation of the actuator is another relevant aspect that is usually not considered for delayed control systems. Note that industrial actuators may have a maximum variation rate, which may lead to unsatisfactory closed-loop performance. For instance, consider the flow control by variable velocity pumps. In such a case the velocity cannot change its value abruptly. The rate saturation can be approached with the tools presented in this chapter by using nested saturation functions. Also, the investigation of nonsymmetric saturating actuators is of particular interest in industrial applications, because the regulation conditions are often nonsymmetric concerning the allowed amplitude variations of the actuators [52].

10.7. Concluding remarks

We have presented a space augmentation-based approach to design state feedback control gains for LPV systems under saturating actuators and time-varying state delay. Additionally, we take into account delay variation between two consecutive instants. The proposed conditions allows us to compute an estimate of the region of attraction in the augmented space, yielding to larger initial conditions than those found in the literature. Moreover, we use the input-to-state stability approach to handle exogenous disturbance with limited energy. Furthermore, we show how the proposed conditions can match quasi-LPV systems, switching systems, and robust control systems. Our proposal is illustrated by comprehensive and didactic examples.

References

[1] F.F. Fowle, Toll telephone traffic an experimental study of the relationship between circuit loads and delay to traffic, Proceedings of the American Institute of Electrical Engineers 33 (1914) 1063–1070.

[2] P.J. Palmer, The behaviour of control systems with definite time delay and random disturbance, Proceedings of the IEE – Part IV: Institution Monographs 100 (1953) 573–575.

[3] J.E. Normey-Rico, E.F. Camacho, Control of Dead-Time Processes, Springer, 2007.

[4] D. Chen, D.E. Seborg, PI/PID controller design based on direct synthesis and disturbance rejection, Industrial & Engineering Chemistry Research 41 (2002) 4807–4822.

[5] D.S. Niculescu, Delay Effects on Stability: A Robust Control Approach, Springer, Germany, 2001.

[6] E. Fridman, Introduction to Time-Delay Systems: Analysis and Control, Birkhäuser, Basel, 2014.

[7] F. Brauer, C. Castillo-Chavez, Mathematical Models in Population Biology and Epidemiology, 2nd. ed., Texts in Applied Mathematics, Springer-Verlag, NY, 2012.

[8] H. Caballero-Barragán, L. Osuna-Ibarra, A. Sanchez, A. Loukianov, Semi-empirical modelling of the mass flow in a pilot-scale tubular reactor and tracking control, Journal of Process Control 69 (2018) 79–85.

[9] S. Tarbouriech, G. Garcia, J.M. Gomes da Silva Jr., I. Queinnec, Stability and Stabilization of Linear Systems With Saturating Actuators, Springer, 2011.

[10] T. Hu, Z. Lin, Control Systems With Actuator Saturation: Analysis and Design, Birkhäuser, Boston, 2001.

[11] P. Pepe, G. Pola, M.D.D. Benedetto, On Lyapunov–Krasovskii characterizations of stability notions for discrete-time systems with uncertain time-varying time-delays, IEEE Transactions on Automatic Control PP (2017) 1–15.

[12] A.N.D. Lopes, V.J.S. Leite, L.F.P. Silva, K. Guelton, Anti-windup TS fuzzy PI-like control for discrete-time nonlinear systems with saturated actuators, International Journal of Fuzzy Systems 518 (2020) 272–285.

[13] A.C. Sousa, V.J.S. Leite, I. Rubio Scola, Affordable control platform with MPC application, Studies in Informatics and Control 27 (2018) 265–274.

[14] J. Chen, S. Xu, X. Jia, B. Zhang, Novel summation inequalities and their applications to stability analysis for systems with time-varying delay, IEEE Transactions on Automatic Control 62 (2017) 2470–2475.

[15] S.Y. Lee, W.I. Lee, P. Park, Polynomials-based summation inequalities and their applications to discrete-time systems with time-varying delays, International Journal of Robust and Nonlinear Control 27 (2017) 3604–3619.

[16] N. Wada, K. Satoh, M. Saeki, An LMI-based controller design method for a discrete-time linear system with time-varying state delays, IMA Journal of Mathematical Control and Information 34 (2017) 405–424.

[17] M. Park, O. Kwon, S. Choi, Stability analysis of discrete-time switched systems with time-varying delays via a new summation inequality, Nonlinear Analysis: Hybrid Systems 23 (2017) 76–90.
[18] A. Seuret, F. Gouaisbaut, E. Fridman, Stability of discrete-time systems with time-varying delays via a novel summation inequality, IEEE Transactions on Automatic Control 60 (2015) 2740–2745.
[19] L. Hetel, J. Daafouz, C. Iung, Equivalence between the Lyapunov–Krasovskii functionals approach for discrete delay systems and that of the stability conditions for switched systems, Nonlinear Analysis: Hybrid Systems 2 (2008) 697–705.
[20] W. Lombardy, S. Olaru, S.-I. Niculescu, L. Hetel, A predictive control scheme for systems with variable time-delay, International Journal of Control 85 (2012) 915–932.
[21] J.V.V. Silva, L.F.P. Silva, I. Rubio Scola, V.J.S. Leite, Robust local stabilization of discrete-time systems with time-varying state delay and saturating actuators, Mathematical Problems in Engineering (2018) 5013056.
[22] C. de Souza, V.J.S. Leite, L.F.P. Silva, E.B. Castelan, ISS robust stabilization of state-delayed discrete-time systems with bounded delay variation and saturating actuators, IEEE Transactions on Automatic Control 64 (2019) 3913–3919.
[23] C. de Souza, V.J.S. Leite, E.B. Castelan, L.F.P. Silva, ISS stabilization of discrete-time LPV systems with interval time-varying state delay and saturating actuators, in: 2nd IFAC Workshop on Linear Parameter Varying Systems LPVS 2018, IFAC-PapersOnLine 51 (2018) 143–148.
[24] K.J. Åström, B. Wittenmark, Computer-Controlled Systems: Theory and Design, Prentice-Hall, 1984.
[25] V. Kapila, W.M. Haddad, Memoryless $\mathcal{H}_\infty$ controllers for discrete-time systems with time delay, Automatica 34 (1998) 1141–1144.
[26] F. Bourahala, K. Guelton, N. Manamanni, F. Khaber, Relaxed controller design conditions for Takagi–Sugeno systems with state time-varying delays, International Journal of Fuzzy Systems 19 (2016) 1406–1416.
[27] B. Zhou, Improved Razumikhin and Krasovskii approaches for discrete-time time-varying time-delay systems, Automatica 91 (2018) 256–269.
[28] R.H. Gielen, M. Lazar, Stabilization of polytopic delay difference inclusions via the Razumikhin approach, Automatica 47 (2011) 2562–2570.
[29] M.-T. Laraba, S. Olaru, S.-I. Niculescu, F. Blanchini, G. Giordano, D. Casagrande, S. Miani, Guide on set invariance for delay difference equations, Annual Reviews in Control 41 (2016) 13–23.
[30] S.-H. Chen, J.-H. Chou, Robust D-stability analysis for linear uncertain discrete singular systems with state delay, Applied Mathematics Letters 19 (2006) 197–205.
[31] L. Hetel, M.B.G. Cloosterman, N. van de Wouw, W.P.M.H. Heemels, J. Daafouz, H. Nijmeijer, Comparison of stability characterisations for networked control systems, in: Joint 48th IEEE Conf. on Decision and Control and 28th Chinese Control Conf., 2009, pp. 7911–7916.
[32] T. Binazadeh, M. Bahmani, Design of robust controller for a class of uncertain discrete-time systems subject to actuator saturation, IEEE Transactions on Automatic Control 62 (2017) 1505–1510.
[33] J.M. Gomes da Silva Jr., D. Limón, T. Alamo, E.F. Camacho, Dynamic output feedback for discrete-time systems under amplitude and rate actuator constraints, IEEE Transactions on Automatic Control 53 (2008) 2367–2372.
[34] B. Zhou, Analysis and design of discrete-time linear systems with nested actuator saturations, Systems & Control Letters 62 (2013) 871–879.
[35] Y. Chen, S. Fei, Y. Li, Robust stabilization for uncertain saturated time-delay systems: a distributed-delay-dependent polytopic approach, IEEE Transactions on Automatic Control 62 (2017) 3455–3460.

[36] Y. Chen, S. Fei, Y. Li, Stabilization of neutral time-delay systems with actuator saturation via auxiliary time-delay feedback, Automatica 52 (2015) 242–247.

[37] E. Fridman, A. Pila, U. Shaked, Regional stabilization and $\mathcal{H}_\infty$ control of time-delay systems with saturating actuators, International Journal of Robust and Nonlinear Control 13 (2003) 885–907.

[38] Y.-Y. Cao, Z. Lin, T. Hu, Stability analysis of linear time-delay systems subject to input saturation, IEEE Transactions on Circuits and Systems. I, Fundamental Theory and Applications 49 (2002) 233–240.

[39] S. Tarbouriech, J.M. Gomes da Silva Jr., Synthesis of controllers for continuous-time delay systems with saturating controls via LMI's, IEEE Transactions on Automatic Control 45 (2000) 105–111.

[40] Y. Chen, S. Fei, K. Zhang, Stabilisation for switched linear systems with time-varying delay and input saturation, International Journal of Systems Science 45 (2014) 532–546.

[41] Y. Chen, S. Fei, Exponential stabilization for discrete-time time-delay systems with actuator saturation, in: Proceedings of the 33rd Chinese Control Conference, 2014, pp. 6130–6135.

[42] J. Lin, Exponential estimates and stabilization of discrete-time singular time-delay systems subject to actuator saturation, Discrete Dynamics in Nature and Society 2012 (2012).

[43] L.F.P. Silva, V.J.S. Leite, E.B. Castelan, M. Klug, K. Guelton, Local stabilization of nonlinear discrete-time systems with time-varying delay in the states and saturating actuators, Information Sciences 518 (2020) 272–285.

[44] V.C. Pal, J. Kumar, R. Negi, A. Pati, $\mathcal{H}_\infty$ control of discrete time delayed systems in presence of actuator saturation, in: 2017 Indian Control Conference (ICC), IEEE 2017 Indian Control Conference (ICC), Guwahati, India, 2017.1.4-2017.1.6, 2017, pp. 295–302.

[45] L.F.P. Silva, V.J.S. Leite, E.B. Castelan, G. Feng, Delay dependent local stabilization conditions for time-delay nonlinear discrete-time systems using Takagi–Sugeno models, International Journal of Control, Automation, and Systems (2018) 1435–1447.

[46] K. Naamane, R. Chaibi, E.H. Tissir, A. Hmamed, Stabilization of discrete-time T-S fuzzy systems with saturating actuators, in: 2017 International Conference on Advanced Technologies for Signal and Image Processing (ATSIP), IEEE, 2017, pp. 1–5.

[47] S. Xu, G. Feng, Y. Zou, J. Huang, Robust controller design of uncertain discrete time-delay systems with input saturation and disturbances, IEEE Transactions on Automatic Control 57 (2012) 2604–2609.

[48] X. Zhang, J. Zhao, G.M. Dimirovski, $\mathcal{L}_2$ gain analysis and control synthesis of uncertain discrete-time switched linear systems with time delay and actuator saturation, International Journal of Control 84 (2011) 1746–1758.

[49] S. Boyd, L.E. Ghaoui, E. Feron, V. Balakrishnan, Linear Matrix Inequalities in Systems and Control Theory, SIAM Studies in Applied Mathematics, SIAM, Philadelphia, PA, 1994.

[50] R. Oliveira, P. Peres, Parameter-dependent LMIs in robust analysis: characterization of homogeneous polynomially parameter-dependent solutions via LMI relaxations, IEEE Transactions on Automatic Control 52 (2007) 1334–1340.

[51] Q. Zhou, C. Wu, P. Shi, Observer-based adaptive fuzzy tracking control of nonlinear systems with time delay and input saturation, Fuzzy Sets and Systems 316 (2017) 49–68.

[52] H. Min, S. Xu, Q. Ma, B. Zhang, Z. Zhang, Composite-observer-based output-feedback control for nonlinear time-delay systems with input saturation and its application, IEEE Transactions on Industrial Electronics 65 (2017) 5856–5863.

CHAPTER ELEVEN

Robust output feedback control of linear state-delayed systems using dynamic IQCs

Xiangming Xue[a], Chengzhi Yuan[b], and Fen Wu[a]
[a]Department of Mechanical and Aerospace Engineering, North Carolina State University, Raleigh, NC, United States
[b]Department of Mechanical, Industrial and Systems Engineering, University of Rhode Island, Kingston, RI, United States

Contents

11.1. Introduction

The past decades have witnessed considerable interests in time-delay systems due to their wide presence in many practical engineering problems, such as industry manufacturing processes, internet environments, nuclear reactors, rolling mills, ecological groups, and so on [1–3]. The stability property of time-delay systems has been investigated extensively, and a great number of results have been reported in the literature (see, e.g., [2–9] and the references therein). In particular, different types of criterions for stability and performance analysis of time-delay systems have been established in [6–8] for different dynamical systems with time delays using different mathematical tools; one common feature of these works is that the authors were seeking to construct sharper Lyapunov–Krasovskii functionals (LKFs) so as to derive tighter analysis results. As observed, less conservative results

Control Strategy for Time-Delay Systems
https://doi.org/10.1016/B978-0-12-820599-0.00016-1

are typically resulted from employing more complicated LKFs. This on one side undoubtedly can improve the analysis accuracy of time-delay systems, whereas it also causes some difficulties for delay controller synthesis/design on the other side, especially for systems with time-varying delays. One of the difficulties is of technical nature and lies in that when the stabilization issue is concerned, most existing analysis results (especially for those associated with complex LKFs) cannot be applied directly to the delay control synthesis problem, because the resulting controller synthesis conditions are usually represented in a nonconvex form (typically, in the form of bilinear matrix inequalities, BMIs), which could be very difficult to solve (see, e.g., [10]). To obtain computationally tractable conditions for the stabilization problem, the most commonly used method is to consider special cases and impose structural constraints on certain decision variables, which however will inevitably introduce some conservatism into the synthesis results. This forms the gap between the analysis and synthesis results in the current literature. It also explains why very limited results for synthesis of delay control systems can be found in spite of the highly rich literature on time-delay systems analysis.

On the other hand, the concept of integral quadratic constraint (IQC) was originally proposed by Megretski and Rantzer [11] to model different types of nonlinearities and uncertainties, such as saturation, dead zone, delay, parametric or unmodeled dynamics, and so on. Later on, it was demonstrated to be successful for stability and stabilization of various uncertain dynamical systems (see, e.g., [5,12–17]). When specifying to time-delay systems, many interesting results have been developed based on the IQC framework. In particular, a library of IQC multipliers have been developed by Kao and Rantzer [5] for continuous-time linear time-invariant (LTI) systems with time-varying delays and by Kao [12] for discrete-time case. In these references the powerfulness of the IQC in time-delay systems analysis has been demonstrated through extensive comparisons against the LKF-based approaches. However, virtually all these results pertain to stability and performance analysis issues, whereas the stabilization issue for time-delay systems under the IQC framework has not yet been addressed adequately. This motivates us exploiting this powerful tool to solve the challenging delay control synthesis problem by providing a systematic and unified design framework that simultaneously renders stringent performance and computational efficacy.

In this chapter, we investigate the robust output-feedback delay control synthesis problem for a class of linear systems with time-varying state delay under the IQC framework. The delay stability and the $\mathcal{L}_2$-gain performance are explored using dynamic IQCs incorporated with quadratic Lyapunov functions. Based on the analysis results, the design of robust dynamic output-feedback controllers that guarantee desired $\mathcal{L}_2$-gain performance is examined. The resulting output-feedback control synthesis conditions are formulated in terms of linear matrix inequalities (LMIs) plus a line search over one scalar variable, which can be solved effectively using existing LMI-based algorithms. The proposed IQC-based robust output-feedback control scheme provides a simple and systematic design framework for synthesis of linear time-delay control systems via convex LMI-based optimization.

The main contributions of this chapter can be summarized from three aspects:

i. Dynamic IQCs are successfully employed for robust output feedback control of systems subject to time-varying state delay with guaranteed stability and optimal $\mathcal{H}_\infty$ controlled performance.

ii. In comparison to existing methods, the new IQC-based delay control design scheme has better controlled performance in terms of both maximum allowable time-delay bounds, better $\mathcal{L}_2$ performance levels, and better design capability in handling cases with fast time-varying delay, especially when the delay rate r in the range of $1 \leq r < 2$.

iii. Synthesis condition are formulated as LMIs plus a line search, which can be solved effectively with convex optimization techniques. As a result, this work significantly simplifies the delay control system design compared to the classical LKF-based and BMIs approaches.

The rest of the chapter is organized as follows. The problem statement and some basic definitions are given in Section 11.2. The main results of the chapter, including the conditions for robust stability analysis and dynamic output-feedback control synthesis, are contained in Section 11.3. Section 11.4 provides numerical studies to illustrate the effectiveness and advantages of the proposed design approach. The conclusions are given in Section 11.5.

Notation. $\mathbb{R}$ and $\mathbb{R}_+$ stand for the sets of real and positive real numbers, respectively. $\mathbb{R}^{m\times n}$ is the set of real $m \times n$ matrices, and $\mathbb{R}^n$ represents the set of real $n \times 1$ vectors. The identity matrix of dimension $n \times n$ is denoted by I_n. $\mathbb{S}^n$ and $\mathbb{S}^n_+$ are used to denote the sets of real symmetric $n \times n$ matrices and positive definite matrices, respectively. A block diagonal matrix with matrices $X_1, X_2, \ldots, X_p$ on its main diagonal is denoted by

$diag\{X_1, X_2, \cdots, X_p\}$. Furthermore, we use the symbol $\star$ in LMIs to denote entries that follow from symmetry. For two integers $k_1 < k_2$, we denote $\mathbf{I}[k_1, k_2] = \{k_1, k_1+1, \ldots, k_2\}$. For a matrix $M \in \mathbb{C}^{m\times n}$, M^T denotes its transpose. For $x \in \mathbb{R}^n$, its norm is defined as $\|x\| := (x^T x)^{1/2}$. L_{2+}^n is the space of functions $u: [0, \infty) \to \mathbb{R}^n$ satisfying $\|u\|_2 := \left(\int_0^\infty u^T(t)u(t)dt\right)^{1/2} < \infty$. Given $u \in L_{2+}^n$, u_T denotes the truncated function $u_T(t) = u(t)$ for $t \le T$ and $u_T(t) = 0$ otherwise. The extended space, denoted as L_{2e+}, is the set of functions u such that $u_T \in L_{2+}$ for all $T \ge 0$.

11.2. Problem statement and preliminaries

Consider a class of linear systems with time-varying state delay described in the following form:

$$\begin{bmatrix} \dot{x}_p \\ e \\ y \end{bmatrix} = \begin{bmatrix} A_p & A_d & B_{p1} & B_{p2} \\ C_{p1} & C_{d1} & D_{p11} & D_{p12} \\ C_{p2} & C_{d2} & D_{p21} & 0 \end{bmatrix} \begin{bmatrix} x_p \\ \mathcal{D}_{\bar{\tau},r}(x_p) \\ d \\ u \end{bmatrix}, \tag{11.1}$$

where $x_p \in \mathbb{R}^{n_p}$ is the plant states, $d \in \mathbb{R}^{n_d}$ is the disturbance, $e \in \mathbb{R}^{n_e}$ is the controlled output, $u \in \mathbb{R}^{n_u}$ is the control input, and $y \in \mathbb{R}^{n_y}$ is the measurement output for feedback control use. Moreover, $\mathcal{D}_{\bar{\tau},r}(x_p)$ denotes the time-varying delay function defined by $\mathcal{D}_{\bar{\tau},r}(x_p) = x_p(t - \tau(t))$ with $\tau(t)$ specifying the delay at time t and satisfying

$$\tau(t) \in \mathcal{T}_{\bar{\tau},r} := \{\tau : \mathbb{R}_+ \to [0, \bar{\tau}], |\dot{\tau}(t)| \le r\}, \tag{11.2}$$

where $\bar{\tau}$ and r represent the maximum delay and its variation bound of the delay $\tau(t)$ for all $t \ge 0$, respectively. For $t \in [-\bar{\tau}, 0]$, we define $x_p(t) = x_{p-}(t)$ with $x_{p-}(t)$ being the initial delay function. We assume that the time-varying delay $\tau(t)$ is unknown but its associated delay parameters $(\bar{\tau}, r)$, together with all the constant system matrices, are available for controller design.

Defining $w = \mathcal{S}_{\bar{\tau},r}(x_p) := \mathcal{D}_{\bar{\tau},r}(x_p) - x_p$, the state-delayed system (11.1) can be rewritten as

$$\begin{aligned} \begin{bmatrix} \dot{x}_p \\ e \\ y \end{bmatrix} &= \begin{bmatrix} A_p + A_d & A_d & B_{p1} & B_{p2} \\ C_{p1} + C_{d1} & C_{d1} & D_{p11} & D_{p12} \\ C_{p2} + C_{d2} & C_{d2} & D_{p21} & 0 \end{bmatrix} \begin{bmatrix} x_p \\ w \\ d \\ u \end{bmatrix}, \\ w &= \mathcal{S}_{\bar{\tau},r}(x_p). \end{aligned} \tag{11.3}$$

In this chapter, we consider the robust $\mathcal{H}_\infty$ delay control problem for system (11.3) via dynamic output feedback under the integral quadratic constraint (IQC) framework. Specifically, we seek to design an output-feedback control law that minimizes the $\mathcal{L}_2$ gain γ from the disturbance d to the error output e (i.e., $\|e\|_2 < \gamma\|d\|_2$) for all uncertain time-varying delays $\tau(t) \in \mathcal{T}_{\bar{\tau},r}$. Before proceeding further, we first recall following definition of time-domain IQC [18].

Definition 11.1. Let Ψ be a stable linear time-invariant (LTI) system, that is, $\Psi \in \mathbb{RH}_\infty^{n_z \times (n_v+n_w)}$, and $W = W^T \in \mathbb{R}^{n_z \times n_z}$. Two signals $v \in L_{2e+}^{n_v}$ and $w \in L_{2e+}^{n_w}$ satisfy the IQC defined by the multiplier Ψ and matrix W if the following inequality holds for all $T \geq 0$:

$$\int_0^T z^T(t) W z(t) dt \geq 0, \tag{11.4}$$

where z is the output of Ψ driven by inputs (v, w) with zero initial conditions. A bounded causal operator $\mathcal{S} : L_{2e+}^{n_v} \to L_{2e+}^{n_w}$ satisfies the IQC defined by (Ψ, W) if condition (11.4) holds for all $v \in L_{2e+}^{n_v}$, $w = \mathcal{S}(v)$, and $T \geq 0$.

IQCs, as a powerful tool for modeling a large variety of nonlinearities, have been demonstrated to be successful for robust stability analysis of various dynamical systems (see, e.g., [5,11,12,18,19]). In this chapter, we solve the robust $\mathcal{H}_\infty$ synthesis problem for system (11.3) by employing dynamic IQCs to characterize the input–output behavior of the time-varying delay nonlinearity $\mathcal{S}_{\bar{\tau},r}(x_p)$. A library of IQC multipliers for time-varying delays can be found in [5] for continuous-time settings and in [12] for discrete-time settings. Due to the space limitation, we refer the readers to these references for more details.

For simplicity, we have the following assumptions on the open-loop plant (11.3).

Assumption 11.1. The pair (A_p, B_{p2}) is robustly stabilizable, and the pair (A_p, C_{p2}) is robustly detectable for the nondelay nominal linear system (11.1).

Assumption 11.2. $\mathcal{S}_{\bar{\tau},r}$ satisfies a collection of IQCs defined by $\{(\Psi_k, W_k)\}_{k=1}^{N_\lambda}$ with $\Psi_k = \begin{bmatrix} \Psi_{11,k} & \Psi_{12,k} \\ 0 & I_{n_p} \end{bmatrix} \in \mathbb{RH}_\infty^{(n_p+n_p)\times(n_p+n_p)}$ and $W_k = \begin{bmatrix} I_{n_p} & 0 \\ 0 & -I_{n_p} \end{bmatrix}$ for all $k \in \mathbf{I}[1, N_\lambda]$.

We stress that these two assumptions do not cause any loss of generality. Assumption 11.1 is necessary to guarantee the existence of an output-feedback stabilizing controller for a nondelayed nominal linear system (11.1). Assumption 11.2 is made to simplify the derivations and can be relaxed via loop transformation [20] and J_{n_p,n_p}-spectral factorization [18] with more extensive and complicated formulae.

With such a system setup, the IQC-induced LTI system $\{\Psi_k\}_{k=1}^{N_\lambda}$ for the delay nonlinearity $\mathcal{S}_{\bar{\tau},r}(x_p)$ in system (11.3) can be described in the following state-space form:

$$\begin{bmatrix} \dot{x}_\psi \\ z_k \end{bmatrix} = \begin{bmatrix} A_\psi & B_{\psi 1} & B_{\psi 2} \\ C_{\psi,k} & D_{\psi 1,k} & D_{\psi 2,k} \end{bmatrix} \begin{bmatrix} x_\psi \\ x_p \\ w \end{bmatrix}, \quad k \in \mathbf{I}[1, N_\lambda], \tag{11.5}$$

where $x_\psi \in \mathbb{R}^{n_\psi}$ denotes the state vector of the operator $\{\Psi_k\}_{k=1}^{N_\lambda}$ with $x_\psi(0) = 0$; $z_k \in \mathbb{R}^{n_z}$ with $n_z = 2n_p$ for all $k \in \mathbf{I}[1, N_\lambda]$ are the operator outputs. In particular, Assumption 11.2 renders the associated output matrices the following structure for all $k \in \mathbf{I}[1, N_\lambda]$:

$$C_{\psi,k} = \begin{bmatrix} \bar{C}_{\psi,k} \\ 0 \end{bmatrix}, \quad D_{\psi 1,k} = \begin{bmatrix} \bar{D}_{\psi 1,k} \\ 0 \end{bmatrix}, \quad D_{\psi 2,k} = \begin{bmatrix} \bar{D}_{\psi 2,k} \\ I_{n_p} \end{bmatrix}.$$

To fulfill the above-mentioned robust output-feedback stabilization objective, we construct the output-feedback controller in the following dynamic form:

$$\begin{bmatrix} \dot{x}_c \\ u \end{bmatrix} = \begin{bmatrix} A_c & B_c \\ C_c & D_c \end{bmatrix} \begin{bmatrix} x_c \\ y \end{bmatrix}, \tag{11.6}$$

where $x_c \in \mathbb{R}^{n_c}$ is the controller state with the order n_c to be determined, and $A_c \in \mathbb{R}^{n_c \times n_c}$, $B_c \in \mathbb{R}^{n_c \times n_y}$, $C_c \in \mathbb{R}^{n_u \times n_c}$, and $D_c \in \mathbb{R}^{n_u \times n_y}$ are controller gain matrices to be designed.

The extended closed-loop system interconnecting the controlled plant (11.3), the IQC-induced system (11.5), and the controller (11.6) can be expressed as

$$\begin{aligned} \begin{bmatrix} \dot{x}_{cl} \\ z_k \\ e \end{bmatrix} &= \begin{bmatrix} A_{cl} & B_{cl1} & B_{cl2} \\ C_{cl1,k} & D_{cl11,k} & D_{cl12,k} \\ C_{cl2} & D_{cl21} & D_{cl22} \end{bmatrix} \begin{bmatrix} x_{cl} \\ w \\ d \end{bmatrix}, \\ w &= \mathcal{S}_{\bar{\tau},r}(x_p), \end{aligned} \tag{11.7}$$

where $x_{cl} := [x_p^T \ x_\psi^T \ x_c^T]^T$ and

$$A_{cl} = \begin{bmatrix} \begin{bmatrix} A_p + A_d & 0 \\ B_{\psi 1} & A_\psi \end{bmatrix} + \begin{bmatrix} B_{p2} \\ 0 \end{bmatrix} D_c \begin{bmatrix} C_{p2} + C_{d2} & 0 \end{bmatrix} & \begin{bmatrix} B_{p2} \\ 0 \end{bmatrix} C_c \\ B_c \begin{bmatrix} C_{p2} + C_{d2} & 0 \end{bmatrix} & A_c \end{bmatrix},$$

$$B_{cl1} = \begin{bmatrix} \begin{bmatrix} A_d \\ B_{\psi 2} \end{bmatrix} + \begin{bmatrix} B_{p2} \\ 0 \end{bmatrix} D_c C_{d2} \\ B_c C_{d2} \end{bmatrix},$$

$$B_{cl2} = \begin{bmatrix} \begin{bmatrix} B_{p1} \\ 0 \end{bmatrix} + \begin{bmatrix} B_{p2} \\ 0 \end{bmatrix} D_c D_{p21} \\ B_c D_{p21} \end{bmatrix},$$

$$C_{cl1,k} = \begin{bmatrix} \begin{bmatrix} D_{\psi 1,k} & C_{\psi,k} \end{bmatrix} & 0 \end{bmatrix},$$

$$C_{cl2} = \begin{bmatrix} \begin{bmatrix} C_{p1} + C_{d1} & 0 \end{bmatrix} + D_{p12} D_c \begin{bmatrix} C_{p2} + C_{d2} & 0 \end{bmatrix} & D_{p12} C_c \end{bmatrix},$$

$$\begin{bmatrix} D_{cl11,k} & D_{cl12,k} \\ D_{cl21} & D_{cl22} \end{bmatrix} = \begin{bmatrix} D_{\psi 2,k} & 0 \\ C_{p1} + D_{p12} D_c C_{d2} & D_{p11} + D_{p12} D_c D_{p21} \end{bmatrix}.$$

In particular, for all $k \in \mathbf{I}[1, N_\lambda]$, we have

$$C_{cl1,k} = \begin{bmatrix} \bar{C}_{cl1,k} \\ 0 \end{bmatrix} := \begin{bmatrix} \begin{bmatrix} \bar{D}_{\psi 1,k} & \bar{C}_{\psi,k} \end{bmatrix} & 0 \\ 0 & 0 \end{bmatrix},$$

$$D_{cl11,k} = \begin{bmatrix} \bar{D}_{cl11,k} \\ I_{n_p} \end{bmatrix} := \begin{bmatrix} \bar{D}_{\psi 2,k} \\ I_{n_p} \end{bmatrix}.$$

11.3. Main results

11.3.1 Robust $\mathcal{L}_2$-gain analysis

In this subsection, we first establish the analysis conditions for robust $\mathcal{L}_2$ stability of the extended closed-loop system (11.7) under the IQC framework.

Theorem 11.1. *Consider the extended closed-loop system* (11.7) *with Assumptions 11.1 and 11.2. Suppose there exist positive definite matrices* $P \in \mathbb{S}_+^{n_p+n_\psi+n_c}$ *and positive scalars* $\gamma, \lambda_k \in \mathbb{R}_+$ *for all* $k \in \mathbf{I}[1, N_\lambda]$ *such that*

$$\begin{bmatrix} He\{PA_{cl}\} & \star & \star \\ B_{cl1}^T P & -\sum_{k=1}^{N_\lambda} \lambda_k I_{n_p} & \star \\ B_{cl2}^T P & 0 & -\gamma I_{n_d} \end{bmatrix} + \frac{1}{\gamma} \begin{bmatrix} C_{cl2}^T \\ D_{cl21}^T \\ D_{cl22}^T \end{bmatrix} \begin{bmatrix} C_{cl2} & D_{cl21} & D_{cl22} \end{bmatrix}$$
$$+ \sum_{k=1}^{N_\lambda} \lambda_k \begin{bmatrix} \bar{C}_{cl1,k}^T \\ \bar{D}_{cl11,k}^T \\ \bar{D}_{cl12,k}^T \end{bmatrix} \begin{bmatrix} \bar{C}_{cl1,k} & \bar{D}_{cl11,k} & \bar{D}_{cl12,k} \end{bmatrix} < 0. \tag{11.8}$$

Then the extended closed-loop system (11.7) *with time-varying state delay is stable and achieves an $\mathcal{L}_2$ gain less than γ for all $\tau(t) \in \mathcal{T}_{\bar{\tau},r}$.*

Proof. Define the Lyapunov function as $V(x_{cl}) = x_{cl}^T P x_{cl}$ for the extended closed-loop system (11.7). Then we have

$$\begin{aligned} \dot{V} &= (A_{cl}x_{cl} + B_{cl1}w + B_{cl2}d)^T P x_{cl} + x_{cl}^T P (A_{cl}x_{cl} + B_{cl1}w + B_{cl2}d) \\ &= \begin{bmatrix} x_{cl}^T & w^T & d^T \end{bmatrix} \begin{bmatrix} He\{PA_{cl}\} & \star & \star \\ B_{cl1}^T P & 0 & \star \\ B_{cl2}^T P & 0 & 0 \end{bmatrix} \begin{bmatrix} x_{cl} \\ w \\ d \end{bmatrix}. \end{aligned}$$

Multiplying the vector $[x_{cl}^T \ w^T \ d^T]^T$ from the right of condition (11.8) and its transpose to the left and using $W_k = diag\{I_{n_p}, -I_{n_p}\}$ for all $k \in \mathbf{I}[1, N_\lambda]$, by Assumption 11.2 we obtain

$$\dot{V} + \frac{1}{\gamma} e^T e - \gamma d^T d + \sum_{k=1}^{N_\lambda} \lambda_k z_k^T W_k z_k < 0.$$

Integrating both sides of this inequality from $t = 0$ to $t = T$ with zero initial conditions, we obtain

$$V(x_{cl}(T)) + \sum_{k=1}^{N_\lambda} \int_0^T z_k^T(t) W_k z_k(t) dt < \gamma \int_0^T d^T(t) d(t) dt - \frac{1}{\gamma} \int_0^T e^T(t) e(t) dt.$$

It follows from the IQC constraint (11.4) and the nonnegativity of the Lyapunov function V that

$$\frac{1}{\gamma} \int_0^T e^T(t) e(t) dt < \gamma \int_0^T d^T(t) d(t) dt.$$

Thus we conclude that the extended closed-loop system (11.7) with time-varying state delay is stable and $\|e\|_2 < \gamma \|d\|_2$ for all $\tau(t) \in \mathcal{T}_{\bar{\tau},r}$. □

Remark 11.1. Note that the robust analysis condition (11.8) reduces to the standard bounded real lemma [20] for system (11.7) without state delay effects.

11.3.2 Robust output-feedback control synthesis

In this subsection, based on the analysis results in Subsection 11.3.1, we address the robust $\mathcal{H}_\infty$ synthesis problem for the output-feedback controller (11.6). The synthesis conditions, together with the controller construction algorithms, are given in the following theorem.

Theorem 11.2. *Consider the linear state-delayed plant* (11.1). *Suppose there exist positive definite matrices* $R, S \in \mathbb{S}_+^{n_p+n_\psi}$, *rectangular matrices* $\hat{A}_c \in \mathbb{R}^{(n_p+n_\psi)\times(n_p+n_\psi)}$, $\hat{B} \in \mathbb{R}^{(n_p+n_\psi)\times n_y}$, $\hat{C}_c \in \mathbb{R}^{n_u\times(n_p+n_\psi)}$, $\hat{D}_c \in \mathbb{R}^{n_u\times n_y}$, *and positive scalars* $\gamma, \hat{\lambda}, \hat{\lambda}_k \in \mathbb{R}_+$ *for all* $k \in \mathbf{I}[1, N_\lambda]$ *such that*

$$
\begin{bmatrix}
He\left\{\begin{bmatrix} A_p + A_d & 0 \\ B_{\psi 1} & A_\psi \end{bmatrix} R + \begin{bmatrix} B_{p2} \\ 0 \end{bmatrix} \hat{C}_c\right\} & \star & \star & \star & \star & \star \\
\hat{A}_c + \begin{bmatrix} A_p + A_d & 0 \\ B_{\psi 1} & A_\psi \end{bmatrix}^T + \begin{bmatrix} C_{p2}^T + C_{d2}^T \\ 0 \end{bmatrix} \hat{D}_c^T \begin{bmatrix} B_{p2} \\ 0 \end{bmatrix}^T & He\left\{\begin{matrix} S\begin{bmatrix} A_p + A_d & 0 \\ B_{\psi 1} & A_\psi \end{bmatrix} \\ +\hat{B}_c\begin{bmatrix} C_{p2} + C_{d2} & 0 \end{bmatrix} \end{matrix}\right\} & \star & \star & \star & \star \\
\hat{\lambda}\begin{bmatrix} A_d \\ B_{\psi 2} \end{bmatrix}^T + \hat{\lambda} C_{p2}^T \hat{D}_c^T \begin{bmatrix} B_{p2} \\ 0 \end{bmatrix}^T & \hat{\lambda}\begin{bmatrix} A_d \\ B_{\psi 2} \end{bmatrix}^T S + \hat{\lambda} C_{d2}^T \hat{B}_c^T & \sum_{k=1}^{N_\lambda}(\hat{\lambda}_k - 2\hat{\lambda}) I_{n_p} & \star & \star & \star \\
\begin{bmatrix} B_{p1} \\ 0 \end{bmatrix}^T + D_{p21}^T \hat{D}_c^T \begin{bmatrix} B_{p2} \\ 0 \end{bmatrix}^T & \begin{bmatrix} B_{p1} \\ 0 \end{bmatrix}^T S + D_{p21}^T \hat{B}_c^T & 0 & -\gamma I_{n_d} & \star & \star \\
\begin{bmatrix} C_{p1} + C_{d1} & 0 \end{bmatrix} R + D_{p12}\hat{C}_c & \left\{\begin{matrix} \begin{bmatrix} C_{p1} + C_{d1} & 0 \end{bmatrix} \\ +D_{p12}\hat{D}_c\begin{bmatrix} C_{p2} + C_{d2} & 0 \end{bmatrix} \end{matrix}\right\} & \hat{\lambda} C_{p1} + \hat{\lambda} D_{p12}\hat{D}_c C_{d2} & D_{p11} + D_{p12}\hat{D}_c D_{p21} & -\gamma I_{n_e} & \star \\
\hat{\Upsilon}_{61} & \hat{\Upsilon}_{62} & \hat{\Upsilon}_{63} & 0 & 0 & -\hat{\Lambda}
\end{bmatrix} < 0, \quad (11.9)
$$

$$
\begin{bmatrix} R & \star \\ I & S \end{bmatrix} > 0, \quad (11.10)
$$

where

$$\begin{bmatrix}\hat{\Upsilon}_{61} & \hat{\Upsilon}_{62} & \hat{\Upsilon}_{63}\end{bmatrix} = \left[\begin{array}{c|c|c} \begin{bmatrix}\bar{D}_{\psi 1,1} & \bar{C}_{\psi,1}\end{bmatrix}R & \begin{bmatrix}\bar{D}_{\psi 1,1} & \bar{C}_{\psi,1}\end{bmatrix} & \hat{\lambda}\bar{D}_{\psi 2,1} \\ \vdots & \vdots & \vdots \\ \begin{bmatrix}\bar{D}_{\psi 1,N_\lambda} & \bar{C}_{\psi,N_\lambda}\end{bmatrix}R & \begin{bmatrix}\bar{D}_{\psi 1,N_\lambda} & \bar{C}_{\psi,N_\lambda}\end{bmatrix} & \hat{\lambda}\bar{D}_{\psi 2,N_\lambda} \end{array}\right],$$

$$\hat{\Lambda} = diag\left\{\hat{\lambda}_1 I_{n_p}, \cdots, \hat{\lambda}_{N_\lambda} I_{n_p}\right\}.$$

Then the closed-loop system (11.7) *with time-varying state delay is stabilized by the output-feedback controller* (11.6) *of order* $n_c = n_p + n_\psi$ *and achieves an* $\mathcal{L}_2$ *gain less than* γ *for all* $\tau(t) \in \mathcal{T}_{\bar{\tau},r}$. *Moreover, the output-feedback controller coefficient matrices are given by*

$$\begin{bmatrix} A_c & B_c \\ C_c & D_c \end{bmatrix} = \begin{bmatrix} N & S\begin{bmatrix} B_{p2} \\ 0 \end{bmatrix} \\ 0 & I_{n_u} \end{bmatrix}^{-1} \begin{bmatrix} \hat{A}_c - S\begin{bmatrix} A_p + A_d & 0 \\ B_{\psi 1} & A_\psi \end{bmatrix} R & \hat{B}_c \\ \hat{C}_c & \hat{D}_c \end{bmatrix} \times \begin{bmatrix} M^T & 0 \\ \begin{bmatrix} C_{p2} + C_{d2} & 0 \end{bmatrix} R & I_{n_y} \end{bmatrix}^{-1}, \tag{11.11}$$

where $MN^T = I - RS$. *The scaling factors corresponding to the IQC multipliers are* $\lambda_k = \hat{\lambda}_k^{-1}$ *for all* $k \in \mathbf{I}[1, \lambda_N]$.

Proof. According to Theorem 11.1, the extended closed-loop system (11.7) is stable and achieves an $\mathcal{L}_2$ gain less than γ if condition (11.8) is satisfied. Then, under Assumption 11.2, through the Schur complement, from condition (11.8) we obtain that

$$\begin{bmatrix} He\{PA_{cl}\} & \star & \star & \star & \star \\ B_{cl1}^T P & -\sum_{k=1}^{N_\lambda} \lambda_k I_{n_p} & \star & \star & \star \\ B_{cl2}^T P & 0 & -\gamma I_{n_d} & \star & \star \\ C_{cl2} & D_{cl21} & D_{cl22} & -\gamma I_{n_e} & \star \\ \Upsilon_{51} & \Upsilon_{52} & 0 & 0 & -\Lambda \end{bmatrix} < 0, \tag{11.12}$$

where

$$\Upsilon_{51} = \begin{bmatrix} \bar{C}_{cl1,1} \\ \vdots \\ \bar{C}_{cl1,N_\lambda} \end{bmatrix}, \quad \Upsilon_{52} = \begin{bmatrix} \bar{D}_{cl11,1} \\ \vdots \\ \bar{D}_{cl11,N_\lambda} \end{bmatrix},$$

$$\Lambda = diag\left\{\frac{1}{\lambda_1}I_{n_p}, \cdots, \frac{1}{\lambda_{N_\lambda}}I_{n_p}\right\}.$$

Denoting $\hat{\lambda}_k = \lambda_k^{-1}$ for all $k \in \mathbf{I}[1, N_\lambda]$ and using the fact that $-\hat{\lambda}_k^{-1} \leq -\hat{\lambda}^{-2}(2\hat{\lambda} - \hat{\lambda}_k)$ for all $k \in \mathbf{I}[1, N_\lambda]$ and any positive scalar $\hat{\lambda}$, it suffices to have a sufficient condition to (11.12) as

$$\begin{bmatrix} He\{PA_{cl}\} & \star & \star & \star & \star \\ B_{cl1}^T P & \sum_{k=1}^{N_\lambda} \hat{\lambda}^{-2}(\hat{\lambda}_k - 2\hat{\lambda})I_{n_p} & \star & \star & \star \\ B_{cl2}^T P & 0 & -\gamma I_{n_d} & \star & \star \\ C_{cl2} & D_{cl21} & D_{cl22} & -\gamma I_{n_e} & \star \\ \Upsilon_{51} & \Upsilon_{52} & 0 & 0 & -\hat{\Lambda} \end{bmatrix} < 0 \quad (11.13)$$

with $\hat{\Lambda} = diag\{\hat{\lambda}_1 I_{n_p}, \cdots, \hat{\lambda}_k I_{n_p}\}$.

Let $n_c = n_p + n_\psi$, and let

$$P = \begin{bmatrix} S & N \\ N^T & X^{-1} \end{bmatrix}$$

be the partition with $S \in \mathbb{S}_+^{n_p+n_\psi}$ and $X^{-1} \in \mathbb{S}_+^{n_p+n_\psi}$. Then define

$$Z_1 = \begin{bmatrix} R & I \\ M^T & 0 \end{bmatrix}, \quad Z_2 = \begin{bmatrix} I & S \\ 0 & N^T \end{bmatrix}$$

such that $PZ_1 = Z_2$ and $MN^T = I - RS$, which implies $X^{-1} = -N^T RM^{-T}$. Based on condition (11.10), we can verify that $Z_1^T PZ_1 = \begin{bmatrix} R & I \\ I & S \end{bmatrix} > 0$; in turn, $P > 0$ as Z_1 is nonsingular.

Then multiplying inequality (11.13) by the matrix $diag\{Z_1, \hat{\lambda} I_{n_p}, I_{n_d}, I_{n_e}, I_{N_\lambda n_p}\}$ from the right and by its transpose from the left, we obtain the following results:

$$Z_1^T PA_{cl}Z_1 = Z_2^T A_{cl}Z_1$$

$$= \begin{bmatrix} \begin{bmatrix} A_p + A_d & 0 \\ B_{\psi 1} & A_\psi \end{bmatrix} R + \begin{bmatrix} B_{p2} \\ 0 \end{bmatrix} \hat{C}_c & \left\{ \begin{matrix} S\begin{bmatrix} A_p + A_d & 0 \\ B_{\psi 1} & A_\psi \end{bmatrix} \\ +\hat{B}_c \begin{bmatrix} C_{p2} + C_{d2} & 0 \end{bmatrix} \end{matrix} \right\} \\ \hat{A}_c & S\begin{bmatrix} A_p + A_d & 0 \\ B_{\psi 1} & A_\psi \end{bmatrix} + \hat{B}_c \begin{bmatrix} C_{p2} + C_{d2} & 0 \end{bmatrix} \end{bmatrix},$$

$$
\begin{aligned}
&\hat{\lambda} B_{cl1}^T P Z_1 = \hat{\lambda} B_{cl1}^T Z_2 \\
&= \begin{bmatrix} \hat{\lambda} \begin{bmatrix} A_d^T & B_{\psi 2}^T \end{bmatrix} + \hat{\lambda} C_{d2}^T \hat{D}_c^T \begin{bmatrix} B_{p2}^T & 0 \end{bmatrix} & \hat{\lambda} \begin{bmatrix} A_d^T & B_{\psi 2}^T \end{bmatrix} S + \hat{\lambda} C_{d2}^T \hat{B}_c^T \end{bmatrix}, \\
&B_{cl2}^T P Z_1 = B_{cl2}^T Z_2 = \begin{bmatrix} \begin{bmatrix} B_{p1}^T & 0 \end{bmatrix} + D_{p21}^T \hat{D}_c^T \begin{bmatrix} B_{p2}^T & 0 \end{bmatrix} & \begin{bmatrix} B_{p1}^T & 0 \end{bmatrix} S + D_{p21}^T \hat{B}_c^T \end{bmatrix}, \\
&C_{cl2} Z_1 = \begin{bmatrix} \left\{ \begin{matrix} \begin{bmatrix} C_{p1} + C_{d1} & 0 \end{bmatrix} R \\ + D_{p12} \hat{C}_c \end{matrix} \right\} & \left\{ \begin{matrix} \begin{bmatrix} C_{p1} + C_{d1} & 0 \end{bmatrix} \\ + D_{p12} \hat{D}_c \begin{bmatrix} C_{p2} + C_{d2} & 0 \end{bmatrix} \end{matrix} \right\} \end{bmatrix}, \\
&\hat{\lambda} D_{cl21} = \hat{\lambda} C_{p11} + \hat{\lambda} D_{p12} \hat{D}_c C_{d2},
\end{aligned}
$$

and for all $k \in \mathbf{I}[1, N_\lambda]$,

$$
\begin{aligned}
\bar{C}_{cl1,k} Z_1 &= \begin{bmatrix} \begin{bmatrix} \bar{D}_{\psi 1,k} & \bar{C}_{\psi,k} \end{bmatrix} R & \begin{bmatrix} \bar{D}_{\psi 1,k} & \bar{C}_{\psi,k} \end{bmatrix} \end{bmatrix}, \\
\hat{\lambda} \bar{D}_{cl11,k} &= \hat{\lambda} \bar{D}_{\psi 2,k},
\end{aligned}
$$

where

$$
\begin{aligned}
\hat{A}_c &= S \begin{bmatrix} A_{p0} + A_{p1} & 0 \\ B_{\psi 1} & A_\psi \end{bmatrix} R + S \begin{bmatrix} B_{p2} \\ 0 \end{bmatrix} D_c \begin{bmatrix} C_{p20} + C_{p21} & 0 \end{bmatrix} R \\
&\quad + N B_c \begin{bmatrix} C_{p20} + C_{p21} & 0 \end{bmatrix} R + S \begin{bmatrix} B_{p2} \\ 0 \end{bmatrix} C_c M^T + N A_c M^T, \\
\hat{B}_c &= S \begin{bmatrix} B_{p2} \\ 0 \end{bmatrix} D_c + N B_c, \\
\hat{C}_c &= D_c \begin{bmatrix} C_{p20} + C_{p21} & 0 \end{bmatrix} R + C_c M^T, \\
\hat{D}_c &= D_c.
\end{aligned} \tag{11.14}
$$

Consequently, we recover condition (11.9) by denoting $\hat{\lambda}_k = \lambda_k^{-1}$ for all $k \in \mathbf{I}[1, N_\lambda]$. The controller formula (11.11) can be further verified by inverting the relations in (11.14). □

MATLAB® program `delayIQC_SFsyn` to solve the IQC-basesd controller is given in Appendix 11.A. Note that the proposed IQC-based robust output-feedback control synthesis conditions are formulated in terms of matrix inequalities plus a scalar variable $\hat{\lambda}$. Such a special type of conditions can still be solved effectively via LMI optimization coupled with a line search over the scalar variable $\hat{\lambda}$. This line search may be effectively achieved through bisection algorithm, implemented in the MATLAB program (see Listing 11.1).

```
%Performing a line search over the scalar variable Lambdahat
ll = 0;      %Lower limit Lambdahat
ul = 10;     %Upper limit Lambdahat
np = 50;     %Number of points
eta = 0.2;   %Interval
cnt = 0;     %Number of searching
%Store the lambdahat and gamma
Lambdahat_store = zeros(np,1);
gamma_store = zeros(np,1);
%Starting bisection line search algrithom
for c1 = 1:np/2
    mp = (ul+ll)/2;
    %Apply the Synthesis condition to obtain the optimal ...
        system performance
    % gamma under given lambdahat
    Lambdahat_b = mp—eta;
    tgamma_b = ...
        delayIQC_SFsyn(Ap,Bp1,Bp2,Ad,Cd1,Cd2,Cp1,Cp2, ...
        Dp11,Dp12,Dp21,Dim,Lambdahat_b,Apsi,Bpsi1,Bpsi2, ...
        barCpsi11,barCpsi12,barDpsi11,barDpsi12, ...
        barDpsi21,barDpsi22);
    Lambdahat_c = mp+eta;
    tgamma_c= ...
        delayIQC_SFsyn(Ap,Bp1,Bp2,Ad,Cd1,Cd2,Cp1,Cp2, ...
        Dp11,Dp12,Dp21,Dim,Lambdahat_c,Apsi,Bpsi1,Bpsi2, ...
        barCpsi11,barCpsi12,barDpsi11,barDpsi12, ...
        barDpsi21,barDpsi22);
    %Update the new lower and upper bound
    if tgamma_b<tgamma_c
        ul = Lambdahat_c;
    else
        ll = Lambdahat_b;
    end
    cnt = cnt + 1;
    Lambdahat_store(cnt) = Lambdahat_b;
    gamma_store(cnt) = tgamma_b;
    cnt = cnt + 1;
    Lambdahat_store(cnt) = Lambdahat_c;
    gamma_store(cnt) = tgamma_c;
end
Lambdahat = Lambdahat_store(end,:);
```

Listing 11.1: mfile: Bisectionlinesearch.m.

With fixed $\hat{\lambda} > 0$ in hand, inequalities (11.9) and (11.10) are expressed in terms of LMIs with respect to all variables. Therefore the optimal control solution with minimal $\mathcal{L}_2$ performance can be obtained by solving the convex optimization problem

$$\begin{aligned} \gamma(\hat{\lambda}) = \min_{R,S,\hat{A}_k,\hat{B}_k,\hat{C}_k,\hat{D}_k,\hat{E}_k,\hat{F}_k,\hat{\lambda}_k} \quad & \gamma \\ \text{s.t.} \quad & (11.9)\text{–}(11.10),\ \forall k \in \mathbf{I}[1, N_{\lambda}]. \end{aligned} \tag{11.15}$$

After solving this problem, the output-feedback controller coefficient matrices are computed by the MATLAB program (see Listing 11.2), which is described in (11.11).

```
%Construct the IQC–based controller from the output of ...
    synthesis condition
%Extract the dimension information from Dim vector
Dim_n = Dim(1);          %number of plant states
Dim_d = Dim(2);          %number of disturbance
Dim_u = Dim(3);          %number of control input
Dim_e = Dim(4);          %number of controlledoutput
Dim_y = Dim(5);          %number of measurement output
Dim_psi  = Dim(6);       %number of IQC states
%Define the augment matrices Aaug and Caug2
Aaug = [Ap+Ad,zeros(Dim_n,Dim_psi);
        Bpsi1 ,Apsi];
Caug2 = [Cp2+Cd2 zeros(Dim_y,Dim_psi)];

%Compute the partition matrix in Lyapunov matrix P
%Define the M matrix as identity matrix
tM = eye(Dim(1)+Dim(6));
tN = (inv(tM)*(eye(Dim(1)+Dim(6))–tR*tS))';

%Compute the controller gain based on Theorem 2
tK = inv([tN tS*[Bp2;zeros(Dim(6),Dim(3))];zeros(Dim(3), ...
    Dim(1)+Dim(6)) eye(Dim(3))])*[tAkhat–tS*Aaug*tR ...
    tBkhat;tCkhat tDkhat]* inv([tM' ...
    zeros(Dim(1)+Dim(6),Dim(5));Caug2*tR eye(Dim(5))]);
Ak = tK(1:Dim_n+Dim_psi,1:Dim_n+Dim_psi);
Bk = tK(1:Dim_n+Dim_psi,Dim_n+Dim_psi+1:Dim_n+Dim_psi+Dim_y);
Ck = tK(Dim_n+Dim_psi+1:Dim_n+Dim_psi+Dim_u,1:Dim_n+Dim_psi);
Dk = tK(Dim_n+Dim_psi+1:Dim_n+Dim_psi+Dim_u,Dim_n+ ...
    Dim_psi+1:Dim_n+Dim_psi+Dim_y);
```

Listing 11.2: mfile: IQCcontroller.m.

Remark 11.2. The proof of Theorem 11.2 involves a relaxation from condition (11.12) to condition (11.13) via $-\hat{\lambda}_k^{-1} \le -\hat{\lambda}^{-2}(2\hat{\lambda} - \hat{\lambda}_k)$ for all $k \in \mathbf{I}[1, N_\lambda]$. It may introduce certain degree of conservatism to the resulting synthesis conditions only when multiple IQC multipliers are employed. Nevertheless, when a single IQC multiplier case (i.e., $N_\lambda = 1$) is applied, the controller synthesis conditions (11.9) and (11.10) are completely equivalent to the analysis one (11.12).

11.4. Examples

In this section, by two examples we demonstrate the effectiveness of the proposed IQC-based robust output-feedback controller.

Example 11.1. Consider system (11.1) with system matrices

$$A_p = \begin{bmatrix} -2 & 1 \\ -1 & 1 \end{bmatrix}, \quad A_d = \begin{bmatrix} 0.2 & 0.1 \\ 0.3 & 0.1 \end{bmatrix}, \quad B_{p1} = \begin{bmatrix} 1 \\ 0 \end{bmatrix}, \quad B_{p2} = \begin{bmatrix} 0 \\ 1 \end{bmatrix},$$
$$C_{p1} = \begin{bmatrix} 1 & 0 \\ 0 & 1 \end{bmatrix}, \quad C_{d1} = \begin{bmatrix} 0 & 0 \\ 0 & 0 \end{bmatrix}, \quad D_{p11} = \begin{bmatrix} 0 \\ 0 \end{bmatrix}, \quad D_{p12} = \begin{bmatrix} 0 \\ 1 \end{bmatrix},$$
$$C_{p2} = \begin{bmatrix} 1 & 3 \end{bmatrix}, \quad C_{d2} = \begin{bmatrix} 0 & 0 \end{bmatrix}, \quad D_{21} = \begin{bmatrix} 1 \end{bmatrix}.$$

The state delay $\tau(t)$ satisfies $|\tau(t)| \le \bar{\tau}$ and $|\dot{\tau}(t)| \le r$. With such a system setup, for IQC-based robust output-feedback synthesis, we pick two dynamic IQC multipliers $\{\Pi_k\}_{k=1}^2$ from [5] to characterize the associated state delay operator $\mathcal{S}_{\bar{\tau},r}(x_p)$, that is,

$$\Pi_1(s) = \begin{bmatrix} |\phi(s)|^2 I_2 & 0 \\ 0 & -I_2 \end{bmatrix},$$
$$\Pi_2(s) = \begin{bmatrix} |\varphi(s)|^2 I_2 & 0 \\ 0 & -I_2 \end{bmatrix}, \tag{11.16}$$

where $\phi(s) = k_1(\frac{\bar{\tau}^2 s^2 + c_1\bar{\tau}s}{\bar{\tau}^2 s^2 + a_1\bar{\tau}s + k_1 c_1}) + \epsilon$ and $\varphi(s) = k_2\frac{\bar{\tau}^2 s + c_2\bar{\tau}s}{\bar{\tau}^2 s^2 + a_2\bar{\tau}s + b_2} + \delta$ with $k_1 = 1 + \frac{1}{\sqrt{1-r}}$, $a_1 = \sqrt{2k_1c_1}$, c_1 is any positive real number such that $c_1 < 2k_1$, $K_2 = \sqrt{8/(2-r)}$, $a_2 = \sqrt{6.5 + 2b_2}$, $b_2 = \sqrt{50}$, $c_2 = \sqrt{12.5}$, and ϵ and δ are two arbitrarily small positive numbers. In this example, we select $c_1 = 1$, $\epsilon = 10^{-7}$, and $\delta = 0.0001$. It is clear that the selected multipliers $\{\Pi_k\}_{k=1}^2$ satisfy Assumption 11.2. By applying the IQC factorization methods in [18] and [19] we obtain J-spectral factorization of the selected IQC multipliers Π_1 and Π_2 in (11.16), respectively:

$$\Psi_1(s) = \begin{bmatrix} \left(\frac{\frac{k_1(c_1-a_1)}{\bar{\tau}} s - k_1^2 c_1/\bar{\tau}^2}{s^2 + \frac{a_1}{\bar{\tau}} + k_1 c_1/\bar{\tau}^2} + k_1 + \epsilon \right) I_2 & 0 \\ 0 & I_2 \end{bmatrix},$$

$$\Psi_2(s) = \begin{bmatrix} \left(\frac{\frac{k_2(c_2-a_2)}{\bar{\tau}} s - k_2 b_2/\bar{\tau}^2}{s^2 + \frac{a_2}{\bar{\tau}} + b_2/\bar{\tau}^2} + k_2 + \delta \right) I_2 & 0 \\ 0 & I_2 \end{bmatrix}. \tag{11.17}$$

The associated IQC-induced dynamics can be expressed in the form of (11.5) with the system matrices

$$A_\psi = \begin{bmatrix} \tilde{A}_\psi & 0 \\ 0 & \tilde{A}_\psi \end{bmatrix}, \quad B_{\psi 1} = \begin{bmatrix} \tilde{B}_{\psi 1} & 0 \\ 0 & \tilde{B}_{\psi 1} \end{bmatrix}, \quad B_{\psi 2} = 0,$$

$$\bar{C}_{\psi,1} = \begin{bmatrix} \tilde{\bar{C}}_{\psi,1} & 0 \\ 0 & \tilde{\bar{C}}_{\psi,1} \end{bmatrix}, \quad \bar{D}_{\psi 1,1} = \begin{bmatrix} k_1+\epsilon & 0 \\ 0 & k_1+\epsilon \end{bmatrix},$$

$$\bar{C}_{\psi,2} = \begin{bmatrix} \tilde{\bar{C}}_{\psi,2} & 0 \\ 0 & \tilde{\bar{C}}_{\psi,2} \end{bmatrix}, \quad \bar{D}_{\psi 1,2} = \begin{bmatrix} k_2+\delta & 0 \\ 0 & k_2+\delta \end{bmatrix},$$

$$\bar{D}_{\psi 2,1} = 0, \quad \bar{D}_{\psi 2,2} = 0$$

with

$$\tilde{A}_\psi = \begin{bmatrix} 0 & 1 & 0 & 0 \\ -\frac{k_1 c_1}{\bar{\tau}^2} & -\frac{a_1}{\bar{\tau}} & 0 & 0 \\ 0 & 0 & 0 & 1 \\ 0 & 0 & -\frac{b_2}{\bar{\tau}^2} & -\frac{a_2}{\bar{\tau}} \end{bmatrix}, \quad \tilde{B}_{\psi 1} = \begin{bmatrix} 0 \\ 1 \\ 0 \\ 1 \end{bmatrix},$$

$$\tilde{\bar{C}}_{\psi,1} = \begin{bmatrix} -\frac{k_1^2 c_1}{\bar{\tau}^2} & \frac{k_1(c_1-a_1)}{\bar{\tau}} & 0 & 0 \end{bmatrix},$$

$$\tilde{\bar{C}}_{\psi,2} = \begin{bmatrix} 0 & 0 & \frac{b_2 k_2}{\bar{\tau}^2} & \frac{k_2(c_2-a_2)}{\bar{\tau}} \end{bmatrix}.$$

MATLAB program to establish the selected IQC multipliers Π_1, Π_2 in (11.16) and the associated IQC induced dynamics (see Listing 11.3).

```
%A script to define the multiplier Pi
%and the associated IQC-induced dynamics
%Define upper bound of delay
taubar =1;
```

```
%Define upper bound of delay derivative
dtau = 0.1;
%Define the parameter in  IQC multipliers Pi1 and Pi2
Δ = 0.0001;
epsilon = 1e-7;
k1 = 1 + 1/sqrt(1-dtau);
c1 = 1;
a1 = sqrt(2*k1*c1);
k2 = sqrt(8/(2-dtau));
b2 = sqrt(50);
a2 = sqrt(6.5+2*b2);
c2 = sqrt(12.5);
%There are three cases in this example, here we define ...
    two multipliers Pi1 and Pi2
%Pi1 is only  valid  for taubar <1
%Pi2 is only  valid  for taubar <2
%If IQC = 1, Using both multipliers Pi1 and Pi2
%If IQC = 2, Using only Pi1
%If IQC = 3, Using only Pi2
IQC = 1;
if IQC == 1  %Using both multipliers Pi1 and Pi2
    Apsi_base1 = [0 1;
        -k1*c1/taubar^2 -a1/taubar];
    Apsi_base2 = [0 1;
        -b2/taubar^2 -a2/taubar];
    tApsi = [Apsi_base1 zeros(2,2);
           zeros(2,2) Apsi_base2];
    Apsi = [tApsi zeros(4,4);
           zeros(4,4) tApsi];
    Bpsi1_base = [0;1;0;1];
    Bpsi1 = [Bpsi1_base zeros(4,1);zeros(4,1) Bpsi1_base];
    Bpsi2_base = zeros(8,2);
    Bpsi2 = Bpsi2_base;
    tbarCpsi1_base = [-k1^2*c1/taubar^2 k1*(c1-a1)/taubar ...
        0 0];
    barCpsi11 = [tbarCpsi1_base zeros(1,4);zeros(1,4) ...
        tbarCpsi1_base];
    tbarCpsi2_base = [0 0 -b2*k2/taubar^2 k2*(c2-a2)/taubar];
    barCpsi12 = [tbarCpsi2_base zeros(1,4);zeros(1,4) ...
        tbarCpsi2_base];
    barDpsi11 = [k1+epsilon 0;0 k1+epsilon];
    barDpsi12 = [k2+Δ 0;0 k2+Δ];
    barDpsi21 = zeros(2,2);
    barDpsi22 = zeros(2,2);
elseif IQC ==2 %Using only Pi1
    Apsi_base1 = [0 1;
        -k1*c1/taubar^2 -a1/taubar];
```

```
    Apsi_base2 = Apsi_base1;
    tApsi = [Apsi_base1 zeros(2,2);
             zeros(2,2) Apsi_base2];
    Apsi = [tApsi zeros(4,4);
            zeros(4,4) tApsi];
    Bpsi1_base = [0;1;0;1];
    Bpsi1 = [Bpsi1_base zeros(4,1);zeros(4,1) Bpsi1_base];
    Bpsi2_base = zeros(8,2);
    Bpsi2 = Bpsi2_base;
    tbarCpsi1_base = [-k1^2*c1/taubar^2 k1*(c1-a1)/taubar ...
        0 0];
    barCpsi11 = [tbarCpsi1_base zeros(1,4);zeros(1,4) ...
        tbarCpsi1_base];
    tbarCpsi2_base = [0 0 -b2*k2/taubar^2 k2*(c2-a2)/taubar];
    barCpsi12 = [tbarCpsi1_base zeros(1,4);zeros(1,4) ...
        tbarCpsi1_base];
    barDpsi11 = [k1+epsilon 0;0 k1+epsilon];
    barDpsi12 = [k1+epsilon 0;0 k1+epsilon];
    barDpsi21 = zeros(2,2);
    barDpsi22 = zeros(2,2);
elseif IQC ==3  %Using Pi2 only
    Apsi_base1 = [0 1;
        -k1*c1/taubar^2 -a1/taubar];
    Apsi_base2 = [0 1;
        -b2/taubar^2 -a2/taubar];
    tApsi = [Apsi_base2 zeros(2,2);
             zeros(2,2) Apsi_base2];
    Apsi = [tApsi zeros(4,4);
            zeros(4,4) tApsi];
    Bpsi1_base = [0;1;0;1];
    Bpsi1 = [Bpsi1_base zeros(4,1);zeros(4,1) Bpsi1_base];
    Bpsi2_base = zeros(8,2);
    Bpsi2 = Bpsi2_base;
    tbarCpsi1_base = [-k1^2*c1/taubar^2 k1*(c1-a1)/taubar ...
        0 0];
    tbarCpsi2_base = [0 0 -b2*k2/taubar^2 k2*(c2-a2)/taubar];
    barCpsi11 = [tbarCpsi2_base zeros(1,4);zeros(1,4) ...
        tbarCpsi2_base];
    barCpsi12 = [tbarCpsi2_base zeros(1,4);zeros(1,4) ...
        tbarCpsi2_base];
    barDpsi11 = [k2+Δ 0;0 k2+Δ];
    barDpsi12 = [k2+Δ 0;0 k2+Δ];
    barDpsi21 = zeros(2,2);
    barDpsi22 = zeros(2,2);
end
```

Listing 11.3: mfile: IQC_multipliers.m.

Table 11.1 Influence of r and $\bar{\tau}$ on $\mathcal{L}_2$-gain performance.

Method	$\gamma \backslash r$ / $\bar{\tau}$	**0**	**0.1**	**0.9**	**1**	**1.9**
Theorem 11.2	0.01	0.6386	0.6387	0.6390	0.6422	0.6721
	0.2	0.8207	0.8259	0.8892	0.9022	1.6033
	2	1.7661	1.8035	2.3477	2.4828	16.3070
	10	1.9761	2.0096	2.4583	2.5652	259.6675
Method in [21]	0.2	0.9293	0.9316	1.2036	infeasible	infeasible
Method in [22]	0.2	0.9505	2.5091	4.8703	infeasible	infeasible

Remark 11.3. Note that Π_1 is only valid for $r < 1$ and Π_2 for $r < 2$. Other appropriate IQCs from [5] can be selected for cases where $r > 2$. For $r < 0.5$, we will apply both IQC multipliers in (11.16) for controller synthesis; when $r > 0.5$, only the IQC multiplier Π_2 will be applied. The above selection of the IQC multipliers is only for illustration purposes; more IQC multipliers can be utilized to achieve the better results.

Since there are two parameters, the delay derivative r and the maximum delay $\bar{\tau}$, involved in the synthesis conditions (11.9) and (11.10), we will first investigate the influence of r and $\bar{\tau}$ on the $\mathcal{L}_2$-gain performance. The results will also be compared with the those obtained by using output-feedback control design proposed in [21] (see Eqs. (23)–(25) in Section III therein) and [22] (see Corollary 1 in Section III therein). Solving the optimization problem (11.15) under different combinations of r and $\bar{\tau}$, we obtain the corresponding $\mathcal{L}_2$-gain performance shown in Table 11.1.

As can be seen, the $\mathcal{L}_2$-gain performances obtained from [21] are worse than those from IQC-based robust output-feedback controller. In particular, for the case of $\bar{\tau} = 0.2$ and $r = 0.9$, the optimal $\mathcal{L}_2$-gain performance obtained from [21] is $\gamma = 1.2036$ using the proposed method, whereas the result was $\gamma = 0.8892$, which is worsened by 35.36%. Moreover, the synthesis conditions of [21] and [22] are only applicable for the case $r < 1$, whereas our approach remains effective.

Further comparisons in terms of $\mathcal{L}_2$-gain performances has been shown in Fig. 11.1. As expected, a larger maximum delay $\bar{\tau}$ and/or delay derivative r in general would lead to worse $\mathcal{L}_2$-gain performances. Therefore better performance can be achieved for the cases with lower bounds on delay and/or delay variation rate. Moreover, these results also demonstrate that the synthesis conditions (11.9) and (11.10) are feasible for small $\bar{\tau}$ and r.

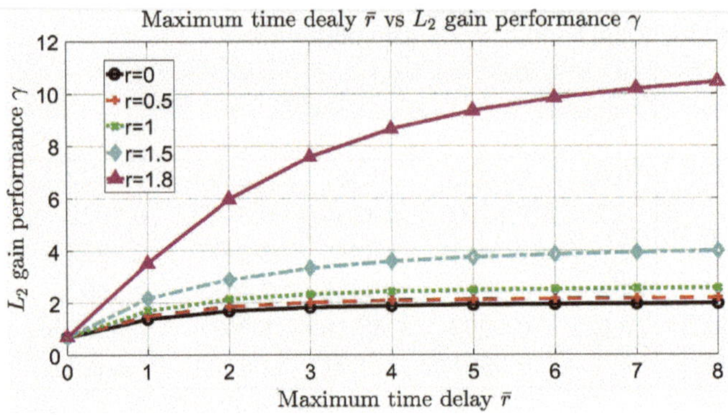

Figure 11.1 Relation between maximum time delay $\bar{\tau}$ and $\mathcal{L}_2$-gain performance γ under different delay derivatives r.

Example 11.2. Consider a practical system: a DC motor with time-varying state delay in measured output, which is given in [23]. The argumented system of the DC motor is given by

$$\begin{aligned}\dot{x}_p &= \begin{bmatrix} -108 & 0 \\ 182B_w & A_w \end{bmatrix} x_p + \begin{bmatrix} 1 \\ 0 \end{bmatrix} d + \begin{bmatrix} 1 \\ 0 \end{bmatrix} u, \\ e &= \begin{bmatrix} 182D_w & C_w \\ 0 & 0 \end{bmatrix} x_p, \\ y &= \begin{bmatrix} -182 & 0 \end{bmatrix} \mathcal{D}_{\bar{\tau},r}(x_p),\end{aligned}$$

where the matrices of the performance weight are

$$\{A_w, B_w, C_w, D_w\} = \left\{-5 \times 10^{-1}, 1, 5 \times 10^{-3}, 0\right\}.$$

The bounded conditions of the time-varying delay in are given by the delay derivative $r = 0.9$ and maximum delay $\bar{\tau} = 0.1$. With such a system setup, for IQC-based robust output-feedback synthesis, only the IQC multiplier Π_2 in (11.16) will be applied.

The associated IQC-induced dynamics can be expressed in the form (11.5) with the system matrices given by

$$A_\psi = \begin{bmatrix} \tilde{A}_\psi & 0 \\ 0 & \tilde{A}_\psi \end{bmatrix}, \quad B_{\psi 1} = \begin{bmatrix} \tilde{B}_{\psi 1} & 0 \\ 0 & \tilde{B}_{\psi 1} \end{bmatrix}, \quad B_{\psi 2} = 0,$$

$$\bar{C}_{\psi,1} = \begin{bmatrix} \tilde{\tilde{C}}_{\psi,1} & 0 \\ 0 & \tilde{\tilde{C}}_{\psi,1} \end{bmatrix}, \quad \bar{D}_{\psi 1,1} = \begin{bmatrix} k_2+\delta & 0 \\ 0 & k_2+\delta \end{bmatrix},$$
$$\bar{C}_{\psi,2} = \begin{bmatrix} \tilde{\tilde{C}}_{\psi,2} & 0 \\ 0 & \tilde{\tilde{C}}_{\psi,2} \end{bmatrix}, \quad \bar{D}_{\psi 1,2} = \begin{bmatrix} k_2+\delta & 0 \\ 0 & k_2+\delta \end{bmatrix},$$
$$\bar{D}_{\psi 2,1} = 0, \quad \bar{D}_{\psi 2,2} = 0$$

with

$$\tilde{A}_{\psi} = \begin{bmatrix} 0 & 1 & 0 & 0 \\ -\frac{b_2}{\bar{\tau}^2} & -\frac{a_2}{\bar{\tau}} & 0 & 0 \\ 0 & 0 & 0 & 1 \\ 0 & 0 & -\frac{b_2}{\bar{\tau}^2} & -\frac{a_2}{\bar{\tau}} \end{bmatrix}, \quad \tilde{B}_{\psi 1} = \begin{bmatrix} 0 \\ 1 \\ 0 \\ 1 \end{bmatrix},$$
$$\tilde{\tilde{C}}_{\psi,1} = \begin{bmatrix} 0 & 0 & \frac{b_2 k_2}{\bar{\tau}^2} & \frac{k_2(c_2-a_2)}{\bar{\tau}} \end{bmatrix},$$
$$\tilde{\tilde{C}}_{\psi,2} = \begin{bmatrix} 0 & 0 & \frac{b_2 k_2}{\bar{\tau}^2} & \frac{k_2(c_2-a_2)}{\bar{\tau}} \end{bmatrix}.$$

The mfile (see Listing 11.4) describes the MATLAB program necessary for IQC-based robust output-feedback controller.

```
% An mfile to setup the parameters and design the IQC ...
    controller
% for the mass-spring-damper systems

%%%%%    Setup System Data    %%%%%
%Define the matricies of the performance weight
Aw = -5*10^-1;
Bw = 1;
Cw = 5*10^-3;
Dw = 0;
%Define the system matrices
Ap = [-108 0;182*Bw Aw];
Bp1 = [1;0];
Bp2 = [1;0];
Cp1 = [182*Dw Cw;0 0];
Cp2 = [0 0];
Dp11 = [0;0];
Dp12 = [0;0];
Dp21 = [0];
%Define the delay matrice
Ad = [0 0;0 0];
Cd1 = [0 0;0 0];
Cd2 = [-182 0];
```

```
%Define extract system dimension
Dim_n = length(Ap(:,1));
Dim_d = length(Bp1(1,:));
Dim_u = length(Bp2(1,:));
Dim_e = Dim_n;
Dim_y = length(Cp2(:,1));
Dim(1) = Dim_n;                          %%%% Dimensions of ...
    Plant State
Dim(2) = Dim_d;                          %%%% Dimensions of ...
    disturbance  matrix
Dim(3) = Dim_u;                          %%%% Dimensions of ...
    control input  matrix
Dim(4) = Dim_e;                          %%%% Dimensions of ...
    measured  matrix
Dim(5) = Dim_y;                          %%%% Dimensions of ...
    measurement output matrix

%Define the system delay information and IQC multipliers with
% mfile IQC_multipliers.m
IQC_multipliers.m
%Performing a line search over the scalar variable ...
    Lambdahat with
% mfile
Bisectionlinesearch.m

%%%%    Output Feedback Controller Synthesis via Solving ...
    Synthesis    %%%%%
% Compute the IQC controller with
% mfile  delayIQCSFsyn.m
[tgamma,tLambdahat1,tLambdahat2,tR,tS,tAkhat,tBkhat, ...
    tCkhat,tDkhat]=delayIQC_SFsyn(Ap,Bp1,Bp2,Ad,Cd1,Cd2, ...
    Cp1,Cp2,Dp11,Dp12,Dp21,Dim,Lambdahat,Apsi,Bpsi1,Bpsi2, ...
    barCpsi11,barCpsi12,barDpsi11,barDpsi12,barDpsi21, ...
    barDpsi22);

%%%%%%%%  Dynamic Output Feedback Controller Construction ...
    %%%%%%%
%Construct the IQC{\chr"E2}{\chr"88}{\chr"92}based ...
    controller from the synthesis condition with
% mfile IQCcontroller.m
IQCcontroller.m
```

Listing 11.4: mfile: generate the IQC controller for DC motor system.

Table 11.2 $\mathcal{L}_2$-gain performance between Theorem 11.2 and [23].

Method	$\gamma \backslash r$ / $\bar{\tau}$	**0**	**0.1**	**0.5**	**0.9**	**1**	**1.9**
Theorem 11.2	0.1	0.0169	0.0169	0.0169	0.0169	0.0169	0.0557
Method in [23]	0.1	2.0024	2.2985	2.8345	3.8127	infeasible	infeasible

The results will also be compared with the controller proposed in [23] (see Theorem 1). Solving the optimization problem (11.15) under different r, we obtain the corresponding $\mathcal{L}_2$-gain performance shown in Table 11.2.

As can be seen, the $\mathcal{L}_2$-gain performances obtained from [23] are worse than those from IQC-based robust output-feedback controller. Utilizing the system data with delay derivative $r = 0.9$ and maximum delay $\bar{\tau} = 0.1$ to solve the optimization problem (11.15), we obtain suboptimal $\mathcal{H}_\infty$ results with minimal $\mathcal{L}_2$ gains $\gamma = 0.0169$ with the corresponding $\hat{\lambda}_1 = 6.3590$ and $\hat{\lambda}_2 = 6.3590$. In addition, the control gains of IQC-based robust output-feedback controller in this case can be obtained and shown as follows:

$$A_k = \begin{bmatrix} 26.3549 & -177.0672 & 0.0010978 & -0.89824 & 0.25118 & -9.9009 & -0.0044213 \\ 59.709 & -177.836 & 0.036642 & -0.67554 & 0.17071 & -6.8021 & -0.070907 \\ 23.3063 & -10.7909 & -39.3075 & 188.067 & 0.14895 & -6.5613 & 0.020924 \\ -0.85017 & -1.4263 & -3.6434 & -47.2123 & -0.0079913 & 0.36107 & -0.00040188 \\ -49238.2254 & -2471.3808 & 17.158 & 665.1141 & 86.2542 & -5222.8911 & -84.9608 \\ -1086.5137 & -39.5315 & 0.39077 & 15.4059 & 2.9761 & -165.4082 & -1.956 \\ -183.2647 & 62.6158 & 0.01692 & -1.0479 & -1.7307 & 76.5887 & -39.4027 \\ 9.7507 & 4.3182 & -0.0013795 & 0.14682 & 0.086953 & -3.882 & -3.6371 \\ 39187.2106 & -88.9346 & -12.4431 & -366.6378 & -71.6417 & 3013.8043 & 59.5902 \\ 886.0095 & -9.7828 & -0.24672 & -7.7733 & -1.0645 & 43.4401 & 1.2174 \end{bmatrix}$$

$$\begin{matrix} -0.32043 & -0.0022064 & 0.12596 \\ -0.14046 & -0.0062797 & 0.31574 \\ -0.34323 & -0.10353 & 4.9045 \\ 0.086283 & -5.2926e-06 & -0.032092 \\ -3063.7132 & 42.4976 & -2370.0136 \\ -71.2774 & 1.2884 & -68.004 \\ 194.1654 & 1.146 & -53.5835 \\ -47.7452 & -0.034324 & 1.7754 \\ 1796.0139 & -156.1144 & 5694.7648 \\ 38.5796 & -2.3777 & 75.9099 \end{matrix}\Bigg], \quad B_k = \begin{bmatrix} 0.00015834 \\ -0.0029668 \\ -0.00026422 \\ -1.7661e-05 \\ -0.41822 \\ -0.0089217 \\ 0.002073 \\ -2.1357e-05 \\ 0.26105 \\ 0.005174 \end{bmatrix},$$

$$C_k = \begin{bmatrix} 282.2977 & -570.1961 & 0.0017931 & -4.436 & 0.42504 & -15.8688 & 0.088269 & 6.9291 & -0.34446 & 13.2088 \end{bmatrix},$$

$$D_k = -0.0074.$$

Now we are ready to simulate the state delay systems under external disturbance. The initial condition is set as $x(0) = \begin{bmatrix} 2 & -2 \end{bmatrix}^T$, The disturbance vector applied to the system specified is the following random like noise signal [24]: $d(t) = (3\sin(9t) + 5\sin(\sqrt{13}t + 7\cos(15t) + 9\cos(19t))/24$ for $t \in$

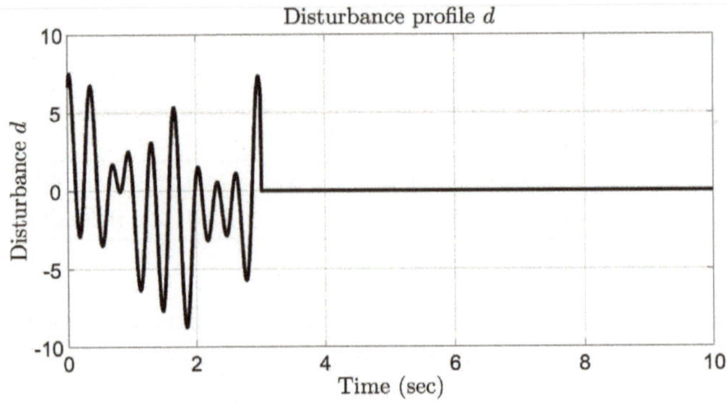

Figure 11.2 Disturbance profile.

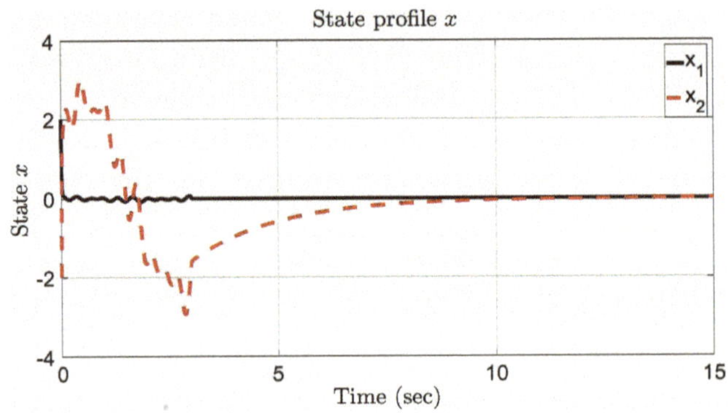

Figure 11.3 State $x_i(t)$ in Example 11.2.

[0, 3] s. This disturbance profile is also plotted in Fig. 11.2. The simulation results under the random like disturbance are provided in Figs. 11.3–11.4. In particular, Fig. 11.3 illustrates the system state information over 15 seconds. Fig. 11.4 shows the corresponding control input u.

As can be seen, the robust $\mathcal{H}_\infty$ is indeed achieved in spite of the presence of the time-varying state delay with a reasonable amount of control effort in 15 seconds.

11.5. Conclusions

In this chapter, we studied the robust dynamic output feedback consensus control problem for a class of linear systems with time-varying state delay subject to external disturbance. Using the proposed IQC-based

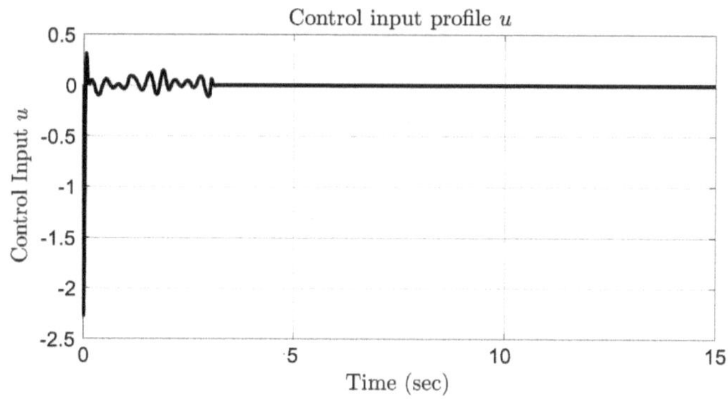

Figure 11.4 Control input u in Example 11.2.

robust output-feedback controller, we are able to tackle the robust $\mathcal{H}_\infty$ performance. We employed dynamic IQCs to characterize the input–output behaviors of the state-delay operator for both robust stability analysis and control synthesis. The synthesis conditions for robust $\mathcal{L}_2$-gain performance were formulated in terms of LMIs plus a scalar variable, which can be solved effectively via LMI optimization coupled with a line search method. Two numerical examples were used to demonstrate the advantages of the IQC-based controller.

The important advantage of the proposed IQC-based synthesis scheme over existing delay control methods lies in its systematic design framework for synthesizing time-delay control systems. The proposed results can be extended to handle the systems with other types of uncertainties or various nonlinearities.

Appendix 11.A. The mfile to solve the IQC-based robust output-feedback control synthesis

The mfile (see Listing 11.5) is the synthesis condition to solve the IQC-based robust output-feedback controller described in Theorem 11.2.

```
function ...
    [tgamma,tLambdahat1,tLambdahat2,tR,tS,tAkhat,tBkhat, ...
    tCkhat,tDkhat]=delayIQC_SFsyn(Ap,Bp1,Bp2,Ad,Cd1,Cd2, ...
    Cp1,Cp2,Dp11,Dp12,Dp21,Dim,Lambdahat,Apsi,Bpsi1, ...
    Bpsi2,barCpsi11,barCpsi12,barDpsi11,barDpsi12, ...
    barDpsi21,barDpsi22)
```

```
%The MATLAB function to solve the IQC-based robust ...
    output-feedback control synthesis
%Input:  Ap,Bp1,Bp2,Cp1,Cp2,Dp11,Dp12,Dp21    System matrices
%        Ad,Cd1,Cd2                            Delay matrices
%        Dim                                   Dimension vector
%        Lambdahat                             Scalar ...
    variable Lambdahat
%        Apsi,Bpsi1,Bpsi2,barCpsi11,
%        barCpsi12,barDpsi11,barDpsi12,
%        barDpsi21,barDpsi22                   IQC-induced ...
    LTI system matrices
%Output: tgamma                                System ...
    performance index
%        tLambdahat1, tLambdahat2              LMI ...
    variable: Lambdahat
%        tR,tS                                 LMI ...
    variable: Lyapunov matrices
%        tAkhat,tBkhat,tCkhat,tDkhat           LMI ...
    variable: Controller gains

%Extract the dinmension information from Dim vector
Dim_n = Dim(1);          %number of plant states
Dim_d = Dim(2);          %number of disturbance
Dim_u = Dim(3);          %number of control input
Dim_e = Dim(4);          %number of controlled output
Dim_y = Dim(5);          %number of measurement output
Dim_psi = Dim(6);        %number of IQC states

%Define the augment matrices in synthesis conditions
Aaug = [Ap+Ad,zeros(Dim_n,Dim_psi);
        Bpsi1 ,Apsi];
Baug2 = [Bp2;zeros(Dim_psi,Dim_u)];
Baug1 = [Bp1;zeros(Dim_psi,Dim_d)];
Adaug = [Ad;Bpsi2];
Caug1 = [Cp1+Cd1 zeros(Dim_n,Dim_psi)];
Caug2 = [Cp2+Cd2 zeros(Dim_y,Dim_psi)];

%%%% Initialize an LMI system
setlmis([]);
%%%%%%%%%%%%%%%%%%%%%%%%%%%%%%%%%%%%%%
%Define the LMI variables
gamma = lmivar(1,[1,0]);
Lambdahat1 = lmivar(1,[1,0]);
Lambdahat2 = lmivar(1,[1,0]);

R = lmivar(1,[Dim_n+Dim_psi,1]);
S = lmivar(1,[Dim_n+Dim_psi,1]);
```

```
Akhat  = lmivar(2,[Dim_n+Dim_psi,Dim_n+Dim_psi]);
Bkhat  = lmivar(2,[Dim_n+Dim_psi,Dim_y]);
Ckhat  = lmivar(2,[Dim_u,Dim_n+Dim_psi]);
Dkhat  = lmivar(2,[Dim_u,Dim_y]);

%%%  Define LMI Terms
    %Define the LMI [R I;I S]>0
    lmiterm([-1,1,1,R],1,1);
    lmiterm([-1,2,1,0],eye(Dim_n+Dim_psi));
    lmiterm([-1,2,2,S],1,1);

    lmiterm([2,1,1,R],Aaug,1,'s');
    lmiterm([2,1,1,Ckhat],Baug2,1,'s');

    lmiterm([2,2,1,Akhat],1,1);
    lmiterm([2,2,1,0],Aaug');
    lmiterm([2,2,1,-Dkhat],Caug2',Baug2');
    lmiterm([2,2,2,S],1,Aaug,'s');
    lmiterm([2,2,2,Bkhat],1,Caug2,'s');

    lmiterm([2,3,1,0],Lambdahat*Adaug');
    lmiterm([2,3,1,-Dkhat],Lambdahat*Cd2',Baug2');
    lmiterm([2,3,2,S],Lambdahat*Adaug',1);
    lmiterm([2,3,2,-Bkhat],Lambdahat*Cd2',1);
    lmiterm([2,3,3,Lambdahat1],1,eye(Dim_n));
    lmiterm([2,3,3,Lambdahat2],1,eye(Dim_n));
    lmiterm([2,3,3,0],-4*Lambdahat*eye(Dim_n));

    lmiterm([2,4,1,0],Baug1');
    lmiterm([2,4,1,-Dkhat],Dp21',Baug2');
    lmiterm([2,4,2,S],Baug1',1);
    lmiterm([2,4,2,-Bkhat],Dp21',1);
    lmiterm([2,4,4,gamma],-eye(Dim_d),1);

    lmiterm([2,5,1,R],Caug1,1);
    lmiterm([2,5,1,Ckhat],Dp12,1);
    lmiterm([2,5,2,0],Caug1);
    lmiterm([2,5,2,Dkhat],Dp12,Caug2);
    lmiterm([2,5,3,0],Lambdahat*Cp1);
    lmiterm([2,5,3,Dkhat],Lambdahat*Dp12,Cd2);
    lmiterm([2,5,4,0],Dp11);
    lmiterm([2,5,4,Dkhat],Dp12,Dp21);
    lmiterm([2,5,5,gamma],-eye(Dim_e),1);

    lmiterm([2,6,1,R],[barDpsi11 barCpsi11],1); %Upsilonhat61
    lmiterm([2,6,2,0],[barDpsi11 barCpsi11]); %Upsilonhat62
```

```
    lmiterm([2,6,3,0],Lambdahat*barDpsi21); %Upsilonhat63
    lmiterm([2,6,6,Lambdahat1],-eye(Dim_n),1);

    lmiterm([2,7,1,R],[barDpsi12 barCpsi12],1); %Upsilonhat61
    lmiterm([2,7,2,0],[barDpsi12 barCpsi12]); %Upsilonhat62
    lmiterm([2,7,3,0],Lambdahat*barDpsi22); %Upsilonhat63
    lmiterm([2,7,7,Lambdahat2],-eye(Dim_n),1);

    lmiterm([-3,1,1,gamma],1,1);
    lmiterm([-4,1,1,Lambdahat1],1,1);
    lmiterm([-5,1,1,Lambdahat2],1,1);

lmisys = getlmis;
%%%%%%%%%%%%%%%%%%%%%%%%%%%%%%%%%%%
n_lmi = lminbr(lmisys);
disp([' Total LMI numbers: ',num2str(n_lmi)])
n_Mat = matnbr(lmisys);
disp([' Total Matrices numbers: ',num2str(n_Mat)])

%%%%  solving optimization problem
%%%%%%%%%%%%%%%%%%%%%%
%%%% decide c for function mincx by using defcx
n_var = decnbr(lmisys);
c_min = zeros(n_var,1);
R_para = 0;
S_para = 0;
gamma_para = 1;
Akhat_para = 0;
Bkhat_para = 0;
Ckhat_para = 0;
Dkhat_para = 0;
Ekhat_para = 0;
Fkhat_para = 0;
for jj = 1:n_var
  [gammaj,tLambdahat1j,tLambdahat2j,Rj,Sj,Akhatj,Bkhatj, ...
      Ckhatj,Dkhatj]=defcx(lmisys,jj,gamma,Lambdahat1, ...
      Lambdahat2,R,S,Akhat,Bkhat,Ckhat,Dkhat);
   c_min(jj) = gamma_para*gammaj;
end

[copt,xopt] = mincx(lmisys,c_min,[0.001,300,1e11,10,0]);
```

```

%%%%%%%%%%%%%%%%%%%%%%%%%% obtaining results ...
    %%%%%%%%%%%%%%%%%%%%%%%%%%%
        tgamma = dec2mat(lmisys,xopt,gamma);
        tLambdahat1 = dec2mat(lmisys,xopt,Lambdahat1);
        tLambdahat2 = dec2mat(lmisys,xopt,Lambdahat2);
        tR = dec2mat(lmisys,xopt,R);
        tS = dec2mat(lmisys,xopt,S);
        tAkhat = dec2mat(lmisys,xopt,Akhat);
        tBkhat = dec2mat(lmisys,xopt,Bkhat);
        tCkhat = dec2mat(lmisys,xopt,Ckhat);
        tDkhat = dec2mat(lmisys,xopt,Dkhat);
%%%%%%%%%%%%%%%%%%%%%%%%%%%%%%%%%%%%%%%%%%%%%%%%%
%%%%%%%%%%%%%%%%%%%%%%%%%%%%%%%%%%%%%%%%%%%%%%%%%
end
```

Listing 11.5: mfile: delayIQC_ SFsyn.m.

References

[1] V.B. Kolmanovskii, A.D. Myshkis, Introduction to the Theory and Application of Functional Differential Equations, Kluwer Academic Publishers, Dordrecht, 1999.

[2] S.-I. Niculescu, Delay Effects on Stability: A Robust Control Approach, Springer, Berlin, 2001.

[3] K. Gu, V.L. Kharitonov, J. Chen, Stability of Time-Delay Systems, Birkhäuser, Boston, 2003.

[4] S. Xu, J. Lam, A survey of linear matrix inequality techniques in stability analysis of delay systems, International Journal of Systems Science 39 (12) (2008) 1095–1113.

[5] C. Kao, A. Rantzer, Stability analysis of systems with uncertain time-varying delays, Automatica 43 (2007) 959–970.

[6] P. Park, A delay-dependent stability criterion for systems with uncertain time-invariant delays, IEEE Transactions on Automatic Control 44 (1999) 876–877.

[7] E. Fridman, U. Shaked, New bounded real lemma representations for time-delay systems and their applications, IEEE Transactions on Automatic Control 46 (2001) 1973–1979.

[8] E. Fridman, Descriptor discretised Lyapunov functional method: analysis and design, IEEE Transactions on Automatic Control 51 (8) (2006) 890–897.

[9] C. Yuan, F. Wu, Exact-memory and memoryless control of linear systems with time-varying input delay using dynamic iqcs, Automatica 77 (2017) 246–253.

[10] C. Briat, O. Sename, J.F. Lafay, Memory-resilient gain-scheduled state-feedback control of uncertain LTI/LPV systems with time-varying delays, Systems & Control Letters 59 (2010) 451–459.

[11] A. Megretski, A. Rantzer, System analysis via integral quadratic constraints, IEEE Transactions on Automatic Control 42 (1997) 819–830.

[12] C. Kao, On stability of discrete-time LTI systems with varying time delays, IEEE Transactions on Automatic Control 57 (5) (2012) 1243–1248.

[13] I.E. Kose, C.W. Scherer, Robust $\mathcal{L}_2$-gain feedforward control of uncertain systems using dynamic IQCs, International Journal of Robust and Nonlinear Control 19 (2009) 1224–1247.

[14] J. Veenman, C.W. Scherer, A synthesis framework for robust gain-scheduling controllers, Automatica 50 (2014) 2799–2812.

[15] C. Yuan, F. Wu, Dynamic IQC-based control of uncertain LFT systems with time-varying state delay, IEEE Transactions on Cybernetics 46 (12) (2015) 3320–3329.

[16] C. Yuan, F. Wu, Delay scheduled impulsive control for networked control systems, IEEE Transactions on Control of Network Systems 4 (3) (2016) 587–597.

[17] C. Yuan, Leader-following consensus control of general linear multi-agent systems with diverse time-varying input delays, Journal of Dynamic Systems, Measurement, and Control 140 (6) (2018) 061010.

[18] P. Seiler, Stability analysis with dissipation inequalities and integral quadratic constraints, IEEE Transactions on Automatic Control 60 (6) (2014) 1704–1709.

[19] H. Pfifer, P. Seiler, Robustness analysis of linear parameter varying systems using integral quadratic constraints, International Journal of Robust and Nonlinear Control 25 (15) (2014) 2843–2864.

[20] K. Zhou, J.C. Doyle, K. Glover, Robust and Optimal Control, Prentice Hall, Englewood Cliffs, NJ, 1996.

[21] E.T. Jeung, J.H. Kim, H.B. Park, $\mathcal{H}^{\infty}$-output feedback controller design for linear systems with time-varying delayed state, IEEE Transactions on Automatic Control 43 (7) (1998) 971–974.

[22] Y. Wang, H. Zhang, D. Yang, M. Dong, H^{∞} output feedback control for uncertain systems with time-varying delays, in: 2010 Chinese Control and Decision Conference, IEEE, 2010, pp. 1570–1575.

[23] M. Nagahara, Y. Suzuki, Y. Uchimura, Controller design based on sum-of-squares for time-varying delay systems, in: 2012 12th IEEE International Workshop on Advanced Motion Control (AMC), IEEE, 2012, pp. 1–6.

[24] C. Yuan, F. Wu, Consensus for multi-agent systems with time-varying input delays, International Journal of Systems Science 48 (14) (2017) 2956–2966.

Index

E

F

G

H

I

L

M

N

T

U

FSC
www.fsc.org
MIX
Papier | Fördert
gute Waldnutzung
FSC® C083411

Kolibri 360 GmbH
Ferdinand-Jühlke-Straße 7
99095 Erfurt, Deutschland
produktsicherheit@kolibri360.de